Engineering-Medicine
Principles and Applications of Engineering in Medicine

Editors

Lawrence S. Chan
Department of Dermatology
University of Illinois College of Medicine
Chicago, IL
USA

William C. Tang
Department of Biomedical Engineering
University of California, Irvine
Irvine, CA
USA

CRC Press
Taylor & Francis Group
Boca Raton London New York

CRC Press is an imprint of the
Taylor & Francis Group, an **informa** business

A SCIENCE PUBLISHERS BOOK

Cover Acknowledgement

Lawrence S. Chan, one of the co-editors, created the cover figure

CRC Press
Taylor & Francis Group
6000 Broken Sound Parkway NW, Suite 300
Boca Raton, FL 33487-2742

© 2019 by Taylor & Francis Group, LLC
CRC Press is an imprint of Taylor & Francis Group, an Informa business

First issued in paperback 2021

No claim to original U.S. Government works

Version Date: 20190101

ISBN 13: 978-0-367-77980-1 (pbk)
ISBN 13: 978-1-138-54087-3 (hbk)

Library of Congress Cataloging-in-Publication Data
Names: Chan, Lawrence S., editor. | Tang, William C., editor.
Title: Engineering-medicine : principles and applications of engineering in medicine / editors, Lawrence S. Chan, William C. Tang.
Description: Boca Raton, FL : CRC Press/Taylor & Francis Group, [2018] | "A Science Publishers book." | Includes bibliographical references and index.
Identifiers: LCCN 2018057448 | ISBN 9781138540873 (hardback : alk. paper)
Subjects: | MESH: Biomedical Engineering | Delivery of Health Care | Education, Medical | Quality Improvement | United States
Classification: LCC R856 | NLM QT 36 | DDC 610.28--dc23
LC record available at https://lccn.loc.gov/2018057448

Visit the Taylor & Francis Web site at
http://www.taylorandfrancis.com

and the CRC Press Web site at
http://www.crcpress.com

Foreword

The interface between engineering and medicine poses both complex challenges and far-reaching opportunities for enhancing healthcare. Among the challenges and opportunities are six major themes:

1. improving healthcare services delivery effectiveness through capitalizing on the rapid and complex impact on social structure and human interaction resulting from engineering developments, most significantly manifested through digital media, big data, and robotics;

2. the optimal balance between the in-person connection of the physician to the patient and the increased access of the patient to the physician through digital media distance connection;

3. the balance between increased efficiency from engineering applications to medicine and the concern about healthcare quality control related to the replacement of some elements of personal patient/physician interaction with elements of robotics and digital connection;

4. ethical issues related to the value and desire for privacy for the individual patient and the risks to privacy from centralized databases and virtual physician/patient connections;

5. enhancement of approaches to patient care through delivery of clinical services by inter-professional teams facilitated by engineering interventions versus continuation of traditional one-on-one personal clinical interactions between the patient and the physician;

6. essential changes in medical and engineering education precipitated by developments and collaboration at the engineering-medicine interface.

This book details the profound opportunities for enhancing the effectiveness, efficiency, and comprehensiveness of healthcare services through the engineering-medicine interface. The chapters address in detail the six major themes as they exist currently and projected into the future.

In essence we face an engineering evolution in medicine. Given the accelerating rate of innovation and development in engineering-medicine one could appropriately consider this trend on the verge of transitioning from evolution to revolution. The challenges currently faced by the American healthcare system—high cost, less than optimal health status outcomes, and projected serious shortages in the healthcare workforce (particularly physician shortages)—powerfully underscore the importance of realizing the potential value inherent in enhanced collaboration at the engineering-medicine interface. The editor and principal author of the book, Dr. Lawrence Chan, initiated the effort based upon his experience as an engineer and physician as well as participation in an innovative executive graduate program focused on the collaboration of senior clinicians and experienced healthcare system administrators. The partnership among senior clinicians, administrators, and engineers in healthcare delivery, leadership, education, and research is essential to realizing the full potential impact of engineering-medicine on the future of healthcare.

Jay Noren
Associate Dean for Leadership Development and
Director of Executive Master of Healthcare Administration
University of Illinois College of Medicine

Acknowledgments

The editors of this book would like to thank all contributors for their hard work and dedication in writing important chapters of this book. Without their help, this book would not be complete. We would also like to thank our family members for tolerating our "absence of duty" during this period of time. In addition, we would like to thank our engineering school alma maters, Massachusetts Institute of Technology (Dr. Chan's) and University of California, Berkeley (Dr. Tang's) for instilling in us a strong basis of logics, engineering principles, and work ethics that sustain us throughout the entire book writing adventure. Most of all, we thank our Creator who is the ultimate Engineer-Physician, who created and heals the world.

Preface

This book is a transforming textbook intended for the students of medicine: the medical students and their teachers, as well as the biomedical engineering students and their teachers.

For the medical students and their teachers, this is the first attempt to write a textbook for teaching them in a novel track of Engineering-Medicine. Since a textbook of this kind has never been written before, this project is truly a pioneering effort. The ultimate aim and central theme of this book is to train our future physicians to "think like an engineer and act as a physician", as we sincerely believe that transforming medical education requires change of thinking and approach, not just merely adding some technology components. This textbook intends to incorporate engineering principles into the teaching of medicine, at the level of undergraduate medical education, i.e., at the level of doctor of medicine (MD) candidate. The primary focuses of this book are to instill the engineering principles and tactics into medical education frameworks, rather than teaching details of individual biomedical subjects. Ultimately, this textbook has one important goal; that is to teach the next generation of physicians to take the profession to a higher level of medicine.

For the students and teachers of the biomedical engineering discipline, this book is our first attempt to provide a broad view on contributions of engineering discipline to improve medicine, not only in diagnosis, treatment, and prevention, but also in efficient healthcare delivery, as they will learn ways their research and design efforts would impact patient care delivery and benefit patients in a very practical fashion. For the biomedical engineering students with a plan to enter medical school, this book serves as a perfect pathway material for their eventual learning of medicine with engineering perspectives.

For the practicing biomedical engineers and the practicing physicians, this book offers a fresh look at the "engineering" way of conducting medicine and medicine-related works, whether it is for the improvement of disease determination, therapeutics, prevention, or healthcare methodology.

In addition to educating the current and future biomedical engineers and physicians, one of the major goals of this textbook is to advocate close working collaboration between engineers and physicians. The combined clinical expertise in physicians and the technical expertise in engineers would exert a powerful force when it comes to designing instrumentation and methods for disease diagnosis, therapy, and prevention.

This textbook is organized in the following manner: After Chapter 1 introduces the topic of engineering-medicine, the following two chapters set the important tone of medical ethics (Chapters 2 & 3), then the principles of engineering and the principles of engineering-medicine are delineated (Chapters 4 & 5). Then following a chapter on the cost-effectiveness analysis of engineering-medicine education (Chapter 6), this book dedicates the subsequent six chapters delineating the quintessential engineering principles and characteristics as applied to medicine; including invention and innovation (Chapter 7), design optimization (Chapter 8), problem-solving (Chapter 9), systems integration (Chapter 10), efficiency (Chapter 11), and precision (Chapter 12). Next, Big Data (Chapter 13), artificial intelligence (Chapter 14), and quality management (Chapter 15) are discussed as they are playing an increasingly important role in engineering-medicine. The next two chapters are devoted to biology subjects (cellular and molecular biology, Chapter 16; systems biology, Chapter 17). Decisively, this book provides advanced biotechnology (Chapter 18), up-to-date biomedical imaging methodology for magnetic resonance imaging (Chapter 19) and molecular imaging (Chapter 20), and state-of-the-art emerging biomedical imaging (optical coherence tomography, Chapter 21 & photoacoustic imaging, Chapter 22) and analytical technology (mass spectrometry, Chapter 23), not only sufficient for today's medicine, but also for its future. This book continues to the subjects of rehabilitation technology (Chapter 24), the role of academic health center and leadership in engineering-medicine (Chapter 25), and finally concludes on environmental protection (Chapter 26).

On the teaching format, this textbook combines various methodologies each intends to achieve the optimum outcomes and they include lectures, group discussions, and individual and group projects. We try to make it as "thought-provoking" as possible. From the aspect of content, this textbook can also serve as a unique "engineering-medicine" curriculum for a medical college which has the courage to test this novel and transformative education methodology. It is also suitable as a teaching textbook of biomedical engineering classes. It is our sincere hope that this book will help bridge the gap between engineering and medicine and contribute to the improvement of healthcare delivery as to fulfill the triple aims of "better care, better health, and lower cost" called for by the Institute of Medicine (now National Academy of Medicine).

As the first of its kind, this "trail-blazing" textbook will unquestionably go through a period of "growing pain". Periodic revision is expected in the future to modify, improve, supplement, and update its contents.

Lawrence S. Chan
William C. Tang

Contents

1

Engineering-Medicine
An Introduction

Lawrence S. Chan

QUOTABLE QUOTES

"Be innovative, challenge the status quo, think out of the box, and make a difference!"

Dr. Victor Dzau, President, the Institute of Medicine (Dzau 2015)

"Whatever we succeed in doing is a transformation of something we have failed to do."

Paul Valery, French Poet (FORBES 2017a)

"History is a race between education and catastrophe."

Hebert G. Wells, English Writer (FORBES 2017b)

"New Year's Day. A fresh start. A new chapter in life waiting to be written. New questions to be asked, embraced, and loved. Answers to be discovered and then lived in this transformative year of delight and self-discovery."

Sarah Ban Breathnach, American Author (BRAINY 2017)

"Engineers can help realize the vision of high-quality, precision, and quantitative medicine while also reducing health care costs. Just as the revolution in medicine created by the advent of molecular biology in the past century, engineering will be the new driving force for the progress of medical research and education in this century and beyond."

Engineering as a new frontier for translational medicine (Chien et al. 2015)

Learning Objectives

The learning objectives of this chapter are to pave the foundation for the rationale of engineering-medicine education and to set the logical foundation for engineering-medicine.

University of Illinois College of Medicine, 808 S Wood Street, R380, Chicago, IL 60612; larrycha@uic.edu

After completing this chapter, the students should:

- Understand the current healthcare and medical education status in the US.
- Understand the rationale for fulfilling the triple aim of better care, better health and lower cost called by Institute of Medicine.
- Understand similar educational goals between engineering and medicine.
- Understand ways engineering education may enhance future physicians' efficiency.
- Understand the potentials of engineering-medicine may improve the overall healthcare deliver.
- Understand the global picture of how engineering was, is, and will be enhancing the quality of healthcare.
- Understand pathways that engineering-medicine may be implemented.

The Current State of US Healthcare and Medical Education

The health of healthcare in the US is presently being assessed in a major way. The current US healthcare faces two major challenges in that while the cost is very high its overall outcome is not excellent. In its 2014 report, the Organization for Economic Co-operation and Development (OECD) informed us that healthcare expenditure of the US costs 16.9% gross domestic products (GDP) in 2012, the highest of all OECD countries (OECD 2014). Yet according to a 2014 study conducted by the Commonwealth Fund, the US is ranked the very last in overall healthcare quality among 11 nations including Australia, Canada, France, Germany, the Netherlands, New Zealand, Norway, Sweden, Switzerland, the United Kingdom and the United States (COMMONWEALTH 2014). Specifically, the US is last or near last on measures of access, efficiency, and equity (COMMONWEALTH 2014). Another major challenge the US healthcare system faces is the increase of aging population as a percentage of overall US population (Colwill et al. 2008). This demographic change thus will increase the demand for healthcare services and will certainly put more burdens to the healthcare providers. Compounding to this challenge is the coming of physician shortage projected by the American Association of Medical Colleges (AAMC), which estimated a substantial physician shortage ranging from 46,000 to 90,000 by the year of 2025 (AAMC 2016). Others also estimated the deficit of 35,000 to 44,000 adult general physicians by the year 2025 (Colwill et al. 2008). Furthermore, the current medical practice is outpaced by the rapid advancement of health science and technology. Moreover, the combined findings of low efficiency and high expenditure of the US health system has prompted the Institute of Medicine (IOM, now National Academy of Medicine) to call for the implementation of three aims of medicine in the future: "better care, better health, and lower costs" (Alper et al. 2013). Since medical practice is by-and-large a reflection of medical education, the improvement of US healthcare state would need to be started at the medical school level.

Having depicted the current state of US healthcare, now let us examine the current state of US medical education. We will pounder the following 4 salient points (efficiency, global perspective, teamwork, and advanced biotechnology) in assessing the current medical education for future physicians:

- Medical students are largely taught to be an individual patient advocate, rather than thinking critically about being steward of the entire health care system, particularly in terms of resources. This issue has prompted the IOM to call for lowering health care cost (Alper et al. 2013). Therefore we need to provide solution by educating students the efficient methods in healthcare delivery.
- Medical students are primarily taught in providing care: in making correct diagnosis and in prescribing right medication or surgical procedure, and not much about other determinants that are also significant for improving health. For example, population studies estimated that medical care contribute only about 10% of the variance in the final outcome of health, whereas a huge percentage (50%) of the outcome is dependent on behavior and social factors (Hershberger et al. 2014). Yet physicians and the healthcare system as a whole are playing little role in the other important contributing factors. This issue has prompted the Association of American Medical Colleges to call for the inclusion of public health in medical education curriculum (AAMC 2015). In fact, the University of Wisconsin School of Medicine has changed its name to "University of Wisconsin School of Medicine and Public

Health" in 2005 to involve students in health promotion and disease prevention. Thus we need to provide medical education, with emphases on global perspective and promote holistic approach and resource stewardship in our future physicians.

- Medical students are predominantly taught to be excellent individual physicians, rather than being a team member of a large health care system. This issue has prompted the American Medical Association (AMA) to call for the promotion of teamwork concept (AMA 2015). To improve on this aspect, our medical curriculum should instill the teamwork concept into the daily practice of our future physicians.

- Rapid advancement of biomedical technology has occurred at a much faster pace than the current medical education and practice (Duda et al. 2014, Rietschel et al. 2015). A recent survey on genetic curricula in US and Canadian medical schools found that only 26% responders reported formal genetic teaching during 3rd and 4th year of school and most responders felt the amount of time spent on genetics was insufficient for future clinical practice in this era of genomic medicine (Plunkett-Rondeau et al. 2015). Not only medical educators felt the current medical education lagged behind in advanced technology training, students at Harvard Medical School expressed similar sentiment (Moskowitz et al. 2015). They felt that "our ability and capacity to train both new and experienced clinicians to manage the tremendous amount of data lag far behind the pace of the data revolution" and that "medical education at all levels must come to address data management and utilization issues as we enter the era of Big Data in the clinical domain" (Moskowitz et al. 2015). In addition, recent studies have pointed out that advanced technology useful to teach undergraduate medical knowledge and skills such as sonography and otolaryngology was underutilized (Day et al. 2015, Fung 2015). The participants of *Translate 2014* meeting in Berlin Germany has reached a consensus on the rate-limiting factor for advancing translational medicine and made an urgent statement: "The pace of basic discoveries in all areas of biomedicine is accelerating. Yet translation of this knowledge into concrete improvements in clinical medicine continues to lag behind the pace of discovery" (Duda et al. 2014, Rietschel et al. 2015). Engineering indeed has great potential to enhance the healthcare quality in many ways (Hwang et al. 2015, Ricci 2017). Three recent biomedical articles published in 2015, randomly chosen as below, illustrate the current state of medical science advancements that are not universally taught in our medical schools:

One example is illustrated by 3-D Printing technology used in the treatment of tracheobronchomalacia. The biomedical engineers at the University of Michigan have successfully implanted customized patient-specific 3-D printed external airway splints in three infants suffered from tracheobronchomalacia, a life-threatening condition of excessive collapse of airways during respiration. As a result, this 3-D printed material eliminated the airway disease, strongly illustrating the biotechnology contribution to medicine (Morrison et al. 2015, Michalski and Ross 2014).

Another example is nanotechnology in gene editing to correct cystic fibrosis mutation. The Yale University scientists have derived a gene editing technique, which a synthetic molecule similar to DNA, called peptide nucleic acids (PNAs), and together with donor DNA are utilized to correct genetic defect in cystic fibrosis. Using biodegradable microscopic nanoparticles to facilitate the delivery of PNA/DNA to the target cells, the Yale researchers were able to trigger the "clamping" of PNA close to the mutation, leading to subsequent DNA repair and recombination pathway in these target cells, resulting the proper gene correction in both human airway cells and mouse nasal cells, as well as in nasal and lung tissues, significantly demonstrating the nanotechnology contribution to medicine (McNeer et al. 2015, Langer and Weissleder 2015).

Further example is a novel mobile phone video microscope utilized in parasitic infection detection. Researchers at the University of California at Berkeley and NIH have together produced a mobile phone automatic microscopic device which can be used by healthcare providers in low-resource areas of Africa to detect filarial parasite L. loa, using whole blood loaded directly into a small glass capillary from a simple finger stick, without the need for conventional sample preparation or staining. Having validated that the automated counts (a mere 2-minute operation) correlated with the manual thick smear counts with 94% specificity and 100% sensitivity, this device can now be used to exclude patients infected with L. loa from life-threatening medication side effects of ivermectin, while the remaining mass population can be safely administered with ivermectin to eliminate the debilitating onchocerciasis and lymphatic filariasis (D'Ambrosio et al. 2015).

Additional evidence that provides indirect support for the notion of fast pace of biomedical technology advancement is the rapid growth of biomedical engineering jobs. The US Bureau of Labor Statistics has projected the employment of biomedical engineers to grow about 20% from 2014 to 2024, much faster than all other occupations (USBLS 2016).

The Call for Medical Education Reform

The American Medical Association (AMA) is keenly aware these educational deficiencies and promotes medical education reform in an article entitled "Accelerating change in medical education" on its website (AMA 2015). The AMA urged seven changes of present-day medical practice in order to adapt to the future health care, through medical education reform (AMA 2015) (Table 1).

Table 1. The goals of education reform in engineering dovetail the goals of education reform in medicine.

Engineering Education Reform[1] {Item No.}		Medicine Education Reform[2] {Item No.}	
Good communication skills (S)	{1}	Rigid time frame in school to flexible competency-based training	{1}
Physical sciences and engineering science fundamentals (K)	{2}	Individual physicians to interprofessional teams	{2}
Skills to identify, formulate, and solve engineering problems (S)	{3}	Low technology utilization to high technology utilization	{3}
Systems integration (K)	{4}	Inpatient focus to outpatient focus	{4}
Curiosity and persistent desire for continuous learning (A)	{5}	Individual patient advocacy to population health outcome	{5}
Self-drive and motivation (A)	{6}	Limited focus on cost to stewardship of resources	{6}
Culture awareness in the broad sense (K)	{7}	Understanding role in healthcare delivery	{7}
Economics and business acumen (K)	{8}		
High ethical standards, integrity, and global, social, intellectual, and technological responsibility (A)	{9}		
Critical thinking (S)	{10}		
Willingness to take calculated risk (A)	{11}		
Efficiency for prioritization (S)	{12}		
Project management (S)	{13}		
Teamwork skills and function on multidisciplinary teams (A)	{14}		
Entrepreneurship and intrapreneurship (A)	{15}		

[1] Goals established by the American Society of Engineering Education (ASEE)
[2] Goals set by the American Medical Association (AMA)

Engineering-Medicine as an Innovative Medical Education Reform

How should we then reform our current medical education? As early as 2005, the National Academy of Engineering (NAE) and IOM have called for an engineering-medicine partnership for the sake of improving healthcare delivery (NAE 2005). In addition, NAE has recognized the important role of engineering in achieving the goals of personalized medicine (precision medicine, NAE 2017). Moreover, National Institutes of Health have realized the great potential of engineering in healthcare improvement and has substantially increase its funding to biomedical engineering and other engineering research projects since 2000, whereas its funding to other scientific research projects remains essentially the same (Chien et al. 2015). Some academicians have argued that engineering is indeed the new frontier of translational medicine (Chien et al. 2015). It has been proposed that "the full understanding of biological processes and the effective management of clinical conditions require quantitative and time-variant considerations, which are the hallmarks of engineering, in addition to feedback control, systems approach, and multiscale modeling" (Chien et al. 2015). Biomedical academicians have started to champion for a new engineering-medicine curriculum for medical education (Chien et al. 2015, Chan 2016, 2018). Now let us examine evidences to support that this proposal would work.

Similar educational Goals for Engineering and Medicine

Comparing these seven changes urged by the AMA and the 15 future engineering education priorities proposed by the American Society of Engineering Education (ASEE) (Table 1), one finds amazingly parallel goals between these two educational organizations (AMA 2015, ASEE 2014). Specifically, item 2, 5, 6, 7 of educational changes supported by the AMA closely correspond to Item 14 (team work), 9 (global perspective), 9 (stewardship), 4 (system practice) of the educational reform priorities proposed by the ASEE, respectively. Item 3 of the change supported by the AMA is the core component of engineering education (high use of technology). Moreover, the engineering education reform goal of good communication skills (item 1) is a basic daily work requirement for patient-physician relationship. In addition, the engineering goal of continuous learning (item 5) is in fact a prerequisite for physician license renewal (CME 2018). Such similarity of educational goals between engineering and medicine should relieve fear that educating medical students through the engineering-based curriculum may lead future physicians into the wrong path of impersonal technicians (Weisberg 2014).

Similar Overall Education Characteristics for Engineering and Medicine

When we give an evaluation of the overall characteristics of engineering and medicine, there are some global similarities. Just as the official motto depicted in the official seal of Massachusetts Institute of Technology, one of the most reputable educational institutions of engineering and technology, depicts this phrase, "*Mens et Manus*", a Latin phrase meaning "minds and hands" in English language. In essence this motto projects the characteristics of engineering as an integration of intellectual pursuit and practical application (MIT 2016) (Fig. 1). Similarly, medical practice, by definition, fits the motto "*Mens et Manus*",

Fig. 1. Official Seal of Massachusetts Institute of Technology.

as physicians utilize the cooperation and coordination of their minds (intellect) and hands (physical ability) to make disease diagnoses and treat patients (Bujak 2008, Carson and Murphey 1996). Neurosurgery is one of the best examples where the surgeon's thorough understanding of the anatomy and physiology of the human brain and nervous system (*Mens*) and the skilled hands (*Manus*) are both critical for the successful performance of brain surgery (Carson and Murphey 1996).

Therefore, since engineering education is so similar in overall characteristics to that of medicine, it could be considered as a good pathway for transforming our future medical education. Since both engineering and medicine educations include hand-on training as their core skills, the future merging engineering-medicine curriculum could include medical students taking part in engineering capstone design project teams so as to gain first-hand experience of how engineering functions. Conversely, biomedical engineering students could participate in medical simulation lab as a pathway to gain better understanding of medicine.

Engineering Education Enhances Efficiency, Global Perspectives, Teamwork, and Technology Utilization

Since efficiency, global perspectives, teamwork, and technology utilization were depicted as needed improvement for future medical education in the session above, we now examine whether engineering-medicine education could fulfill these needs.

First, since efficiency is always a key emphasis in engineering education, engineering-medicine would go a long way in improving medical efficiency (Table 1).

To instill global perspectives, engineering-medicine education should definitely fulfill its mission, as global perspective has depicted it as one of their goals by ASEE (Table 1).

For teamwork, a key characteristic of engineering education, would support the suitability of engineering-medicine in teamwork training (Table 1).

In terms of advanced technology utilization, engineering is synonymous with technology and engineering-medicine education thus should improve the application of advanced technology in its medical training (Table 1). Some academicians further argued that "Moore's Law", meaning functional improvement is accompanied by adjusted lower costs and increased usage, is applicable here—that the engineering-led technological advancement should enhance quality, reduce cost, and democratize healthcare delivery (Chien et al. 2015).

Steps in Engineering-medicine Education

Since undergraduate medical education is the exclusive physician pipeline and the gateway to the future of medicine, what we educate the medical students today would have a profound influence what the physicians will practice in the 50 years ahead. The most effective way to transform medicine of the future is, therefore, through medical education reform. In his commencement speech delivered at University of Illinois College of Medicine on May 8th, 2015, Dr. Victor Dzau, President of IOM, encouraged the medical graduates to "be innovative, challenge the status quo, think out of the box, and make a difference!" (Dzau 2015) The news of a proposed Engineering-based medical school, Carle-University of Illinois College of Medicine, to be established at the University of Illinois Urbana/Champagne Campus for enrollment in 2018, has triggered various reactions within academic medicine community (Cohen 2015, ENGINEERING 2017). Furthermore, Texas A & M University also established a special EnMed track for medical students (TEXAS 2017). An engineering-medicine education curriculum, which intends to promote innovation, creativity, and efficiency, could indeed be a roadmap to the future medicine. In the subsequent chapters of this book, an overview of engineering principles will be described (Chapter 4), followed by an attempt to solidify an overview of engineering-medicine principles and the rationale for the engineering-medicine education (Chapter 5), by an economic evaluation of engineering-medicine education (Chapter 6), and then by discussions of several important engineering-medicine educational components derived from key engineering characteristics: Invention and Innovation (Chapter 7), Design Optimization (Chapter 8), Problem-Solving (Chapter 9), Systems Integration (Chapter 10), Efficiency (Chapter 11), and Precision (Chapter 12). Although a developing physician goes through two stages of training, the medical school education followed by a post-graduate residency, this book focuses more on the former. Nevertheless, a successful transformation would require a reinforcing of this concept into the post-graduate level of training in residency. For the biomedical engineering students, this book intends to provide them with a greater perspective on potential contributions in medicine with their engineering skills and to pave a path for their future career consideration, whether it is in medicine, engineering, or both. Moreover, the cross-fertilization of cultures of engineering and medicine, the most important goal of this textbook, will not only enhance physicians' appreciation of quantitative and systems aspects of biomedical research and education, but will also promote a better recognition of the critical and challenging problems in clinical medicine by the biomedical engineers (Chien et al. 2015). Collectively, an engineering- medicine education would fulfill the medical education reforms called for by the IOM, AMA, and AAMC. The timing may be ripe now to test this concept in a small number of institutions.

Summary

In this introductory chapter, the main purpose was to lay the ground works for this textbook. Specifically, this chapter provides the "why's" before introducing the "what's" of this new concept of engineering-medicine. The current states of US medical practice and education are first examined, followed by discussion on the concept of engineering-medicine and the possible benefits of engineering-medicine education for helping the development of future physicians to be more efficiency- and quality-oriented. To convince a very conservative medical profession for considering a "radical" transformation like engineering-medicine will be a tall order. The huge challenge of this transformative change notwithstanding, this book represents our deep dedication to start the trailblazing works. This chapter also provides the rationale and a broad perspective for biomedical engineering students on the goal of engineering-medicine. We shall now begin the detailed discussions.

Acknowledgment

The author would like to acknowledge the academic support, for this and other chapters, provided by the University of Illinois College of Medicine (in which the author is currently the Dr. Orville J. Stone Endowed Professor of Dermatology) and the Captain James Lovell Federal Health Care Center (in which the author is currently an attending staff physician).

References

[AAMC] Association of American Medical Colleges. 2015. Public health in medical education online community of practice. [www.aamc.org] accessed June 25, 2015.

[AAMC] Association of American Medical Colleges. 2016. Physician supply and demand through 2025: key findings. AAMC [https://www.aamc.org] accessed June 25, 2016.

Alper, J., J. Sanders and R. Saunders. Rapporteurs. 2013. Core measurement needs for better care, better health, and lower costs: Counting what counts-workshop summary. Institute of Medicine of the National Academies. National Academy Press, Washington, DC USA.

[AMA] American Medical Association. 2015. Accelerating change in medical education. [www.ama-assn.org] accessed June 25, 2015.

[ASEE] American Society for Engineering Education. 2014. Transforming undergraduate education in engineering. [www.asee.org] accessed June 25, 2015.

[BRAINY] Brainy Quotes. 2017. Transformative Quotes. [https://www.brainyquote.com/authors/topics/transformative] accessed November 11, 2017.

Bujak, J.S. 2008. Inside the Physician Mind: Finding Common Ground with Doctors (ACHE Management). 1st Edition. Health Administration Press. IL: Chicago, USA.

Carson, B. and C. Murphey. 1996. Gifted Hands: The Ben Carson Story. Zondervan. Michigan: Grand Rapid. USA.

Chan, L.S. 2016. Building an engineering-based medical college: Is the timing ripe for the picking? Med Sci Edu 26: 185–190.

Chan, L.S. 2018. Engineering-medicine as a transforming medical education: A proposed curriculum and a cost-effectiveness analysis. Biol Eng Med 3: 1–10. doi: 10.15761/BEM.1000142.

Chien, S., R. Bashir, R.M. Nerem and R. Pettigrew. 2015. Engineering as a new frontier for translational medicine. Sci Transl Med 7: 281fs13. Doi:10.1126/scitranslmed.aaa4325.

[CME] 2018. State required CME/CE. My CME [https://www.mycme.com/state-required/section/5320/?DCMP=ILC-mycme_us_myCME_Nav_State_market] accessed August 25, 2018.

Cohen, J.S. 2015. U. of I. trustees approve new medical school. Chicago Tribune. March 12, 2015. [www.chicagotribune.com] accessed June 25, 2015.

Colwill, J.M., J.M. Cultice and R.L. Kruse. 2008. Will generalist physician supply meet demands of an increasing and aging population? Health Affairs 27: w232–241.

[COMMONWEALTH] Commonwealth Fund. 2014. Mirror, mirror on the wall. Update: How the U.S. health care system compares internationally [www.commonwealthfund.org/publications] accessed September 13, 2015.

D'Ambrosio, M.V., M. Bakalar, S. Vennuru, C. Reber, A. Skandarajah, L. Nilsson et al. 2015. Point-of-care quantification of blood-borne filarial parasites with a mobile phone microscope. Sci Transl Med 7: 286re4.

Day, J., J. Davis, L.A. Riesenberg, D. Heil, K. Berg, R. Davis et al. 2015. Integrating sonography training into undergraduate medical education: A study of the previous exposure of one institution's incoming residents. J Ultrasound Med 34: 1253–1257.

Duda, G.N., D.W. Grainger, M.L. Frisk, L. Bruckner-Tuderman, A. Carr, U. Dirnagl et al. 2014. Changing the mindset in life sciences toward translation: a consensus. Sci Transl Med 6:264cm12. doi: 10.1126/scitranslmed.aaa0599.

Dzau, V. Commencement speech delivered at University of Illinois College of Medicine on May 8th, 2015.

[ENGINEERING] Engineering-based medicine. 2017. A new era of medicine: How engineering will revolutionize the future of medicine [https://engineering.illinois.edu/initiatives/medicine/index.html] accessed November 14, 2017.

[FORBES] Forbes Quotes. 2017a. [https://www.forbes.com/quotes/search/transformation/] accessed November 9, 2017.

[FORBES] Forbes Quotes. 2017b. [https://www.forbes.com/quotes/search/education/] accessed November 9, 2017.

Fung, K. 2015. Otolaryngology-head and neck surgery in undergraduate medical education: advances and innovations. Laryngoscope 125: Suppl 2: S1–S14.

Hershberger, P.J. and D.A. Bricker. 2014. Who determines physician effectiveness? JAMA 312: 2613–2614.

Hwang, J., Y. Jeong, J.M. Park, K.H. Lee, J.W. Hong and J. Choi. 2015. Biomimetics: forecasting the future of science, engineering, and medicine. Int J Nanomedicine 10: 5701–5713.

Langer, R. and R. Weissleder. 2015. Scientific discovery and the future of medicine. Nanotechnology. JAMA 313: 135–136.

McNeer, N.A., K. Anandalingam, R.J. Fields, C. Caputo, S. Kopic, A. Gupta et al. 2015. Nanoparticles that deliver triplex-forming peptide nucleic acid molecules correct F508del CFTR in airway epithelium. Nat Commun 6: 6952.

Michalski, M.H. and J.S. Ross. 2014. The shape of things to come. 3D printing in medicine. JAMA 312: 2213–2214.

[MIT] MIT Libraries. 2016. Seal of the Massachusetts Institute of Technology [http://libraries.mit.edu/mithistory/institute/seal-of-the-massachusetts-institute-of-technology/] accessed July 31, 2016.

Morrison, R.J., S.J. Hollister, M.F. Niedner, M.G. Mahani, A.H. Park, D.K. Mehta et al. 2015. Mitigation of tracheobronchomalacia with 3D-printed personalized medical devices in pediatric patients. Sci Transl Med 7: 285ra64.

Moskowitz, A., J. McSparron, D.J. Stone and L.S. Celi. 2015. Prepare a new generation of clinicians for the era of Big Data. Harvard Medical Student Review. 2015 January 3 [www.hmsreview.org] accessed August 24, 2015.

[NAE] National Academy of Engineering & Institute of Medicine. 2005. Building a better delivery system: a new engineering/health care partnership [https://nae.edu/Publications/Reports/25657.aspx] accessed November 25, 2017.

[NAE] National Academy of Engineering. 2017. Grand challenges-engineer better medicines. [http://www.engineeringchallenges.org/challenges/medicines.aspx] accessed November 14, 2017.

[OECD] OECD health statistics. 2014. How does the United States compare? [www.oecd.org/unitedstates] accessed September 13, 2015.

Plunkett-Rondeau, J., K. Hyland and S. Dasgupta. 2015. Training future physicians in the era of genomic medicine: trends in undergraduate medical genetic education. Genet Med 17: 927–934.

Ricci, T. 2017. How bioengineers are enhancing the quality of healthcare. American Society of Mechanical Engineers [https://www.asme.org/engineering-topics/articles/bioengineers-are-enhancing-quality-of-healthcare] accessed November 14, 2017.

Rietschel, E.T., L. Bruckner-Tuderman, G. Schutte and G. Wess. 2015. Translation. Moving medicine forward faster. Sci Transl Med 7: p.277ed2.

[TEXAS] Texas A & M University. 2017. EnMed. Engineering & Medicine [https://enmed.tamu.edu] accessed November 25, 2017.

[USBLS] US Bureau of Labor Statistics. 2016. Biomedical Engineers: Occupational outlook handbook [www.bls.gov] accessed July 5, 2016.

Weisberg, D.F. 2014. Science in the service of patients: lessons from the past in the moral battle for the future medical education. Yale J Biol Med 87: 79–89.

QUESTIONS

1. What can medical students learn from cross-fertilization of the cultures of engineering and medicine?

2. What can biomedical engineering students learn from cross-fertilization the cultures of engineering and medicine?

3. What perspectives would engineering-medicine provide to the biomedical engineering students?

4. What are the similarities between the educational reform goals of engineering and that of medicine?

5. What are the deficiencies or perceived deficiencies in the traditional medical education?

6. How could the cross-fertilization of the cultures of engineering and medicine help fulfilling the call for the IOM's triple aims of "better care, better health, and lower costs"?

7. What are the ethical concerns relating to the cross-fertilization of the cultures of engineering and medicine?

PROBLEMS

Each student will be asked to brainstorm and bring a real-life medical problem, for which the student has not been able to solve. The student will then think about ways that engineering method may help solving the problem and prepare a class presentation of how engineering may help resolving the problem.

2

General Ethics in Engineering-Medicine

Lawrence S. Chan

QUOTABLE QUOTES

"Leges sine Moribus vanae"

Motto of University of Pennsylvania (UPENN 2016)

"The human spirit must prevail over technology"

Albert Einstein, Scientist (Szczerba 2015)

"The real danger is not that computers will begin think like men, but men will begin to think like computers."

Sydney Harris, Journalist (Szczerba 2015)

Learning Objectives

The learning objectives of this chapter are to familiarize the students the ethical considerations for the novel curriculum of engineering-medicine, including the ethical concerns of curriculum changes, the concern of patient privacy, the concern of compromising quality and safety for efficiency sake, and the concern of over utilizing technology at the expense of physician-patient relationship.
After completing this chapter, the students should:

- Understand the importance of general ethics in medicine.
- Understand ethical consideration of changing learning curriculum.
- Able to use proper methods to guard against patient privacy compromise.
- Understand the ethical concern of compromising safety and quality in the process of improving healthcare efficiency.
- Understand the ethical perspective on increasing technology utilization.
- Understand the ethical perspective in utilizing artificial intelligence in medicine.
- Able to use sound ethical principles in dealing with healthcare issues.

University of Illinois College of Medicine, 808 S. Wood Street, R380, Chicago, IL 60612; larrycha@uic.edu

Introduction to Medical Ethics

At the founding of University of Pennsylvania, a motto was established by the first Provost William Smith in Latin language "Leges sine Moribus vanae", which translates into English language as "Laws without morals are useless" (UPENN 2016). If ethical principles are so essential in guiding the practice of law, they would be more important in the application of medicine. Medicine deals directly with human beings and it intimately affects all of us in a physical, emotional, and psychological ways. Although certain laws and regulations are present to govern how we practice medicine on a global level, it still depends on the individuals to carry out the implementation of the ethical principles. It depends on physicians who daily encounter patients and medical issue, and the biomedical engineers who design specific medical devices or computer algorithm. Therefore, when we propose a novel engineering-medicine teaching curriculum, this new curriculum must be examined at the outset under the microscope of ethics so as to ensure this innovative teaching does not in any way harm the ethical characteristics of medicine. This is for these concerns that this chapter and the following chapter are designed to address.

Ethical Considerations in Engineering-Medicine Curriculum

The potential to improve efficiency through this engineering-medicine curriculum notwithstanding (Chan 2016, 2018, NEW 2015), this kind of new curriculum raises new ethical concerns. Five particular areas of concerns need to be addressed from the teaching institution's perspectives: (1) The ethical consideration of change in teaching curriculum; (2) The ethical consideration of patient privacy protection; (3) The ethical consideration of quality and safety compromise with improving efficiency; (4) The ethical consideration of increased technology utilization; and (5) The ethical considering of utilizing artificial intelligence in medicine. These considerations are important for both medical and biomedical engineering students.

The Ethical Consideration of Changing Teaching Curriculum

The simple fact of changing medical curriculum merits its unique ethical consideration. Since medical students have a fixed amount of four years in learning time, a new content (and time slot) added to the curriculum must be simultaneously balanced by elimination or shortening of an existing content (and time slot). By altering its original curriculum content, one must consider if the "new" teaching materials replacing the "eliminated" materials are ethically justified in a way that the medical students will not miss the opportunity to acquire some knowledge and skill essential to their future medical career. To analyze and attempt to release the concern on this perspective, it will be prudent to ask these questions based on the traditional two-and-two medical curriculum: How much time do the medical students need to devote the first two years of undergraduate medical education to the traditional basic science curriculum? And how much time do they need to dedicate the last two years of undergraduate medical education to the clinical curriculum (clinical clerkship)? Since there is no consensus agreement among medical educators on this perspective and Liaison Committee on Medical Education (LCME) offers no strict requirement guideline, the next best way may be to examine currently how much time do medical students actually spend in these categories of learning at their respective medical schools. It is indeed surprisingly interesting to discover that there is a big variation among US medical colleges in this regard. Furthermore, at the present time many medical schools are undergoing substantial curriculum revision and the clear boundary between the traditional class room-type basic science teaching (first 2 years) and bed-side teaching (clinical clerkship, last two years) has been blurred (YALE 2016, CORNELL 2016, UMICH 2016, COLUMBIA 2016, CASE 2016). Yale University School of Medicine, for example, has initiated, by Fall of 2015, a new integrated basic science/clinical curriculum of one and a half year, followed by a 12-month clinical clerkship, leaving the final 18 months for students to engaged in sub-internship, clinical electives, or biomedical research (YALE 2016). Similarly, Weill Cornell University Medical College established a combined basic/clinical science curriculum for the first 18 months of medical school, followed by a clinical clerkship time period of 12 months, and then followed by the final 18 months devoting to electives and sub-internship (CORNELL 2016). Another example is observed in the curriculum of University of Michigan's MD program, which offered the first 15 months of medical school for basic science-concentrated teaching, following by core clinical rotations to be completed at the remaining of the second year. Dramatically it devotes the entire 3rd and 4th years for students to have the flexibility to earn a dual degree, participate domestic and international clinical electives, develop a primary research project, or to engage in other academic and health-related activities (UMICH 2016). Additional example is Columbia University College of Physicians and Surgeons, which has in its third and final phase of curriculum, dedicating 14 months to electives and scholarly projects (COLUMBIA 2016). Moreover, the School of Medicine at Case Western Reserve University has a curriculum that allows a 10-month time-frame at the fourth year for flexible scheduling of advanced clinical and scientific studies (CASE 2016). Currently the medical schools of Yale University, Columbia University, University of Michigan, Weill Cornell University, and Case Western Reserve University are ranked, among the 141 LCME-accredited US mainland MD-granting medical schools, Nos. 7, 8, 10, 18, and 24 in the 2016 nationwide survey, according to US News and World Report. And there is no apparent deficiency of their medical teachings (USNWR 2016). Thus, from these data, one may conclude that setting aside approximately 9–12 months of time from a 4-year medical school curriculum (up to about 25% total medical school time) for engineering-medicine curriculum by the medical college administration could be ethnically justified, since it would not short change the essential medical training of the medical students, not counting the potential benefits of improving future physicians' efficiency. Having stated that, it is utmost important for the medical educator to ensure that the remaining medical curriculum provides the traditionally needed medical training for the medical students in this special curriculum.

The Ethical Consideration of Patient Privacy Protection

The tendency of utilizing patients' personal information in the application of Engineering-Medicine obviously will raise concern for the privacy protection. One particular subject of engineering-medicine, precision medicine, which utilizes patients' genomic information, is an obvious example of heightened ethical concern (Plunkett-Rondeau et al. 2015). In the following chapter (Chapter 3), a detailed and comprehensive ethical discussion will be dedicated to cover this area and other areas of ethical consideration, in particular to precision medicine, for which extensive discussions are designated to Chapter 12 of this book.

The Ethical Consideration of Quality and Safety Compromised with Improving Efficiency

Engineering-medicine values efficiency (Chapter 11). While efficiency is desirable in bringing the costs down, it may at the same time lead to lower quality and/or safety, therefore raises the ethical concern of trading quality and safety for the sake of efficiency (Vogus et al. 2010). It is, therefore, an imperative for the engineering-medicine college to instill a culture of safety to their students. Vogus and colleagues champion a safety health care culture through three essential processes of "enabling, enacting, and elaborating" (Vogus et al. 2010). These authors defined the "enabling" process as "To enable is to single out and draw attention to safety-relevant aspects of the larger organizational culture, and to create contexts that make it possible for people to translate these aspects into meaningful activities in their local health care routines". Vogus et al. further defined that "enacting" a culture of safety requires "highlighting and accurately representing latent and manifest threats to safety and acting to reduce them." Latent and manifest threats are defined as system error hidden in the design or process and personal errors, respectively, that can potentially lead to safety compromise. To "elaborate" a culture of safety, Vogus et al. stated, is to "evolve, expand, and enlarge the initial set of safety practices that were extracted from the safety climates", i.e., a refinement process (Vogus et al. 2010). The medical school could formulate such teaching about a culture of safety into a class of medical ethics, which may include a series of lectures and group discussions. The medical students participated in their group discussions may be challenged with prudent questions for their discussions. Some of these questions, listed at the end of this chapter, would be excellent thought-stimulations for both medical and biomedical engineering students. Obviously, while undergraduate medical education (medical school) can pave the foundation of such safety culture, this culture needs to be reinforced through the post-graduate medical education (residency) and beyond when the daily practices of the culture of safety actually occur. To encourage the consideration of healthcare quality, a separate chapter in this book (Chapter 15) is devoted for the discussions of quality management.

The Ethical Consideration of Increased Technology Utilization

The engineering-medicine curriculum would, by its very nature, encourage optimal utilization of available technology (Chapters 18–Chapter 24). Such curriculum increases the teaching in technology utilization and carry along with it a new ethical concern (Weisberg 2014, Chan 2016, 2018, Morrison 2016). In addition to the ethical concern of patient privacy due to the increase usage of electronic health record and Big Data, the increase of technology utilization raises additional ethical concerns of erosion of physician-patient relationship, patient mistreatment, reduction of affordability, and impersonal medicine (Weisberg 2014). Thus, it is imperative from the administrative perspective that an engineering-medicine curriculum would include teaching to reinforce a culture of strong physician-patient relationship (Emanuel and Emanuel 1992, Borza et al. 2015, Wachter 2015), respect human dignity (Badcott and Leget 2013, Bailey et al. 2015, Caplan et al. 2015, Chan 2015, Ishii 2015, Baltimore et al. 2015, Foster 2015, Sugarman 2015, Henry et al. 2015, Michael 2014, Akozer and Akozer 2016), technology and patient safety (Vogus et al. 2010, Geiger and Hirshi 2015, Capozzi and Rhodes 2015), and stewardship (Leon-Carlyle et al. 2015, Detsky and Verma 2012, Cooke 2010, Weinberger 2011, Wolfson et al. 2014). Some of the recommendations are discussed below. Again, it would be important that these areas of ethical teaching in the medical school level be reinforced through the post-graduate medical education and beyond when the daily practices of medicine occur.

Building a Culture of Physician-Patient Relationship

Emanuel and Emanuel, in their classic 1992 paper in JAMA, described four models (types) of physician-patient relationship: paternalistic, informative, interpretive, and deliberative (Emanuel and Emanuel 1992). They proposed the best model being the "deliberative" model, especially in cases of chronic disease management (Emanuel and Emanuel 1992, Borza et al. 2015). The objective of this "deliberative" model is best described as "to help the patient determine and choose the best health-related values that can be realized in the clinical situation". Specifically, the physician would delineate the information on the patient's situation, elucidate the patient's values embodied in the available options, suggesting why certain values are more worthy to pursue, thus helping the patient to make the final decision on method of therapy (Emanuel and Emanuel 1992). In this model, physicians act as teacher or friend of their patients. Physician-patient relationship is undoubtedly changing, for example, most physicians nowadays have moved away from the "paternalistic" model, where patient's values are largely excluded from consideration (Emanuel and Emanuel 1992, Borza et al. 2015). Furthermore, patient education has been added to this relationship equation (Borza et al. 2015). While "deliberative" model is a desirable form of physician-patient relationship, it is also a time-consuming process (Emanuel and Emanuel 1992, Borza et al. 2015). Cultivating medical students in the direction of building "deliberative" physician-patient relationship would be best handled in the bed-side teaching format, where medical school faculty members showed their students the practice of "deliberative" relation with patients while mentoring their students in a clinical setting. As the medical educator Robert Wachter pointed out in his beautifully illuminating essay on "Reviving the doctor-patient relationship", "The iPatient can be useful as a way of representing a set of facts and problems, and big data can help us analyze them and better appreciate our choices. But ultimately, only the real patient counts, and only the real patient is worthy of our full attention" (Wachter 2015). On the other extreme, the physician-patient relationship can also become an ethical concern if patients become emotionally dependent on physicians (Ryle 1987).

Building a Culture of Respecting Human Dignity

The concept of human dignity is not easy to define. Bioethicist Adam Schulman defines human dignity as "the essential and inviolable core of our humanity" (Badcott and Leget 2013) and in the famed German philosopher Immanuel Kant's (i.e., Kantian philosophy), the human dignity definition is "the absolute inner worth of being human" (Akozer and Akozer 2016). Though there is not an overwhelming consensus on the definition of human dignity by experts in the field as evident from a recent meeting of the President's Council on Bioethics in 2008, there is a general public recognition when there is a flagrant violation of respect of human dignity (Badcott and Leget 2013). Thus to some extent, this area of ethical education may be better taught in relationship to law, as the federal government has issued and considered many regulations in areas concerning the technology utilization and human dignity. For example, US Food and Drug Administration (FDA) has already provided some regulatory oversight on gene and cell therapy products (Bailey et al. 2015). Many areas of regulations are being actively debated among the academicians (Caplan et al. 2015, Chan 2015, Ishii 2015, Baltimore et al. 2015). A class of "Medical Ethics and Law" included in the curriculum of an engineering-medicine college would serve this purpose well (Foster 2015). Other educational contents could also improve physician's expressions of respect for human dignity. Physicians' attitudes, actions, and behaviors can, in a conceptual model study, either contribute to or detract from conveying expressions of respect for patient dignity (Sugarman 2015, Henry et al. 2015). Including such teaching materials in the engineering-medicine curriculum may yield fruits. Furthermore, conveying dignity in the fundamental framework of human right may also help instilling this essential concept in the young mind of medical and biomedical engineering students (Michael 2014).

In addition, to balance the increase curriculum load in technology, an engineering-medicine education would be better served by including courses of humanity. For example, a first year course termed "Profession of Medicine" taught at the Geisel School of Medicine at Dartmouth University introduces its students to some of the complex human issues in medicine, such as disability and end-of-life care with real patient encounter, aiming to teach students to deal with patients compassionately (Wiencke 2014). A Medical Arts program established in 2009 at the University of Michigan School of Medicine, introduces the students to the humanities and arts, might also be a good model (Crawford 2015).

Building a Culture of Patient Safety in Healthcare Technology

Ethical issues involved healthcare are particularly keen when new technology or technique is first introduced (Geiger and Hirschi 2015). Among different issues are timing of deployment of the technology, method of informing patients regarding potential side effects before they are known, method of proper evaluation of technology, and responsibilities of the patients vs. that of the society (Geiger and Hirschi 2015). Besides safety, other ethical issues involved with new technology are the selection of new technology investment under budget constrain, decision on patient selection, payer for the technology utilization, and conflict of interest (Morrison 2016, Capozzi and Rhodes 2015). Thus, the ethics course in the engineering-medicine curriculum also needs to include discussions about the safety in the use of technology as well other ethical considerations on technology and medicine. It has been strongly suggested that clearly delineated policies regarding who and when to utilize new technology should be developed at the healthcare administration level before they become available, so as to establish a fair utilization formula and to avoid negative community image and potential legal battle (Morrison 2016).

Building a Culture of Cost Consciousness

As medical community and the society as a whole recognize the importance of resource stewardship, so should such recognition be a part of the formal medical education (Morrison 2016, Leon-Carlyle et al. 2015, Detsky and Verma 2012, Cooke 2010, Weinberger 2011, Wolfson et al. 2014). In the past decades, resource stewardship is not one of priority subjects in the medical education curriculum and the other good qualities such as thoroughness, curiosity, and "zebra" (rare and unique case) search were celebrated, rather than restraint (Leon-Carlyle et al. 2015, Detsky and Verma 2012). Evidence-based teaching can be an excellent way of teaching medical students and physicians on the issue of resource stewardship. And in fact, some clinical studies have provided strong arguments that under certain clinical conditions, utilizing less expensive or less invasive treatment options would result in equally good or better clinical outcomes (Carroll 2015, Bachur et al. 2015, Salminen et al. 2015). To ensure the resource stewardship becoming a normal practice pattern, some authors urged, this teaching needs to start early in medical education (Leon-

Carlyle et al. 2015). By integrating stewardship into the lectures and seminars in each of medical school years would likely help students in developing early in their career training a culture cost-consciousness that could be sustained into the future practice (Leon-Carlyle et al. 2015). The same principles of resource consciousness could also be applied to the biomedical engineering education.

The Ethical Consideration of Artificial Intelligence (AI)

Although AI is a technological advancement, the utilization of AI merits a separate discussion in ethical consideration. AI has been utilized to help solving the computation challenge in the big data analytics needed for precision medicine (Chapters 12 and 13) and a detailed discussion is dedicated in Chapter 14. Although most physicians are comfortable with the introduction of new diagnostic techniques and new treatment options, for the sake of improving patient care, some may have variant degrees of uneasiness when it comes to AI. This concern is certainly understandable because the very word of AI invokes a keen sense of "impersonal" feeling. Furthermore, some physicians may feel that the introduction of AI may lead to their loss of control of medical decision making as the "intelligence" of AI may take over physicians' autonomy as healers. Although application of AI in medicine has existed since 1980s (Szolovits 1982), the speed of current AI development is very rapid indeed. In 2016 alone, more than $20 billion has been invested in AI development globally (Bughin et al. 2017). Dr. Reif, President of Massachusetts Institute of Technology (MIT), the very institute that nurtured the early development of AI, has made a decision to commit major resources to this area of research. With more than 200 principal investigators already present in its computer science and artificial intelligence research laboratory, MIT just announced on February 1, 2018 to launch a new institute-wide initiative to advance human and machine intelligence research (Dizikes 2018). At the same time Dr. Reif also raised the awareness that AI will have major impact on our society and that human should be in control of AI destination with his recent opinion article in Boston Globe (Reif 2017). Dr. Reif further urged that the utilization of AI should be aiming for the benefits of all people and that the society must act now to ensure those desirable outcomes (Reif 2017).

Therefore, it is both timely and relevant that AI and its potential impact to be placed in the forefront of discussions in both biomedical engineering and medicine professions. If AI is utilized in our medical decision making, what are the ethical considerations and how then can we remedy those concerns? One of major concerns is whether the introduction of AI to medical education will lead to lose empathy in medical students. As it is there is already a concern of medical students losing empathy in medical school primarily due to the inherent stressful work load and desensitization environment encountered during clinical rotations (Brown 2017). Will the "invasion" of AI further contribute to this erosion of empathy in medical students? Another ethical consideration is whether the introduction of AI into medicine will lead to loss of physicians' job? This concern reflects a general feeling of the population at large about AI, as a recent survey conducted by PEW Foundation found that 72% of US population had major concern that AI might result in job loss (Reif 2017). Even if the introduction of AI will not lead to physicians' job loss, the concern of whether AI will lead to loss of physicians' abilities to make medical decision is real and relevant one.

While there are many ways we could consider in remedying the above mentioned ethical concerns, one good way may be to educate biomedical engineers and students, medical professionals, and medical students about the principles, the limitations, and the potential benefits of AI. A quote from Dr. Reif's Op-ed article here will be appropriate to conclude our discussion: "Automation will transform our work, our lives, our society. Whether the outcome Is inclusive or exclusive, fair of laissez-faire, is up to us. Getting this right is among the most important and inspiring challenges of our time-and it should be a priority for everyone who hopes to enjoy the benefits of a society that's healthy and stable, because it offers opportunity for all" (Reif 2017). Rather than waiting for the whole weight of AI to load upon us suddenly and caught us unprepared and scrambled to respond, academic institutions and professional societies should probably make a proactive move to provide comprehensive information for our students and colleagues regarding this coming phenomenon. Knowledge is the key, as the pioneering microbiologist Louis Pasteur famously said, "chance favors the prepared mind" (FORBES 2018).

Summary

Since an engineering-medicine education would, through innovative, technological, and solution-oriented approaches, teach biomedical engineering students and medical students (and future physicians) the concept and skills to become more efficient in healthcare delivery, this kind of curriculum could provide a positive transformation of medical education in the future. However, this kind of new curriculum also raises new ethical concerns and it is imperative that such ethical considerations are addressed explicitly by the engineering and medical school administrations through formal education curriculum. Some suggestions have been outlined in this chapter and the following one in the hope of generating vigorous discussion in the academic medicine community.

References

Akozer, M. and E. Akozer. 2016. Basing science ethics on respect for human dignity. Sci Eng Ethics 22: 1627–1647.

Bachur, R.G., J.A. Levy, M.J. Callahan, S.J. Rangel and M.C. Monuteaux. 2015. Effect of reduction in the use of computed tomography on clinical outcomes of appendicitis. JAMA Pediatr 169: 755–760.

Badcott, D. and C. Leget. 2013. In pursuit of human dignity. Med Health Care Philos 10: 933–936.

Bailey, A.M., J. Arcidiacono, K.A. Benton, Z. Taraporewala and S. Winitsky. 2015. United States Food and Drug Administration regulation of gene and cell therapies. Adv Exp Med Biol 871: 1–29.

Baltimore, D., P. Berg, M. Botchan, D. Carroll. R. Alta Charo and G. Church. 2015. Biotechnology. A prudent path forward for genomic engineering and germline gene modification. Science 348: 36–38.

Borza, L.R., C. Gavrilovici and R. Stockman. 2015. Ethical models of physician-patient relationship revisited with regard to patient autonomy, values and patient education. Rev Med Chir Soc Med Nat Isai 119: 496–501.

Brown, R.S. 2017. Do students lose empathy in medical school? The Pharos. Autumn 2017. PP23–26.

Bughin, J., E. Hazan, S. Ramaswamy, M. Chui, T. Allas, P. Dahlstrom et al. 2017. Artificial intelligence: the next digital frontier? McKinsey Global Institute.

Caplan, A.L., B. Parent, M. Shen and C. Plunkett. 2015. No time to waste-the ethical challenges created by CRISPR: CRISPR/Cas, being an efficient, simple, and cheap technology to edit the genome of any organism, raises many ethical and regulatory issues beyond the use to manipulate human germ line cells. EMBO Rep 16: 1421–1426.

Capozzi, J.D. and R. Rhodes. 2015. Ethical challenges in orthopedic surgery. Curr Rev Musculoskelet Med 8: 139–144.

Carroll, A.E. 2015. Doing less is sometimes enough. JAMA 314: 2069–2070.

[CASE] Case Western Reserve University School of Medicine Curriculum. 2016. [https://case.edu/medicine/admissions/programs/university-program/curriculum/] accessed 30, Jan. 2016.

Chan, D.K. 2015. The concept of human dignity in the ethics of genetic research. Bioethics 29: 274–282.

Chan, L.S. 2016. Building an engineering-based medical college: Is the timing ripe for the picking? Med Sci Educ 26: 185–190.

Chan, L.S. 2018. Engineering-medicine as a transforming medical education: A proposed curriculum and a cost-effectiveness analysis. Biol Eng Med 3(2): 1–10. doi: 10.15761/BEM.1000142.

[COLUMBIA] Columbia University College of Physicians and Surgeons Curriculum. 2016. [http://ps.columia.edu/education/curriculum/medical-school-currculum/differentiation-in] accessed 29, Jan. 2016.

Cooke, M. 2010. Cost consciousness in patient care-what is medical education's responsibility? N Engl J Med 362: 1253–1255.

[CORNELL] Weill Cornell Medical College Curriculum. 2016. [http://weill.cornell.edu/education/curriculum/] accessed 29, Jan. 2016.

Crawford, L. 2015. The art of medicine. Medicine at Michigan. Summer 2015. [www.medicineatmichigan.org/magazine/2015/summer/art-medicine] accessed September 19, 2015.

Detsky, A.S. and A.A. Verma. 2012. A new model for medical education: Celebrating restraint. JAMA 308: 1329–1330.

Dizikes, P. 2018. MIT News. MIT IQ. Institute launches the MIT intelligence quest: New institute-wide initiative will advance human and machine intelligence research. February 1, 2018. [http://news.mit.edu/2018/mit-launches-intelligence-quest-0201] accessed February 2, 2018.

Emanuel, E.J. and L.L. Emanuel. 1992. Four models of the physician-patient relationship. JAMA 267: 2221–2226.

[FORBES]. 2018. Forbes Quotes. Thoughts on the Business of Life. [https://www.forbes.com/quotes/author/louis-pasteur/] accessed February 21, 2018.

Foster, C. 2015. Hman dignity in bioethics and law. J Med Ethics 41: 935.

Geiger, J.D. and R.B. Hirschi. 2015. Innovation in surgical technology and techniques: Challenges and ethical issues. Semin Pediatr Surg 24: 115–121.

Henry, L.M., C. Rushton, M.C. Beach and R. Faden. 2015. Respect and dignity: A conceptual model for patients in the intensive care unit. Narrat Inq Bioeth 5: 5A–14A.

Ishii, T. 2015. Germline genome-editing research and its socioethical implications. Trends Mol Med 21: 473–481.

Leon-Carlyle, M., R. Srivastava and W. Levinson. 2015. Choosing wisely Canada: integrating stewardship in medical education. Acad Med 90: 1430.

Michael, L. 2014. Defining dignity and its place in human rights. New Bioeth 20: 12–34.

Morrison, E.E. 2016. Ethics in Health Administration. A Practical Approach for Decision Makers. 3rd ed. Jones & Bartlett Learning. Barlington, MA, USA, pp. 165–187.

[NEW] New College. New Medicine. 2015. The first college of medicine specifically designed at the intersection of engineering and medicine. [https://medicine.illinois.edu/news.html] accessed September 17, 2015.

Plunkett-Rondeau, J., K. Hyland and S. Dasgupta. 2015. Training future physicians in the era of genomic medicine: trends in undergraduate medical genetic education. Genet Med. 2015 Feb 12. doi: 10.1038/gim.2014.208.

Reif, L.R. 2017. Transforming automation is coming: The impact is up to us. Boston Globe. Op-ed, November 10, 2017. [https://www.bostonglobe.com/opinion/2017/11/10/transformative-automation-coming-the-impact/az0qppTvsUu5VUKJyQvoSN/story.html] accessed February 2, 2018.

Ryle, A. 1987. Problem of patients' dependency on doctors: discussion paper. J R Soc Med 80: 25–26.

Salminen, P., H. Paajanen, T. Rautio, P. Norstrom, M. Aarnio, T. Rantanen et al. 2015. Antibiotic therapy vs. appendectomy for treatment of uncomplicated appendicitis. JAMA 313: 2340–2048.

Sugarman, J. 2015. Toward treatment with respect and dignity in the intensive care unit. Narrat Inq Bioeth 5: 1A–4A.

Szczerba, R.J. 2015. 20 great technology quotes to inspire, amaze, and amuse. February 9, 2015. [www.forbes.com] accessed July 22, 2016.

Szolovits, P. 1982. Artificial intelligence and medicine. Westview Press, Colorado, USA.

[UMICH] University of Michigan Medical School Curriculum. 2016. [https://medicine.umich.edu/medschool/education/md-program/curriculum/] accessed 30, Jan. 2016.

[UPENN] University of Pennsylvania. 2016. University history. A guide to the usage of the seal and arms of the University of Pennsylvania. [http://www.archives.upenn.edu] accessed June 23, 2016.

[USNWR] US News and World Report. 2016. Best Medical Schools: Research. [http://grad-schools.usnews.rankingsandreviews.com/best-graduate-schools/top-medical-schools] accessed 30, Jan. 2016.

Vogus, T.J., K.M. Sutcliffe and K.E. Weick. 2010. Doing no harm: enabling, enacting, and elaborating a culture of safety in health care. Academy management Perspectives 24: 60–77.

Wachter, R. 2015. Reviving the doctor-patient relationship. The Pennsylvania Gazette May–June: 16–17.

Weinberger, S.E. 2011. Providing high-value, cost-conscious care: A critical seventh general competency for physicians. Ann Intern Med 155: 386–388.

Weisberg, D.F. 2014. Science in the service of patients: lessons from the past in the moral battle for the future of medical education. Yale J Biol Med 87: 79–89.

Wiencke, M.C. 2014. Teaching the intangibles. Dartmouth Medicine. Fall: 32–5.

Wolfson, D., J. Santa and L. Slass. 2014. Engaging physicians and consumers in conversations about treatment overuse and waste. A short history of choosing wisely campaign. Acad Med 89: 990–995.

[YALE] Yale University School of Medicine Curriculum. 2016. [medicine.yale.edu/education/rebuild/index.sapx] accessed 29, Jan. 2016.

QUESTIONS

1. Why is patient safety essential in healthcare?
2. What is the relationship between safety and quality in healthcare?
3. What are the scenarios do you think in health care where emphasizing efficiency may compromise safety or quality of care?
4. Could improving efficiency and quality of healthcare occur at the same time?
5. Would you provide clinical examples where efficiency and quality of care can be enhanced simultaneously?
6. How would you guard against patient privacy compromise in the era of electronic healthcare record and Big Data?
7. How would you define human dignity?
8. Could you describe a recent clinical situation where human dignity was compromised?
9. Do you feel that heavy utilization of electronic health record system contributes to a weaken physician-patient relationship? Is so, how would you modify the medical note recording process for a better physician-patient relationship?
10. Could you give an example where advanced technology may interfere physician's relationship with patients? How could you improve?
11. What are the potential concerns with regard to utilizing artificial intelligence in medical decision?

PROBLEMS

As a final project of this course, students are requested to report one major ethically challenging encounter during their training, and provide detailed discussion on how does the encounter affect the students, what ethical principles do the students utilize to navigate the situation, what were the outcomes of the encounter, and what the students wish he or she would have done differently.

3

Ethics in the Era of Precision Medicine

Michael J. Sleasman

QUOTABLE QUOTES

"The right drug for the right patient at the right time is the mantra of personalized medicine."

Edward Abrahams, Executive Director, Personalized Medicine Coalition (Abrahams 2008)

"So if we combine all these emerging technologies, if we focus them and make sure that the connections are made, then the possibility of discovering new cures, the possibility of applying medicines more efficiently and more effectively so that the success rates are higher, so that there's less waste in the system, which then means more resources to help more people—the possibilities are boundless."

Barack Obama, U.S. President, at the White House, January 30, 2015 (Obama 2015)

"Each generation exercises power over its successors: and each, in so far as it modifies the environment bequeathed to it and rebels against tradition, resists and limits the power of its predecessors. ... In reality, of course, if any one age really attains, by eugenics and scientific education, the power to make its descendants what it pleases, all men who live after it are the patients of that power."

C.S. Lewis, British Author (Lewis 1996)

Learning Objectives

The learning objectives of this chapter are to familiarize the students with the ethical considerations and other related considerations relating to precision medicine.

After completing this chapter, the students should:

- Understand the history and development of precision medicine.
- Understand the potential ethical, legal, and social implications of precision medicine.
- Understand the role of government in the regulation and ethics of precision medicine.
- Understand the role of physician in the ethics of precision medicine treatment.
- Be able to apply the ethical analysis of precision medicine in real-life practice.

The Center for Bioethics & Human Dignity, Trinity International University, 2065 Half Day Road, Deerfield, IL 60015; msleasman@cbhd.org

Introduction to the Ethics of Precision Medicine

Precision medicine seeks to tailor healthcare delivery to the individual patient through the use of genetic, genomic, and other 'omic' (e.g., proteomics, metabolomics, microbiomics, etc.) technologies and information to further patient-centered healthcare. As the latest development in patient-centered healthcare, precision medicine advances previous efforts in individualized and personalized medicine utilizing advances in DNA sequencing and analysis, combined with advances in Big Data and bioinformatics to guide clinical decisions in health prevention, diagnosis, and therapeutic intervention.

Precision medicine holds out the prospect of reducing adverse drug reactions through advances in pharmacogenomics (such as targeted therapeutics tailored to the specific genomic signature of a particular cancer) and creating personalized or patient-specific health protocols based on genetic markers and individualized assessment of disease risk (Collins and Varmus 2015, Chawla and Davis 2013, Jain 2009). Proponents claim that precision medicine will promote increased effectiveness in healthcare delivery due to reductions in overall healthcare expenditures that will result from increases in the use of effective interventions (and the correlate reduction in ineffective interventions). Further improvements will result from the reduction in patient care burden that adverse drug reactions generate (Jain 2009). In short, precision medicine will promote "the right treatment to the right patient at the right time" (IMI 2014), shifting away from the "one size fits all" model of traditional medicine that relied upon a trial-and-error or blockbuster approach to diagnosis and treatment (Yousif et al. 2016).

Those more critical of the prospects for precision medicine argue that medicine has historically promoted the best interest of the individual patient and sought personalized care (Feiler et al. 2017). Rather than a revolution in medicine, the developments of precision medicine are merely an evolutionary step in increasingly sophisticated patient observation that builds upon, but will not replace, successful aspects of traditional medicine (Jain 2009). Furthermore, critics challenge several of the core assumptions of the anticipated benefits of precision medicine, questioning the extent to which: (1) genetic risk information is clinically useful and will generate actual health gains; (2) genetic information will influence preventive behavioral changes; and (3) precision medicine will actually yield projected savings given the increased costs to deploy the diagnostic technologies necessary to leverage genetic and genomic information (Laberge and Burke 2008, Coote and Joyner 2015, Jameson and Longo 2015). Laberge and Burke (2008) also point out that genetic diagnostics alone cannot make healthcare more personalized, but rather, it is "relationships with health care providers who know [the patient], value their perspective, and engage in shared decision-making about health care choices." For precision medicine to be successful, critics charge that it must be incorporated within the broader context of the physician-patient relationship and a commitment to whole person medicine.

Historical Development & Context of Precision Medicine

From the early days of the Human Genome Project (HGP) in the 1990s, the ethical, legal, and social implications (ELSI) of genetics have been a significant area of consideration accompanying the technical scientific research and its translational application in medicine. Alongside the establishment of the National Center for Human Genome Research, the U.S. Congress mandated through the National Institutes of Health Revitalization Act of 1993 that "'not less than' 5% of the NIH Human Genome Project budget be set aside for research on the ethical, legal, and social implications of genomic science" (McEwen et al. 2014). Such ELSI considerations were raised not only from the momentous nature of what was being undertaken in the HGP, but also from residual concerns related to the eugenics sentiments and practices of the first half of the 20th Century. Eugenic beliefs are often most noted for motivating practices of the Nazi Holocaust (including not only the mass extermination of specific people groups, but also euthanizing the elderly and the disabled who were deemed 'unfit for life'). Eugenic beliefs and practices, however, extended to a variety of other countries including the U.S. and Britain resulting in forced sterilizations and "Fit Family" contests. Such concerns regarding eugenic beliefs and practices were compounded with developments in genetics and reproductive technologies of the 1950s–1970s that led to the possibility of genetic engineering. The mapping of the human genome led to legal questions regarding genetic property rights and gene patents and the use of genetic information by law enforcement (e.g., forensics), employers, insurers, and beyond (Andrews et al. 2015).

The prospects of a range of potential genetic interventions raised a unique set of ethical questions that are beyond the scope of this present volume, but generally involve issues of risk assessment, unintended consequences, the permissibility of interventions to the germline, and a discussion of naturalness among other important considerations. The topical discussion of such issues involve the spectrum of applications from genetically modified organisms to gene drives for the potential elimination of mosquito-borne diseases, from gene therapy for inherited disorders to human enhancement, and from gene editing and CRISPR to reprogenetics and the prospect of designer babies (Berry 2007, Green 2007, Kilner et al. 1997, Skene and Thompson 2008).

While the HGP resulted in numerous developments in genetic and genomic science that have permitted advances in direct genetic interventions such as gene therapy and gene editing, one of the key developments involved the increasingly sophisticated technologies and methodologies for DNA sequencing, and the evolution of subsequent generations of diagnostic and interpretive technologies. In contrast to the ethical issues of direct genetic interventions, precision medicine emerges from the diagnostic aspects of the genetic revolution coupled with significant advances in information technology that have involved exponential increases in processing (computing power) and data storage capacity. These advances in technological capability led to exponential decreases in the cost of sequencing a single human genome from more than $40 million in 2003 to the prospect of sequencing whole exomes or genomes for less than $1000 (Jameson and Longo 2015, NRC 2011).

Further developments in the interpretive power of developments in bioinformatics and Big Data converged with a renewed focus on patient-centered healthcare as promoted by proponents of individualized medicine (Chawla and Davis 2013). The convergence of these developments led to rising interest in utilizing genetic and genomic information through stratification of patient subpopulations to tailor preventive and therapeutic interventions (commonly referred to as targeted therapeutics) through personalized (and later precision) medicine initiatives (Feiler et al. 2017). Distinct from the ethical issues raised by direct genetic interventions, the ethics of precision medicine focus primarily on considerations raised by the use of genetic testing and information management (e.g., privacy and informed consent), as well as issues related to healthcare allocation more generally and reflections on the nature of medicine, the increasing use of technology in the physician-patient relationship, and the relationship of human identity and a patient's genetic information.

Defining Terms: The Evolution of Precision Medicine

Despite the novelty of precision medicine with its incorporation of DNA-based patient information and the accompanying critique of the "one size fits all" model of traditional medicine, many point out that care focused on the individual patient has always been the goal of modern medicine (Abrahams and Silver 2011, Collins and Varmus 2015, Vogenberg et al. 2010a), regardless of the sophistication of the diagnostic tools and therapeutic pathways involved. Hippocratic medicine distinguished itself from cultic "medical" practices in the Greco-Roman world by its focus on observational medicine and the classification of disease, a practice underscored by the Hippocratic physician Galen in the second century CE. The ensuing tradition of medical philosophy with this emphasis on patient and disease observation culminates in the empirical and evidence-based approaches of modern medicine.

The terms 'precision medicine' and 'personalized medicine' are often used interchangeably (Jameson and Longo 2015), with notable individuals and organizations retaining the use of personalized medicine. As noted earlier, precision medicine emerged directly from the personalized medicine movement that rose to prominence in the late 1990s and accompanied the Human Genome Project. Personalized medicine likewise drew upon earlier emphases of the individualized medicine movement that dates several decades prior (Jain 2009).

In 2011, the U.S. National Research Council published a report that advocated for the shift in terminology toward "precision medicine" to avoid potential confusion that 'personalized' may mean "that each patient will be treated differently from every other patient" (NRC 2011). By preferring the term 'precision medicine', the National Research Council emphasized the importance of patient stratification, specifically, "the ability to classify individuals into subpopulations that differ in their susceptibility to a particular disease, in the biology and/or prognosis of those diseases they may develop, or in their response to a specific treatment" (NRC 2011). Despite such calls, the terms continue to be used interchangeably in the research literature.

The Future of Precision Medicine

As an emerging area of medical research and clinical translation, the potential benefits and prospects of precision medicine continue to evolve. Advocates have identified at least five areas where research is demonstrating the most potential: (1) diagnosis/prognosis by assessing particular subtypes of a disease; (2) treatment prediction—analyzing likely response to a certain treatment; (3) determining appropriate dosing; (4) safety—"anticipating adverse treatment reactions in certain subpopulations"; and (5) monitoring patient response to treatment to assist in making adjustments to treatment regimen (PMC 2013). The ability of genetic and genomic testing for patient stratification—categorizing patients into subpopulations—of heightened risk factors and the insight into potential clinical effectiveness and reaction to certain treatment regimens, offers the ability to increasingly tailor patient-centered healthcare through targeted therapeutics.

A particular area of attention is the continued improvement in the accuracy and precision of biomarkers, including the potential of CRISPR screens for target discovery and targeted therapeutics (Guo et al. 2017). Furthermore, such research is leading to increasingly precise classification, understanding not just subpopulations of human genetic variation and disease risk, but also opening the possibility for "a more accurate and precise 'taxonomy of disease'" (NRC 2011). The most direct benefit of a more precise classification has occurred in oncology and cancer research, though significant challenges still exist particularly with respect to tumor heterogeneity and clonal evolution (Maughan 2017). Another area of promise for an improved taxonomy of disease may be in mental health and psychiatry, particularly if genetic tests could provide clinically useful information in the disease classification, diagnosis, and/or treatment of brain disorders (Gordon and Koslow 2011).

Future prospects for precision medicine also involve developments in artificial intelligence and drug development. Contemporary developments in medical imaging and electronic health records (EHRs) as well as developments in machine learning and artificial intelligence (AI) may offer even further opportunities for precision medicine to create a multi-faceted assessment of patient health and appropriate preventive and therapeutic pathways (Jameson and Longo 2015). Other developments in machine learning and AI are likely to be used for advanced diagnostic aids and clinical decision support (CDS) systems to be used in tandem with EHRs (Bolouri 2010). Efforts are also underway to better integrate genetic research into the drug development process itself, building upon previous work in pharmacogenetics and pharmacogenomics. Such methods may involve identifying specific drug targets for accelerated development and also may be used to better identify research subjects from specific patient subpopulations for targeted drug trials (Boname et al. 2017).

To achieve future prospects for precision medicine, the number of patients participating in DNA sequencing will need to grow such that researchers will have access to larger pools of medical records and patient genetic information with which to mine for risk factors (Bolouri 2010). At the outset of the U.S. Precision Medicine Initiative, Collins and Varmus (2015) suggested the need for "a longitudinal 'cohort' of 1 million or more Americans" to volunteer for such research, which was later formalized as the All of Us Research Initiative. Similarly, in 2012, the British National Health Service (NHS) launched the 100,000 Genomes project as a key step in their efforts to promote genomic medicine.

Ethical Concerns & Other Challenges for Precision Medicine

The Hype and Promise of Precision Medicine

Particularly within Western countries, the current academic research and funding environment has resulted in criticisms regarding the 'hype" of emerging areas of research in medicine and biotechnology (e.g., stem cell research, nanotechnology, gene therapies, synthetic biology, artificial intelligence, etc.). For precision medicine one particular area of concern involves the optimism with which proponents argue it will generate significant improvements in cost-effectiveness and overall patient health and reductions in patient burden. As has been noted in other areas of emerging research, optimistic language has become a staple of individual research applications and budget proposals for entire fields of inquiry as a result of the increasingly competitive grant procurement process and the overall funding environment (Maughan 2017). Such optimistic language may take multiple forms, such as exaggerating "a project's feasibility, likely results or significance" with regard to benefits (McGinn 2010) and may also lead to media distortion in the coverage of emerging areas of technology development and medical research, and this is no less true for precision medicine. While media distortion is not solely the result of hype within the research community, as McGinn (2010) notes continued "researcher participation in or endorsement of media coverage of scientific or engineering developments that turns [sic] out to be distorted can dilute public trust and foster public misunderstanding of science and engineering." Such activities become counter-productive to active public engagement by impeding important ethical and social considerations necessary for careful public deliberation.

Trends toward Medicalization

Nicol et al. (2016; see also Bolouri 2010, Maughan 2017) have noted that while a purported benefit of precision medicine will be the reduction in unnecessary care both in fiscal terms and with respect to patient burden, there is a real risk the opposite might occur. "[W]ithout appropriate safeguards", Nicol et al. (2016) argue, "precision medicine may drive defensive medical practices, shift standards of care to expect more rather than less intervention, and produce extraneous information of uncertain clinical utility. In turn, this could lead to a growing cohort of the 'worried well'." In short, they suggest that without proper attention, precision medicine could exacerbate existing broader trends to medicalization and the over-treatment of society.

Medicalization describes a conceptual shift which perceives, diagnoses, and treats the individual—particularly their nonmedical problems—from a medical perspective, and identifies the subsequent dangers of overtreatment that result from such a perspective (Hadler 2004, Conrad 2007). Clifton Meador (1994) in a provocative commentary on medicalization describes a fictitious patient encounter set in the near future, where the protagonist is a middle-aged, former stockbroker who retires from his second profession to give full attention to the demands of preventive care and medical testing in his constant pursuit of health. To the patient's chagrin nothing can be found to be wrong with him, but this is the last person for whom this was true. While Meador's satirical reflection predates the rise of both personalized and precision medicine, the concern about the trend toward medicalization has continued. Nicol et al. (2016) describe similar concerns in actual patient care referring to the emerging phenomena of "patients in waiting".

Rising awareness of genetic risk factors may also lead to increasing interventions among previously well populations. Such concerns have already been raised by some healthcare professionals with respect to expansions in the categories of pre-diabetes, prehypertension/borderline hypertension/hypertension, and cholesterol screening. Recent attention has been given to the rise in prescription treatment of borderline hypertension with blood pressure medication and the expansion of the population eligible for statin use for cholesterol. While such concerns may be unrelated to the outcomes of precision medicine, nevertheless, appropriate attention should be directed to the potential "harms of over-diagnosis and unnecessary healthcare interventions" (Nicol et al. 2016).

Paradigm Changes in Medicine

Precision medicine also raises a number of ethical issues surrounding the nature and practice of medicine. One such transition involves the increasing presence of technology in the physician-patient encounter and relationship (e.g., EHRs, patient web portals, patient kiosks in the medical office, email communication, and developments in telemedicine such as teleconsultation). The increasing use of technology within the clinical context has led to concerns about technological distance and the perception of medicine as an increasingly technical discipline focused on patient data at the loss of the art of medicine in the care of the whole person. Haack (2015a) expresses concerns regarding the depersonalizing influence of EHRs in the replacement of physician notes with preset dropdown menus, removing the personal narrative of the patient and the story of the physician's relationship with them. While DNA-based information may give physicians an ever greater understanding of the individual patient and their risk of disease, such information also has the potential for reductionist interpretations that replace other important aspects of whole person care. Attention must be given to balancing these interpretations such that the traditional goals of medicine are not replaced by an exclusive emphasis on cost-effectiveness and efficiency (Haack 2015a). These temptations are not limited to the development and deployment of precision medicine, but rather place such developments within and compounded by already existing tensions in contemporary medical practice.

A second paradigm shift in medicine facilitated by precision medicine is the rise of the patient as research contributor. Proponents note the importance of patient participation for the future success of precision medicine, but much of this participation is predicated on shifting assumptions regarding the role of patients in medical research (Prainsack 2017). Such a transition is driven by the need for population level genetic information from which to derive patient stratification and subpopulation risk assessment, and was most formally initiated at a public level through the incorporation of non-genetic patient information in EHRs as part of meaningful use requirements to improve health outcomes. Patient participation is further complicated with the increasing use of commercial direct-to-consumer (DTC) genetic testing and the anticipated development of cloud-connected patient monitoring devices performing on-demand telemonitoring. Ethical challenges exist regarding the blurring of traditional distinctions between medicine and research science, particularly when the goals and likely outcomes of population health research may be unknown and the role of the patient as participant is open-ended and/or unclear (Prainsack 2017). As noted above, the attention given to ELSI considerations throughout the Human Genome Project anticipated to some degree such ethical, legal, and social issues. Nonetheless, the shifting paradigm will require additional attention to patient and research protections that address conceptual and regulatory ambiguities between medical practice and scientific research, and the ideas of patient empowerment.

Privacy and Security

Chief among the concerns for precision medicine are those that emerge from its information technology roots, principally data privacy and security (Boname et al. 2017) and subsequent issues of consent and data sharing. Data security presents challenges at the hardware and software level in order to prevent data manipulation and corruption, unauthorized disclosures, and other breaches of security. Relatedly, data privacy involves protection from unauthorized disclosures of an individual's personal genetic profile, and the more heightened concern over the possibility of re-identification of genetic data. While a given genetic sequence is typically de-identified as a standard practice of anonymizing data for the purposes of research, Bolouri (2010) and others note the surprising ease of identifying "an individual uniquely and with high confidence" with a modest number of DNA markers (between 30–80).

Informed Consent

Another important area of ethical consideration is the realm of consent, particularly as it relates to the participation in population level research and ongoing access to patient information by primary care providers and third party researchers. While such considerations are standard for many areas of medical research and particularly for those involving biological samples and genetic information, as noted earlier the model advanced by precision medicine assumes large patient participation and ongoing sharing and access to relevant medical data. Ethical issues of genetic testing in general have created questions surrounding disclosure of incidental findings to patients and/or family members, posthumous legal rights regarding genetic information, difficulties in subsequent decisions to opt-out of research, and the previously mentioned concern about re-identification of genetic information. Ongoing disputes in research ethics and bioethics examine the appropriateness of broad consent practices that would authorize all future research in a single consent as contrasted with the more traditional model of narrow consent which only approves specific known uses. Most of the stakeholders involved in precision medicine recognize the importance of the issues for broad patient participation and discuss the importance of trust and collaboration with patients. Successful practices will need to involve more comprehensive patient education regarding the prospects and concerns for involvement in precision medicine initiatives.

Diverse and Representative Patient Stratification

As with many areas of clinical and scientific research racial and ethnic minorities are significantly underrepresented in genome-wide association studies, with the majority of participants of European descent (Boname et al. 2017). This lack of diverse patient information becomes particularly crucial among specific minority subpopulations where the burden of a particular disease is much higher, such as with diabetes or asthma (Boname et al. 2017). A key factor in the potential realization of the societal benefits of precision medicine will be the incorporation of diversified patient populations to facilitate genome-wide association studies that apply more comprehensively to a given society. Others have noted that to realize the full potential of precision medicine more expansive data sets involving socioeconomic status and environmental factors may be necessary (Boname et al. 2017), raising further questions regarding patient privacy and the scope of patient information that should be incorporated within the scope of medical analysis.

The Role of Government in the Regulation and Ethics of Precision Medicine

Given the potential for innovation in healthcare delivery models, it is no surprise that many governments are taking a leading role in the promotion and analysis of precision medicine. One of the government's key roles in the deployment of precision medicine, at least in the U.S., is primarily a regulatory function directed predominantly towards the diagnostic tools used to identify subpopulations of risk and susceptibility, as well as those used in the context of testing the genetic and genomic profile of the individual patient. Regulation of genetic diagnostic tools has been a source of some ambiguity given overlapping oversight roles by different governmental entities (e.g., FDA, CMS). Further ambiguity results from the multiple classifications of a given diagnostic tool as an *in vitro* diagnostic (IVD), a laboratory developed test (LDT), and/or a companion diagnostic medical device.

As an example, the U.S. Food & Drug Administration (FDA) regulates IVDs as a product "used to collect specimens, or to prepare or examine specimens . . . after they are removed from the human body" (FDA 2010). Though it has selectively enforced such diagnostics in the past and exempted others from premarket approval, nonetheless the FDA categorizes companion diagnostics and LDTs as IVDs which fall under its jurisdiction (Kwon 2016, Knowles et al. 2017). The focus of FDA regulatory oversight of IVDs is of "the material or mechanism as a medical device; it does not monitor how a QC [quality control] component is used within a laboratory" (FDA 2018). Meanwhile, the Centers for Medicare & Medicaid Services (CMS) held traditional jurisdiction over the analytical validity of LDTs since the passage of the Clinical Laboratory Improvement Amendments of 1998 (CLIA) which regulated "federal standards applicable to all U.S. facilities or sites that test human specimens for health assessment or to diagnose, prevent, or treat disease" (CDC 2018, Kwon 2016, Knowles et al. 2017). It is notable that while CMS regulates analytic validity of clinical genetic tests, there is no federal oversight of the clinical validity. As such the FDA has proposed policies to expand oversight in this area (NHGRI 2018, Knowles et al. 2017).

Primary regulatory concern surrounding the use of IVDs emphasizes the issues of quality control and the potential for serious, life-threatening risk due to "[m]isdiagnosis and/or error in treatment caused by inaccurate test results" or those that could result in functional impairment or damage to the body (FDA 2010). Such errors may result from false positives, which could also result in additional unnecessary testing and treatment that may cause unnecessary physical and/or psychological harms, or false negatives, which could delay critical care in a timely manner. Aside from concern for quality control, other regulatory interests surrounding IVDs and precision medicine generally follow the traditional considerations to minimize risk in human subject research.

The U.S. federal government also regulates personal health information through the privacy protections of the Health Insurance Portability and Accountability Act of 1996 (HIPAA). HIPAA restricts access to personal health information by requiring consent for the disclosure of personal health information even to immediate family members and to authorize access to any third parties. Furthermore, HIPAA includes provisions monitoring the security and confidentiality of personal health information. Subsequent legislation in 2008 and 2010 extended such protections to include electronic health records, breach-notification obligations, and may prevent covered entities from "remuneration for constructing databases containing a limited set of 'anonymized' patient information" (Vogenberg et al. 2010b). However, despite these expansions, HIPAA regulations apply only "to health plans, healthcare clearinghouses, and healthcare providers" and do not apply to other kinds of organizations that may handle sensitive patient data (Bolouri 2010).

Other regulatory activities by the government include privacy protections and protections against genetic discrimination. As noted above the concern of genetic discrimination would be the misuse of personal genetic information resulting in prejudicial treatment of the individual on the basis of their genetic risk factors. Such misuse could be envisioned in the work sector by prospective employers or in the realm of health or other insurance coverage. The Genetic Information Nondiscrimination Act (GINA) of 2008 was intended to protect individuals from such discrimination, but was limited to a prohibition of such discrimination by employers and health insurers (Abrahams and Silver 2011, Vogenberg et al. 2010b, Bolouri 2010). GINA does not comprehensively prevent potential discrimination from other forms of insurance, and recent attention has been given to long-term insurance and other insurance products that may be exempt from the protections of GINA. Similarly, changes in healthcare regulation have opened the door to the possible use of behavioral incentives (offered in the form of rebates or discounts) to existing insurance customers for their participation in genetic testing or screening activities. While such practices would not directly threaten potential coverage, pricing models for insurance could increasingly incorporate financial disincentives to those who do not wish to participate in such activities regardless of justification.

An area of evolving governmental regulation involves the role of the emerging direct-to-consumer (DTC) genetic testing market. Functioning outside of traditional research and clinical arenas, DTC genetic testing presents challenges both at the level of quality control and patient protections regarding privacy and consent (Knowles et al. 2017, Bolouri 2010). As discussed earlier, the FDA and CMS hold primary authority to regulate genetic tests, including DTC genetic testing. However, given the commercial activities of DTC companies, the U.S. Federal Trade Commission (FTC) has also played a regulatory role investigating select companies accused of false advertising and/or misleading product claims. Meanwhile industry groups and proponents of DTC genetic tests have advocated that consumers should have access to buy and use new tests even if they are not approved by the FDA under right to access or right to know their own genetic information type of arguments (Vogenberg et al. 2010c).

Additional regulatory challenges have emerged as commercial DTC companies have begun using their databases of genetic information to engage in research programs in the private and public sector (Nicol et al. 2016). Such developments raise significant questions of oversight of research and broader considerations for the ethics of research at the convergence of commercial interests, research interests, consumer interests, and human subjects protections. The landscape of governmental regulation of DTC continues to evolve along with the commercial genetic testing industry, and will require constructive participation of industry, regulatory, and consumer advocacy groups in the coming years, particularly as the FDA seeks to expand regulatory oversight in this area (Nicol et al. 2016, Knowles et al. 2017).

While the focus of this section has been on the regulatory environment within the U.S., similar complexities and ambiguities exist within the European Union. One such complexity regards the application of the EU's General Data Protection Regulation (GDPR), which offers strict compliance regarding personal information (including health information) that goes well beyond traditional HIPAA privacy rules and involves both explicit consent, patient access to personal data, and the right to be forgotten (ESR 2017, Horgan et al. 2015). As noted earlier, the success of precision medicine depends upon the creation of large population samples to data mine for clinically significant risk factors. While the GDPR was largely supported as a strong move by privacy advocates, such policies may further complicate efforts in large patient population recruitment. While such efforts need not directly conflict it points to the importance of navigating the interests of patients with public interests in the pursuit of precision medicine.

One such example of private-public collaboration, the Precision Medicine Initiative (PMI) was launched In January 2015 by U.S. President Barack Obama to accelerate discoveries in genetic and genomic medicine and diagnostics, as well as guide the broad deployment of precision medicine. A key aspect of the PMI involved the Privacy and Trust Principles which sought to "engage individuals as active collaborators—not just as patients and research subjects" by providing guidelines in the areas of: governance; transparency; respecting participant preferences; participant empowerment through access to information; data sharing, access, and use; and data quality and integrity (WH 2015). In 2016, the PMI unveiled an additional set of principles focused on data security (WH 2016). These sets of principles set out prospective areas of potential regulatory and legislative action surrounding the collection, storage, and use of genetic and genomic patient information, and were intended to guide decision-making by organizations involved in such activities.

One final area of governmental work involves continued funding of ELSI-like activities assessing the broader implications of developments in 'omic' technologies and their application to healthcare. Such work may involve commissioning reports and ethics bodies dedicated to examining the broader concerns of bioethics and inquiries focused on the specific ethical, legal, and social considerations raised by precision medicine. Prominent examples include the reports *Privacy and Progress in Whole Genome Sequencing* and *Anticipate and Communicate* published by the Presidential Commission for the Study of Bioethical Issues in 2012 and 2013 respectively relevant to genetic testing (PCSBI 2012, 2013). As federal priority is given to various precision medicine initiatives, efforts should be made to insure that ongoing consideration is given to the broader ELSI concerns of such developments.

Role of Physician in the Ethics of Precision Medicine

Finally, a key area in the ethics of precision medicine involves the role of physicians and other healthcare professionals. As Bolouri (2010) notes, "Clinicians cannot be expected to perform ad hoc bioinformatics analyses" and as such cannot be expected to be involved in the detailed scrutiny of "every variant within a patient's genome." Bolouri goes on to note the particular challenges of the increasing complexity of medical diagnosis with the growing volume of patient information and data available through EHRs, a challenge that precision medicine will further exacerbate. While such developments will complicate the actual practice of medicine, they are also accompanied by important ethical concerns as noted previously. Medical education and clinical training must evolve to address the needs of increased physician reliance upon the diagnostics and therapeutics of precision medicine, incorporating more background training in DNA-based information, genome sequencing, and genetic risk factors (Gronowicz 2016), but it must also evolve to incorporate discussion of the ethical issues raised by precision medicine and the rising use of genetic patient information.

The management of all this general and specialized data will present difficulties for the individual physician, but developments in bioinformatics and artificial-intelligence assisted diagnostic aids should provide opportunities for increasing use of CDS systems that work in tandem with EHRs to assist physicians to deploy targeted therapeutics and identify genetic risks for individual patients more readily (Bolouri 2010). While such developments may be welcomed with respect for their potential to manage the increasing complexity of medical diagnosis and subsequent therapeutic options, nonetheless it presents a change to the classic model of physician as medical expert and will further contribute to trends of increasing technological use in the bedside or clinical encounter. Bolouri (2010; see also Haack 2015a) notes that such "sophisticated medical informatics tools" should not be "intended to replace physicians' good practice habits" and should be evaluated by the extent to which they aid such practice habits.

Furthermore, while genetic risk information is likely to offer a number of improvements in areas such as proper dosage and adverse drug reaction, and may assist in the reduction of ineffective therapeutics, such information must also be balanced with other non-genetic risk factors that previously have been demonstrated as effective for guiding healthcare for a given patient (Laberge and Burke 2008). A 2008 study suggested that it was the combination of high risk genetic and non-genetic profiles that saw a substantially elevated odds ratio for coronary infarction (Trichopoulou et al. 2008). Furthermore, while genetic risk information may discourage some clinical practices as a waste of healthcare resources due to lack of personal risk for a given condition, nevertheless some of these practices may be important for overall health. One practice that generated debate involved the recent proposal to eliminate regular cervical screening, seemingly pitting the interests of population health against those of individual care (Haack 2015b). Physicians will need to demonstrate increasing wisdom in balancing these interests as greater levels of emphasis is placed on clinical utility, cost-effectiveness, and genetic risk factors.

One final area of physician engagement with precision medicine involves the previously discussed availability of DTC genetic testing. Given their direct-to-consumer status these tests do not require the involvement of a physician and, thus, remain an area of emerging research as to the clinical validity and utility of such services, as well as the psychological, behavioral, and clinical effects for consumers/patients (Bloss et al. 2011). In one early study of more than 3600 subjects, only 10.4% reported discussing their results with a board-certified genetic counselor employed by the testing company, and only 26.5% reported sharing their results with their physicians (Bloss et al. 2011). Some studies have indicated "no significant effects of communicating DNA based risk estimates" for a wide variety of disease prevention behaviors (Hollands et al. 2016), while others have indicated a positive relationship between physician consultation of results and uptake of preventive behavioral changes (Hayashi et al. 2018).

Questions surrounding likely behavioral influence resulting from knowledge of genetic risk factors are not limited to DTC genetic testing alone, as the evidence for such behavioral modification remains unclear. There are noted areas where such preventive behavioral influence, however, does appear to be effective. One such example is women who obtain results of higher risk for breast cancer being more likely to get regular mammography screening (Laberge and Burke 2008). Regardless, for the benefits of precision medicine to be fully realized more attention needs to be given to relationship between genetic risk information and behavioral change, and the role of the physician in both communicating such information and encouraging preventive behaviors.

Summary

The goals of precision medicine to improve patient health and the cost-effectiveness of healthcare delivery through the use of genetic and genomic patient information are laudable ones. While precision medicine is best seen as an evolution of traditional medical practice through employment of patient stratification and genetic risk factors, it is also important that it not abandon previously successful population health strategies and the historic emphases of traditional medicine on the importance of the physician-patient relationship and whole patient care. As public involvement in precision medicine gains momentum through broader patient participation, significant attention will need to be given to examining the ethical, legal, and social implications and instituting proper regulatory and professional mechanisms for insuring appropriate patient protections, privacy, and appropriately informed models of consent and research authorization. Furthermore, medical schools and continuing medical education will need to evolve to accommodate emerging demands of the technical competencies necessary for the deployment of precision medicine, as well as attendant competencies necessary for robust patient care in the context of expanded use of DNA-based patient information.

References

Abrahams, E. 2008. Right drug—right patient—right time: personalized medicine coalition. Clin Transl Sci 1: 11–12. [https://dx.doi.org/10.1111%2Fj.1752-8062.2008.00003.x] accessed June 26, 2018.

Abrahams, E. and M. Silver. 2011. The history of personalized medicine. pp. 3–16. *In*: Gordon, E. and S. Koslow [eds.]. Integrative Neuroscience and Personalized Medicine. Oxford University Press, New York.

Andrews., L., M. Mehlman and M. Rothstein. 2015. Genetics: Ethics, Law, and Policy. 4th ed. West Academic, St. Paul, MN.

Berry, R. 2007. The Ethics of Genetic Engineering. Routledge, New York.

Bloss, C., N. Schork and E. Topol. 2011. Effect of direct-to-consumer genomewide profiling to assess disease risk. N Engl J Med 364(6): 524–534.

Bolouri, H. 2010. Personal Genomics and Personalized Medicine. Imperial College Press, Hackensack, NJ.

Boname, M., A.W. Gee, T. Wizemann, S. Addie and S.H. Beachy. 2017. Enabling Precision Medicine: The Role of Genetics in Clinical Drug Development. National Academics Press, Washington, D.C.

[CDC] Centers for Disease Control and Prevention. 2018. CLIA law & regulations. [https://wwwn.cdc.gov/clia/Regulatory/default.aspx] accessed July 31, 2018.

Chawla, N. and D. Davis. 2013. Bringing big data to personalized healthcare: a patient-centered framework. J Gen Intern Med 28: 660–665.

Collins, F. and H. Varmus. 2015. A new initiative on precision medicine. N Engl J Med 372 (9): 793–795.

Conrad, P. 2007. The Medicalization of Society: On the Transformation of Human Conditions into Treatable Disorders. The Johns Hopkins University Press, Baltimore, MD.

Coote, J. and M. Joyner. 2015. Is precision medicine the route to a healthy world? The Lancet 385(9978): 1617. [https://doi.org/10.1016/S0140-6736(15)60786-3].

[ESR] European Society of Radiology. 2017. The new EU general data protection regulation: what the radiologist should know. Insights Imaging 8(3): 295–299.

[FDA] U.S. Food & Drug Administration. 2010. Guidance for industry and FDA staff: in vitro diagnostic (IVD) device studies—frequently asked questions. [https://www.fda.gov/downloads/medicaldevices/deviceregulationandguidance/guidancedocuments/ucm071230.pdf] accessed July 31, 2018.

[FDA] U.S. Food & Drug Administration. 2018. Overview of IVD regulation: how does FDA look at quality control. [https://www.fda.gov/medicaldevices/deviceregulationandguidance/ivdregulatoryassistance/ucm123682.htm] accessed July 31, 2018.

Feiler, T., K. Gaitskell, T. Maughan and J. Hordern. 2017. Personalized medicine: the promise, the hype, and the pitfalls. The New Bioethics 23(1): 1–12.

Gordon, E. and S. Koslow [eds.]. 2011. Integrative Neuroscience and Personalized Medicine. Oxford University Press, New York.

Green, R. 2007. Babies by Design: The Ethics of Genetic Choice. Yale University Press, New Haven, CT.

Gronowicz, G. 2016. Personalized Medicine: Promises and Pitfalls. CRC Press, Boca Raton, FL.

Guo, X., P. Chitale and N. Sanjana. 2017. Target discovery for precision medicine using high-throughput genome engineering. pp. 123–145. *In*: Tsang, S. [ed.]. Precision Medicine, CRISPR, and Genome Engineering: Moving from Association to Biology and Therapeutics.

Haack, S. 2015a. The promises and perils of technological progress in healthcare. Dignitas 22(3): 8–11.

Haack, S. 2015b. Transitions in women's healthcare: the impact of the new population paradigm. Dignitas 22(1): 1, 4–5, 7.

Hadler, N. 2004. The Last Well Person: How to Stay Well Despite the Health-Care System. McGill Queens University Press, Montreal.

Hayashi, M., A. Watanabe, M. Muramatsu and N. Yamashita. 2018. Effectiveness of personal genomic testing for disease-prevention behavior when combined with careful consultation with a physician: a preliminary study. BMC Research Notes 22: 223 [https://doi.org/10.1186/s13104-018-3330-9] accessed August 9, 2018.

Hollands, G., D. French, S. Griffin, T. Preost, S. Sutton, S. King et al. 2016. The impact of communicating genetic risks of disease on risk-reducing health behavior: systematic review with meta-analysis. BMJ 352 [https://doi.org/10.1136/bmj.i1102] accessed August 9, 2018.

Horgan, D., A. Pardiso, G. McVie, I. Banks, T. Van der Wal, A. Brand et al. 2015. Is precision medicine the route to a healthy world? The Lancet 386(9991): 336–337. [https://doi.org/10.1016/S0140-6736(15)61404-0].

[IMI] Innovative Medicines Initiative. 2014. The right prevention and treatment for the right patient at the right time: strategic research agenda for innovative medicines initiative 2. [https://www.imi.europa.eu/sites/default/files/uploads/documents/reference-documents/IMI2_SRA_March2014.pdf] accessed June 26, 2018.

Laberge, A.M. and W. Burke. 2008. Personalized medicine and genomics. pp. 133–136. *In*: Mary Crowley [ed.]. From Birth to Death and Bench to Clinic: The Hastings Center Bioethics Briefing Book for Journalists, Policymakers, and Campaigns. The Hastings Center, New York.

Jain, K. 2009. Textbook of Personalized Medicine. Springer, New York.

Jameson, J.L. and D. Longo. 2015. Precision medicine—personalized, problematic, and promising. N Engl J Med 372(23): 2229–2234.

Kilner, J., R. Pentz and F. Young [eds.]. 1997. Genetic Ethics: Do the Ends Justify the Genes? Eerdmans, Grand Rapids, MI.

Knowles, L., L. Westerly and T. Bubela. 2017. Paving the road to personalized medicine: recommendations on regulatory, intellectual property and reimbursement challenges. Journal of Law and the Biosciences 4(3): 453–506.

Kwon, S. 2016. Regulating personalized medicine. Berkeley Technology Law Journal 31(2): 931–960.

Lewis, C.S. 1996. The Abolition of Man. Touchstone Books, New York.

Maughan, T. 2017. The promise and the hype of 'personalized medicine'. The New Bioethics 23(1): 13–20.

McEwen, J., J. Boyer, K. Sun, K. Rothenberg, N. Lockhart and M. Guyer. 2014. The ethical, legal, and social implications program of the National Human Genome Research Institute: reflections on an ongoing experiment. Annual Review of Genomics and Human Genetics 15: 481–505 [https://doi.org/10.1146/annurev-genom-090413-025327] accessed May 18, 2018.

McGinn, R. 2010. Ethical responsibilities of nanotechnology researchers: A short guide. Nanoethics 4(1): 1–12.

Meador, C. 1994. The last well person. N Engl J Med 330(6): 440–441.

[NHGRI] National Human Genome Research Institute. 2018. Regulation of genetic tests. [https://www.genome.gov/10002335/regulation-of-genetic-tests/] accessed on July 18, 2018.

[NRC] National Research Council Committee on A Framework for Developing a New Taxonomy of Disease. 2011. Toward Precision Medicine: Building a Knowledge Network for Biomedical Research and a New Taxonomy of Disease. National Academies Press, Washington, D.C.

Nicol, D., T. Bubela, D. Chalmers, J. Charbonneau, C. Critchley, J. Dickinson et al. 2016. Precision medicine: drowning in a regulatory soup? Journal of Law and the Biosciences 3(2): 281–303.

Obama, B. 2015. Remarks by the president on precision medicine. The White House. [https://obamawhitehouse.archives.gov/the-press-office/2015/01/30/remarks-president-precision-medicine] accessed July 25, 2018.

[PMC] Personalized Medicine Coalition. 2013. Pathways for oversight of diagnostics. [http://www.personalizedmedicinecoalition.org/Userfiles/PMC-Corporate/file/pmc_pathways_for_oversight_diagnostics1.pdf] accessed May 18, 2018.

Prainsack, B. 2017. Personalized Medicine: Empowered Patients in the 21st Century? New York Press, New York.

[PCSBI] Presidential Commission for the Study of Bioethical Issues. 2013. Anticipate and Communicate: Ethical Management of Incidental and Secondary Findings in the Clinical, Research, and Direct-to-Consumer Contexts. Washington, D.C.

[PCSBI] Presidential Commission for the Study of Bioethical Issues. 2012. Privacy and Progress in Whole Genome Sequencing. Washington, D.C.
Skene, L. and Janna Thompson. 2008. The Sorting Society: The Ethics of Genetic Screening and Therapy. Cambridge University Press, New York.
Trichopoulou, A., N. Yiannakouris, C. Bamia, V. Benetou, D. Trichopoulous and J. Ordovas. 2008. Genetic predisposition, nongenetic risk factors, and coronary infarct. Arch Intern Med 168(8): 891–896.
Vogenberg, F.R., C.I. Barash and M. Pursel. 2010a. Personalized medicine: part 1: evolution and development into theranostics. Pharmacy and Therapeutics 35(10): 560–562, 565–567, 576.
Vogenberg, F.R., C.I. Barash and M. Pursel. 2010b. Personalized medicine: part 2: ethical, legal, and regulatory issues. Pharmacy and Therapeutics 35(12): 670–671, 673–675.
Vogenberg, F.R., C.I. Barash and M. Pursel. 2010c. Personalized medicine: part 3: challenges facing health care plans in implementing coverage policies for pharmacogenomic and genetic testing. Pharmacy and Therapeutics 35(11): 628–631, 642.
[WH] The White House. 2015. Precision medicine initiative: privacy and trust principles. [https://obamawhitehouse.archives.gov/sites/default/files/microsites/finalpmiprivacyandtrustprinciples.pdf] accessed July 15, 2018.
[WH] The White House. 2016. Precision medicine initiative: data security policy principles and framework. [https://obamawhitehouse.archives.gov/sites/obamawhitehouse.archives.gov/files/documents/PMI_Security_Principles_Framework_v2.pdf] accessed July 31, 2018.
Yousif, T., K. Bizanti and B. Elnazir. 2016. Uses of personalized medicine in current pediatrics. International Journal of Clinical Pediatrics 5: 1–5. [http://dx.doi.org/10.14740/ijcp241w].

QUESTIONS

1. How is precision medicine related to traditional medical care?

2. What is the developments that resulted in the emergence of personalized and precision medicine?

3. What are some of the potential benefits of precision medicine over traditional medical care?

4. What are some of the ethical considerations of precision medicine?

5. How would you define informed consent in the context of precision medicine?

6. Could you describe a clinical situation in which informed consent might be compromised?

7. What are some of the regulatory issues for precision medicine?

8. Do you feel that incorporation of genetic information into the electronic health record system and the use of genetic diagnostic aids will improve patient care? What effect might this have on the clinician's practice and the physician-patient relationship?

9. Do you feel that over-diagnosis and unnecessary medical treatment are a likely consequence of precision medicine? If so, what steps might be taken to prevent such results?

PROBLEMS

As a part of class discussion, students are requested to explore the use of direct-to-consumer genetic testing by patients and to determine the extent which such results should be incorporated into clinical guidance or included within the patient's care history. Discuss how to counsel patients regarding the use of DTC testing. Finally, discuss the merits and areas of concern with promoting broad patient participation in genetic testing such as is involved with the All of Us Research Initiative.

4

Engineering Principles Overview

William C. Tang

QUOTABLE QUOTES

"Scientists dream about doing great things. Engineers do them."

James A. Michener, American Author (BRAINYQUOTE 2018a)

"Creativity is not just for artists. It's for businesspeople looking for a new way to close a sale; it's for engineers trying to solve a problem; it's for parents who want their children to see the world in more than one way."

Twyla Tharp, American Dancer, Choreographer, and Author (BRAINYQUOTE 2018a)

"At its heart, engineering is about using science to find creative, practical solutions. It is a noble profession."

Queen Elizabeth II (BRAINYQUOTE 2018b)

"I have always been an engineer devoted to the potential of advanced technologies. Like most engineers, I have a keen sense of curiosity and a deep desire to learn."

Min Kao, Engineer and Entrepreneur, Co-founder of GPS maker Garmin (BRAINYQUOTE 2018a)

"Engineers ... are not mere technicians and should not approve or lend their name to any project that does not promise to be beneficent to man and the advancement of civilization."

John Fowler, British Engineer (WORKFLOWMAX 2018).

"It is a great profession. There is the satisfaction of watching a figment of the imagination emerge through the aid of science to a plan on paper. Then it moves to realization in stone or metal or energy. Then it brings job and homes to men. Then it elevates the standards of living and adds to the comforts of life. That is the engineers' high privilege."

Herbert Hoover, American Engineer and 31st President of US (WORKFLOWMAX 2018).

Learning Objectives

The learning objectives of this chapter are to help the students to acquire the overall concept and perspective of the broad and interdisciplinary field of engineering.

University of California, Irvine, Department of Biomedical Engineering, 3120 Natural Sciences II, Zot 2715, Irvine, CA 92697-2715; wctang@uci.edu

After completing this chapter, students are expected to:

- Understand the essence of engineering and how it serves as the backbone of human development.
- Understand the most fundamental principle of engineering to benefit humankind.
- Understand the process principle of Identify → Invent → Implement.
- Able to utilize the principle of engineering to approach broadly defined problem and seek viable and practical solutions with expedience, cost effectiveness, and societal and ethical awareness.

Engineering Defined

In the simplest of terms, engineering is to make things. While science studies what happens in the natural world, engineering solves problems by creating new things. There are two qualifiers in this definition: the action of making is intentional and the things being made do not yet exist or are not readily available in nature. For example, a person builds a bridge over a river with the intention of efficiently bringing something to the other side without getting wet. The assumptions are that a means to cross the river without getting wet does not yet exist and that the person does not build the bridge accidentally or for no good reason. To examine the definition of engineering more closely, it is central to point out that the word "engineering" is derived from the Latin *ingenium*, which means something like a brilliant idea or a flash of genius (IAE 2018). As a matter of fact, the very core of engineering cannot be separated from purposeful invention and innovation. Since the earliest traceable existence of human beings through archeological discoveries, engineered structures, tools, utensils, etc., mark the defining major eras. The Stone Age refers to the broad period when human made structures and tools out of stone. For 3.4 million years, our ancestors had been engaging their inventive creativity and the dexterity of their hands to engineer devices and structures to first and foremost enhance survivability of our species, a very practical and vital purpose (Harmand et al. 2015). Ko suggested that human intelligence (our inventive abilities) and tool making (engineering) are tightly linked in the long history of human development (Ko 2016). There are also archeological evidences that engineered structures serve not only the practical purpose to survive, but also to express our artistic and metaphysical thoughts. For example, some of the world's oldest free-standing structures are the Ġgantija temples in Gozo, Malta, dated back to circa 3,600 BC. These limestone structures that are older than the pyramids of Egypt are believed to be places of worship. The Stone Age gradually transitioned to the Bronze Age between 8,700 BC and 2,000 BC, which is then followed by the Iron Age. This three-age Stone-Bronze-Iron system, as proposed by Christian Jürgensen Thomsen, uses the hallmarks of what we engineered to define and classify ancient human society (Heizer 1962). Engineering as a profession is built upon the bedrock of scientific discoveries and mathematical skills, motivated by the desire to benefit society, and powered by our mental capacity to invent and innovate. The broad definition of engineering, therefore, consists of four elements:

- Engage the human ingenuity
- Utilize scientific and mathematical knowledge
- Build and operate a structure, a process, a system, or an organization
- For the purpose of benefiting humankind

Biomedical engineering, being the newest of all engineering sub-disciplines, also follow these four essential elements, with the main purpose of benefiting humankind in the area of medicine, to provide knowledges, methods, and equipment for the diagnosis, treatment, and prevention of diseases.

Engineering Principles Defined

Having delineated the definition of engineering, we now focus on the principles of engineering. The principles of engineering are a set of processes and guidelines by which knowledge and skills are efficiently and effectively employed to devise novel solutions to a problem, which, when solved, will promote welfare to society.

As mentioned earlier, the Latin origin of the word engineering means a brilliant idea or a flash of genius. However, there is no consensus among neural scientists on how our thoughts are formed and far less on how a person acquires a brilliant idea (Kandel et al. 2012). At the same time, human experience in the field of engineering culminated into a set of principles that facilitate the generation of novel ideas to solve complex problems with a purpose. It should be emphasized that the best novel ideas are those generated with a purpose. The more compelling and urgent the purpose, the more novelty we will produce. Dr. Thomas Fogarty is a cardiovascular surgeon who is generally recognized as one of the most prolific medical device inventors in history, with many of his technologies in active use across a wide spectrum of patient care. He has founded or co-founded over 30 companies and was inducted into the National Inventors Hall of Fame in 2001. Based on his unparalleled wealth of experience in innovating medical devices, Dr. Fogarty asserted that "the most important principle is that we innovate to improve the lives of patients. Commitments to ourselves, the institution we serve, and others are secondary" (Yock et al. 2015). Thus, the first and foremost principle in engineering, in the case of engineering for medicine, is the unwavering purpose of improving the lives of patients.

Having established the foundational principle in engineering for medicine, we shall now examine the process principles. The Biodesign Program established at Stanford University in early 2000 offers one of the most effective process in medical technology innovation (Yock 2015). Using numerous case studies and concepts based on hands-on experience, they established a three-phase biodesign process:

Identify → Invent → Implement

Identify

The most important starting point of any engineering endeavor must be to identify a compelling need. This need must be clearly defined that is significant and of the right scope. The end goal of the *Identify* step is a clear and succinct statement of the need.

Significance

Before committing time and resources to an engineering project, one must first ascertain the significance of the project. The project's important impact naturally leads to passion and enthusiasm of the participants. In fact, it is the passion to solve a particular problem that sustains the engineer through obstacles and failures along the way. This same passion also translates into tremendous joy and reward for the engineer when the end goal is finally achieved. It is important to note that different persons are attracted to different projects for different reasons, and therefore it will be a good idea that a careful personal inventory be carried out before assigning project. The purpose of the inventory is to identify the mission of the individual or team, as well as their strengths and weaknesses (Yock et al. 2015). Having a clear self-awareness of strengths and weaknesses and clearly defined personal mission, the engineer can quickly narrow down to personally relevant and significant projects.

The Right Scope

There are many unmet needs in the world, particularly in the medical field. We must pick and choose the right unmet need and the right scope of that need to work on to optimize the use of limited resources and the chance of success. One should pick the proverbial low-hanging fruit first, if you will. This low-hanging fruit should have most of the following characteristics:

- The relevant scientific background and mathematical foundations are sufficiently well understood.
- There have been demonstrated engineering solutions to similar (but not identical) or lesser unmet needs.
- The unmet need affects a large portion of the population or community.
- The unmet need is particularly urgent.
- The benefit-to-cost ratio can be estimated, and is potentially high.
- The solution to the unmet need is financially rewarding.

None of the items in the above list is in itself absolutely needed, and their relative priority is highly dependent on the engineering team. Therefore, again, the engineers must invest the time in developing an accurate personal inventory, which will help clarify how to prioritize the above characteristics, which then leads to a sensible and reasoned decision in choosing the unmet need to pursue. It should be noted that the choice is an iterative process requiring patience and persistence in objectively investigating each of the 6 characteristics as related to the unmet need. Sometimes, it may be necessary to abandon an unmet need that may at first looks attractive and start all over again. It is much easier and far less costly to change our minds at this stage than to do so in later stages.

Invent

Although how our brains work in generating ideas and novelties is not fully understood, we have accumulated enough experience to figure out how to facilitate the generation of ideas, or ideation. A two-step approach has been proposed and its effectiveness has been widely demonstrated (Yock et al. 2015): Concept Generation and Concept Screening.

Concept Generation

A powerful tool that facilitates the proliferation of ideas in a team setting is brainstorming. This approach was first described and practiced in business development by Alex Osborn (Osborn 1979) more than 60 years ago, and is still widely practiced today. There are variations in how a brainstorming session is conducted for different context and purposes. In general, a typical brainstorming session may include 6 to 10 participants with different backgrounds and trainings, but share the same goal of addressing the unmet need. The difference in backgrounds among the participants greatly facilitate the cross-pollination of fresh ideas from different perspectives. One of the participants will serve as the facilitator to start the process, keep the process going, and conclude the session. A good brainstorming session should last about an hour with the end goal of generating as many as 100 ideas. Props, figures, drawings, etc., that are relevant to the unmet needs should be brought into the room to stimulate creative thinking. Other than that, items that may distract the process should be minimized. During the session, all ideas are welcome and are succinctly written on a piece of sticky note and posted on sizable white board(s). Evaluation and criticism on the ideas must be deferred. The purpose of the session is to generate ideas. Any judgment on an idea will stifle further idea generation. Evaluation of these ideas should come later. This process is succinctly summarized into the following 7 rules (Kelley 2001):

- Defer judgment—don't dismiss any ideas.
- Encourage wild ideas—think "outside the box."
- Build on the ideas of others—no "buts," only "ands."
- Go for quantity—aim for 100 ideas in 60 minutes.
- One conversation at a time—let people have their say.
- Stay focused on the topic—keep the discussion on target.
- Be visual—use objects and toys to stimulate ideas.

Concept Screening

After a broad range of ideas has been generated from the brainstorming sessions, a disciplined process proceeds in screening and choosing the most suitable solution. This step involves comparing all of the generated ideas against the defined need specification to evaluate how well they satisfy the need (Yock et al. 2015). The steps to follow in concept screening include:

- Grouping and organizing ideas—the reason of using sticky notes to write down the generated ideas is to allow grouping and organizing by moving them around. They can be organized according to the mechanism of action or technical feasibility. This allows the engineers to visually identify gaps, biases, and synergies among ideas.

- Identifying gaps—this will become obvious after the organization is done carefully. The identified gaps will become topics of further brainstorming to uncover missed opportunities.
- Uncovering biases—if the grouped ideas tend to concentrate on certain particular mechanisms of action, then additional participants from a different background and training may be invited to the brainstorming session.
- Discovering synergies—when two or more groups of ideas present synergies, then they can be combined into a synergistic approach.
- Differentiating general vs. concrete—when certain group of ideas tend to be mostly general approaches and lacking in concrete ideas, it will also serve as topics for further brainstorming.

It is very rare that the process of *Invent* can be completed without iterating between Concept generation and Concept screening. After several rounds of these two steps, the engineers will prepare a concept map that visually represents the different clusters of possible solutions to the unmet need. The engineers will now proceed to the following step.

Comparing ideas against the need—each of the concept should be evaluated rigorously, consistently and objectively against the specifications from the unmet need. This will shed invaluable light on the "low-hanging fruit" among the different possible solutions.

Implement

Implementing the chosen solution is by far the longest, most complex, and costliest stage. Even after we have put in the best efforts in *Identify* and *Invent*, success in *Implement* is not guaranteed. In the medical field, implementation of a great idea involves serious efforts in development strategy and painstaking planning. All the following aspects must be carefully considered simultaneously:

- Intellectual property and protection
- Regulatory strategy including process management and quality system
- Reimbursement strategy
- Business strategy including marketing, sales, and distribution, and
- Combining all assets and strategies to develop a sustainable competitive business advantage

These aspects must be considered simultaneously because they interact with each other in a complex and intricate way (Yock et al. 2015). More details on these aspects are discussed in the Invention and Innovation Chapter (Chapter 7).

Summary

Engineering is to make things creatively. The creative process of engineering is backbone of human development and the chief reason for the survival of our species. The most important principle in engineering is that it must serve a benevolent purpose for humankind. The process principles can be summarized with a three-step innovation process of Identify → Invent → Implement.

References

[BRAINYQUOTE] 2018a. Brainy Quote. Engineers Quotes. [https://www.brainyquote.com/topics/engineers] accessed August 11, 2018.

BRAINYQUOTE] 2018b. Brainy Quote. Engineers Quotes. [https://www.brainyquote.com/authors/queen_elizabeth_ii] accessed August 15, 2018.

Harmand, S., J.E. Lewis, C.S. Feibel, C.J. Lepre, S. Prat, A. Lenoble et al. 2015. 3.3-million-year-old stone tools from Lomekwi 3, West Turkana, Kenya. Nature 521: 310–315.

Heizer, R. 1962. The background of Thomsen's three-age system. Tech Cult 3: 259–266.

[IAE]. International Association of Engineers. 2018. [www.iaeng.org] accessed August 10, 2018.

Kandel, E.R., J.H. Schwartz, T.M. Jessell, S.A. Siegelbaum and A.J. Hudspeth. 2012. Principles of neural sciences, 5th Ed., McGraw-Hill Education, New York, USA.

Kelley, T. 2001. The art of innovation. Currency/Doubleday, New York, USA.

Ko, K.H. 2016. Origins of human intelligence: The chain of tool-making and brain evolution. Anthro Notebooks 22: 5–22.

Osborn, A.F. 1979. Applied imagination: Principles and procedures of creative problem-solving, 3rd Ed., Scribner, New York, USA.

[WORKFLOWMAX]. 2018. The unconventional guide to work. [https://www.workflowmax.com/blog/101-engineering-quotes-from-the-minds-of innovators] accessed August 11, 2018.

Yock, P.G., S. Zenios, J. Makower, T.J. Brinton, U.N. Kumar, F.T.J. Watkins et al. 2015. Biodesign: The process of innovating medical technologies, 2nd Ed., Cambridge University Press, Cambridge, UK.

QUESTIONS

1. What is the key difference between the discipline of science and that of engineering?
2. Why is engineering important for our human society?
3. What are the three important steps of engineering process principle?
4. Why judgement should be avoided during the "concept generation" step?
5. Why is the "significance" step of engineering project so important?
6. Why is unmet need an essential element of the project scope discussion?

PROBLEM

Students are grouped into 5 or more people in each. Within each group, students will first brainstorm to identify a real-life medical conditions for which the students need to find answers either to make correct diagnosis or to determine best treatment option. The students will together as a group, try to solve the problem by utilizing the 3 essential engineering process principle: Identify, Invent, and implement. Finally, they will present their works to the entire class.

5

Engineering-Medicine Principles Overview

Lawrence S. Chan

QUOTABLE QUOTES

"Think like an engineer, act as a physician."

Dr. Lawrence S. Chan, Medical Researcher and Educator, Editor of Medical Textbooks (Chan 2018).

"Courage is rightly esteemed the first of human qualities because it is the quality which guarantees all others."

Winston Churchill, Prime Minister of UK (BRAINY QUOTES 2018)

"In parallel with the incorporation of engineering in biomedical research, medical education should also integrate engineering principles."

Engineering as a new frontier for translational medicine (Chien et al. 2015)

"We might possess every technological resource...but if our language is inadequate, our vision remains formless, our thinking and feeling are still running in the old cycles, our process may be 'revolutionary' but not transformative."

Adrienne Rich, American Poet (BRAINY QUOTES 2017)

Learning Objectives

The learning objectives for this chapter are to define and detail the general principles of engineering-medicine.

After completing this chapter, the students should:

- Understand the definition of the transformative engineering-medicine.
- Understand what are the necessary actions for such transformation.
- Understand the principles of engineering-medicine.
- Understand the key of engineering-medicine is to "think like an engineer and act as a physician".
- Understand the thinking framework of engineers as applicable to medicine discipline.

University of Illinois College of Medicine, 808 S Wood Street, R380, Chicago, IL 60612; larrycha@uic.edu

Engineering-Medicine Defined

Having examined engineering principles in Chapter 4, we would now extend our discussion to delineate the general concept of engineering-medicine. But before we discuss the general concept, we should first define what engineering-medicine is. As of today, there is no general consensus in the academic medicine community in defining "engineering-medicine". Nevertheless, we may gain some insights from three related educational domains: We will first view the definition of "biomedical engineering". Secondly, we will examine the mission statements of education institutes with combined goals of engineering and medicine. Thirdly, we will look at the educational statement posted by a proposed new medical college aiming to teach "engineering-based medicine".

The US Bureau of Labor Statistics offers this definition of "biomedical engineering" in terms of what biomedical engineers do (USBLS 2016). It states that "Biomedical engineers combine engineering principles with medical and biological sciences to design and create equipment, devices, computer system, and software used in healthcare."

Another definition of biomedical engineering comes from Meriam-Weber Dictionary that was established since 1828 (MERRIAM 2016). It defines "bioengineering" as "biological or medical application of engineering principles or engineering equipment—called also biomedical engineering."

Established in 1995 at the Massachusetts General Hospital, with collaboration partners of Massachusetts Institute of Technology the Harvard Medical School, and the Shriners Burns Hospital, The Center for Engineering in Medicine declared "The mission of the Center for Engineering in Medicine is twofold: first, to train MDs, PhDs, and predoctoral students in the fundamentals of biomedical engineering; and second, to bring the principles and tools of biomedical engineering to the forefront of biomedical research and patient care" (CEM 2016).

The Engineering in Medicine Program coordinated by the Thayer School of Engineering and the Geisel School of Medicine at Dartmouth University offering both combined MD/MS and MD/PhD programs in medicine and bioengineering, describes its education and research goals this way: "Engineering in medicine research addresses today's technology-driven healthcare system. Advances depend not only on clinical expertise, but also on those trained to look at the technical side of patient care" (THAYER 2016).

Created in 1996, the Institute for Medicine and Engineering at the University of Pennsylvania provides this mission statement "The mission of the institute for Medicine and Engineering (IME) is to stimulate fundamental research at the interface between biomedicine and engineering/physical/computational sciences leading to innovative applications in biomedical research and clinical practice" (IME 2016).

Similarly, the Biomedical Engineering Department at Yale School of Engineering and Applied Science posted this educational goal: "Biomedical Engineering at Yale has two related goals: first, the use of the tools and methods of engineering to better understand human physiology and disease; second, the development of new technologies for diagnosis, treatment, and prevention of disease" (YALE 2016).

Another way to characterize engineering-based medical education is to view the two educational goals depicted by the national first proposed "engineering-based" medical college, Carle-UI College of Medicine (NEW 2015). These two goals are "(1) Re-invent health care around revolutionary advances in engineering and technology to further research, education, and health care delivery. (2) Transform health care education of physicians through the development of team-based innovative approaches to achieving improved health care outcomes through the continuum of care: preventive medicine, chronic disease management, acute care, rehabilitative medicine, and end of life care" (NEW 2015). Instead of the traditional discipline-based approach (like anatomy, biochemistry, etc.), this curriculum would converge medicine and engineering, quantitative and computer sciences, and technology to teach human body as an integrated system that is critical to the development of analytical and problem-solving skills. Engineering technologies and approaches would be incorporated throughout the entire 4-year curriculum. For example, in the course of microbiology, its curriculum will teach an understanding of microbes and engineering approaches to the alteration of genomes and biological circuits. In the computer science course, its curriculum will educate students the data mining methods and patient care guideline development. And in the clinical training, its curriculum will incorporate 3D printing, bioreactor, and advanced analysis techniques (NEW 2015). Accordingly, the emphases of this "engineering-based" curriculum seem to be engineering technology and team-based approach.

The new EnMed track (a medical sub-track that combines engineering and medicine) of Texas A & M University stated that "EnMed would be an innovative engineering medical school option created by Texas A & M University at Houston Methodist Hospital to educate a new kind of doctor who will create transformative technology for health care" (ENMED 2017). Technology seems to be the focus for this EnMed program.

With the above statements examined, an engineering-medicine education may be characterized as "an educational system that aims to teach and train medical students not only the traditional theory and practice of patient care, but also the principles of biomedical engineering in innovative and efficient approaches to health care delivery and in state-of-the-art technology utilization for understanding, diagnosis, treatment, and prevention of human diseases" (Chan 2016). Likewise, engineering-medicine will teach students of biomedical engineering not only the engineering aspects of biology and medicine, but also a broad perspective of how the principles of engineering can help elevating medicine to be more efficient, precise, innovative, and higher quality.

Engineering-Medicine Principles Defined

In a succinct statement, it is the principle goal of engineering-medicine to train our current and future biomedical engineers and physicians to "think like an engineer and act as a physician" (Chan 2018). As the "mind and hands" concept depicted by the motto on the seal of Massachusetts Institute of Technology indicates, the engineering process requires the intellectual direction to govern the physical activities (MIT 2018). Similarly, the proper engineering-medicine applications need the engineering-trained minds to direct the medical applications. Thus, the most important training for the engineering-medicine curriculum would be the principles of engineering thinking and approach. The American Poet Adrienne Rich succinctly pointed out the important link between changing our way of thinking and transformation, "We might possess every technological resource…but if our language is inadequate, our vision remains formless, our thinking and feeling are still running in the old cycles, our process may be 'revolutionary' but not transformative" (BRAINY QUOTES 2017). In other word, simply adding advanced biomedical technology into medical education will never be able to fulfill the transforming goals of engineering-medicine. To think like an engineer, physicians and biomedical engineers educated through the engineering-medicine curriculum are expected to approach medical problems in these manners:

- Think **logically** and mechanistically, rather than merely fact memorization.
- Think whole patient as an **interconnected**, coordinated, and integrated system, rather than isolated body parts.
- Think the medical care of patient in a coordinated and **integrated** manner, rather than fragmented fashion.
- Think invention and **innovation** of healthcare delivery, rather than simply follow the current format and method.
- Think **efficiency** of the entire system, rather than isolated parts.
- Think healthcare delivery **design and optimization**, rather than simply follow the tradition.
- Think in terms of healthcare **problem-solving**, rather than simply problem-identification.
- Think to achieve the best **quality** in medical care and medical management.
- Think in terms of **environmental** impact.
- Think in terms of **precision** in medical applications.

In short, engineering-medicine is about principles and applications of engineering in medicine.

Accordingly, this textbook is arranged to teach the medical students and biomedical engineering students the following thinking processes as depicted in the following table:

Engineering-Medicine Principle	Course of Teaching
Logical thinking	Engineering Principle
Systematic thinking	Systems Biology, Big Data
Team thinking	Systems integration
Innovative thinking	Invention and Innovation
Efficiency thinking	Efficiency, Design Optimization
Problem-solving thinking	Problem-solving
Quality thinking	Precision, Quality Management
Environmental thinking	Environmental Protection
Technology thinking	Advance Biotechnology, Biomedical imaging, Artificial Intelligence
Precision thinking	Precision, Big Data

In the following chapters, the engineering thinking and approaches, as applied to medicine, will be discussed in greater details. After setting the important tone of medical ethics in Chapters 2 and 3 and providing an economic evaluation of engineering-medicine education in Chapter 6, this book will cover the engineering-medicine subjects with quintessential engineering characteristics (Chapters 7–12), Big Data (Chapter 13), artificial intelligence (Chapter 14), quality management (Chapter 15), human biology (Chapter 16), systems biology (Chapter 17), and then subjects of biotechnology (Chapters 18–24). In addition, we will discuss the implementation of engineering-medicine from academic health center and leadership's perspectives (Chapter 25). To conclude, the subject of environmental protection was included to raise the awareness of this essential issue (Chapter 26). Medical examples will be provided to illustrate the application of engineering-medicine principles in the medical teaching and learning process. As emphasized in the preface regarding the major aims of this textbook, we hope to train future physicians and biomedical engineers to achieve the thinking and approaches of an engineer, rather than providing details on the specific medical subject matters. In the final analysis, it is the thinking that guides the actions.

Summary

Engineering-medicine is aiming to train future physicians and biomedical engineers, first and foremost, the principles of engineering as applicable to medicine. Engineering-medicine is training future physicians and biomedical engineers to think as engineers think, logically, efficiently, innovatively, precisely, globally, systematically, solution-oriented, integrated, and in terms of quality and environmental impact. Without educating the engineering thinking and approach, the simple addition of advanced biomedical technology will not fulfill the goals of engineering-medicine and will not be transforming. With a clear understanding of engineering principles in thinking and approach, engineering-medicine will guide the training of advanced biotechnology applications, for the engineers and physicians, in the most optimal manner, to best serve our patients.

References

[BRAINY QUOTE] Brainy Quote. 2017. Adrienne Rich Quotes. [https://www.brainyquote.com/authors/adrienne_rich] accessed November 11, 2017.
[BRAINY QUOTE] Brainy Quote. 2018. Winston Churchill Quotes. [https://www.brainyquote.com/authors/winston_churchill] accessed May 11, 2018.
[CEM] The Center for Engineering in Medicine. 2016. [http://cem.sbi.org] accessed July 5, 2016.
Chan, L.S. 2016. Building an engineering-based medical college: Is the timing ripe for the picking? Med Sci Edu 26: 185–90.
Chan, L.S. 2018. Engineering-medicine as a transforming medical education: A proposed curriculum and a cost-effectiveness analysis. Biol Eng Med 3: 1–10. Doi: 10.15761/BEM.1000142.

Chien, S., R. Bashir, R.M. Nerem and R. Pittigrew. 2015. Engineering as a new frontier for translational medicine. Sci Transl Med April 1; 7: 281fs13. Doi: 10.1126/scitranslmed.aaa4325.

[ENMED] Engineering and Medicine. 2017. Creating the future of health. EnMed. https://enmed.tamu.edu/] accessed June 30, 2017.

[IME] Institute for Medicine and Engineering. 2016. About the IME. University of Pennsylvania. [http://hosting.med.upenn.edu/ime/] accessed July 5, 2016.

[MERRIAM] Merriam-Webster Dictionary. 2016. Bioengineering. [http://www.merriam-webster.com] accessed July 5, 2016.

[MIT] 2018. Seal of the Massachusetts Institute of Technology. MIT History. MIT Libraries. [https://libraries.mit.edu/mithistory/institute/seal-of the Massachusetts-institute-of-technology/] accessed August 6, 2018.

[NEW] New College. New Medicine. 2015. The first college of medicine specifically designed at the intersection of engineering and medicine. [https://medicine.illinois.edu/news.html] accessed September 17, 2015.

[THAYER] Thayer School of Engineering at Dartmouth. 2016. Engineering in Medicine. [http://engineering.dartmouth.edu] accessed July 5, 2016.

[USBLS] US Bureau of Labor Statistics. 2016. Biomedical Engineers: Occupational outlook handbook. [www.bls.gov] accessed July 5, 2016.

[YALE] Yale School of Engineering and Applied Science. 2016. Biomedical Engineering. [http://seas.yale.edu/departments/biomedical-engineering] accessed July 5, 2016.

QUESTIONS

1. How would you define "Engineering-Medicine"?

2. Why engineering-medicine is not simply adding advanced biomedical technology to medicine?

3. What is the most important education aspect of engineering-medicine?

4. What do you understand as "thinking like an engineer"?

5. What are the characteristics of engineering approach as applied to medicine?

6. How would "thinking like an engineer" help improving healthcare delivery?

7. What are some of the ethical concern with regard to medical education transformation by engineering-medicine?

PROBLEMS

Students are grouped into 3–5 people each. In each group, students will brainstorm and come up with one patient care issue for the purpose of exercise. Based on this patient care issue, the students will derive two different approaches to the issue: one method tries to solve with traditional medical methodology and the other technique attempts to resolve with engineering-medicine principles and document the full participations of each student. Students will discuss the difference of each approach and the pros and cons of each approach.

6

Economy of Engineering-Medicine Education
A Cost-Effectiveness Analysis

Lawrence S. Chan

QUOTABLE QUOTES

"The Internet is the first technology since the printing press which could lower the cost of a great education and, in doing so, make that cost-benefit analysis much easier for most students. It could allow American schools to service twice as many students as they do now, and in ways that are both effective and cost-effective."

John Katzman, Author and Educationist (BRAINYQUOTE 2018)

"By using our renewable resources, energy can be generated in a carbon-responsible, cost-effective manner that also creates jobs and mitigates climate change, providing a workable and affordable solution that will keep our planet healthy."

Tulsi Tanti, Entrepreneur (BRAINYQUOTE 2018)

"Vaccines are the most cost-effective health care interventions there are. A dollar spent on a childhood vaccination not only helps save a life, but greatly reduces spending on future healthcare."

Ezekiel Emanuel, American Physician, Medical Educator, and Medical Ethicist (BRAINYQUOTE 2018)

Learning Objectives

The learning objectives of this chapter are to introduce to the students the rationale, method, and results of an economic evaluation of the transforming engineering-medicine education. A brief cost-effectiveness analysis on such an engineering-medicine education, from a societal perspective and under a set of assumptions, results in a positive cost-effective outcome.

University of Illinois College of Medicine, 808 S Wood, Suite 380, Chicago, IL USA; Larrycha@uic.edu

After completing this chapter, the students should:

- Understand the rationale of conducting an economic evaluation for a new educational endeavor.
- Understand the rationale of conducting a cost-effectiveness analysis for engineering-medicine education.
- Understand the method of cost-effectiveness analysis.
- Understand the essential elements of cost-effectiveness analysis.
- Understand the limitation of cost-effectiveness analysis.
- Able to perform cost-effectiveness analysis.

Rationale and Methods of Cost-Effectiveness Analysis on Engineering-Medicine Education

In the first chapter of this book, we delineated the rationale for a transforming Engineering-Medicine education and its potential impact on healthcare (AMA 2015, Chan 2016). Having stated the importance of Engineering-Medicine education in the future medical practice from an educational perspective, we now turn to another important question: Can we determine if such education is cost-effective from the health economic perspective? In the US, perhaps in the whole world, only two schools have similar programs. An EnMed program established by the Texas A & M University will start its new medical education program in 2017 for a portion of their students (TEXAS 2017) and a new engineering-based medical school will start its inaugural class in 2018 (CARLE 2017). There is no successfully tested program one can use to evaluate its cost-effectiveness. Therefore, in order to conduct such economic evaluation, we have to rely on other available data and some assumptions. Part of the following analysis has been published in an open access journal (Chan 2018).

Let us first define the key elements of this particular cost-effectiveness analysis. Cost-effectiveness analysis in healthcare is a scientific method conducted in a systematic manner with the purpose to assist leaders in making logical decisions on interventions or programs meant to improve health (Drummond et al. 2015, Muennig and Bounthavong 2016). Such methods will help healthcare managers focus on what works and reduce the chance of decision error or waste. In short, the "whole point of cost-effectiveness, after all, is to examine the optimal course of action when there is considerable uncertainty" (Muennig and Bounthavong 2016). Since uncertainty is one element of cost-effectiveness analysis, there is no absolute guarantee for its outcome. Accordingly, the key elements of cost-effectiveness analysis include the followings (Drummond et al. 2015, Muennig and Bounthavong 2016):

- *Perspective*: Before going about the cost-effectiveness analysis, it is imperative that the perspective of the analysis is delineated, whether it is from the perspective of the institution or from the perspective of the society, since the benefits to these entities are different. For this project, the perspective will be from the society as a whole.
- *Competing alternative*: Although this point seems to be obvious, it needs to be clarified. To make a decision on a new intervention or program, we need to compare both the cost and effectiveness of the new one with that of the existing one (conventional or traditional one). For this specific calculation, the competing alternative for engineering-medicine education is the traditional medical education. It should be clear for all concerned that there is a mutual exclusivity between the new proposal and the traditional mode, as this is an "either/or" decision.
- *Determine the costs*: It is essential that the new intervention or program will not be cost prohibitive even if it is more effective. In this project, the costs used for comparison will include the physical facility, information technology (IT) infrastructure and maintenance, faculty, and administrative staff. We will estimate such costs in dollar terms.
- *Determine the effectiveness*: This point is front and center of the analysis, as the new intervention or program should be at the minimum as effective as the existing one, if not more effective. The obstacle here is that there is no school of engineering-medicine that have successfully produced physicians so

that we can examine their "educational products". In the absence of such data, the best effort is made to collect existing studies analyzing interventions or programs utilizing engineering principles that resulted in improving healthcare effectiveness. The assumption is that these interventions or programs, if incorporated into medical education, would have the similarly effective outcomes. We will define the effectiveness as cost saving for the healthcare system in dollar terms.

- *Calculate the Cost-effectiveness Ratio*: Once the cost and effect are determined, this ratio will provide substantial help to the decision-making process. The formula of this ratio is the difference of cost (between the new one and the competing alternative) divided by the effectiveness (between the new one and the competing alternative). Since both cost and effectiveness are determined in dollar terms, a ratio smaller than one (1.0) would indicate that the increase of effectiveness outweighs the increase of cost, thus indicating the program in consideration is cost-effective.

- *One-way sensitivity analysis*: In this analysis, we will test the sensitivity of projected outcomes when we vary the input assumptions (one variable) in different directions.

- *Two-way sensitivity analysis*: Finally, we will test the sensitivity of projected outcomes by varying both the cost side and the effectiveness side of the equation.

Cost Analysis

First, let us delineate our analytical assumptions. For the purpose of this analysis, we will assume the new medical college, regardless if it is an engineering-medicine college or a traditional one, will have an inaugural class size of 50.

With our assumptions set, the cost data were collected by the following manner:

Physical Facility Costs: Traditional vs. Engineering-Medicine

To estimate the cost for a new medical school education building, we will examine the costs of similar buildings in some recently completed facilities. For example, University of North Dakota, built a new educational building in 2013 with a total of 325,000 square feet space that cost $125 Million (UND 2016). Another example is the new educational building of Cooper Medical School of Rowan University which cost $139 million, with the capacity of 200,000 square feet to accommodate 100 students per class (ROWAN 2016). Another new medical education building to be completed in 2016 for University of Texas in Rio Grande Valley cost $54 million with the capacity of 88,250 square feet, including classrooms, conference rooms, study rooms, faculty offices, simulation center, digital anatomy lab, an auditorium, a library/learning center, and a student lounge (AAMC 2016). For a new medical school with a projected class size of 50, it is estimated that we need an educational building of 100,000 square feet, sufficient to contain some small class rooms, an auditorium with the size to accommodate 250 attendees (50 students/ class X 4 classes +50 faculty and others), some laboratories, and a simulation center, an anatomy lab, as well as some study rooms for students and office spaces for faculty and staff. An estimate of $80 million is considered sufficient for a conventional school. Let us further assume that the more technology-intense Engineering-Medicine curriculum will require a facility that is relatively larger and equipped with more advanced instrumentation than the traditional medical college. We will make an assumption that 15% more a price tag is required for the education building of the engineering-medicine school of the same size, equating to $92 million. If we assume a 5% interest loan amortize the building expense to a 25-year usage, the annual expense for the medical education building will be $5,676,196 and $6,527,626, for the traditional and engineering-medicine school, respectively (Table 1).

Information Technology (IT) Costs: Traditional vs. Engineering-Medicine

Let us also assume that IT cost in the more technology-intense Engineering-Medicine curriculum will require an expense in IT infrastructure and maintenance that is 10% higher than the conventional medical college. If we set the annual IT cost for traditional school to be $500,000, the corresponding cost for engineering-medicine school will be $550,000 (Table 1).

Table 1. Annual cost comparison between conventional and engineering-medicine school.

Cost Item	Conventional School	Engineering-Medicine School
Information Technology	$500,000	$500,000 × 1.1 = $550,000
Faculty		
Basic Science	$67,932 × 65 = $4,415,580	$4,415,580
Clinical	$249,000 × 195 = $48,555,000	$48,555,000
Engineering	0	$97,023 × 13 = $1,261,299
Staff	$60,000 × 100 = $6,000,000	$6,000,000
Subtotal	$59,470,580	$60,781,879
Education Building	($80 million loan, 5% interest)	($92 million loan, 5% interest)
(Amortized to annual cost)	$5,676,196	$6,527,626
Total	$65,146,776	$67,309,505
Nation-wide total	$65,146,776 × 1,496 =	$67,309,505 × 1,496
(total students)	$97,459,576,896	$100,695,019,480
Differential:		+$3,235,442,584

Faculty Costs: Traditional vs. Engineering-Medicine

On the faculty equation, the engineering-medicine curriculum will obviously require a new set of engineering faculty, in addition to the medical faculty (UFL 2013, BLS 2016, HIGHEREDU 2016). In terms of number of faculty relative to number of students, currently there is no standard to follow and there is a wide variation among medical schools in the US. A review of published literature revealed some interesting findings. Harvard University, ranked No. 1 Best Research-oriented Medical School by the U.S. News & Report this year, has a full-time faculty to student ratio of 13.3:1. Other medical schools on this list showed that University of Pennsylvania (ranked No. 3), University of Michigan (ranked No. 11), University of North Carolina at Chapel Hill (ranked No. 22), Case Western University (ranked No. 25), Brown University (ranked No. 35), and Medical College of Wisconsin (ranked No. 55), Drexel University (ranked No. 82), and University of Central Florida (ranked No. 88) have full-time faculty to student ratio of 4.7:1, 2.6:1, 1.8:1, 3.0:1, 1.4:1, 1.91, 0.6:1, and 0.8:1, respectively (USNEWS 2016). Obviously, there is no good correlation between the faculty to student ratio and the quality of education per se, but the better schools tend to have at least a 1.3:1 ratio. It is not clear, however, to what extent these faculty members participate in direct medical student teaching. Thus the exact direct teaching contribution of these faculty members is not defined. For the purpose of this analysis, we will use a ratio of 1.3:1. Thus for the full capacity of student body of 200, we will aim for the faculty members of 260, including both basic science and clinical faculty, for the traditional medical school calculation. In terms of ratio of basic science to clinical faculty, there is also no standard. We will make the assumption of 25% basic science and 75% clinical faculty. Thus we will need 65 basic science and 195 clinical faculty members for a traditional medical school. Using salary data from University of Florida, the average annual salary pooled from 18 specialties of clinical faculty was approximated to be $249,000 (UFL 2016). For the engineering-medicine school, we will calculate additional 13 engineering faculty members (of Associate Professor level), one for each of the engineering subjects depicted in the curriculum in the section above. According to a higher education survey, the average biological science and engineering faculty annual salary (2015–16) at the Associate Professor level are $67,932, and $97,023, respectively (HIGHEREDU 2016). Thus an additional $1,261,299 annual cost for engineering faculty members will be calculated into the engineering-medicine equation. The detailed calculation of cost will be depicted in Table 1.

Administration Costs: Traditional vs. Engineering-Medicine

For this item, the assumption is that there will be no increase of cost between traditional and the engineering-medicine schools. We estimated that 100 staff members are needed (Table 1).

Comparing Total Costs: Traditional vs. Engineering-Medicine

According to a recent (2015) data, the total number of medical graduates per class in the United States is 18,705 (KFF 2016). This total graduate number multiplied by 4 and then divided by 50 will give a factor of 1,496, which will be used to multiply by the annual cost of a 50-student-per-class school to obtain the nation-wide total cost (Table 1). Finally, we determined by extrapolation that the engineering-medicine school, if operated for the entire United States, will cost $3,235,442,584 ($3.235 billion) more than the traditional medical school annually (Table 1). Having determined the cost differential, we now move to examine the effectiveness.

Effectiveness Analysis

As we did for the cost analysis, we will also define the following analytical assumptions: Effectiveness will be defined by the healthcare saving in dollar terms, whether it is accomplished by increased efficiency or by reducing cost (Reid et al. 2005).

With the assumptions set, the next step will be data collection. Since circulatory, musculoskeletal, respiratory, and endocrine group of diseases, along with ill-defined conditions, account for the top 5 disease groups where the US health system is spending its largest sum of money, it is logical to collect as much data in relation to these diseases, for the purpose of this cost-effectiveness analysis. In 2012, one organization estimated that US national expenditures were estimated to be $241 Billion, $186 Billion, $157 Billion, and $138 Billion, for circulatory, musculoskeletal, respiratory, and endocrine diseases, respectively (TRACKER 2016).

Healthcare System Savings on Musculoskeletal Diseases

Geriatric hip fracture is a rather common medical problem among senior citizens (Lau et al. 2013, Lau et al. 2017). It will be prudent to consider ways to improve clinical outcomes and to reduce cost in these clinical encounters. Utilizing the engineering concept of integration, an implementation of integrated, collaborative, standard treatment protocol called the Geriatric Hip Fracture Clinical Pathway (GHFCP) resulted in improvement of clinical outcomes and in reduction of provider manpower utilized for each fracture occurrence. Specifically, the length of hospital stays in the surgery and recovery were reduced by 6.1 days and 14.2 days respectively (an overall 50% reduction). In addition, the post-operative pneumonia infection rate was reduced from 1.25% to 0.25% (a 1% reduction) (Lau et al. 2017). According to a study, the medical expenditures for osteoporotic fractures in the US in 1995 was estimated to be $13.8 billion, with about $8.5 billion spent for hip fractures, for which the hospital costs were about 65% (Ray et al. 1997). Thus, the total costs of in-patient hip fracture care will be about $5.5 billion each year. For a 50% reduction of hospital stay, the healthcare saving could be near *$2.75 billion* annually for the US, even without counting the potential saving from reduction of post-operative pneumonia by this new pathway, assuming no increase of cost in the implementation of this integrated system of GHFCP.

Healthcare System Savings for Diabetes

Diabetes is a major endocrine disease where the engineering principle of integration could help reducing the cost, which was estimated between $100 and $245 billion in the US for the year of 2012 (TRACKER 2016, ADA 2016a, Grant and Chika-Ezeriche 2014). The estimated expenditure in physician office visits is 9%, accounting for $15.5 billion (0.09 × $175 billion, which is used as the mid figure between $100 and $245 billion). Among the common non-acute diabetes complications and their required corresponding primary and specialty physician office visits are: high blood pressure and stroke (primary care & neurology), hyper- and hypo-glycemia (endocrinology), heart diseases (cardiology), neuropathy (dermatology & podiatry), retinopathy (ophthalmology), gastroparesis (gastroenterology), and kidney malfunction (nephrology) (ADA 2016b). If we use a conservative estimation that on average a patient with diabetes will encounter 50% of these complications, then a patient would need to see 5 different physicians to control their disease co-morbidities. Without integrative care, these patients would need to set up appointment and commute to see

5 different physicians in 5 separated times. During physician office visits, each of these 5 physicians would need to take a history, perform physician examinations, order diagnostic tests, and prescribe appropriate treatments. Engineering integration could help improve effectiveness by transforming the care delivery to an integrated diabetes center. Like that of a cancer center where patients with cancer will get all the necessary and coordinated cancer-related care in one place, an innovative diabetes center will provide all the necessary and coordinated diabetes-related care in one place. Logistically, patients with diabetes will be able to make appointments with all 5 physician office visits in an integrated manner. During the coordinated office visits, the patient will first see a primary care physician, who will take a comprehensive history and physical examination, order common laboratory tests, and prescribe non-specialty treatments, then send the patient to be seen by the first specialty physician within the diabetes center in the same day. The first specialty physician will then utilize the medical record completed by the primary care physician (that contains most essential history, physical findings, lab results, and treatments), perform a focused specialty-related physician examination, order specialty-related lab tests and treatments, and then send the patient to the second specialty physicians also located in the same diabetes center, and so on down to the 5th physician. This kind of engineering-based integrative care will not only save patients' time for arranging and commuting to 5 physician visits, it will importantly, also save physicians' manpower. If we conservatively estimate that on the average 30% of a physician visit is spent in taking the history, we will save equivalent physician manpower of 1.2 office visit (0.3 visit × 4) for each diabetes patient we care for, or overall a physician manpower saving of 24% ((0.3 × 4)/5). Extrapolating this saving to a nation-wide equation, we could potentially save *$3.72 billion* ($15.5 billion × 0.24) annually, assuming no cost increase will be needed to implement this integrated care system. Other potential saving from this integrated diabetes care will be the reduction of the expenditures on duplicated lab tests. In addition, a societal benefit will be the reduction of non-productive time and energy of the diabetes patients. If we assume each of the 24 million diabetic patients in the US will have one annual visit to their respective physicians (ADA 2016c), and if we conservatively estimate that each physician visit will cost a patient non-productive time of 1.25 hour (30 minute in commute, 45 minutes in office visit), the total annual cost will be 6.25 hours. On the other hand, the integrated visit will cost non-productive time of 3.35 hours (30 minutes in commute, 171 minutes or 2.85 hour (45 + (45 × 0.7 × 4)) in office visits), we will potentially save 69.6 million hours of productive time (6.25 − 3.35) × 24 million) for diabetes patients annually. Using a minimum wage of $15/hour, we could easily save the US society *$1.04 billion* annually. Other potential savings to the society include reduction of expenses on gasoline, automobile repair and depreciation.

Healthcare System Savings for Heart Diseases

Cardiovascular diseases as a group has the second highest costs of healthcare in the US. Among this group of diseases, heart failure has a substantial cost: In 2012, the direct cost of heart failure in the US was $21 billion, 80% of which was the cost of hospitalization, accounting for $16.8 billion annually (HEART 2016). The mean cost of a single congestive heart failure readmission is $13,000, with a 25.1% readmission rate (CMS 2016). In a study published in 2011, healthcare systems in the State of Maine leveraged an integrated care system, which was able to reduce the heart failure readmission by 5.83% (from 18.5% to 12.67%) (Cawley and Grantham 2011). If readmission accounts for 25% of total hospital admissions of heart failure, this integrated system could potentially reduce the cost of hospital readmission of congestive heart failure by *$245 million* ($16.8 billion × 0.25 × 0.0583) annually in the US, assuming there is no increase cost in implementing this integrated heart failure care system.

Cost-effectiveness Ratio

{Ratio} = (Cost of Engineering-Medicine − Cost of Traditional School)/(Effectiveness of Engineering-Medicine − Effectiveness of Traditional School) = ($3.235 billions)/($2.75 + $3.72 + 1.04 + $0.245) billion = $3.235 billion/$7.755 billion = 0.42.

According to the above calculation with the said assumptions, the engineering-medicine school would be cost-effective.

One-way Sensitivity Analysis

- Assuming there is a 30% reduction on the side of effectiveness due to increased cost in implementing the clinical integration:

$$Ratio = \frac{\$3.235 \text{ billion}}{\$7.755 \text{ billion} \times 0.70} = \frac{3.325}{5.4285} = 0.61$$

Cost-effectiveness is still achieved by the engineering-medicine school.

- Assuming 75% efficiency of the physician efficiency:

$$Ratio = \frac{\$3.235 \text{ billion}}{\$7.755 \text{ billion} \times 0.75} = \frac{3.325}{5.816} = 0.57$$

With 75% efficiency, the engineering-medicine school remains highly cost-effective.

- For 50% efficiency of the physician efficiency:

$$Ratio = \frac{\$3.235 \text{ billion}}{\$7.755 \text{ billion} \times 0.5} = \frac{3.325}{3.8775} = 0.86$$

Even with 50% efficiency, Engineering-medicine remains cost-effective.

- Breakeven % efficiency of the physician effectiveness (Ratio = 1)

$$1 = Ratio = \frac{\$3.235 \text{ billion}}{\$7.755 \text{ billion} \times \% \text{ physician efficiency}}$$

{Breakeven point} = $3.235/$7.755 = 42% physician efficiency.

Two-way Sensitivity Analysis

- Assuming 75% efficiency of the physician effectiveness

  Assuming also the cost of engineering-medicine school will be more than initially projected: The IT cost is 20% above the traditional medical school, staff requirement for the engineering-medicine school is increased by 10%, and the engineering faculty requirement is increased by 30% to 17 (Table 2). With the above modifications, we still find engineering-medicine school to be cost-effective:

Table 2. Two-way Sensitivity Analysis with increased costs in Engineering-Medicine School & reduced physician efficiency to 75%.

Cost Item	Conventional School	Engineering-Medicine School
Information Technology	$500,000	$500,000 × 1.2 = $600,000
Faculty		
Basic Science	$67,932 × 65 = $4,415,580	$4,415,580
Clinical	$249,000 × 195 = $48,555,000	$48,555,000
Engineering	0	$97,023 × 17 = $1,649,391
Staff	$60,000 × 100 = $6,000,000	$60,000 × 110 = $6,600,000
Subtotal	$59,470,580	$61,819,971
Education Building	($80 million loan, 5% interest)	($92 million loan, 5% interest)
(Amortized to annual cost)	$5,676,196	$6,527,626
Total	$65,146,776	$68,347,597
Nation-wide total	$65,146,776 × 1,496 =	$68,347,597 × 1,496 =
(total students)	$97,459,576,896	$102,248,005,112
Differential:		+$4,788,428,216
{Cost-effectiveness Ratio} = $4,788 billion/($7.755 × 0.75) billion = 0.82 (still cost-effective)		

$$Ratio = \frac{\$4.788 \text{ billion}}{\$7.755 \text{ billion} \times 0.75} = 0.82$$

- Assuming the efficiency of physician effectiveness is reduced further to 50% and the costs are also increased the same amount as above.

{Cost-effectiveness Ratio} = $4,788 billion/($7.755 × 0.50) billion = 1.23

Now the engineering-medicine school is no longer cost-effective.

Summary

Through data collected or estimated for three major diseases where engineering principles could save healthcare dollars, we showed, based on limited available data, that engineering-medicine education has the potential to help generate healthcare savings and it could be cost-effective from the societal perspective. This analysis revealed relative insensitivity to variation of efficiency of physician effectiveness alone in a one-way sensitivity method, but demonstrated relative sensitivity to combined variation of cost and efficiency of physician effectiveness in a two-way sensitivity analysis. Having determined its potential cost-effectiveness, engineering-medicine education should be appropriate to conduct in a small segment of undergraduate medical schools as a proof of concept project. If successful, it could prove to be a good model for future medical education reform. Since the cost-effectiveness is demonstrated from a societal perspective and not necessarily from a medical college (institutional) perspective, the additional cost for training these engineering-physicians should probably be bored by society which would stand to benefit from this novel path of education.

References

[AAMC] 2016. New Buildings. [https://www.aamc.org] accessed December 5, 2016.

[ADA] 2016a. The cost of diabetes. American Diabetes Association. [http://www.diabetes.org/advocacy/news-events/cost-of-diabetes.html] accessed December 4, 2016.

[ADA] 2016b. Complications. American Diabetes Association. [http://www.diabetes.org/living-with-diabetes/complications/] accessed December 4, 2016.

[ADA] 2016c. Diabetes prevalence expected to double in next 25 years. American Diabetes Association. [http://www.diabetes.org/newsroom/press-releases/2009/diabetes-prevalence-expected-to-double.html] accessed December 4, 2016.

[AMA] 2015. Accelerating change in medical education. American Medical Association. [www.ama-assn.org] accessed June 25, 2015.

[BLS] 2016. Biomedical Engineers: Occupational outlook handbook. US Bureau of Labor Statistics. [www.bls.gov] accessed July 5, 2016.

[BRAINYQUOTE] 2018. Brainy Quote. Cost-effective quotes. [https://www.brainyquote.com/topics/cost-effective] accessed August 5, 2018.

[CARLE] 2017. Carle Illinois College of Medicine. The first engineering-based college of medicine. [https://medicine.illinois.edu] accessed April 21, 2017.

Cawley, J. and C.C. Grantham. 2011. Building a system of care: Integrating across the heart failure continuum. Perm J 15: 37–42.

Chan, L.S. 2016. Building an engineering-based medical college: Is the timing right for the picking? Med Sci Edu 26: 185–90.

Chan, L.S. 2018. Engineering-medicine as a transforming medical education: A proposed curriculum and a cost-effectiveness analysis. Biology, Engineering and Medicine 3(2): 1–10. Doi: 10.15761/BEM.1000142.

[CMS] 2016. Six (6) stats on the cost of readmission for CMS-tracked conditions. Becker's Infection control & clinical conditions. [http://www.beckershospitalreview.com/quality/6-stats-on-the-cost-of-readmission-for-cms-tracked-conditions.html] accessed December 4, 2016.

Drummond, M.F., M.J. Sculpher, K. Claxton, G.L. Stoddart and G.W. Torrance. 2015. Methods for the economic evaluation of health care programmes. Oxford University Press. Oxford, UK.

Grant, P. and I. Chika-Ezeriche. 2014. Evaluating diabetes integrated care pathway. Practical Diabetes 31: 319–22.

[HEART] 2016. What is the cost of heart failure on the economy? Economic Impact. [http://www.heartfailure.com/hcp/heart-failure-cost.jsp] accessed December 4, 2016.

[HIGHEREDU] 2016. Tenured/Tenure-track faculty salaries. HigherEd Jobs. [https://www.higheredujobs.com/salary/SalaryDisplay.cfm?SurveyID=37] accessed December 7, 2016.

[KFF] 2016. Total number medical graduates. KFF.org. [http://kff.org/other/state-indicator/total-medical-school-graduates/?currentTimeframe=0] accessed December 10, 2016.

Lau, T.W., C. Fang and F. Leung. 2013. The effectiveness of a geriatric hip fracture clinical pathway in reducing hospital and rehabilitation length of stay and improving short-term mortality rates. Geriatr Orthop Surg Rehabil 4: 3–9.

Lau, T.W., C. Fang and F. Leung. 2017. The effectiveness of a multidisciplinary hip fracture care model in improving the clinical outcome and the average cost of manpower. Osteoporosis Int 28: 791–798.

Muennig, P. and M. Bounthavong. 2016. Cost-effectiveness analysis in health: a practical approach. Third Ed. Jossey-Bass. San Francisco, CA, USA.

Ray, N.F., J.K. Chan, M. Thamer and J. Melton. 1997. Medical expenditures for the treatment of osteoporotic fractures in the United States in 1995: Report from the National Osteoporosis Foundation. J Bone Mineral Res 12: 24–35.

Reid, P.P., W. Dale-Compton, J.H. Grossman and G. Fanjiang. 2005. Building a Better Delivery System: A new engineering/health care partnership. National Academies Press, Washington, DC, USA.

[ROWAN] 2016. Six-story Medical School Building Rise in Camden. Cooper Medical School of Rowan University. [http://www.rowan.edu/coopermed/about/news/] accessed December 5, 2016.

[TEXAS] 2017. EnMed. Engineering & Medicine. Texas A & M University. [https://enmed.tamu.edu/] accessed April 21, 2017.

[TEACKER] 2016. How much does the U.S. spend to treat different diseases? Peterson-Kaiser Health System Tracker. [http://www.healthsystemtracker.org/chart-collection/how-much-does-the-u-s-spend-to-treat-different-diseases/] accessed December 4, 2016.

[UFL] 2013. University of Florida College of Medicine Faculty Compensation Plan. July 1, 2013 to June 30, 2014. [https://connect.ufl.edu/comfs/ComPlan/] accessed March 9, 2016.

[UND] 2016. New Building. School of Medicine & Health Sciences. UND: University of North Dakota. [http://www.med.und.edu/construction/] accessed December 5, 2016.

[USNEWS] 2016. U.S. News & World Report. Education. [http://grad-schools.usnews.rankingsandreviews.com/] accessed December 7, 2016.

QUESTIONS

1. What is the purpose of cost-effectiveness analysis?

2. What are the key elements of cost-effectiveness analysis?

3. Why is there a need to define the perspective of the analysis?

4. Why is more appropriate to place the perspective on the society rather than on the institution, for this particular subject?

5. What, if any, can the results of cost-effectiveness analysis be guaranteed?

PROBLEMS

The students will be grouped into 3–5 people each. Within each group, students will brainstorm to define a real-time medical issue for which a cost-effectiveness analysis is needed. Collaboratively, the students within each group will collect data for cost and effectiveness and perform a cost-effectiveness analysis and present their results to the class.

7

Invention and Innovation

Lawrence S. Chan[1],* and *Sonali S. Srivastava*[2]

QUOTABLE QUOTES

"Imagination is more important than knowledge. Knowledge is limited. Imagination encircles the world."

Albert Einstein, Scientist, at Smithsonian, February 1979 (Benna and Baer 2016)

"The common strands that seemed to transcend all creative fields was an openness to one's inner life, a preference for complexity and ambiguity, an unusually high tolerance for disorder and disarray, the ability to extract order from chaos, independence, unconventionality, and a willingness to take risk."

Scott B. Kaufman and Carolyn Gregoire, Authors (Kaufman and Gregoire 2015)

"Creativity is infectious"

Scott B. Kaufman and Carolyn Gregoire, Authors (Kaufman and Gregoire 2015)

"Physicians are a highly selected group of individuals who have intelligence, drive, problem-solving ability, and, at times, inventiveness."

Dr. John A. Pacey, Surgeon, inventor (Pacey 2015)

"Two days ago, the USPTO celebrated the issuance of patent 10 million with President Trump at a signing ceremony in the Oval Office. This is a truly remarkable accomplishment and a testament to American imagination and perseverance that has changed the world. It is an accomplishment that far exceeds the patent activity of any other country, and demonstrates that the U.S. patent and trademark system is not only the backbone of our economy, but is also at the core of innovation, invention, entrepreneurship, and human progress."

Andrei Iancu, Director of US Patent and Trademark Office (USPTO 2018)

Learning Objectives

The learning objectives are to help students to understand the process of invention and innovation and to develop intellectual and management abilities to achieve marketable invention and innovation.

---

[1] University of Illinois College of Medicine, 808 S Wood Street, R380, Chicago, IL 60612.
[2] Patent Department, Abbvie, Inc. 1 N Waukegan Road, North Chicago, IL 60015; sonali.srivastava@abbvie.com
* Corresponding author: larrycha@uic.edu

After completing this chapter, the students should:

- Understand the definition of discovery, invention and innovation and the difference amongst them.
- Understand the importance of invention and innovation in engineering.
- Understand the significant impact of invention and innovation in medicine.
- Understand the major obstacles for invention and innovation in medicine.
- Understand ways invention and innovation can be nurtured in an engineering-medicine education curriculum.
- Recognize the essential issues associated with invention: idea development, pre-invention market analysis, intellectual property protection, and product approval processes of the U.S. Food and Drug Administration (FDA).
- Able to carry out successfully a senior project that will enable the student to develop inventive/ innovative talent to solve a real-life biomedical engineering problem.
- Obtain the knowledge of business planning with regard to successful invention or innovation.

Introduction to Discovery, Invention and Innovation

Since engineering works by creating new products or new processes out of nature, therefore, scientific discovery, invention and innovation are a part of its DNA makeup. Moreover, scientific discovery, inventions and innovations are meaningfully different. While discovery is identification of pre-existing scientific facts, such as discovery of ultrasonic waves and chemical biomarkers, by contrast invention and innovation relates to use of these scientific principles for making new and useful products, such as an ultrasound machine based on the principles of ultrasound waves and a diagnostic kit for identification of diseases from the discovery of biomarkers.

All inventions and innovations, however, start with a fundamental understanding of principles of science, or discovery that remain the fundamental truth underlying the invention or the innovation. Here we focus on invention and innovation.

There is a constant confusion about what constitutes discovery and what separately constitutes a patentable innovation or invention, especially in the field of medicine. Perhaps a good way to distinguish is an example regarding genetics. Recognition of a unique genetic defects, such as the BRCA gene is deemed to be a discovery, not suitable for patents, whereas, DNA chips that would identify the defective BRCA gene would be elegant engineering solutions, well worthy of patents. *Mayo v. Prometheus*, 566 U.S. 66 (2012).

Furthermore, the words of "invention" and "innovation" are commonly used in parallel setting, there are differences between these two processes. Invention is the "creation of a product or introduction of a process for the first time" and Thomas Edison, who invented light bulbs that never existed before, was an example of inventor. Innovation occurs when someone "improves on or makes a significant contribution" to something that has already been invented and Steve Jobs, the formal Apple CEO who improved mobile phone applications and user friendliness, was an example of innovator (Bhasin 2012). There is another way to look at this subtle yet important difference, "Invention creates an ability but innovation takes that ability and allows it to scale and create some kind of a market impact" (Morgan 2015). Another expert in the field describes the difference between invention and innovation in this manner: "Invention, ... is what happens when you translate a thought into a thing. Innovation is what happens afterward." Innovation, according to this author, is the act of turning a creative idea into a business opportunity, just like what Steve Jobs had successfully accomplished for his apple computer products (Kennedy 2016). While invention and innovation are different in their scopes of the changing spectrum, both aim at improving human living conditions. Whereas inventors seek opportunities to create something totally new to better serve humankind, innovators look for ways to improve existing products or processes to make things even better. And both invention and innovation, by all means, require creativity to accomplish. In medicine, invention and innovation aim to generate novel and improve existing healthcare equipment, medication, process, respectively, so as to improve the efficiency and quality of healthcare.

In either invention or innovation situation, the most important characteristics that propel the success of inventor or innovator is the mind of openness, the mindset of not content with the current status quo, the vision of "seeing" something better or much better, the will to initiate the first step of change, and the perseverance to see the entire process come to a fruitful end, all while learning from failures. Since invention is to generate something new, the process requires creativity, which in turn needs a creative individual to achieve. Moreover, the process requires resilience, the ability to make iterative improvements while learning from each failure. When Thomas Edison failed in hundreds (some say thousands) of ways of making the light bulb, it is said that he spoke eloquently about it from a creative process, and had stated that he had in fact not failed a thousand times but had successfully learned the thousand ways, to not make the light bulb. "I have not failed. I've just found 10,000 ways that won't work" (AZ QUOTES).

Invention and Innovation in the History of Medicine

The history of medicine has recorded and documented many significant inventions and innovations that have help propelling the medical field forward. In some occasions, they literally pushed the medical field in a leap and bound manner. In the early dates, the use of cowpox to immunize people against small pox infection by Edward Jenner in 1796 (Riedel 2005); the generation of an synthetic anesthetic by German physician and botanist Valerius Cordus {1515–1544} who synthesizes diethyl ether by distilling ethanol and sulphuric acid and the subsequent use of anesthetics for surgical procedures (WLMA 2016); and the initial discovery of Penicillium mold inhibiting bacterial growth by Dr. Alexander Fleming of St. Mary Hospital of London, the subsequent extraction and proving clinical efficacy of penicillin in mice by the Oxford University scientists, and the subsequent use of this naturally occurring antibiotic penicillin to treat infection (Markel 2013) are some of the major medical inventions and innovations that have significant impact on medicine and are still benefiting us today. In more recent times, there are many more outstanding medical inventions and innovations. For example, computer axial tomography (CT scanner) was developed in 1960s by a UK engineer Godfrey Hounsfield from EMI Laboratories in England and a South Africa-born US physicist Allan Cormack of Tufts University in Massachusetts, USA. From its first test scan conducted on a mouse in 1967, to the first patient scan performed in 1971, to current medical practice, the CT scanner has become a core imaging tool in thoracic and other disease diagnosis. In 1979, only 8 years later, Hounsfield and Cormack were awarded a Nobel Prize in Physics and Medicine, a well-deserved honor, for their transformative invention. Chest imaging nowadays accounts for about 30% of all CT scanner usage (Goodman 2010, HBT 2016). Similarly, the Magnetic Resonance Imaging (MRI scanner), discovered some 30 years ago and first documented for its clinical usefulness by a Physician-scientist named Dr. Raymond Damadian in 1970s, has become the image of choice for brain scanning for various disease diagnoses, since no harmful x-rays is used in this diagnostic methodology (Ai et al. 2012, Bellis 2017). CT and MRI scanners revolutionize our abilities to look at human body slice by slice in a transverse fashion or in 3-dimensional construct, unlike the prior technique of X-ray films, which could only look at human body in one-dimension. The earliest MRI equipment in healthcare were made available to the hospitals at the beginning of the 1980s. In terms of its impact, worldwide, approximately 22,000 MRI scanners were in use and a record of more than 60 million MRI examinations were performed in the year of 2002 alone (Bellis 2017). Another example of excellent medical invention is molecular breast imaging (MBI) that has been developed recently (HBT 2016, O'Connor et al. 2009). Over many years, medical professionals had lots of frustration in using mammography in certain patients. While mammography has been useful in most patients, it has difficulties in detecting breast cancers in dense breast tissues. MBI was developed in response to this frustration (HBT 2016). MBI is a nuclear medicine scanning technique which performs 5 minutes after a radioactive material is injected into patient's vein. It has achieved a 90% sensitivity and an 82% sensitivity in detecting all breast tumors and breast tumors of 10 mm or less in size respectively, and has a two to three times higher sensitivity in detecting breast tumor in dense tissues and a slightly higher specificity than the conventional mammography (O'Connor et al. 2009). More example of recent medical invention includes the Active Bionic Prosthesis (Wearable Robotic Devices), developed by a group of Mayo Clinic's biomedical engineers in 2012. With this new device, prosthesis is not just a piece of motionless cosmetic cover up, rather, this medical invention allows patients to replicate the action of a person's tendons and muscles to imitate a variety of natural human body motions. Utilizing microprocessors, battery-powered motors, and Bluetooth technology, this device enables a person to adjust settings simply using a smart phone to ensure natural and consistent motions (HBT 2016).

In the digital age, medical innovation has taken patient care to a newer, higher platform. Case in point is the Medtronic Carelink for remotely monitoring a patients' pace maker from the comfort of their home, sometimes while the patient is sleeping. If a patient has an implantable heart device, the care does not end with the surgical implant of the device. Remote monitoring is a way for the implanted heart device to communicate with the physician, while using a small bed-side monitor. The performance of the pace-maker and the device information is sent back to the physician for routine monitoring and observation via a secure network. More details about the functions of this device can be viewed at: http://www.medtronic.com/us-en/patients/treatments-therapies/remote-monitoring.html. The new and emerging field of 3-D printing leaves many open chapters to be written. The ability to print, on-demand, parts and pieces of given material with a desired density is likely to reshape how future inventions are made. Imagine the ability to print human tissue and human skin on demand to repair damaged tissue or repair deep wounds. More of these 3-d printings can be viewed at: http://www.3ders.org/articles/20180503-uot-develops-handheld-3d-skin-printer-for-fast-repair-of-deep-wounds.html.

Invention and Innovation in Engineering-Medicine

At the first thought of invention and innovation as an educational subject, one may start to wonder if the subject matter of invention and innovation, which has such an innate quality of human beings, can be taught in the engineering or medical school. Can we really train future engineers or physicians to be inventors and innovators? Can creativity be taught, trained, encouraged, or otherwise nurtured? That is a very challenging question indeed. The traditional medical education, which places heavy emphasis on structured learning and conformity, has essentially put creativity in our medical students to a complete stop. Our classroom lectures for a long time have our students to committed to the daily grind of fact memorization of the highest level. Worse yet, the nation-wide administered medical examinations that qualify our students for promotion and determine their abilities to practice medicine in the future predominantly test their abilities to recall isolated data and phenomena and very little on their capacities to think for logical solution, let alone to create new solution. Thus, our current medical examination system for most part strongly encourages simple fact memorization and conformity, and at the same time provide no incentive for logical thinking or creative solution. How then could we revive the human creative potential and restore the innovative energy in our medical students? Perhaps we could examine examples of creative members of our society in recent history and observe what they do differently, so that we may gain some insights on ways to cultivate the "spirit" or "habit" of invention and innovation (Kaufman and Gregoire 2015). Similarly, we could also learn from them how to improve and nurture the "creativity" among biomedical engineering students. It has been recognized by experts in the field that there are things those highly creative people do differently than those people in the general population, including the following ten activities. If generally speaking the outcomes of successful creativities could be derived from these ten "things" depicted below, these 10 character traits would be very helpful for engineering and medical school admission offices in seeking out the best engineering and medical student candidates for their engineering-medicine curriculum (Kaufman and Gregoire 2015):

- Imaginative Play
- Passion
- Daydreaming
- Solitude
- Intuition
- Openness to Experience
- Mindfulness

- Sensitivity
- Turning Adversity into Advantage
- Thinking Differently

Most of all, deep and holistic understanding the unmet need would go a long way. Necessity, after all, is the mother of inventions. Therefore, having delineated the common traits of creative people, the next question would then be could we somehow help encouraging, mentoring, nurturing, and otherwise coaching our young engineering and medical students to be the future medical inventors and innovators? The answer is a resounding yes and there are real-life examples to prove this argument (Kennedy 2016). As proposed by experts in the field, "we are all, in some way, wired to create and that everyday life presents myriad opportunities to exercise and express that creativity. This can take the form of approaching a problem in a new way, seeking out beauty, developing and sticking to our own opinions (even if they're unpopular), challenging social norms, taking risks, or expressing ourselves through personal style" (Kaufman and Gregoire 2015). Here is one case exemplifying the notion that creativity can be encouraged to develop and be successfully nurtured with the right environment. In 1990s, Dr. Neil Gershenfeld, a professor at the Massachusetts Institute of Technology decided to teach a class termed "How to make (almost) everything" and allowed his students to have free access of his Research & Development Laboratory and he was surprised to found that his class was in such a high demand and that the students who attended the class were able to accomplish many creativity works by translating their fantasy vision into invention. Furthermore, his students are not just from the school of engineering as he initially designed for, but are also from the schools of art and architecture (Kennedy 2016). Looking from a different angle on the measurability of creativity, we also see that creativity itself can be enhanced. Here is another example. Not long ago researchers in the field have proven that the score of a special kind of examination termed "Spatial-reasoning test" is a useful predictor of creative achievement, fully documented by experimental evidence (Kell et al. 2013). As it turns out, this "Spatial Reasoning", a great predictor of creative achievement, can also be learned and its scores can be improved by training (Uttal et al. 2013). Another question is what environment or condition could the educational institute provide to optimize the opportunity of successful invention? To this regard, some experts have concluded that: Users that have repeatedly encountered the same problem over a long period of time have the greater driver to develop creative solution to resolve the frustration that the problem generated (Kennedy 2016). In the context of undergraduate biomedical and medical education, this project is best conducted during the students' senior year, when sufficient problem encounter and frustration cumulated in their first three years of engineering and clinical training would provide a strong motivational force for creative solution. Hence our decision to term this senior project.

As experts in the field pointed out, invention is no small task and when tackling a big question, the young inventor must project a daring spirit, dissent the "dogma", question the traditional doctrine, and leap passing the ordinary boundary, and the inventor will inevitably face self-doubt, rejection from peers, ridicule from uninformed individuals, and opposition from competitors. And there is when we need "empowerment" to sustain the invention process (Kennedy 2016). Medical invention is no exception. Not surprisingly, a report in 1930s published by Joseph Rossman, a patent inspector by trade, pointed out that nearly 70 percent of surveyed inventors provided the answer "perseverance" to the question of "what are the characteristics of a successful inventor?" (Kennedy 2016). How to train, encourage, and help our young inventor to persevere is the utmost important, as an expert in the field pointed out, "fortitude might be the most important ingredient in any inventive effort—without it, an idea remains just a thought or a scribble" (Kennedy 2016). At the end, those who succeed would need to have the perseverance to learn how not to make the light bulb, a thousand times, just like Edison, before the elusive success arrives.

So if we can nurture creativity, what method or methods then should we employ in order to obtain the optimum outcomes? As many educators concluded, the best way to nurture creative outcomes is through involving the people in creative projects, through a type of "project-based learning" (Kennedy 2016). A more recent example of innovative program that can be incorporated into the engineering-medicine curriculum is a field-tested medical technology design project (Loftus et al. 2015). In this 6 months-long program, medical students were partnered with business, law, design and engineering students to form interdisciplinary teams aiming to develop practical solution for unmet medical needs and were provided with prototyping fund, access to clinical and industry mentors, development facilities, and didactic teachings in ideation, design, intellectual property, FDA regulation, prototyping, market analysis, business planning, and capital acquisition. After a period of 4 years, the 396 participants have developed 91 novel medical devices, and helped launching 24 new companies (Loftus et al. 2015). This kind of project would likely teach our future biomedical engineers and physicians lessons on creative solution, teamwork, innovation, and solution-oriented engineering and medical practices.

Processes of Invention

Generally speaking, a successful medical invention would go through a series of processes:

- *Generating invention idea*: through thoughtful thinking of solving a medicine-related problem that has substantial impact.
- *Discovering unmet need:* for invention consideration.
- *Confirming invention idea*: through patent search, pre-invention marketing analysis.
- *Preparing the invention instrumentation and methodology*: through literature and book review, internet search, and consultation.
- *Securing the invention funding*: through personal saving or public support (in which case the patent right will not be assigned solely to the inventor).
- *Generating the invention*: through many hours, days, weeks, months, or even years of hard works and overcoming the failure of attempts.
- *Testing the invention*: through verifying the usefulness of the invention in real-life medical conditions.
- *Protecting the invention*: through patent filing and defense.
- *Approving the invention*: through filing the product with FDA.
- *Commercializing the invention*: through licensing or developing the products fully.
- *Funding the invention commercialization*: through traditional or venture capital funding sources.
- *Building sustainable commercial value models for the invention:* by understanding the health economics and outcomes research associated with commercializing the invention and setting up for financial success by understanding the medical product reimbursement landscape. Best products without reimbursement are likely to stagnate and fail.

Senior Invention Project

The senior invention project can be conducted by the students on an individual basis or team-based. The students will prepare a senior invention/innovation project that will be supervised and mentored by engineering-medicine faculty members. In addition to individual faculty mentoring, the students will receive group lectures on essential subject matters related to invention and innovation: invention and innovation idea cultivation, essence of invention and innovation design, intellectual property protection and patent filing, FDA regulation of medicine and medical device, prototype development, market analysis of invention and innovation products, business planning for product development, and capital acquisition for product development support (Loftus et al. 2015). The proposed project should fulfill the following basic requirements:

- A problem the student has faced during the first three years of engineering or medical school education.
- A problem the student has faced repeatedly.
- The solution of the problem in theory would have high impact in the patient care delivery or healthcare in general.
- The solution of the problem in theory would help fulfilling the Institute of Medicine's triple aims: better care, better health, and lower cost.
- The project has been discussed and approved by the student's Senior Project Advisors.

Management of Invention and Innovation

Intellectual Property Protection and Patent Filing

The message of invention is incomplete without the discussion of intellectual property protection, the very purpose of which is to protect the fruits of invention. Since substantial amount of hard work and financial resources could be required to achieve a marketable invention, our legal systems provide an avenue to protect the inventor from other competitors on making, using, selling the invention. Recently, ten-millionth US Patent was granted and the Under Secretary of Commerce for Intellectual Property and Director of US Patent and Trademark Office, Andre Iancu made this comment, "On June 19, the U.S. Patent and Trademark Office will issue **patent number 10 million**—a remarkable achievement for the United States of America and our agency. More than just a number, this patent represents one of ten million steps on a continuum of human accomplishment launched when our Founding Fathers provided for intellectual property protection in Article 1, Section 8, Clause 8 of our Constitution" (COMMERCE 2018). Director Iancu also commented "Some of the greatest leaps of humanity has made, have been fueled by our greatest inventors, Americans who have changed the course of history with their brilliance and dogged perseverance."

In the US, the patent filing is through the US Patent and Trademark Office (USPTO 2016a). In this government website, it offers links to perform step-by-step search for existing patents (USPTO 2016b) and instruction for patent filing online (USPTO 2016c). To file protection overseas, which is optional, the inventor could file international patent protection through an international agreement called Patent Cooperation Treaty (PCT), which significantly streamlines the patent application process to multiple countries. By filing one patent application with the US Patent and Trademark Office, a US inventor can simultaneously, by extension, file a centralized patent application in participating 140 countries who are members of this treaty (USPTO 2016d). The students should understand, however, the enforcement of patent right in foreign countries may not be as vigorous as that occurred in the US. In seeking intellectual property protection, the key is to file the appropriate application as early as technically possible, because similar patent filed by others just one day prior to your application filing will likely mean the loss of your invention protection. Students, thus, are encouraged to become familiar with the patent filing process and the patent agents and lawyers in their academic institutions in case their inventions reveal the potential to become commercial products.

The patent application process, which eventually leads to the grant of a patent can be a long laborious process that involves serious financial support and time. The process generally starts with the provisional application, which is then filed as a PCT application, prior to 12 months of time. This PCT application then goes through a process of centralized examination and review, where the patent examiner forms a preliminary opinion on patentability. It appears discouraging very quickly, as the examiners will find the closest art, and reject the invention as is. Creativity is invaluable in responding to the Patent Examiner's review. Generally, prior to the 30-month deadline, the PCT application is Nationalized in each country, where you truly want to have your patent. After careful selection of countries, the patent application is then reviewed by each specific country's patent examining body, before either receiving a final thumbs-up or thumbs-down. Eventually each country grants its own patent, and as patents are jurisdictional, a US patent protects the inventor from others using, selling or manufacturing the product in the United States, and so on and so forth.

Since securing patents is an expensive and time-consuming effort, clear strategic foresight and constant review of the art should guide a patent filing strategy, both offensively and defensively. In certain countries, after the granting of a patent, third parties may continue to challenge the patent through post-grant oppositions, additional caution is necessary as one navigates the patent landscape.

FDA Regulation of Medication and Medical Device

Because medicine significantly affects all human lives, the government has a rightful interest in regulating all of its contents and methods to ensure the efficacy of proposed medical products (drugs and devices) and to protect public from harm by defected or otherwise non-functional medical products. In the US, the regulatory agency is called the U.S. Food and Drug Administration (FDA), which derives its legal authority to regulate both medical devices and electronic radiation-emitting products from the Federal Food Drug & Cosmetic Act

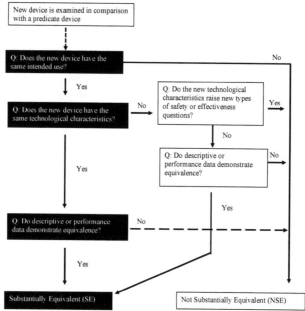

Fig. 1. FDA Approval Process for New Medical Devices (GAO 2009).

(FD&C Act) (FDA 2016a). To accomplish the provisions specified by the FD&C Act, FDA develops, publishes, and implements the appropriate regulations (FDA 2016a). The proposed rules of FDA are initially published for public comments in Federal Register (FR). Once the FDA rules become final, they will be codified in the Code of Federal Regulations (CFR). For most parts, medical device and radiation-emitting product regulations under FDA are covered by Title 21 CFR Parts 800–1299, covering various aspects of device design, namely, clinical assessment, manufacturing, product labeling and packaging, post-marketing surveillance, as well as standards and product reports of radiation-emitting products (FDA 2016a). At the FDA, the proposed new medical devices go through several evaluation processes to analyze if the new devices demonstrate the same intended use purpose, the same technical characteristics, and the same performance data as compared to an existing device. The evaluation helps decide whether or not a new device is "substantially equivalent (SE, approval)" or "Not substantially equivalent (NSE, non-approval)" to an existing device (GAO 2009, Fig. 1). To ensure drug efficacy and safety, FDA places newly developed drugs through very vigorous process of evaluation (FDA 2016b). Table 1 details the steps for the new drug approval process published by FDA (FDA 2016b). In addition to medical devices and drugs, FDA also regulates the efficacy and safety of other medicine-related products such as vaccine, blood products, and biologics (FDA 2016b). According to a study by the Independent Institute, the average time requirement, costs, and overall probability of success for phase I, II, and III clinical trials are 22, 26, 31 (months); $15, $23, $87 (million dollars); and 30%, 14%, 9% (success rate), respectively. Collectively about 8% of new drugs that gone through the vigorous FDA approval process were finally approved for sale (FDAR 2016). According to PhRMA, "the average cost to research and develop each successful drug is estimated to be $2.6 billion. This number incorporates the cost of failures—of the thousands and sometimes millions of compounds that may be screened and assessed early in the R&D process, only a few of which will ultimately receive approval. The overall probability of clinical success (the likelihood that a drug entering clinical testing will eventually be approved) is estimated to be less than 12%" (PHRMA 2018). By contrast, the approval speed and approval rate of new medical devices through the US FDA is much higher than that for the new drug development. According to the US government's General Accounting Office (GAO) report on medical devices evaluated between 2000 and 2011, the final approval rates ranged from 56% (2009) to 93% (2011). And the average time from application submission to final decision ranged from about 400 days (2001) to 1,100 days (2007) (GAO 2012). Basic knowledge on how medical devices and drugs are regulated by FDA in the US (FDA 2016b), Medicine & Healthcare Products Regulatory Agency (MHPRA) in UK (MHPRA 2016), and other government agencies (in other countries) will provide useful background information to guide our future medical inventors.

Table 1. Steps of new drug approval and regulation by FDA (FDA 2016b).

Step Stage	New Drug Developer	FDA
1. Pre-clinical	Performs laboratory & animal tests to learn how the proposed drug works and its safety to test in humans	Reviews for approval prior to human test
2. Clinical	Conducts three phases of clinical trials on humans	
	Phase I: Conducts trials predominantly For safety purpose	Monitors and approves for Phase II clinical trial
	Phase II: Conducts trials for safety, Dosing, and efficacy	Monitors and approves for Phase III clinical trial
	Phase III: Conducts trials for safety, Efficacy, and side effects	Monitors and approves for new drug application
3. Post-Market Surveillance		Monitors new drug's safety through MedWatch and also Review other databases 1. Issue communication to providers if risks detected 2. Add special label to raise safety concern 3. Withdraws drug out of market if serious risks determined

Pre-invention Market Analysis

Prior to creating a desired product, it is an important starting point to create a desired product profile that defines the metes and bounds of the invention or the innovation. A well-defined problem or thesis is more likely to produce a definitive solution, whereas an undefined desired product is likely to languish between, wishful attributes that may not be solved or realized. Once a target product profile is created, a pre-invention market analysis become possible. While the approval of FDA (in US), MHPRA (in UK) and other government agencies (in other countries) determines whether a new medical product (device or drug) can be legally marketed for patient use, the essence of market analysis provides an estimation on how well the new medical product may fare in the open healthcare market system. In addition, a deep and holistic understanding of the unmet need will pave an excellent way to a market-winning strategy. How to achieve deep and holistic understanding of the unmet need? One may consider the following disciplined approaches:

- Careful observation and inquiry into the current practice.
- Careful inventory of the inventors' passion, skills, and life mission.
- Objective analysis of the pervasiveness and market demands of the unmet need.

The right unmet need to pursue is likely within the overlap of the above 3 approaches.

With a few rare exceptions where a new medical product is one-of-a-kind without peer, competition is the name of the game. A new medical product will therefore likely be compared and competed against existing products for efficacy, safety, and price in various ways. Therefore, a simple market analysis of the proposed invention idea, prior to all the intellectual, time, and financial capitals are deployed, may be very helpful for the future medical inventors, i.e., our biomedical engineering and engineering-medicine students, to set the right targets for their invention projects. To perform a pre-invention market analysis is probably considered a brief analysis of environment, similar to that used in healthcare strategic planning: one performs an internal and an external assessment of competitiveness of the products of the new invention (Zuckerman 2012). Once the assessment data are collected, they can be visually displayed on a SWOT graph, which displays data according to their categories: Strength, Weakness, Opportunities, and Threats, in the upper left, upper right, lower left, and lower right quadrants of the graph, respectively (Zuckerman 2012, Fig. 2). Based on the relative strength of four quadrants in his SWOT analysis, the medical inventors would need

- Newly patent technology or drug?
- More effective?
- More efficient method?
- Less side effect?
- More accuracy?
- Higher technology?
- More sensitive technique?
- More specificity?

- Higher cost?
- Less convenient?
- High capital investment?
- Higher learning curve?

Strength   **Weakness**

Opportunities   **Threats**

- New Treatment paradigm?
- New treatment indication?
- New treatment pathway?
- New healthcare market?

- Similar products on the market?
- Emerging new products?
- More well-funded competitors?
- Similar products are not reimbursed through medical insurance system?

Fig. 2. Pre-invention Market Analysis Display by SWOT Model.

to make an educated guess as if the invention would go forward or to be terminated. Again, we are dealing with uncertainty in this pre-invention market analysis, and the final outcome cannot be guaranteed.

Capital Funding

Once the inventor develops a prototype and obtains a patent, the inventor will then have the options to further develop the product by forming a small business venture or simply pass onto other investors by selling or licensing its invention right. The advantage for selling or licensing its invention right at the earliest stages is that the inventor will receive a lump sum of monetary reward (or a continuous loyalty) without further effort required. The disadvantage for selling or licensing its invention right at the earliest stages is that its selling or licensing value is much less than a more mature product since the full product potential may not be apparent to the investors at that point. For those inventors who want to develop the product fully through a small business venture or who want to develop the product to a more mature stage before passing onto other investors, capital funding, other than additional efforts, is therefore needed to move forward, unless the inventor is independently wealthy to support the product development.

To obtain capital funding at these early stages of invention, the best option will be through "Venture Capital", since at the early stages of product development, the inventor would not be able to obtain capital funding from traditional sources such as public markets and banks (VC 2017, SBA 2017a). Venture capital investments are usually obtained by the inventor as cash in exchange for shares and active management role in the investment partners (VC 2017, SBA 2017a). It is also important for the inventors to understand the different funding criteria between "Venture Capital" funding and traditional capital funding (SBA 2017a, Table 2). Moreover, the inventors should understand that it is likely that some equity cushion or security (collateral) are required by the "Venture Capitalist" before they can obtain "Venture Capital" fund. Furthermore, since "Venture Capital" investors have a low priority in claim against the assets of inventor's company (should the new company fail), a higher rate of return on investment (ROI) is required, in comparison to that of traditional capital funding (SBA 2017a). Generally speaking, inventor can obtain capital funding from two major avenues of "Venture Capital": high net worth individuals (HNWIs, also known as "angel investors") or venture capital firms (VC 2017, SBA 2017a). The US government's Small

Table 2. Different Funding Characteristics between Traditional Capital and Venture Capital (SBA 2017a).

Characteristics	Traditional Capital Funding	Venture Capital Funding
Focus	Stable growth	Young, high-growth potential
Invest format	Debt	Equity capital
Risk tolerance	Low to medium risk	High risk, potentially high return
Investment Horizon	Short to medium term	Long-term
Participation	Passive	Actively participate in board of director, capital structure, governance, strategic marketing
Priority in claim against borrower's assets	First	Last
Required rate of return on investment (ROI)	Low to medium	High

Business Administration (SBA) has developed a Small Business Investment Company (SBIC) Program which works with private investment funds licensed as SBICs to provide seed money to US small businesses (SBA 2017b). While SBICs are regulated by SBA, they are profit-seeking private investment firms that make independent investment decisions (SBA 2017b). In addition to SBIC Program, SBA also provide small loans for starting businesses directly (SBA 2017c). In order to obtain SBA loans, the inventor would first visit a local bank or financial institution that participate in SBA programs and submit the SBA-specific loan application.

To the qualified applicants, SBA would guarantee the loan, meaning SBA will repay the lender the remaining portion of the loan in the case that the inventor default on the loan payments (SBA 2017c). Three types of common SBA loans are available for starting business purpose: (1). Microloan Program that provides a maximum of $50,000; (2). Certified Development Company (CDC) 504 Loan Program that provides long-term fixed-rate loan for major fixed assets such as land and buildings; (3). Basic 7(a) Loan Program that provides basic funding for start-up business (SBA 2017c). Another funding source the inventor could look into would be the National Venture Capital Association (NVCA), a venture capital organization, which has comprised a list of hundreds of venture capital firms (NVCA 2017).

Business Planning

Regardless of which type of funding the inventor would seek, a well-developed business plan is essential for such high-risk development (SBA 2017d). In addition to being a "roadmap" for the success of a starting business, a business plan is the very first document a "Venture Capital" investor will examine in detail in the determination of whether a new invention is worth of its capital investment (VP 2017, SBA 2017a). Other benefits of such strategic business planning include better priorities setting, more accurate resources allocation, timely success measurement and strategy revision, generation of commitment, and improved coordination of the new enterprise (Zuckerman 2012). In addition, business plan should provide the inventor the opportunity to realize the substantial personal investment he or she will be needed for the success and prepare a logical exit strategy should the business fails. Thus business plan is necessary for capital acquisition. In the SBA website, the inventor could register and login a password-protected account to build the business plan online, simply by following the step-by-step guide provided by the website. The inventor could save the generated business plan in a pdf file, which would be revised and updated as the inventor sees fit. And the inventor's business plan could be stored at the government website up to six months from the last login date (BSA 2017e). In general, a business plan should include the following elements (SBA 2017d):

- *An executive summary*: A concise description of business plan as a whole, this summary should highlight the strength of its overall plan, commonly for a five-year period. It generally includes a mission statement (central purpose of the new enterprise), a brief description of the new enterprise, a short statement of the new product (or service) and its growth potential, and a summary of the future plans. For starting business, the focus should be the inventor's experience, background, and decision leading to the new enterprise. Importantly, it should include the statement that the plan is supported by the result of market analysis and the proposed new product is unique in filling an unmet need. Although this summary appears first in the business plan, it should be written the last as it summarizes all the important information of the entire plan.

- *A description of the new company*: The specific product or service of the new company, its difference from others, its competitive advantages, and its market targets of the new company.

- *A market analysis*: The analysis of the position of new product in the market, its target market size, its unique characteristics to fill unmet needs, its potential regulatory restriction, its potential market share, its pricing structure, and its competition: strength, weakness, opportunities, and threats (SWOT analysis).

- *An organization and management structure*: The overall reporting system and authority line (organizational chart), ownership information, and board of directors' data.

- *A description of the product or service*: These include details of the new product or service, how it may meet the need and benefit the targeted customers, its unique advantages, its life cycle (the period of time a product is developed, marketed, matured, declined, and eventually terminated), its intellectual property data, and its research and development (R & D) activities.

- *A marketing and selling strategy*: The marketing strategies would include a market penetration strategy, a market growth strategy, a distribution channel strategy, and a communication strategy. The selling strategies would include a sale force strategy and a sale activity strategy.

- *A funding request*: These would include specific amounts needed, both current and future (up to five years); the intended use of the fund, and a strategic financial plan for the future: buyout, debt repayment, acquisition, or selling.

- *A financial projection*: These statements should match the funding request. The specific prospective financial data would include forecasted income statements, balance sheets, cash flow statements, and capital expenditure budgets. The methods for formulating these statements can be found in this reference (Gapenski and Reiter 2016). For the first year, these statements should be formatted on a monthly basis.

- *Appendix*: These would include the product pictures and illustrations, resumes, licenses, permits, contracts, leases, the inventor's credit history, relevant publications, letters of references, and other legal and supporting documents.

Health Economics and Outcomes Research and (HEOR) Medical Product Reimbursement Landscape

In the business of medicine, it is important for decision-makers at a national or state level to make decisions based on sound economic value of each innovation in medicine. For example, a medicine or device that includes data showing decrease in absenteeism, or increase in work productivity because of better pain management, would be favored over another similar device that does not possess the health economics and outcomes data.

Decisions are made not only at the national level, where major purchasers like Medicare and Medicaid are trying to assess the value of each innovation, but also at a private level where the health insurers are trying to justify benefit and cost to employers. In all measures, such health economics and outcomes data is valuable. Where decision makers are asked to select from many ranges of innovation, from economically thrifty to economically extravagant, both the decision maker and the seller has to rely on sound economic data to support rationale choice and selection of prices. The cost-price of innovation and medical device needs to be finely balanced by the economic benefit derived from each innovation. At the end, it is about, "is the price right for the benefit the device brings to the patient?"

Therefore, evidence-based medicine is no longer limited to safety and efficacy or clinical-effectiveness of the product, but has expanded to cost-effectiveness, or HEOR value provided by the innovation. "Health

Technology Assessment" entities, such as NICE for United Kingdom has emerged as a key decision-maker that can make an approved product, either get reimbursed, or not. NICE was originally set up in 1999 as the National Institute for Clinical Excellence, a special health authority, to reduce variation in the availability and quality of NHS treatments and care. NICE currently assess the clinical and cost-effectiveness of both pharmaceutical products and medical devices, amongst others (NICE 2018). The main health outcome measure that NICE uses is the quality-adjusted life year (QALY). A QALY is a unit that combines both quantity (length) of life and health-related quality of life into a single measure of health gain. Therein lies the line between a commercially successful medical product, a blockbuster, or an innovation that did not make its mark.

Similar to the European institutions, the US and State governmental agencies such as Centers for Medicare & Medicaid Services (CMS), the Veterans Affairs, Tricare (for active duty military) and Children's Health Insurance Program (CHIP) carry a lot of negotiating power in product reimbursement, based on number of lives covered. Any product not covered or reimbursed by these entities would have to decide its financial fate in the private insurers market, and may not emerge to be successful product. The federal healthcare programs are designed with many objectives, including preventing overutilization of health care services or items. Therefore, an overstatement of benefits would result in overutilization of improper products or services by physicians.

Summary

Invention and innovation, the heart and soul of engineering principles, have contributed huge beneficial impact to our human society historically. By incorporating these principles and putting them into real-life medical practices as depicted in this course work, we are providing outstanding tools to our future biomedical engineers and engineer-physicians for their future success in moving medicine forward to be higher in efficiency and quality.

References

Ai, T., J.N. Morelli, X. Hu, D. Hao, F.L. Goerner, B. Ager et al. 2012. A historical overview of magnetic resonance imaging, focusing on technological innovations. Invest Radiol 47: 725–41.

[AZ QUOTE] 2018. Thomas A. Edison. [http://www.azquotes.com/author/4358-Thomas_A_Edison] accessed June 30, 2018.

Bellis, M. 2017. Magnetic Resonance Imaging MRI. About.com. April 20, 2017. [http://inventors.about.com/od/mstartinventions/a/MRI.htm] accessed October 29, 2017.

Benna, S. and D. Baer. 2016. 25 quotes that take you inside Albert Einstein's revolutionary mind. August 26, 2015 [www.businessinsider.com] Accessed July 4, 2016.

Bhasin, R. 2012. This is the difference between invention and innovation. April 2, 2012 [www.businessinsider.com] Accessed July 4, 2016.

[COMMERCE] 2018. 10 Millions Patents: A celebration of American Innovation. Commerce.gov. [https://www.commerce.gov/news/blog/2018/06/10-million-patents-celebration-american-innovation] accessed June 30, 2018.

[FDA] U.S. Food and Drug Administration. 2016a. Code of Federal Regulation. [http://www.fda.gov/MedicalDevices/DeviceRegulationandGuidance/Overview/ucm1344 99.htm] accessed December 22, 2016.

[FDA] U.S. Food and Drug Administration. 2016b. How FDA evaluates regulated products: Drugs. [http://www.fda.gov/AboutFDA/Transparency/Basics/ucm269834.htm] accessed December 22, 2016.

[FDAR] FDAReview.org. 2016. The drug development and approval process. [http://www.fdareview.org/03_drug_development_php] accessed December 23, 2016.

[GAO] United States Government Accountability Office. 2009. Medical devices. FDA should take steps to ensure that high-risk device types are approved through the most stringent premarket review process. Report to Congressional Addressees. GAO-09-190. January 2009. [http://www.gao.gov/news.items/d09190.pdf] accessed December 23, 2016.

[GAO] United States Government Accountability Office. 2012. Medical devices. FDA has met most performance goals but device reviews are taking longer. Report to Congressional Requesters. GAO-12-418. February 2012. [http://www.gao.gov/assets/590/588969.pdf] accessed December 23, 2016.

Gapenski, L.C. and K.L. Reiter. 2016. Healthcare Finance: An introduction to accounting & financial management. 6th Ed. Health Administration Press, Chicago, USA.

Goodman, L.R. 2010. The Beatles, the Nobel Prize, and CT scanning of the chest. Radiol Clin North Am 48: 1–7.

[HBT]. Healthcare Business & Technology. 2016. The 10 Greatest Medical Inventions of the last 50 years. [http://www.healthcarebusinesstech.com/the-10-greatest-medical-inventions-of-the-last-50-years/] accessed July 29, 2016.

Kaufman, S.B. and C. Gregoire. 2015. Wired to create: Unraveling the mysteries of the creative mind. Perigee Book, New York, USA.

Kell, H.J., D. Lubinski, C.P. Benbow and J.H. Steiger. 2013. Creativity and technical innovation: Spatial ability's unique role. Psychol Sci 24: 1831–6.

Kennedy, P. 2016. Inventology: How we dream up things that change the world. Hough\ton Mifflin Harcourt, Boston, USA.

Loftus, P.D., C.T. Elder, T. D'Ambrosio and J.T. Langell. 2015. Addressing challenges of training a new generation of clinician-innovators through an interdisciplinary medical technology design program: Bench-to-Bedside. Clin Transl Med 19: 4–15.

Markel, H. 2013. The Real Story behind Penicillin. September 27, 2013. PBS.

Newshour. [http://www.pbs.org/newshour/rundown/the-real-story-behind-the-worlds-first-antibiotic/] accessed July 29, 2016.

[MHPRA] Medicine and Healthcare Products Regulatory Agency. 2016. Government of UK. [https://www.gov.uk/government/organisations/medicines-and-healthcare-products-regulatory-agency] accessed December 22, 2016.

Morgan, J. 2015. What's the difference between invention and innovation? September 10, 2015 [www.forbes.com] Accessed July 4, 2016.

[NICE] National Institute for Health and Care Excellence. UK. 2018. [https://www.nice.org.uk] accessed June 30, 2018.

[NVCA] The National Venture Capital Association. 2017. [http://nvca.org/] accessed January 19, 2017.

O'Connor, M., D. Rhodes and C. Hruska. 2009. Molecular breast imaging, Expert Rev Anticancer Ther 9: 1073–1080.

Pacey, J.A. 2015. Life-changing medical invention: Build a successful enterprise and a new world. Advantage, Charleston, USA.

[PHRMA] 2018. PhRMA. Biopharmaceutical Research and Development Overview. [http://phrma-docs.phrma.org/sites/default/files/pdf/rd_brochure_022307.pdf] accessed June 30, 2018.

Riedel, S. 2005. Edward Jenner and the history of smallpox and vaccination. Proc (Bayl Univ Med Cent). 18: 21–25.

[SBA] US Small Business Administration. 2017a. Starting and Managing. Venture Capital. [https://www.sba.gov/starting-business/finance-your-business/venture-capital/venture-capital] accessed January 21, 2017.

[SBA] US Small Business Administration. 2017b. SBIC. Directory of SBIC Licensees. [https://www.sba.gov/sbic/financing-your-small-business/directory-sbic-business] accessed January 21, 2017.

[SBA] US Small Business Administration. 2017c. Starting and Managing. SBA Loans. [https://www.sba.gov/starting-business/financing-your-business/loans/sba-loans] accessed January 21, 2017.

[SBA] US Small Business Administration. 2017d. Starting and Managing. Write Your Business Plan. [https://www.sba.gov/starting-business/write-your-business-plan] accessed January 21, 2017.

[SBA] US Small Business Administration. 2017e. Build Your Business Plan. [https://www.sba.gov/tools/business-plan/?interiorpage2015] accessed January 22, 2017.

[USPTO] United States Patent and Trademark Office. 2016a. [https://www.uspto.gov/] accessed December 22, 2016.

[USPTO] United States Patent and Trademark Office. 2016b. Search for Patents. [https://www.uspto.gov/patents-application-process/search-patents] accessed December 22, 2016.

[USPTO] United States Patent and Trademark Office. 2016c. File Online. [https://www.uspto.gov/patents-application-process/file-online] accessed December 22, 2016.

[USPTO] United States Patent and Trademark Office. 2016d. Protecting Intellectual Property Rights Overseas. [https://www.uspto.gov/patents-getting-started/international-protection/protecting-intellectual-property-rights-ipr] accessed December 22, 2016.

[USPTO] United States Patent and Trademark Office. 2018. Remarks by Director Andrei Iancu at the American Enterprise Institute. June 21, 2018. [https://www.uspto.gov/about-us/news-updates/remarks-director-andrei-iancu-american-enterprise-institute] accessed June 30, 2018.

Uttal, D.H., N.G. Meadow, T. Tipton, L.L. Hand, A.R. Alden, C. Warren et al. 2013. The malleability of spatial skills: A meta-analysis of training studies. Psychol Bulletin 139: 352–402.

[VC] Venture Capital. 2017. Investopedia.
[http://www.investopedia.com/terms/v/venturecapital.asp] accessed January 19, 2017.

[WLMA] The Wood Library-Museum of Anesthesiology. 2016. History of Anesthesia. [http://www.woodlibrarymuseum.org/history-of-anesthesia/] accessed July 29, 2016.

Zuckerman, A.M. 2012. Healthcare strategic planning. 3rd ed. Health Administration Press, Chicago, USA.

QUESTIONS

1. Determine which of the followings are inventions and which are innovations:

   - Dr. Smith modifies a surgical instrument to make it more user friendly.
   - Medical student Dennis generates a medical device useful to detect blood sugar level in nasal swap that has never been known to exist.

- Biomedical engineering student Amy develops a novel diagnostic technique useful to identify a kind of brain infection in peripheral blood sample without the need of lumbar puncture (an invasive method to obtain central nervous system fluid).
- Biomedical engineer Dr. Gold improves the usefulness of ultrasonography by increasing its visual resolution power.
- Hospital CEO Dr. Brown makes a change in healthcare delivery resulting in increase of efficiency and reduction of waste.

2. Generally speaking, what are the potential obstacles an inventor may encounter during the course of generating invention? What can you do to overcome those obstacles if occurred?
3. Can creativity be nurtured? If so, how?
4. How does FDA determine the substantially equivalent in approving medical devices?
5. How important is pre-invention market analysis? Why?
6. What is SWOT analysis?
7. Which is more important for creating invention: the intellectual ability of getting the right idea or the technical ability of testing the idea?

PROBLEMS

In preparation for the senior invention project, you are encouraged to give thorough considerations of the followings. After thoughtful consideration, prepare in writing for the following project steps and arrange a discussion session with your project advisor:

- What are the practical medicine-related problems that you have encountered for multiple times during your training thus far?
- For which of these practical medical problems (above) do you think its resolution would produce the greatest impact on healthcare and why?
- For which of these practical medical problems (above) do you think its solution (medical device or medication) would have the greatest market value?
- What instruments and experimental methods do you intend to use in order to generate the resolution of the problem you have selected to conquer?
- What intellectual mentoring and technical assistances do you need for the project?

8

Design Optimization

William C. Tang

QUOTABLE QUOTES

"We are all tasked to balance and optimize ourselves."

Mae Jemison, American Engineer, Physician, and Astronaut (BRAINYQUOTE 2018)

"If you optimize for money too early, you will be minimizing for learning, almost without exception."

Tim Ferriss, American Author and Entrepreneur (BRAINYQUOTE 2018)

"Machines have the ability to assemble things faster than any human ever could, but humans possess the analytics, domain expertise, and valuable knowledge required to solve problems and optimize factory floor production."

Joe Kaeser, CEO of Siemens (BRAINYQUOTE 2018)

"Your genome sequence will become a vital part of your medical record, thereby providing critical information about how to optimize your wellness."

Leroy Hood, American Biologist and National Medal of Science Awardee (2011) (BRAINYQUOTE 2018)

Learning Objectives

The learning objectives of this chapter are to help the students to acquire the overall concept and essential skills to perform the design and optimization process in engineering.

After completing this chapter, students are expected to:

- Understand the essence of design optimization.
- Understand the relationship between *Design Variables, Objective Functions,* and *Constraints.*
- Able to utilize the knowledge of design optimization and multidisciplinary design optimization to approach biomedical engineering problem.

University of California, Irvine, Department of Biomedical Engineering, 3120 Natural Sciences II, Zot 2715, Irvine, CA 92697-2715; wctang@uci.edu

Design Optimization Defined

Questions as simple as how to cross a river without getting wet to as difficult as how to land a man on the moon and returning him safely to the Earth stimulate and challenge us to engage our creative minds to seek solutions—not just any solution, but the best solution. We will not settle for any way to cross the river except the easiest way if we are to cross the river many times. Most certainly we will not choose any solution unless it can deliver the most likelihood of success in landing a man on the moon and returning him safely to the Earth. In other words, we are asked to perform the process of design optimization.

When confronted with the task of designing the best solution to a problem, the first question an engineer would ask is what is considered as the best, that is, what criteria are being used to evaluate any solution as good, better, or best. For the river-crossing example, the criteria may include any or all of the following: (1) requiring the least effort from the users (those who need to cross the river without getting wet), (2) little to no training needed for using the solution, and (3) the solution is readily accessible anytime, anywhere along the bank of the river. For the moon landing example, the best solution is not the easiest, but the one with maximum chance of success. In this case, the criteria may include: (1) least uncertain outcomes in operating the solution, (2) most robust against user errors, (3) most tolerant of hostile operation environment. We call this set of criteria the *Objective Function* in design optimization.

The second question an engineer would ask is what resources and time are available to construct the solution. Level of resources and amount of time may be traded to certain extent. President John F. Kennedy gave the nation less than 9 years to go to the moon, believing that "we possess all the resources and talents necessary" (JFKLIBRARY 2018). We succeeded in landing a man on the moon and returning him safely to the Earth with the given time of less than 9 years. However, the entire Apollo program cost $20.4 billion. We could probably achieve the same goal with less budget, but we could not have done it in 9 years. The available means for the design optimization are called *Constraints* or *Design Requirements*.

The engineer then proceeds to examine different designs to find out how each of them change the *Objective Function* and whether the *Constraints* are satisfied. The different designs are called *Design Variables*. Optimization is achieved when the *Design Variables* are found that maximize (or minimize) the *Objective Function* while satisfying the *Constraints*.

Mathematical Example

The following is a simple example where there is only one *Design Variable, x.* There is also only one *Objective Function*, given by $f(x)$. The only *Constraint* is $g(x) \leq 100$. The optimization task is to maximize $f(x)$ subject to $g(x) \leq 100$.

Let $f(x) = 18x - x^2$ and $g(x) = 8x$

Solving for the *Constraint* yields $g(x) = 8x \leq 100 \Rightarrow x \leq 12.5$

Differentiating $f(x)$ with respect to x: $\dfrac{d}{dx} f(x) = \dfrac{d}{dx}\left(18x - x^2\right) = 18 - 2x$

Setting the result equal to 0 gives the maximum at

$\dfrac{d}{dx} f(x) = 0 \Rightarrow 18 - 2x = 0 \Rightarrow x = 9$ which satisfies the *Constraint* of $x \leq 12.5$,

Therefore, the optimum design is <u>$x = 9$</u> resulting in maximum *Objective Function* of $f(x) = 18x - x^2 = 162 - 81 = \underline{\underline{81}}$ while satisfying the *Constraint* of $g(x) \leq 100$.

In practice, one of the *Constraints* often is cost, while the *Objective Function* can be a certain aspect of the performance. The *Design Variable* is what is controllable and is at the disposal of the engineer. The above numerical example illustrates a scenario where the "best" design from all possible *Design Variable x* happens to also satisfy the *Constraint*. In reality, the *Constraint*, particularly cost constraints, limit the *Design Variable* to a range that is financially viable. The engineer will then seek what is "best" within that range, i.e., maximizing the *Objective Function* subject to *Constraint*.

Suppose, as often the case in practice, there are two or more *Objective Functions* in a given optimization task. A common example is to design a car that is (1) as light as possible to save fuel, and (2) as fast as possible to satisfy customer demand. The engineer must then decide the relative priority of these two *Objective Functions* and assign a *Weighing Factor* to each of them based on the priority. The *Weighing Factors* can be numerical constants that sum up to be 1, such as 0.6 to represent a higher priority for a particular *Objective Function* than the other that gets assigned 0.4. Suppose the size of the engine is the *Design Variable, x,* that the engineer can use to optimize the solution. Let $f(x)$ be the *Objective Function* for fuel economy, and $h(x)$ for acceleration. A small engine will increase $f(x)$ but will decrease $h(x)$. Suppose also that the engineer was told that fuel economy is a higher priority than acceleration, and that the relative priority is 0.6 to 0.4. Then the engineer will solve for the following differential equation:

$$\frac{d}{dx}\left[0.6f(x) + 0.4h(x)\right] = 0$$

Similarly, practical optimization problems almost always involve multiple *Constraints. Constraints* are fixed and must be satisfied. Examples include passing the safety regulations for a car, maximum allowable manufacturing cost, and maximum allowable time for the development cycle. To optimize within these *Constraints* would often mean that the differential equation cannot be solved to obtain the value zero. The engineer will then seek the highest value of $[0.6 f(x) + 0.4h(x)]$ along the boundaries of the *Constraints.*

Multidisciplinary Design Optimization

Having defined and illustrated Design Optimization with a simple mathematical model involving one variable, we should now turn to a real-life situation where multiple *Design Variables* simultaneously change multiple *Objective Functions* subject to multiple *Constraints.* The complexity of these multivariable problems quickly becomes intractable, particularly when the *Objective Functions* and *Constraints* can only be expressed in implicit forms. Fortunately, the rapid advancement and availability of powerful computation tools allow these kinds of problems to be approached with numerical methods. The latest development is in Multidisciplinary Design Optimization (MDO) (Martins and Lambe 2013), in which all relevant disciplines are incorporated simultaneously to solve a problem. The optimization results are superior to the design found by optimizing each discipline sequentially, which misses the crucial and intricate interactions among disciplines. However, seeking optimization with simultaneous multidisciplinary approach increases the complexity exponentially, which can be approached only with the latest powerful computational platforms. MDO is most often employed in aerospace engineering. For example, the Boeing Blended Wing Body (BWB) aircraft concept was developed with heavy uses of MDO to incorporate aerodynamics, structural analysis, propulsion, control theory, and economics (Viana et al. 2014). Each of these disciplines presents hundreds of *Objective Functions* at different hierarchies with different weighing factors. The complete optimization model may involve thousands of *Design Variables* and thousands of *Constraints.*

Design Optimization in Biomedical Engineering

In biomedical engineering, the *Constraints* include those commonly encountered in engineering design tasks. In addition, the entire list from the *Implement* step as discussed in previous chapter should also be included, which is repeated here for convenience:

- Intellectual property and protection
- Regulatory strategy including process management and quality system
- Reimbursement strategy
- Business strategy including marketing, sales, and distribution, and
- Combining all assets and strategies to develop a sustainable competitive business advantage

The above list of Constraints is not easily quantifiable, and yet to make sense of the design optimization steps, best efforts must be paid to quantify them. The biomedical engineer quickly realizes that these *Constraints* demand the highest level of creativity and "out-of-the-box" thinking in reaching design optimization as well as the use of sophisticated computational tools.

Optimization in Engineering-Medicine

Having described optimization in engineering process, we should use some length of this chapter to briefly discuss optimization in engineering-medicine. Broadly speaking, engineering-medicine optimization could utilize the above-described engineering optimization method, which could be applied to two fronts: globally to the entire healthcare system and individually to the specific patients:

- *Optimization of the delivery for global healthcare system*: To accomplish this objective, the healthcare system could aim at system integration and focus on efficiency. To assist physicians in this pursuit, this book has provided a chapter on system integration (Chapter 10) and a chapter in efficiency (Chapter 11).

- *Optimization of the care for individual patient*: One good way for optimizing care for individual patient could be reliance on the principles of precision medicine, for which an entire chapter is devoted (Chapter 12). Since the central objective of precision medicine is to provide the right medicine to the right person at the right time, it is naturally the most optimal way of care for individual patients. In addition, the chapter on Big Data (Chapter 13), artificial intelligence (Chapter 14), advance biotechnology (Chapter 18), and biomedical imaging and diagnostic technology (Chapters 19–23) could provide additional intellectual and practical support, as the ultimate goals of these chapters are oriented towards achieving precision in medicine, whether it is for diagnosis, treatment, or prevention.

Summary

There are usually multiple solutions to a given engineering problem. Design optimization is to seek the "best" solution. How one defines as "best" solution given certain resource limitations and design requirements influence how one chooses the design. Real-life design optimization problems involve multiple *Objective Functions, Constraints,* and *Design Variables*, which can be approached with powerful computational tools.

References

[BRAINYQUOTE] 2018. Brainy Quote. [https://www.brainyquote.com/topics/optimize] accessed August 15, 2018.
[JFKLIBRARY] 2018. John F. Kennedy Presidential Library and Museum. [www.jfklibrary.org] accessed August 12, 2018.
Martins, J.R.R.A. and A.B. Lambe. 2013. Multidisciplinary design optimization: A Survey of architectures. AIAA Journal 51(9): 2013. DOI: 10.2514/1.J051895.
Viana, F.A.C., T.W. Simpson, V. Balabanov and V. Toropov. 2014. Metamodeling in multidisciplinary design optimization: How far have we really come? AIAA Journal 52(4): 670–690 (DOI: 10.2514/1.J052375).

QUESTIONS

1. How do you define "the best solution" in engineering term?
2. What is **Design Variable** in design optimization?
3. What is **Objective Function** in design optimization?
4. What is **Constraint** in design optimization?
5. What is the relationship between **Design Variable, Objective Function,** and **Constraint**?
6. What is **Weighing Factor** and how do engineers use it?

PROBLEM

The students will be grouped into 3–5 people per group. Within each group, the students will brainstorm for a real-life medical problem that demands a design optimization. The students will together deliberate to decide how to use the engineering principle of design optimization to generate the "best" outcome, and present their works to the entire class.

9
Problem-solving

Lawrence S. Chan

QUOTABLE QUOTES

"Engineers solve all sorts of problems, and one of their most important tools is their own creativity."

National Academy of Engineering (NAE 2016)

"We cannot solve our problems with the same thinking we used when we created them."

Albert Einstein, Theoretical Physicist and Nobel Laureate in physics, 1921 (ENTREPRENEUR 2018)

"Problems are not stop signs, they are guidelines."

Robert H. Schuller, American Evangelist and Author (ENTREPRENEUR 2018)

"All problems become smaller when you confront them instead of dodging them."

William F. Halsey, American Admiral (ENTREPRENEUR 2018)

"Problem-solving leaders have one thing in common: a faith that there's always a better way."

Gerald M. Weinberg, American Computer Scientist and Author (ENTREPRENEUR 2018)

"Each problem that I solved became a rule, which served afterwards to solve other problems."

Rene Descartes, French philosopher and Scientist (ENTREPRENEUR 2018)

"You can increase your problem-solving skills by honing your question-asking ability."

Michael J. Gelb, Author and Public Speaker (ENTREPRENEUR 2018)

Learning Objectives

The learning objectives of this chapter are to help students to understand the ways engineers solve the problems that could be substantially different from methods used in other field.
After completing this chapter, the students should:

- Understand the general principles of engineering problem-solving.
- Understand methods of engineering problem-solving, including the traditional method, the Thayer School's Problem-solving Cycle and Problem-solving Matrix.
- Understand the general principles of problem-solving in medicine.
- Able to apply the engineering method of problem-solving in real-life medical situation.

University of Illinois College of Medicine, 808 S Wood Street, R380, Chicago, IL 60612; larrycha@uic.edu

Problem-solving as a Cornerstone of Engineering and Engineering Education

One of the key ingredients of engineering education and ultimately one the key characteristics of engineering practice is to solve problems, whether the problems are large or small (NAE 2016). When engineers are called to solve the problems, they may create a brand new way to the solution or they may modify and improve the existing way to reach for the solution. In another word, the engineers can solve the problems through the invention of a new equipment or a novel methodology. Alternatively, they may solve the problems innovatively by improving the efficiency or effectiveness of an existing equipment or method. While problem-solving is essentially a daily task for many engineers, there are many different methods to carry out their problem-solving activities. Engineers, in turn, during formal school years learned their problem-solving methods which are considered as the cornerstone of engineering abilities, as employers seek out engineering school graduates who have the ability to solve open-ended problems. In addition, accreditation agencies for engineering education require teaching institution to provide documented effectiveness of instruction on problem-solving. Moreover, educators in engineering have advocated the need to document the effectiveness of existing methods in teaching problem-solving in engineering schools (Sharp 1991, Sobek and Jain 2004). In the following discussion, the general principles of engineering problem-solving will be define. Several methods utilized by several engineering schools will be detailed including Memorial University of Newfoundland, the Thayer School of Engineering at Dartmouth University (with problem-solving cycle and the problem-solving matrix). The rationale these examples are used is that they seem to be easy to understand by engineering and non-engineering students alike (Sharp 1991, THAYER 2016).

General Principles of Engineering Problem-solving

For most educators of professionally oriented programs that include engineering and medicine, adequate educational curriculum commonly includes three essential components (Sharp 1991):

- Fundamental knowledge acquisition and understanding
- Practical, not theoretical, problem-solving application
- Professional skill development for real-life practice

Thus, problem-solving is a key educational goal for engineering school and is a must-have ability for any and all engineers. Furthermore, problem-solving is a practical ability, rather than a theoretical knowledge. Therefore, problem-solving ability needs to be acquired not simply by lecture-type of instruction but through the right-kind of hand-on practice, as the famed American football player, coach, and executive of National Football League Vince Lombardi was quoted saying "only perfect practice makes it perfect" (BRAINYQUOTE 2018).

The Traditional Engineering Problem-solving Model

Traditionally, engineering problem-solving model include the following sequence of steps and condition requirements for each step (Lewis 1974, Jewel 1986, Sharp 1991):

Sequential Steps	Condition Requirement
Recognizing a particular need	Understanding symptom may not be the true problem for solving
Define problem, objectives, & constraints	Insight and understanding
Gathering data and information	Diligent and thorough
Generation of alternative solutions	Creativity and innovation
Evaluation of consequence of different solutions	Analytical techniques
Deciding and specifying "best" solution	Analytical techniques

The Problem-solving Cycle

The approach to engineering problem solving at Dartmouth's Thayer School of Engineering is through bringing "real world" into the classroom, i.e., the problems they themselves encounter outside of the classroom. And the students then make effort to solve those problems using the problem-solving cycle or the problem-solving matrix methodology (THAYER 2016, Figs. 1 and 2). This problem-solving cycle, as such it means to continue cycle until problem resolved, consists the following sequential steps:

- *State the problem*: by clearly identifying the problem to be solved.
- *Redefine the problem*: by simplifying and defining a solvable problem.
- *Identify constraints and set general specifications*: by determining the specific areas and issues to be resolved.
- *Identify alternative solutions*: by brainstorming multiple options for solution.
- *Analyze the alternative*: by examining the pros and cons of the various possible solution approaches.
- *Select the alternative*: by deciding on the approach to the solution.
- *Check the solution vs. the problem*: by measuring the result of decision in comparison to the resolution of the problem.
- Iterate the cycle until finding a definite solution for the problem.

The students would proceed through the problem-solving cycle with careful step-by-step approach and document the works for each of the steps. When encountering unviable solution, they would return to examine the paper trial and move back only as far as they need to find alternative solution. That way, they would not waste time and energy to start all over from the beginning when a solution is blocked at a given step (Fig. 1). As the students complete the full round of the problem-solving cycle, they would examine the solution against the original problem as to ensure their solution is sufficiently specific for the stated problem. Otherwise the students would iterate the cycle for a better solution.

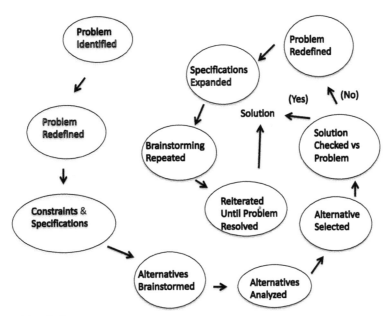

Fig. 1. Problem-solving Cycle.

Problem-Solving Matrix

	Scores for Spec 1	Scores for Spec 2	Scores for Spec 3	Scores for Spec 4	Total Scores
Alternative Solution Idea A					
Alternative Solution Idea B					
Alternative Solution Idea C					
Alternative Solution Idea D					

Fig. 2. Problem-solving Matrix.

The Problem-solving Matrix

The students at Thayer School of Engineering of Dartmouth also utilize the Problem-solving matrix approach, which aims to result in the best possible solution framed by a series of problem-solving matrices. The format of matrix consists of columns of specifications and of rows of ideas for the alternative solutions. The best possible solution, by definition, will be the one that results the most points by satisfying the most numbers of specifications (THAYER 2016, Fig. 2).

Other Engineering Problem-solving Strategies

Besides purely technical skills, other engineer offered very useful problem-solving strategies that include the followings (Buie 2018):

- Great ideas that solve the problems are results of persistent hard works.
- Clear communications with user-friendly text, table, and graphics are critical for idea sharing and development, and ultimately problem-solving.
- Essential to characterize uncertainty, which is present in any real-life problem-solving situation.
- Minimize uncertainty introduced through unintentional experimental variation by following solid written protocol, robust experimental planning, good experimental practice and training, and regular instrument maintenance and inspection.
- Reduce uncertainty introduced via systematic variation in experimental results by ensuring measurement equipment to be accurately calibrated, sensitive, and characterized and the variation measured by further improvement.
- Decrease uncertainty introduced by random variation in experimental results by ways of critical thinking and avoiding acknowledge random pattern as systematic pattern.
- Design experiments with multiple variables changed at the same time allow for the capturing of complex interactions between variables, in fewer experiments.
- Planning and reviewing the plans are equally important.

Problem-solving in Engineering-Medicine

Having delineate the principles and applications of engineering methods in problem-solving, we now turn the attention to the consideration of problem-solving in medicine. First we will delineate the rationale of needing problem-solving in medicine, and then consider some engineering problem-solving principles and techniques might be useful for medicine.

The Rationale of Problem-solving in Medicine

What is the rationale of emphasizing problem-solving in medicine? Like engineers, physicians also tend to think of themselves as problem-solvers (UMASS 2018). In fact, essentially every single daily activity of a practice physician is a problem-solving activity. The physicians' problem-solving activities are commonly occurred at two levels: Individual physician and group physicians.

Individual Physician

Every day individual physicians try to solve the problems of their patients' complaints (symptoms) by listening, inquiring, examining, testing, and finally determining a diagnosis from that complaint. Physicians then move next to solve the diagnosed problems by deliberating, consulting, and finally deciding on appropriate therapeutic plans for their patients. Following the initiation of the treatment, physicians also need, from time to time, to solve the problems of treatment non-responses by investigating the reasons, by consulting with other healthcare colleagues, and then selecting alternative treatments for those patients who do not respond to the initial treatments. In addition, physicians need, sometimes, to solve the problems of treatment side effects by seeking out cause, by reaching out to colleagues of other medical specialties, and by searching for equally effective replacement options. With all these problem-solving needs, an effective method will be very helpful. Will the engineering problem-solving methods provide any assistance in this regard?

Group Physicians

In certain medical situations, the physicians of several specialties gather together to perform problem-solving activities as a team effort. One good example is Tumor Board, a team of oncology experts come together to determine the best option for diagnosis and treatment of a given cancer. The traditional tumor board usually consisted of a medical oncologist, a pathologist, a radiologist, and an oncological surgeon. According to National Cancer Institute, a tumor board is "A treatment planning approach in which a number of doctors who are experts in different specialties (disciplines) review and discuss the medical condition and treatment options of a patient. In cancer treatment, a tumor board review may include that of a medical oncologist (who provides cancer treatment with drugs), a surgical oncologist (who provides cancer treatment with surgery), and a radiation oncologist (who provides cancer treatment with radiation). A tumor board is also called a panel of 'multidisciplinary opinion" (NCI 2018). More recently, a new kind of precision medicine-oriented Molecular Tumor Board is utilized for cancer treatment and its members could include a medical oncologist, an oncological surgeon, and a geneticist, who will provide the cancer genomic data to the board for treatment option discussion (van der Velden et al. 2017). Under these medical situations, groups of physicians perform problem-solving together.

The Principles of Engineering Problem-solving in Medicine

In addition to engineering principles for problem-solving, when it comes to medicine, the principles must include one additional and very essential item: *Do No Harm* (NLM 2018). Unlike most engineering processes, every step of problem-solving process in medicine the consideration of *"Do No Harm"* should always be kept in mind by the problem-solver, the physicians and the biomedical engineers.

The Applications of Engineering Problem-solving in Medicine

The methods of problem-solving offered by engineering profession, could potentially be utilized in medicine, provided some modification can be done. Now we will consider one typical medical example that individual physicians could encounter and we will think through how the engineering method could be utilized to solve the problem. The step-by-step utilization of the problem-solving cycle is depicted below:

- *Patient complaint*: abdomen pain onset in the early morning (note: symptom does not equal the true problem).
- *Problem identified*: symptom suggests an upper gastrointestinal problem or a cardiovascular problem.
- *Problem redefined*: Electrocardiogram (EKG) and blood cardiac enzyme tests ruled out myocardial infarction (heart attack). Empirical treatment with sublingual (under the tongue) nitroglycerine did not improve the symptom but treatment with antacid improve symptom suggest an esophagitis (inflammation of esophagus) and not a cardiac problem.
- *Constraints & Specification*: although upper gastrointestinal endoscope to examine the esophagus and stomach is the best approach, the patient cannot commit the time for the procedure.
- *Alternatives brainstorming*: Consider treatment with antacid or pain relief medication.
- *Alternatives analyzed*: Antacid is preferred since pain relief medication will mask the cause and will render future problem analysis difficult.
- *Alternative decision*: Patient treated with antacid during meal time.
- *Solution checked vs. problem*: No. There is no resolution of abdomen pain.
- *Problem redefined*: The abdomen pain may be a problem in stomach and not in esophagus.
- *Specification expanded*: In order to identify the problem accurately, we need to get a direct visualization of the problem.
- *Brainstorming repeated*: Upper gastrointestinal endoscopy or radiographic examination.
- *Alternatives analyzed*: Upper gastrointestinal endoscopy is preferred due to its ability in direct visualization of the problem, although it requires sedation with valium.
- *Alternative decision*: Upper gastrointestinal endoscopy performed and stomach ulcer identified as the true problem. Patient systematically treated with stomach acid-production inhibitor and oral antibiotics.
- *Solution checked vs. problem*: Yes. There is a resolution of abdomen pain.

Although the above exercise of engineering problem-solving cycle in medicine was conducted for individual physicians, the same method can be utilized for medical situations where group physicians participated such as Tumor Board or Molecular Tumor Board. In the group physician situation, the brainstorming, alternative analysis, and decision making will be more involved and less straight forward, as there are many individual physicians involved in the process. The particular challenge will be in the decision making step where different and divergent opinions could possibly be expressed by members of the Tumor Board or Molecular Tumor Board. Obviously, this engineering problem-solving cycle method should be applicable for biomedical engineering projects.

Summary

This chapter first focuses on defining the engineering principles and methodology of problem-solving, a cornerstone ability for all engineers. The principles of engineering problem-solving method in medicine was discussed and its application was illustrated with an exercise using a common medical example. Utilization of engineering problem-solving techniques in real-life medical problems is feasible, provided that the provision of "Do No Harm" is kept in the consideration for each step of the problem-solving process.

References

[BRAINYQUOTE] 2018. Practice makes perfect. Briany Quote. [https://www.brainyquote.com/topics/practice_makes_perfect] accessed May 27, 2018.

Buie, M. 2018. Problem Solving for New Engineers: What every engineering manager wants you to know. CRC Press, Boca Raton, FL.

[ENTREPRENEUR] 2018. Problem solving: 27 quotes to change how you think about problems. Entrepreneur. [https://www.entrepreneur.com/article/288957] accessed May 21, 2018.

Jewel, T.K. 1986. A Systems Approach to Civil Engineering Planning and Design. Harper and Row, New York, USA.

Lewis, W.P. 1974. Observations of problem-solving by engineering students. Australia Journal of Education 18: 172–183.

[NAE] 2016. Engineers are creative problem-solvers. National Academy of Engineering. [www.engineeringmessages.org] Accessed July 4, 2016.

[NCI] 2018. Tumor Board Review. National Cancer Institute. NIH. [https://www.cancer.gov/publications/dictionaries/cancer-terms/def/tumor-board-review] accessed May 27, 2018.

[NLM] 2018. Greek Medicine. National Library of Medicine. NIH. [https://www.nlm.nih.gov/hmd/greek/greek_oath.html] accessed May 27, 2018.

Sharp, J.J. 1991. Methodologies for problem solving: An engineering approach. The Vocational Aspect of Education 114: 147–157.

Sobek, D.K., II and V.K. Jain. 2004. The engineering problem-solving process: Good for students? Proceedings of the 2004 Society for Engineering Education Conference & Exposition. Session 1331.

[THAYER] 2016. What is engineering problem solving? Thayer School of Engineering at Dartmouth. [http://thayer.dartmouth.edu/teps/what.html] accessed July 6, 2016.

[UMASS] 2018. Physician as clinical problem solver. University of Massachusetts Medical School. [https://umassmed.edu/oume/oume/umms-competencies-for-medical-education1/physician-as-clinical-problem-solver/] accessed May 21, 2018.

van der Velden, D.L., C.M.L. van Herpen, H.W.M. van Laarhoven, E.F. Smit, H.J.M. Groen, S.M. Willems et al. 2017. Molecular tumor boards: Current practice and future trends. Annuals Oncology 28: 3070–3075.

QUESTIONS

- Why do you consider engineers as problem-solvers?
- Why do you consider physicians as problem-solvers?
- Is there a fundamental difference between problem-solving in engineering and problem-solving in medicine?
- What are the potential difficulties or challenges when applying the engineering problem-solving method in medicine?
- Why is "Do No Harm" so important in the problem-solving process in medicine?

PROBLEM

Form groups of 3–5 students and have the students in each group to discuss a typical medical symptom that the students can formulate a problem-solving process. When utilizing the engineering problem-solving cycle method, the students will go through each step of the entire problem-solving cycle in order to have a complete understanding and working knowledge of engineering problem-solving method in a real-life medical situation.

10

Systems Integration

Lawrence S. Chan

QUOTABLE QUOTES

"Whether for academics or for anyone in society (regardless of role), integration (along with discovery, application, and teaching) is the key determinate of success in everyday life."

Gary O. Langford, Author on Engineering Systems Integration (Langford 2012)

"Wellness is the complete integration of body, mind and spirit—The realization that everything we do, think, feel and believe has an effect on our state of well-being."

Greg Anderson, Author and Wellness Promoter (WELLNESS 2018)

"Wholeness is not achieved by cutting off a portion of one's being, but by integration of the contraries."

Carl Jung, Swiss Psychiatrist and Founder of Analytical Psychiatry (AZ QUOTES 2018)

Learning Objectives

The learning objectives of this chapter are to familiarize the students the engineering concept of systems integration and the potential applications and challenges of systems integration in medical practices. After completing this chapter, the student should:

- Understand the definition of systems integration and its importance in engineering practice.
- Understand the current state of medical practice with respect to systems integration.
- Understand ways systems integration can be applied in medicine to improve health outcome.
- Able to construct real-life medical examples where systems integration could help enhancing healthcare delivery.

Systems Integration in Engineering

Systems integration is an important engineering concept and practice in the field of engineering (Langford 2012). One way to look at the effect of systems integration is using team sport as an illustrative example. When a coordinated team is competing with a group of individuals acting independently, some individuals in the independence-acting group may be superior in some aspects than the coordinated team. But ultimately, the coordinated team will end up winning the most points. Likewise, "integration links related (and often

University of Illinois College of Medicine, 808 S Wood Street, R380, Chicago, IL 60612; larrycha@uic.edu

integral) objects into the same context to provide an overall management of effort that benefits the team through efficiencies of communication, planning, organization, directing actions, controlling their positions, and exhibiting structured teamwork" (Langford 2012). State another way, systems integration put the individuals of the coordinated teamwork together in a collaborative manner to achieve greater objectives. It increases efficiencies and reduced overall costs (Langford 2012). To sum up, "Fundamentally, integration is a method that facilitates outcomes that are beyond what an individual object can do either individually or by a number of objects acting independently, that is, makes things happen that would otherwise not happen. The whole is crucially greater than the sum of its parts" (Langford 2012). Perhaps an even better example to illustrate engineering integration will be a complex engineered system, like an automobile. The engine, the transmission, the suspension, the steering, the wheels, the breaks, and other coordinating parts cannot be independently designed and created by separately by different engineering teams without actual coordination, and then just assembled together—and hope that will work. This will never work. These separate subsystems must be co-designed and co-developed together as one integrated system. Having delineated the functions of engineering systems integration, we could now turn our attention to its application to medicine by way of medical education.

Systems Integration in Engineering-Medicine

Current Status of Medicine Practice

Before we go into the discussion of applying systems integration to the medical education and medical practice, it is proper to spend sometimes to first examine the current status of medicine practices with regard to the aspect of "integration". Since current medical practice pattern reflects what were taught in medical school, improvement of future practice depends at least to some extent the teaching occurred at the undergraduate medical education. When a patient comes to a physician's office for a visit, the physician will obtain the medical history, perform a physical examination, and then may order some laboratory tests to confirm or rule out diseases or conditions suspected from the information acquired from the medical history and physical examination. According to the combined results from medical history, physical examination, and laboratory tests, the physician could then reach a final diagnosis of the illness and prescribe medical and/or surgical treatments accordingly. The medical history part itself traditionally consists many components, including the followings (Bickley and Szilagyi 2013):

- *Identification data & source of history*: also including the history reliability.
- *Chief complaint*: what the patient comes to physician's office for, usually the symptom or concern that causes the patient to seek care.
- *History of present illness*: When and where the symptoms and physical signs of illness start, characteristics of these symptoms and signs, and the progression of the illness, and remediation with medication or therapy.
- *Past medical history*: Patients' major medical illness occurred in the past, including medical, surgical, psychiatric, and obstetric/gynecological (for female), immunization, frequency of preventative care.
- *Family medical history*: Related illness occurred in patient's family members including grandparents, parents, and siblings.
- *Current medication*: Medications taken at the time of office visit.
- *Allergic history*: History of allergy to drug, food, and other items.
- *Personal and social history*: Social aspect of the patient, including education level, family of origin, household situation, personal hobby, and life style.
- *Review of systems*: presence or absence of common symptoms of each major body systems: HEENT (head, eye, ear, nose, throat), respiratory, cardiovascular, musculoskeletal, neurologic, cutaneous, gastrointestinal, urogenital, renal, endocrinal, hematologic, and psychiatric.

As depicted above, this composition of history, usually taken by primary care physician like internist, is a very comprehensive piece of medical document, as it covers the entire body systems. Based on this history taken by primary care physicians, other physicians would have a very good idea all medical problems the patient has. As such, it provides a solid basis for the consideration of integration in patient care delivery below.

Integration in Patient Care Delivery

To better understand how we can incorporate systems integration to engineering-medicine education, a real-life clinical example will be utilized to illustrate this point. Similar to a complex engineered system with many interacting subsystems, the human body behave the same way. Illnesses that struck a patient always affect the whole person even if the disease appears localized. The cited example of diabetes below is an excellent illustration for this point. Diabetes mellitus, one of the most common chronic disorders, affected millions of patients in the US (ADA 2013). According to American Diabetes Association, one of the most important education institutions on diabetes, the prevalence of diabetes in 2012 was estimated to be 20 million of the US population, or about 9% of the total population, with another approximately 9 million undiagnosed patients (ADA 2013). In such a commonly affected disease like diabetes, the economic cost on the US is enormous, with total cost in the year of 2012 summing up to $245 billion, which consists $176 billion for direct medical care and $69 billion due to loss or reduce productivity (ADA 2013). Thus, chronic disease management is the greatest issue of healthcare where systems integration can really make a big difference in improving efficiency and reducing cost. The reason is that there are multiple complications and co-morbidity conditions occurred in patients affected by diabetes that require care provided by various medical specialty physicians. Specifically, diabetic patients have complications and co-morbid conditions of hypoglycemia, hypertension, cardiovascular diseases, stroke, kidney disease, and diabetic retinopathy, that require the medical cares of endocrinologist, general internist, cardiologist, neurologist, nephrologist, and ophthalmologist, respectively (ADA 2013). Based on the current model of healthcare delivery, if a patient has several of these complications/co-morbid conditions, he or she will need to make appointment and to visit each of these specialty physicians separately. On each of these visits, physicians will need to obtain a set of history, perform physical examinations, and then order a set of laboratory tests. The total costs of medical care and loss of productivity in these combined visits will be very high. Now let us examine how systems integration could make a big difference. A proposal for integrated care system would be something like establishing a "diabetes care center", where primary care physicians like general internists and all the medical specialties commonly encountered in diabetes care will be located. When a new patient visits the care center, the patient will first be seen by a general internist, who will obtain a comprehensive set of history, perform general physical examinations, and order a set of laboratory tests to include the commonly needed tests from all other specialty physicians. The internist will check the laboratory test results while the patient is waiting at the patient waiting-room. If the general internist found the needs to consult an endocrinologist for better blood sugar control, an ophthalmologist for ruling out retinopathy progression, a cardiologist for monitoring cardiovascular disease, and a nephrologist for monitoring kidney dysfunctions, the internist will then immediately transfer the care to these specialists at the same facility on the same day. Since a comprehensive medical history set has been completed and on the electronic medical record, there is no need for repeating it, so the specialty physicians can simply review the already-written medical history by the internist, obtain a brief specialty-specific history, and then perform brief specialty-specific examinations and specialty-specific tests to carry out the care, done in a sequential way and on the same day. The savings from this integrated care scenario will include reduction of physician man powers in taking unnecessarily repeated history, the reduction of costs for repeated general laboratory tests, reduced complications to the patients (as all the co-morbid conditions are taken care on the same day rather than the possibility of delayed care that can lead to complication), and reduction of loss productivity to the patients (as all of his needed medical cares are accomplished in one coordinated visit, one location, and at a reduced total amount of time). On the subsequent follow-up visit, the care coordination will be even better in terms of integration. Since the specialty needs of a given patient will be known by then, the patient's care will then be pre-arranged through the appointment system in a sequential manner as illustrated below (assuming the patient needs the following medical specialty care):

- *General internist*: comprehensive history, general examination, laboratory test, clinical decision and referral to first specialist-Endocrinologist.

- *Endocrinologist*: brief history, examination, specialty test, clinical decision & send to next specialist-Ophthalmologist.
- *Ophthalmologist*: brief history, examination, specialty test, clinical decision & send to next specialist-Cardiologist.
- *Cardiologist*: brief history, examination, specialty test, clinical decision & send to next specialist-Nephrologist.
- *Nephrologist*: brief history, examination, specialty test, clinical decision & send patient home.

If a very conservative estimation of 15% cost saving (physician manpower, tests, and loss of productivity) could be achieved in one patient, these can be translated into a huge cost saving for the entire healthcare system, a more than $35 billion saving per year in the US alone. Thus, to effectively and efficiently deal with the diabetic illnesses, a team of physicians would better coordinately act in a way similar to a team of engineers who co-design and co-develop an automobile, i.e., to co-locate in a diabetes care center and work with the "systems integration" approach for better healthcare delivery. But our ability to integrate patient care delivery is far beyond diabetic care. Potential integration and cost saving could occur in several other disease categories that consumed the greatest amount of resources in dollar term, including circulatory diseases, musculoskeletal diseases, and respiratory diseases, which consumes approximately $240 billion, $180 billion, and $160 billion, respectively in year 2012 (Chan 2018). The use of advanced information technology, like big data and artificial intelligence, could be the center piece of systems integration when multiple specialists are working together to deal with one illness with multiple complications. The use of these technologies will greatly facilitate the collaborative and integrative efforts, in which the specialists are both the contributors and the users of the big data, with the help of artificial intelligence.

Integration of Science and Clinical Practice

Just as engineers integrate state-of-the-art scientific knowledges into a new design or a new technology, physicians should also integrate up-to-date medical discoveries and technology advancement into their diagnostic and therapeutic tools. In a recent roundtable workshop on value and science-driven healthcare organized by the Institute of Medicine, the topic of integrating research and practice was the center of the discussion (Alper and Crossman 2015). It is estimated that by the year 2020, clinical supports in ninety percent of all clinical practices will be enhanced by evidence-based accurate, nearly real-time, and up-to-date information. But having information by itself is not sufficient to improve the patient care, and the patient care could be improved if and only if the useful clinical information was put into practice, that is, to integrate the science into the daily medical practice (Alper and Crossman 2015). Several workshop speakers provided the following key points that are essential in integrating clinical research and practice:

- *A successful research partnership*: (Alper and Crossman 2015, Huang et al. 2013).
- *A user-friendly data*: The need to avoid overburdening healthcare providers by utilizing data already routinely collected through the electronic health record system (Alper and Crossman 2015, Gerhardt et al. 2007).
- *A healthcare system stabilizing innovation*: The need to avoid the healthcare system being destabilized by disruptive innovation (Alper and Crossman 2015, Grossman and Kemper 2016).
- *An ethical motivation to "do the right thing"*: The need to promote high moral standard in the integration process (Havranek et al. 2003, Alper and Crossman 2015, Moore et al. 2016).
- *A cooperative reimbursement system*: The need to avoid the impediment of integration by the uncooperative reimbursement system (Alper and Crossman 2015, Havranek et al. 2003).

Integration of Medical Care with Other Health Determinants

Traditionally, medical education focuses almost entirely on the instruction of scientific knowledges of normal body function and pathophysiology of disease and on the development of professional skills in providing medical care to patients. The medical community is now coming to acknowledge the important roles of other "health determinants" play in the totality of health. In fact, one report claims that the medical care provided by healthcare providers (physicians, nurses, pharmacists) account for only about 10% of impact of health (Hershberger and Bricker 2014, Chan 2016). The National Academy of Medicine (formerly Institute of Medicine) has called for the triple aims of "better care, lower costs, and better health" as to emphasize "better health" is one big step beyond "better care" (Bisognano 2013, Chan 2016). In fact, University of Wisconsin School of Medicine has actually changed its name to "University of Wisconsin School of Medicine and Public Health" in 2005 to reflect its commitment to integrate other determinant to total medicine (Chan 2016). Because the other health determinants include environmental, economic, social and behavior factors, the challenges for integrating these determinants into medical school teaching and medical care are very big and will need to involve not only individual institutions, but will need to have a national dialog, debate, and strategy, if our aim is to achieve a big health impact (Hershberger and Bricker 2014).

Challenges of Systems Integration in Medicine

There are several challenges facing the systems integration in medicine. However, there may be ways that these road blocks be overcome.

Individual Preference

In the area of patient care delivery, one difficulty in the integration will be patients' preference. If a patient who has diabetes prefers a certain specialty physician outside of the diabetes care center where integration occurs, then the integration process might not work for this individual patient, at least not to the full extent. For this reason, the Veterans Affairs (VA) Medical Center could be the best place in implementing systems integration in patient care delivery, since patient preference of physician in not a policy there, if timely proper care is available within the VA system.

Medical-legal and Medical Billing Issues

Also in patient care delivery, systems integration could face a challenge in terms of medical record documentation in relationship to legal requirement and medical billing. As the current medical billing process requires a set of document including history, physical examination, diagnosis and treatment plan, how would the specialty physicians bill their services if the history is mainly based on that taken by the primary care physicians? In addition, how would this systems integration-generated medical record, in which the specialist's history document is based primarily on that taken by primary care physician, stands the court of law?

Funding Issue

In the integration of medical school teaching and medical care with other health determinants, the biggest challenge is perhaps the funding, as those determinants are large and implementation would require large sum of funding. Individual institution would not be able to shoulder all the financial burdens and society as a whole would need to decide how to provide the needed fund, knowing that end result of systems integration will be cost-saving for the entire society.

Summary

In this chapter, we provided the engineering concept of systems integration and its potential applications in medicine. With its goal of improving efficiency and cost saving, systems integration could potentially provide the needed improvement for the healthcare in the US. While challenges will be encountered, the medical community, along with a supporting society, would be able to overcome the difficulties to achieve the objectives.

References

[ADA] 2013. American Diabetes Association. March 6, 2013. [http://www.diabetes.org/newsroom/press-releases/2013/annual-costs-of-diabetes-2013.html] accessed May 28, 2018.

Alper, J. and C. Crossman. 2015. Integrating research and practice: Health system leaders working toward high-value care. Roundtable on value & science-driven healthcare. The Institute of Medicine. The National Academies Press, Washington DC, USA.

[AZ QUOTES] 2018. AZ Quotes. Integration. [http://www.azquotes.com/quotes/topics/integration.html] accessed May 28, 2018.

Bickley, L.S. and P.G. Szilagyi. 2013. Bate's guide to physical examination and history taking. 11th ed. Nolter Kluwer & Lippincott Williams & Wilkins. New York, NY.

Bisognano, M. 2013. The vision and importance of measuring the three-part aim: Core metrics for better care, lower costs, and better health. An institute of medicine workshop. National Academy of Medicine. December 5, 2013. [https://nam.edu/wp-content/uploads/2017/12/Bisognano.pdf] accessed May 28, 2018.

Chan, L.S. 2016. Building an engineering-based medical college: Is the timing ripe for the picking? Med Sci Edu 26: 185–190.

Chan, L.S. 2018. Engineering-medicine as a transforming medical education: A proposed curriculum and a cost-effectiveness analysis. Biol. Eng. Med. 3:1-10/ doi: 10.15761/BEM.1000142.

Gerhardt, W.E., P.J. Schoeltker, E.F. Donovan, U.R. Kotagal and E.S. Muething. 2007. Putting evidence-based clinical practice guidelines into practice: an academic pediatric center's experience. Jt. Comm. J. Qual. Patient Saf. 33: 226–235.

Grossman, D.C. and A.R. Kemper. 2016. Confronting the need for evidence regarding prevention. Pediatrics 137: e20153332. Doi: 10.1542/peds. 2015-3332. Epub 2016 Jan 12.

Havranek, E.P., H.M. Krumholz, R.A. Dudley, K. Adams, D. Gregory, S. Lampert et al. 2003. Aligning quality and payment for heart failure care: defining the challenges. J Car Fail 9: 251–254.

Hershberger, P.J. and D.A. Bricker. 2014. Who determines physician effectiveness? JAMA 312: 2613–2614.

Huang, S.S., E. Septimus, K. Kleinman, J. Moody, J. Hickok, T.R. Avery et al. 2013. Targeted versus universal decolonization to prevent ICU infection. N Engl J Med 368: 2255–2265.

Langford, G.O. 2012. Engineering systems integration: theory, metrics, and methods. CRC Press, New York, NY.

Moore, S.L., I. Fischer and E.P. Havranek. 2016. Translating health services research into practice in the safety net. Health Serv Res 51: 16–31.

[WELLNESS] 2018. Guiding Wellness. [http://guidingwellness.com/wellness-and-holism/what-is-wellness/] accessed May 26, 2018.

QUESTIONS

1. What is your understanding of the concept of "engineering systems integration"?
2. Could you in one or two sentences, describe your understanding of systems integration in your own words?
3. Why is systems integration important for healthcare?
4. What areas of healthcare would you consider as feasible for systems integration?
5. What are the potential challenges in implementing systems integration in medicine?

PROBLEM

Students will be arranged in small groups of 3–5 people each in each group, the students will be assigned a systems integration project to deliver optimum healthcare outcome with lowest costs to 500 diabetes patients living within 15 miles radius of a medical center equipped with both outpatient clinics and inpatient hospital.

11
Efficiency

Lawrence S. Chan

QUOTABLE QUOTES

"First, efficiency is a central value in engineering"

J.K. Alexander, Engineer and Author (Alexander 2009).

"Efficiency is doing things right; effectiveness is doing the right things."

Peter Drucker, American Management Consultant (BRAINY QUOTE 2018a)

"The first rule of any technology used in a business is that automation applied to an efficient operation will magnify the efficiency. The second is that automation applied to an inefficient operation will magnify the inefficiency."

Bill Gates, Founder of Microsoft (BRAINY QUOTE 2018b)

"The great thing in life is efficiency. If you amount to anything in the world, your time is valuable, your energy precious. They are your success capital, and you cannot afford to heedlessly throw them away or trifle with them."

Orison Swett Marden, American Physician and Author (BRAINY QUOTE 2018c)

Learning Objectives

The learning objectives of this chapter are to familiarize the students the engineering concept on efficiency, importance of efficiency in medicine, and practical ways of applying engineering efficiency in medicine. After completing this chapter, the students should:

- Understand the engineering concept of efficiency.
- Understand the engineering approach to efficiency.
- Understand importance of efficiency in medicine.
- Able to apply engineering principle of efficiency in medicine.

University of Illinois College of Medicine, 808 S. Wood Street, R380, Chicago, IL 60612; larrycha@uic.edu

Engineering Concept of Efficiency

When one thinks about the characteristics of engineering products, "efficiency" always surfaces as an obvious one. An airplane can bring people across the oceans thousands miles away in just few hours. A computer can make an accurate 3-dimensional design in few minutes. A global positioning system (GPS) device can instantly locate your destination and guide you step-by-step to the desired locale. These are just a few commonly encountered examples of engineering efficiency and there are many more such examples. In engineering term, "The conversion, of fuel to motion or work, is at the heart of engineering efficiency" (Alexandar 2009). This means that engineering efficiency is the ratio of output to input. For example, if an engineer takes 100 units of fuel (input) in a machine to generate 70 units of works (output), then the efficiency of this machine will be 70%, with the other 30% fuel being wasted in the process of generating the desired works. The goal of engineering efficiency, obviously, is to make a machine with percentage of efficiency as high as possible, so that little energy is wasted in the process of generating the desirable works. Engineers utilize different methods to improve efficiency. For example, to increase an automobile energy efficiency, the engineers may try to modify the size of the engine, the type of engine, the exterior shape of the automobile, the weight of the automobile, the acceleration control, the electronics of the automobile, or even the type of fuel. Engineers will make modification, test, and retest until the desired efficiency level is achieved. Understandably, engineers' abilities to achieve efficiency are also under specific constraints. For example, in the case of automobile, engineers cannot replace important structural components that are made from metal with certain structural strength requirement by light weight plastic that does not provide the same strength, for obvious safety reason. In addition, engineers are also under quality constraints and thus cannot aim to achieve fuel efficiency at the expenses of automobile quality. Likewise, safety and quality cannot be compromised in pursuing efficiency in medicine, as we will now discuss.

Efficiency Concept in Medicine

As efficiency is important in engineering design and optimization, it is also essential for medicine. The equation of efficiency is taking up a more significant meaning lately when the US healthcare is comparing with other nations of similar economic category, revealing that the US healthcare is costing more but providing less (OECD 2014, MIRROR 2014). In light of the forecast of physician shortage and the high costs of healthcare in the US, the Institute of Medicine (National Academy of Medicine) has called for a triple aim in medicine: better care, better health, and lower costs (IOM 2015, AAMC 2015). Increase healthcare efficiency is certainly in the forefront of discussion (Fraser et al. 2008, Hussey et al. 2009, Hershberger and Bricker 2014, Russo and Adler 2015). However, when it comes to medicine, particularly in day-to-day patient care, the physicians must see their patients one at a time, how would engineering principle and the teaching of engineering-medicine help increasing the future physicians' efficiency?

Before we go into the detailed discussion on improving medicine efficiency, it is proper to clarify the slight conceptual difference between engineering efficiency and economic efficiency. Although engineering efficiency does affect economic efficiency, economic efficiency takes into account other important factors, such as opportunity costs (Drummond et al. 2015). Economic efficiency concerns about value and it is what we are about to discuss, since medicine, after all, is about value we provide for the patients, not just about some kind of numerical ratio.

Efficiency, in economic term, measures "how well resources are being used to promote social welfare" (Henderson 2015). Naturally, efficient use of scarce resources improves social wellbeing whereas inefficient utilization of resources is wasteful. Technically, economic efficiency can be defined by either maximizing production (output) with the same amount of resources (input or effort) or by minimizing the use of resources in the generation of the same amount product (Henderson 2015). These two technical aspects of economic efficiency will be discussed in the sections below.

When efficiency is in the consideration, technology can be helpful and technology teaching in medical school or residency could be enhanced. In fact, both faculty and students of the current medical schools felt that their education is outpaced by the rapid advancement of health science and technology (Plunkett-Rondeau et al. 2015, Moskowitz et al. 2015, Day et al. 2015, Fung 2015). Due to the fact that undergraduate medical education is the exclusive physician pipeline and the gateway to the future of medicine, what we educate the medical students today would have a profound influence what the physicians will practice in the 50 years ahead. The most effective way to transform medicine of the future may in fact be, therefore, through undergraduate medical education reform. The establishment of an engineering-based medical school, Carle-University of Illinois College of Medicine (Carle-UI) at the University of Illinois Urbana/ Champagne Campus has aimed to transform undergraduate medical education by incorporating engineering training into medical education (Cohen 2015, NEW 2015). In this chapter, I will first discuss in theory how an engineering-medicine education may help improving the future physicians' efficiency, and then followed by examining literature evidences in support of the efficiency-improving ability of engineering-medicine curriculum. An engineering-medicine education may be characterized as an educational system aims to teach and train medical students not only the traditional theory and practice of patient care, but also the principles of biomedical engineering in innovative and efficient approaches to health care delivery and in state-of-the-art technology utilization for understanding, diagnosis, treatment, and prevention of human diseases (Chan 2016, 2018).

While the engineering-medicine education may improve healthcare in many ways, this article concentrates on the issue of improving delivery efficiency. Although the focus of this paper will be on undergraduate medical education, it is indeed essential that such training be reinforced at the post-graduate medical education level.

Improving Future Physicians' Efficiency in Healthcare Delivery through an Engineering-Medicine Education

Since currently there is no established and tested engineering-medicine curriculum, a recently proposed curriculum of such program will serve as the basis for the discussions below (Chan 2018).

Increase Efficiency by Increasing Output with same Amount of Input Effort

Theoretical Consideration

Indeed, a course entitled "Efficiency" has been included in the proposed Engineering-medicine curriculum, which introduces students to the engineering concept of efficiency and the application of engineering principles to improve medicine efficiency (Chan 2018). Furthermore, a course entitled "Design and Optimization", also in the proposed curriculum, will introduce students to the engineering concept of design and optimization, its potential role in medicine, and its applications in medicine (Chan 2018). Stated by one expert in the engineering optimization field, "the process of determining the best design is called optimization" (Parkinson et al. 2013). Defined by other expert, "The purpose of optimization is to achieve the 'best design' relative to a set of prioritized criteria and constraints. These include maximizing factors such as productivity, strength, reliability, longevity, efficiency, and utilization" (Merrill et al. 2008). Thus, engineering's design and optimization aim for improvement of precision, function, quality, and efficiency. Similarly, this course aims to optimize healthcare value by improving healthcare efficiency. Moreover, other courses included in this proposed curriculum such as "Advanced Biotechnology" and "Biomedical Imaging" will teach the students on technology-driven healthcare devices (Chan 2018). These tech-oriented courses, would likely increase the efficiency of future physicians by shaping the future physicians to be more technology savvy, more intelligent in optimal technology utilization, and with more ability to adopt to the upcoming Big Data-driven healthcare service, thus would increase their efficiency in patient care (higher output). In addition, these courses will help biomedical engineering students to further crystalize their important technology roles in medicine and healthcare. Moreover, a course teaching "Precision medicine" in this proposed curriculum (Chan 2018) would enhance the practice of providing the right medicine to the right patient at the right time, thus will increase the medical efficiency in this regard (higher outcome with same input).

Supporting Evidence

Having discussed theoretically some areas of engineering-medicine education may help improving physicians' efficiency, can we then find evidence from existing literature in support of such theoretical benefits?

In the proposed curriculum, the teaching of new tools and technology for patient encounters and care is included, and such teaching would provide innovative solution to help increasing clinical practice efficiency (Chan 2018). This potential effect is supported by evidence in a recent study where a tech-driven "healthcare delivery science" improved the hospital emergency department efficiency that is sustained over time, as measured by near elimination the occurrence of patients "left without being seen", and improvement on waiting times from entering door to be seen by physician, patient satisfaction, and total length of stay (DeFlitch et al. 2015).

Increase Efficiency by Producing Same Amount of Output with Reduced Input Effort

Theoretical Consideration

In the proposed engineering-medicine curriculum, a course different from the traditional curriculum, termed "Systems Integration" is included (Chan 2018). This way, the students would likely learn to conduct patient care with a mind-set of team-approach, thus would help shaping the future physicians in performing care in a coordinated manner. As we know that many patients with chronic inflammatory diseases like diabetes require cares from multiple physicians of several specialties such as endocrinologist, ophthalmologist, dermatologist, and nephrologist. A coordinated care service, which coordinate all the diabetes-needed specialty cares in one place and serve the patient in one coordinated clinic visit will help reducing duplicating office visits, redundant lab tests, conflicting treatments, and medical errors, thus will provide better care outcome with reduced man power costs (lower input effort). In addition, this proposed curriculum teaches a course "Introduction to System Biology", in addition to an organ-based format (Chan 2018). This format would likely reinforce medical students in understanding that pathology in one body component affects the entire human body, and not just a single organ, thus would help shaping the future physicians to care for their patients in a more coordinated fashion. A coordinated healthcare will result in substantial saving of healthcare providers man power while providing the same level of care (Chan 2018).

Supporting Evidence

A recent study from pediatric specialty examined whether a reduction of utilizing computed tomography (CT scan, more sensitive and more expensive) in pediatric patients as a result of switching to ultrasonography (US, less sensitive and less expensive) would adversely affect the clinical outcomes (measured for negative appendectomy, appendiceal perforation, and 3-day emergency room revisit (Bachur et al. 2015). Although CT scan is generally recognized as a more sensitive diagnostic methodology, this study showed no difference between patients whom the diagnoses were made by CT scan and US in the measures of appendiceal perforation or 3-day emergency room revisit (Bachur et al. 2015). There is a slight decline in negative appendectomy (defined as appendectomy performed in patients without true appendicitis) in patients diagnosed by US (Bachur et al. 2015). Thus, this kind of evidence-based medicine training would be able help future physicians to achieve identical healthcare outputs with lower cost (less input effort) (Carroll 2015). If this kind of evidence-based information gathering and sharing are assisted by Big Data analytic, one can achieve an even better outcome of efficiency (see Chapter 13). Furthermore, the elimination of potential side effects of ionized radiation with CT scan on children if the diagnoses were made by US would also be considered a cost-reduction to the entire health system, although this aspect of cost saving would be difficult to measure (Carroll 2015).

Another supporting evidence is illustrated by a study of using antibiotics as first-line treatment for acute appendicitis in adults (Salminen et al. 2015). A large investigative trial enrolled 530 patients, age ranged from 18 to 60, with uncomplicated appendicitis confirmed by CT scan. The patients were randomly assigned to receive antibiotics or undergo appendectomy. The success rate of controlling appendicitis was higher for the group assigned to surgery (99.6%), compared to 73% in the antibiotic group. After one year, for the remaining 27% patients in the antibiotic group, who's appendicitis were not controlled and ultimately received surgery, 83% had uncomplicated appendectomy, 7% had negative appendectomy, and 10% had complications (3% of the entire antibiotic group). However, no patients developed abdominal abscesses or other major complications in the antibiotic group (Salminen et al. 2015). Armed with the evidence from this study, physicians can discuss the treatment options with their patients and these informational discussions could result in cost saving if antibiotic is chosen as the first-line treatment (Salminen et al. 2015). To be sure, physicians would be much better informed if such evidence-based information collection and sharing are directed by Big Data analytics (see Chapter 13).

Further evidence is illustrated in a primary care setting, where a study has performed to measure the impact of a "Clinical Decision Support" (CDS) system, a bioinformatics data based on Big Data resource, on the time spent by physicians on preventive services and chronic disease management (Wagholikar et al. 2015). 30 patients from a primary care clinic were randomly selected and assigned to 10 physicians, who performed chart review to decide on preventive services and chronic disease management. Done in a randomized crossover manner, each patient received 2 sets of recommendations: one from a physician with CDS assistance and another from a physician without CDS assistance. The results showed that on the average, physicians completed the recommendations in 1 minute 44 seconds and 5 minutes, if they were assisted or unassisted by CDS, respectively (Wagholikar et al. 2015). This study results demonstrate that teaching medical students appropriate use of information technology could achieve the identical outcome while substantially reduce physicians' manpower usage (lower input effort).

Summary

Therefore, since an engineering- medicine education would, through innovative, technological, and solution-oriented approaches, teach medical students (and future physicians) the concept and skills to become more efficient in healthcare delivery, this kind of curriculum could provide a positive transformation of medical education for the future. However, this review does not deal with the question of whether such increase of efficiency would parallel an increase healthcare quality. On a preliminary basis, a cost-effectiveness analysis with a set of assumptions, gave a positive cost-effective result (Chan 2018). A future, rigorous, cost-effectiveness analysis, which is beyond the scope of the current course, could provide a strong and economically beneficial rationale to launch an engineering-medicine college, if the evaluation is supportive. For the biomedical engineering students, this chapter will help crystalizing their important roles in making medicine efficient. In addition, this chapter will help guiding them in their future biomedical instrument design and biomedical algorithm development to achieve medical efficiency.

References

[AAMC] 2015. Physician supply and demand through 2025: Key findings. [www.AAMC.org] accessed December 12, 2015.

Alexander, J.K. 2009. The concept of efficiency: an historical analysis. *In*: Meijers, A. [ed.]. Philosophy of Technology and Engineering Sciences. Vol. 9. Elsevier. Amsterdam, The Netherlands.

Bachur, R.G., L.A. Levy, M.J. Callahan, S.J. Rangel and M.C. Monuteaux. 2015. Effect of reduction in the use of computed tomography on clinical outcomes of appendicitis. JAMA Pediatr 169: 755–760.

[BRAINY QUOTE] 2018a. Brainy Quote. Efficiency Quotes. [https://www. Brainyquote.com/authors/peter_drucker] accessed July 23, 2018.

[BRAINY QUOTE] 2018b. Brainy Quote. Efficiency Quotes. [https://www. Brainyquote.com/authors/bill_gates] accessed July 23, 2018.

[BRAINY QUOTE] 2018c. Brainy Quote. Efficiency Quotes. [https://www. Brainyquote.com/topics/efficiency] accessed July 23, 2018.

Carroll, A.E. 2015. Doing less is sometimes enough. JAMA 314: 2069–2070.

Chan, L.S. 2016. Building an engineering-based medical college: Is the timing ripe for the picking? Med Sci Edu 26: 185–190.

Chan, L.S. 2018. Engineering-medicine as a transforming medical education: A proposed curriculum and a cost-effectiveness analysis. Biology, Engineering and Medicine 3(2): 1–10. Doi > 10.15761/BEM.1000142.

Cohen, J.S. 2015. U. of I. trustees approve new medical school. Chicago Tribune. March 12, 2015. [www.chicagotribune.com] accessed June 25, 2015.

Day, J., J. Davis, L.A. Riesenberg, D. Heil, K. Berg, R. Davis et al. 2015. Integrating sonography training into undergraduate medical education: A study of the previous exposure of one institution's incoming residents. J Ultrasound Med 34: 1253–7.

DeFlitch, C., G. Geeting and H.L. Paz. 2015. Reinventing emergency department flow via healthcare delivery science. HERD 8: 105–115.

Drummond, M.F., M.J. Sculpher, K. Claxton, G.L. Stoddart and G.W. Torrance. 2015. Methods for the economic evaluation of health care programmes. 4th Ed. Oxford University Press, Oxford, UK.

Fraser, I., W. Encinosa and S. Glied. 2008. Improving efficiency and value in healthcare: Introduction. Health Serv Res 43: 1781–1786.

Fung, K. 2015. Otolaryngology-head and neck surgery in undergraduate medical education: advances and innovations. Laryngoscope 125: Suppl 2: S1–14.

Henderson, J.W. 2015. Health economics and policy. Cengage Learning. Stamford, CT.

Hershberger, P.J. and D.A. Bricker. 2014. Who determines physician effectiveness? JAMA 312: 2613–2614.

Hussey, P.S., H. de Vries, J. Romley, M.C. Wang, S.S. Chen, P.G. Shekelle et al. 2009. A systematic review of health care efficiency measure. Health Serv Res 44: 784–805.

[IOM] 2015. Core measurement needs for better care, better health, and lower costs: Counting what counts-workshop summary. Institute of Medicine. June 24, 2013. [www.iom.edu] accessed June 25, 2015.

Merrill, C., R.L. Custer, J. Daugherty, M. Westrick and Y. Zeng. 2008. Delivering Core Engineering Concepts to Secondary Level Students. J Technology Edu 20: 48–64.

[MIRROR] 2014. Mirror, mirror on the wall, 2014 update: How the U.S. health care system compares internationally. [www.commonweathfund.org/publications] accessed September 13, 2015.

Moskowitz, A., J. McSparron, D.J. Stone and L.A. Celi. 2015. Prepare a new generation of clinicians for the era of Big Data. Harvard Medical Student Review. 2015 January 3. [www.hmsreview.org] accessed August 24, 2015.

[NEW] 2015. New College. New Medicine. The first college of medicine specifically designed at the intersection of engineering and medicine. [https://medicine.illinois.edu/news.html] accessed September 17, 2015.

[OECD] 2014. OECD health statistics 2014. How does the United States compare? [www.oecd.org/unitedstates] accessed September 13, 2015.

Parkinson, A.R., R.J. Balling and J.D. Hedengren. 2013. Optimization Methods for Engineering Design: Applications and theory. Copy Right by Brigham Young University. [apmonitor.com/me575/uploads/Main/optimization_book.pdf] accessed August 2, 2018.

Plunkett-Rondeau, J., K. Hyland and S. Dasgupta. 2015. Training future physicians in the era of genomic medicine: trends in undergraduate medical genetic education. Genet Med 2015 Feb 12. doi: 10.1038/gim.2014.208.

Russo, P. and A. Adler. 2015. Health care efficiency: Measuring the cost associated with quality. Managed Care July: 38–44.

Salminen, P., H. Paajanen, T. Rautio et al. 2015. Antibiotic therapy vs. appendectomy for treatment of uncomplicated appendicitis. JAMA 313: 2340–2348.

Wagholikar, K.B., R.A. Hankey, L.K. Decker, S.S. Cha, R.A. Greenes, H. Liu et al. 2015. Evaluation of the effect of decision support on the efficiency of primary are providers in the outpatient practice. J Prim Care Community Health 6: 54–60.

QUESTIONS

1. How do you define engineering efficiency?
2. What is the difference between engineering efficiency and economic efficiency?
3. Why is healthcare efficiency important?
4. What are the two major forms of efficiency improvement in health care?
5. What characteristics of healthcare cannot be compromised in achieving healthcare efficiency?

PROBLEM

Students are grouped into 3–5 people groups. Within each group, students will brainstorm to come up with a real-life healthcare problem for which the students can discuss and decide on ways to improve its efficiency. Each group will be represented by one student, who will discuss their chosen topic in the class, in terms of how they will make the healthcare process more efficient as well as precaution on potential compromise of safety and quality in pursuing efficiency.

12

Precision

Lawrence S. Chan

QUOTABLE QUOTES

"Everything is about your movements and precision and timing, which is what gymnastics is about."

Shawn Johnson, 2008 Olympic balance beam gold medalist (BRAINY 2016)

"Be precise. A lack of precision is dangerous when the margin of error is small."

Donald Rumsfeld, Secretary of Defense under President George W. Bush (AZQUOTES 2016)

"Precision medicine is about matching the right drug to the right patient."

Precision medicine for cancer with next-generation functional diagnostics (Friedman et al. 2015)

"Advances in tumor biology, coupled with a rapid decline in the cost of multiplex tumor genomic testing of human samples, has ushered in the era of precision cancer medicine."

The landscape of precision cancer medicine clinical trials in the United States (Roper et al. 2015)

"Personalized molecularly based medicine is driving the major revolution of your time. We are all pioneers in a brave new world."

Pieter Cullis, PhD, Scientist and Author (Cullis 2015)

Learning Objectives

The learning objectives of this chapter is to familiarize the students the engineering concept of precision and the concept of precision medicine. The chapter will further delineate the importance of precision medicine illustrated with real-life medical applications including next-generation of sequencing, immunotherapy, functional cancer diagnostic, mass spectrometry-based proteomic, structure-based precision vaccine design, genome editing, and the role of artificial intelligence in precision medicine.

After completing this chapter, the student should:

- Understand the engineering principle of precision.
- Understand the concept of precision in engineering-medicine.
- Understand the rationale for the development of precision medicine.
- Obtain the key elements of precision medicine.

University of Illinois College of Medicine, 808 S. Wood Street, R380, Chicago, IL 60612; larrycha@uic.edu

- Understand why precision medicine may impact our healthcare delivery in the coming years.
- Become familiar with next generation of sequencing, next generation of functional diagnostics, next generation of vaccine design, mass spectrometry for proteomic profiling, cell-based immunotherapy for cancers, and gene therapy.
- Able to embraces the precision medicine culture.
- Capable to think healthcare delivery in a precision manner.
- Able to apply principles of precision medicine to solve real medical problem.

Precision: An Essential Engineering Characteristic

Engineering, in many ways, is synonymous with precision. Imaging what happens if the time of your watch is off 10 minutes every day, if the speedometer of your automobile is off by 10 miles per hour, if your computer keyboards do not produce the same alphabets each time you type a word, or if you arrange a trip to South America but the airplane takes you to South Africa instead due to aviation instrument failure. Notwithstanding that we generally take for grant the precision those products of engineering have provided to us on a daily basis, precision is one of key important functions of engineering. It can be simply stated that one of the characteristics of engineering is that it makes things precise and accurate.

According to the American Society of Precision Engineering, precision engineering is an engineering discipline that "focuses on many areas of interests that are important in the research, design, development, manufacture and measurement of high accuracy components and systems" (ASPE 2016). These scopes of precision engineering technologies would include, but are not limited to, the following industries: microelectronics, automotive, optics and photonics, microelectromechanical systems (MEMS), nanotechnology, defense and machine tools (ASPE 2016), as the members of this society represent a variety of technical areas—from engineering fields (mechanical, electrical, optical and industrial) to scientific fields (materials science, physics, chemistry, mathematics and computer science) (ASPE 2016). In addition, the precision engineering principle is also widely applied in agricultural industry (Whelan and Taylor 2013, Zhang 2016). In engineering term, precision and accuracy have carried slightly different meanings. But for the purpose of this textbook, these two terms are used interchangeably to mean utilizing methods of high fidelity, accuracy, and consistence in solving problems.

Precision in Engineering-Medicine

The importance of precision in medicine cannot be overemphasized. Just as Donald Rumsfeld, the formal Secretary of Defense served under President George W. Bush, indicated, the critical situation where precision is needed the most is when the margin of error is small, medicine has no shortage of situations where the margin of error is small, both in the diagnostic processes and in the therapeutic courses (AZ QUOTES 2016). When a general surgeon operates to remove a tumor in a patient's thyroid gland within few millimeters from the important recurrent laryngeal nerve that controls the movement of patient's voice box (larynx); when an anatomic pathologist makes a determination of whether a surgically removed pigmented skin lesion is or is not a melanoma, a deadly skin cancer that will require a follow-up wide-margined second-round surgery, comprehensive survey for metastasis, and perhaps even additional chemotherapy or immunotherapy; when an ophthalmologist applied laser treatment to a diabetic patient's retina to stop retinal blood vessel bleeding that occurred within few millimeters of fovea, the extremely important center of visual acuity; the small margin of error is keenly manifested, and thus the intense importance of precision and accuracy. The importance of precision in medicine is applicable not just to the field of surgery where the individual anatomical variation demands high precision and accuracy, it is also applicable to non-surgical areas. In fact, precision is taking up a whole new meaning in the diagnosis of diseases and non-surgical treatments of disease, as we should examine it in the following sections. The term of "precision medicine" was introduced in a report of the National Academies of Sciences, Engineering, and Medicine which called for creation of a "biomedical research and patient data network for more accurate classification of disease, move toward 'precision medicine'" and stated that this emerging "new taxonomy" would define diseases by their underlying molecular causes and other factors in addition to their traditional physical signs and symptoms and that this new data network would also improve biomedical research by providing scientists' access to clinical information (NATIONAL ACADEMIES 2011).

Precision Medicine: Rationale

Over the last few decades, medical professionals have slowly come to acknowledge that the current ways of practicing medicine are not quite right. One of the major challenges the medical profession currently face is this general approach to patient care in a kind of "one-size-fits-all" fashion (Cullis 2015). Certainly, when a new medication is developed, it usually becomes effective for most patients, but some patients simply did not receive positive outcomes when they received the same medication at the identical dose. In reality, the percent patients responding to a particular medication in post-marketing general population is lower than that in the pre-marketing clinical trial population (Cullis 2015). It is well documented that the difference in effectiveness and risk of a given medication between clinical trial population and the general population could be as big as 60% (Martin et al. 2004). How then does the medical profession proceed with this subset of patients who do not respond? One of the major factors contributing to this drug responsiveness variation is racial/ethnicity (Burroughs et al. 2002). This racial/ethnic factor is in turn depends heavily on genetic variation (Burroughs et al. 2002). One other major drawback of this "one-size-fits-all" approach is keenly revealed in the area of adverse drug reactions (Cullis 2015, Budnitz et al. 2007, Sultana et al. 2013). As a practicing dermatology physician, I have personally witnessed this huge problem. Many patients who were admitted to the hospitals for surgeries and developed severe or even life-threatening adverse reactions to drug they received as a measure to prevent post-surgical bacterial infection or seizure. Many more patients were admitted to the hospitals for adverse drug reactions after they developed allergic reactions to medications they received as outpatients. In fact, it is estimated that adverse drug reactions cost the US $30 billion dollars each year in economic loss (Sultana et al. 2013). Furthermore, it produces a heavy burden to the manpower of the medical professionals as it was estimated that between 11% and 35% of all visits to hospital emergency departments are due to adverse drug reactions (Budnitz et al. 2007). It is true that adverse drug reactions were monitored and reported in the pre-marketing period of time, i.e., the reports of findings from the clinical trials. However, pre-marketing evaluation does not provide a complete picture of the likelihood of adverse drug reaction due to several inherent limitations of clinical trials: First, it is limited by its scope, since clinical trials could go so far in terms of number of patients they test. Secondly, it is limited by the lack of heterogeneity in the clinical trial population, as certain subgroups of patients are commonly excluded from a given trial. One should also note that the inclusion criteria of clinical trials are not always published. Thirdly, it is limited by its short duration (Cullis 2015, Martin et al. 2004, Sultana et al. 2013). Thus, to fulfill the IOM's triple aims of "better care, better health, and lower cost", a superior way to predict and prevent adverse drug reaction, to the individual patient level, must be included as the integral part of discussion for the future of medicine. Prediction, in turn, would be the key to success, as prevention is dependent on the ability for accurate prediction. "Precision Medicine", in essence, was born to provide the precision of prediction for accurate diagnosis and treatment to a unique individual at the molecular level, as one author defined "Precision Medicine" this way, "the concept of precision medicine, an approach for disease prevention and treatment that is personalized to an individual's specific pattern of genetic variability, environment and lifestyle factors, has emerged" (Reitz 2016). For that reason, precision medicine is also referred to as "molecular medicine", "genomic medicine", or "personalized medicine" (Cullis 2015). Nevertheless, the term "precision medicine" will be used throughout this textbook, for three major reasons: first, "precision medicine" is a more appropriate engineering term for an engineering-medicine curriculum than the term "personalized medicine", "genomic medicine", or "molecular medicine". Second, some physicians objected to the use of the term "personalized medicine" that may imply that they are currently providing "impersonal care" to the patients and are therefore required to transform to "personalized medicine". Thirdly, the more recently utilized precision medicine is going beyond genomic data and utilizes proteomic and functional data to diagnostic and therapeutic methodology. Therefore, the term "genomic medicine" does not cover all areas of "precision medicine" as it excludes the contribution of other "omics" such as proteomics in the healthcare of patients (Jain 2016). Moreover, "precision medicine" is a better term in conveying the idea of precision or accuracy in medical practice in an intuitively clear manner, whereas "personalized medicine", "genomic medicine", or "molecular medicine" does not provide a connotation of precision.

Precision Medicine: Key Elements

The National Institute of Health has defined precision medicine as "an emerging approach for disease treatment and prevention that takes into account individual variability in genes, environment, and lifestyle for each person" (USNLM 2018). How then could precision medicine accomplish its goals of provide the best medicine with the abilities to predict clinical efficacy and potential adverse reaction? It is due to the timely convergence of two recently developed areas of major advancements, namely genomic and other molecular biological technology, and computer science and other digital technology. And within the first major component, the key elements are consisted of genome, proteome, metabolome, and microbiome (Cullis 2015). Artificial intelligence, particularly machine learning, is the most important element of the second major component. In the following sections these 5 key elements will be discussed in greater details.

Genome

Genome is the entire genes make up of an individual person. The genome is a kind like the blue print of a human body as it directs the production of proteins that control every aspect of human structure and function in every single and all biological systems, organs, tissues, all the way down to a single cell level. The human genome is very large and contains about 20,000 different genes and it has the same contents in every single cell. On the first encounter, the thought of sequencing every patient's genome seems to be huge undertaking. The good news is that the cost for a complete human genome sequencing is coming down extremely fast, from $2.5 billion (the first ever genome sequenced) to $1 million in 2008, to about $1,000 in 2015 due to the ability of sequencing by electronic method rather than the traditional chemical method. The advance of technology also speeds up the sequencing process from first genome sequence that completed in 15 years to that completed in just a few days in 2015 (Cullis 2015, NHGRI 2016). A commercial company now prices only $399 to sequence the entire coding DNA sequence (whole exome) portion of human genome (GENOS 2016). The cost is going to go down even more in the not-too-distant future. Once we have our genome sequenced, we can expect the potential usefulness in several key health issues:

- *Ability to predict drug responsiveness in an individual patient*: For a starter, pharmacogenetic information now can predict with some degrees of certainty that patients with certain genomic biomarkers would or would not respond to certain medications (Cullis 2015, FDA 2016).

- *Ability to predict if a given drug would likely to induce adverse drug reaction in an individual patient. (pharmacogenomic application)*: This is extremely important from both an ethical standpoint and from a medical economy perspective (Cullis 2015). As mentioned above, every single year, the US collectively wastes $30 billion dollars to take care patients who have developed an adverse drug reaction and most of these incidences can be prevented if pharmacogenetic approach can provide a clue for the medical professions (Sultana et al. 2013).

- *Accurate Diagnosis of Disease*: The presence or absence of certain mutated genes would help making accurate diagnoses in some diseases.

- *Ability to predict disease risks in an individual patient*: As genomic information can help making disease diagnosis, the presence or absence of certain mutated genes could help access the disease risk in certain individuals.

- *Ability to treat diseases more effectively*: This ability was indeed the impetus for the development of precision medicine in the first place.

Proteome

Proteome is the entire protein make up of an individual person. Whereas the genome is the blue print of the human body, proteome is the workhorse of the human body. While the human genome is essential in forming the basis of human body structure and function and the genomic information has very useful effects in healthcare as described above, it is not the whole story. With the exception of genetic diseases, the proteins, or the translational results of the genes, are actually the keys to the formation of normal structures and the performance of normal functions that sustain and maintain all aspects of human beings, and by extension, the keys to determine the malfunctions of human body, i.e., the disease process. The major reasons are that not all genes are translated into proteins, not all proteins are translated into the same amount of proteins, not all translated proteins are fully functional, and not all functional proteins are stable in a given individual. Therefore, we will also need to understand the conditions of entire protein make-up of the individual person. Unlike the analysis for human genome, the study of the human proteome has not advanced to the same degree. The challenges are three folds:

- Protein data is much more complex than the simple DNA data, which consisted of only 4 different bases: A, T. C, G. Protein, on the other hand, is consisted of 20 different amino acid basic building blocks (Hillis et al. 2014).
- Protein is not as stable as DNA and can be broken down by proteases easily.
- Protein is not as easy to determine as for DNA as it requires antibody detection until the recent new development of mass spectrometry.

Many commercially available services currently offer proteomic analyses using mass spectrometry in conjunction with liquid chromatography (DCBIOSCIENCES 2018).

Metabolome

Metabolome is the entire metabolite make up of an individual person. Most of the metabolites are measurable in the human blood. Metabolites are mostly proteins in nature and therefore face the same analytic challenge as proteins. Essentially, our metabolites reflect what we consumed, what broke down in our bodies, and what our current health states are, and is like an instantaneous snapshot of the entire physiology of a human being. In combination with genomic, proteomic, and microbiomic studies, metabolomics can provide a comprehensive insight into our entire biological processes. Mass spectrometry has been a key technology for the new generation of metabolomics studies and commercially available services often utilize mass spectrometry in combination with liquid chromatography (LC-MS), gas chromatography (GC-MS), and ion chromatography (IC-MS) (THERMO 2018). While LC-MS has the wildest coverage for metabolite analysis, GC-MS is suitable for analyzing rare volatile metabolites. IC-MS is an alternative method to LC-MS for analyzing highly polar compounds (THERMO 2018).

Microbiome

Microbiome is entire make up of microorganism that live inside and on the surface of an individual person. Having been neglected for their importance in human wellbeing, the microorganisms that live inside our body or on our surface have recently been recognized for their significant contributions. Recognized as "the Father of the Microbiome", Professor Jeffery I. Gordon of Washington University in St. Louis has been a pioneer in investigating the relationship between microbes and their human hosts in a variety of health and disease conditions. Dr. Gordon "often speaks of the gut's nonhuman residents—the microbiota—as a microbial organ. Like the body's other organs, Gordon has shown, the microbiota performs specific and vital functions" (Strait 2017). For example, studies have documented that our body's microbiome make up influence not only our health, but also our response to therapeutic treatments during our disease state (Gopalakrishnan et al. 2018). Gene sequencing technology has allowed us to examine the presence, relative abundance, and function of our gut microorganisms with reasonable costs, by simply proving a stool sample (GENOTEK 2018).

Artificial Intelligence

Recent medical literatures have pointed out an important role of artificial intelligence (AI) in fulfilling the mission of precision medicine (Yu et al. 2012, Sela-Culang et al. 2015, Bahl et al. 2018). Attending a recent

Precision Medicine World Conference Silicon Valley held in the Computer History Museum in Mountain View, California, USA (January 22–24, 2018) has allowed this author to appreciate the rapid moving field of AI and its very important roles in assisting precision medicine on fulfilling its mission of providing the right medicine for the right patient (PMWC 2018). Although AI and AI in medicine has been initiated in early 1950s and 1980s, respectively, the recent development of AI has been very rapid and could have much significant impact on automation and transformation of labor forces (Szolovits 1982). Realizing this huge transforming force, Dr. Reif, the President of Massachusetts Institute of Technology which nurtured the very development of AI, has written a newspaper article, calling for society debate and preparation for this coming technology wave (Reif 2017). Parallel to the general development of AI which saw $20 billion investment in 2016 globally (Bughin et al. 2017), the number of medicine-related AI publications have skyrocketed in recent years. On a global perspective, a 2017 report published by McKinsey Global Institute (Bughin et al. 2017), the number one private-sector think tank ranked by the University of Pennsylvania's Lauder Institute, stated that AI would benefit healthcare in the following aspects:

- AI would optimize healthcare task, make diagnosis quicker and provide better treatment options.
- AI could forecast disease epidemic and offer preventative medical solution.
- AI will enhance diagnostic accuracy and improve healthcare productivity.
- AI would provide data support for personalized medicine.

A chapter independently and exclusively devoted to AI is included in this book and students will be able to learn more about AI in that chapter (Chapter 14). To illustrate the role of AI in precision medicine, a real-life clinical example is described below.

A recent scientific report regarding AI in medicine is the enhancement of accuracy for early breast cancer detection. The key advancement achieved was the decline of unnecessary surgeries for a group of patients having received a "high-risk lesion" pathology mark by needle biopsy after mammogram revealed a "suspicious" lesion. Usually, for needle biopsy performed on "suspicious" lesions spotted by mammography, 20% results turn out to be malignant, 70% results are benign, and the remaining 10% are determined to be "high-risk". The majority of these "high-risk" lesions ultimately were diagnosed to be benign when surgery is actually performed and the end results therefore are many unnecessary operations. In seeking for a better way to improve the management for this 10% patients with "high-risk" lesions, MIT engineers and Harvard physicians utilized an AI method called "random-forest classifier" to "train" the computer with 600 existing "high-risk lesions" obtained from needle biopsy so as to identify patterns, in combination with other different data elements including demographics, family history, past biopsies, and pathology reports. The "trained" computer were then utilized to examine an additional set of 335 "unknown" "high-risk" lesions and were able to reach correct diagnosis of 97% breast cancers as malignant in this set of "unknowns". As a result, there could be a decrease of more than 30% unnecessary surgeries (Bahl et al. 2018, Conner-Simons 2017) and Harvard physicians would soon employ this AI model to assist their actual patient care.

Precision Medicine: Applications

Having delineated the key elements of precision medicine and the basic methodology of precision medicine that can be beneficial to medicine, let us now look at several recent real-life clinical examples of how precision medicine can significantly contribute to our healthcare (Friedman et al. 2015, Flores-Morales and Iglesias-Gato 2017, FOUNDATION 2017, Crowe et al. 2001, Crowe 2017, Roper et al. 2015).

Next Generation of Sequencing

The National Cancer Institute delivered the promise of precision medicine in this statement, "the hope of precision medicine is that treatment will one day be tailored to the genetic changes in each person's cancer. Scientists see a future when genetic tests will help decide which treatments a patient's tumor is most likely to respond to, sparing the patient from receiving treatments that are not likely to help" (NCI 2018a). The next generation of sequencing (NGS) technique is now available for faster and larger scale of gene sequencing and analysis. On November 30, 2017, the FDA has approved a new NGS-based *in vitro* diagnostic test commercially available by Foundation Medicine. This new test, termed "FoundationOne CDx" (F1CDx) can test 324 different genes and identify actionable mutations across 5 solid tumor types (non-small cell lung cancer, melanoma, breast cancer, colorectal cancer, and ovarian cancer) and detect gene alterations in EGFR, KRAS, BRAF, BRCA1/2, ALK, and several others (FDA 2017, FOUNDATION 2017). With one run of this test, healthcare providers can now have the data to make better informed medical decisions to tailor cancer treatment in improving care outcomes and potentially reduce costs. Furthermore, the tissues submitted for regular histology (formalin-fixed paraffin-embedded specimen) are sufficient for this test, without the need for extra effort of additional tissue collection (FOUNDATION 2017). Moreover, Center for Medicare and Medicaid Services simultaneously approved this usage, thus facilitating the medical use substantially for patients with Medicare insurance coverage (FDA 2017).

Quantitative Mass Spectrometry-based Proteomic Profiling

Being one of the most commonly diagnosed male neoplasms in the developed nations and the second most common cancer in men of the US, prostate cancer, is known to occur in 170,000 new patients annually in the US, according to the latest available statistics of 2014 (CDC 2017). However, currently there is significantly inefficiency in the care of these patients. The key point of this inefficiency is that these patients as a group were massively over-treated. Due to the slow-growing nature of prostate tumors, many of these patients may not require aggressive treatment and suffer the significant treatment side-effects but were treated nevertheless due to our inability to identify this subgroup of patients (Schroder et al. 2009). Thus, we need to delineate biomarkers for patient stratification based on prognostic risk and for more precise treatment options (Flores-Morales and Iglesias-Gato 2017). Here is where precision medicine can be very beneficial. While the recent large scale of genomic studies on prostate cancer have provided advanced understanding of the molecular mechanism driving the tumor development, our understanding of the progression of prostate cancer remains restricted (Grasso et al. 2012, CGARN 2015, Robinson et al. 2015, Fraser et al. 2017, Flores-Morales and Iglesias-Gato 2017). The recent development of the method of using mass spectrometry enables the detection and quantification of thousands of proteins and post-translational modifications from small amounts of biological materials such as fresh frozen biopsy, formalin-fixed paraffin-embedded tissues, blood, and urine (Flores-Morales and Iglesias-Gato 2017). This mass spectrometry-based proteomic profiling thus forms the basis for the prostate cancer precision medicine. One of the new tools that facilitates this proteomic profiling is ITRAQ, abbreviations for isobaric tag for relative and absolute quantitation. The ITRAQ method is based on the derivatization of primary amino acid groups in intact proteins or peptides and utilization of isobaric tag for quantitation. Since the isobaric mass design reagents do not alter the mass of the tagged proteins or peptides, the samples (tagged or untagged) would appear as single peaks in mass spectrometry analysis, thus allowing accurate protein quantitation (Wiese et al. 2007). Figure 1 schematically depicts the common approaches for this proteomic

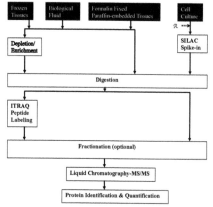

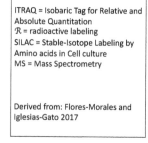

ITRAQ = Isobaric Tag for Relative and Absolute Quantitation
$\mathcal{R}$ = radioactive labeling
SILAC = Stable-Isotope Labeling by Amino acids in Cell culture
MS = Mass Spectrometry

Derived from: Flores-Morales and Iglesias-Gato 2017

Fig. 1. Mass spectrometry-based proteomic profiling.

profiling methodology. This technology will significantly facilitate the determination of protein profiles in given diseases much speedier than the traditional antibody-based testing method, and would provide healthcare providers with much more information they needed to make medical decisions. Advance mass spectrometry, in addition to proteomic profiling, is now able to perform intraoperative cancer margin identification in nearly real-time and this book has devoted an independent chapter on that development.

Immunotherapy

Immunotherapy is another treatment avenue that precision medicine can make a big impact. Immune treatment, by its very nature, is by definition a very precise one, since immune cells and antibodies have to recognize their "targeted enemies" through a very specific (and precise) immune binding process. Specifically, a strong immune reaction depends on a perfect conformational match between the antibody and the target antigen (for antibody reaction) and a perfect match between T cells and the MHC molecule on tumor cells (for cellular reaction) (Hillis et al. 2014). The followings depict some state-of-the-art immunotherapy methods currently being used or further investigated for potential cancer therapeutics.

T Cell-based Immunotherapy

Since T-cell response depends on recognition of antigen presented to them and co-stimulation by antigen presenting cells, researchers have studied ways to promote this T-cell and antigen presenting cells interaction by artificially engineered antigen presenting cells, with either human-based or nanoscale cells (Butler and Hirano 2014, Perica et al. 2014). In one study, investigators tested two nanoscale particle platforms, biocompatible iron dextran paramagnetic particles (50–100 nm diameter) and avidin-coated quantum dot nanocrystals (30 nm) and documented that these nanoparticles induced T cells proliferation *in vitro* and enhanced tumor rejection in mouse model of skin melanoma (Perica et al. 2014). Using human-based cells, investigators have looked into the effectiveness of engineered artificial antigen presenting cells for antigen-specific CD8+ T cells and CD4+ T cells (Butler and Hirano 2014).

At the 2018 Precision Medicine World Conference Silicon Valley held in California, USA, an overview for Immuno-oncology as a new frontier of precision medicine was presented by Ira Mellman, Vice President of Cancer Immunology at Genentech, South San Francisco, California and the following is the excerpt of the presentation (PMWC 2018):

The force that propels the development of immune-oncology is that treatment through cancer genomic has led the way for the last two decades but it is now clear that genome-based treatment is not as effective as we have hope for. Melanoma is an excellent demonstrating case. Although oncogene helped initial management, some tumors returned after treatment. There are several challenges for future therapies going forward. Regarding the treatment-resistant tumor, combinatory drug therapies can be considered but toxicity increases with additional drug. Furthermore, low frequency tumor genotype (as in rare cancers) makes drug development difficult to justify economically from pharmaceutical companies' perspective. Therefore, cancer Immunotherapy becomes the next frontier. Typically, immunotherapies, if T cell-mediated, require patients to go through two phases of treatment, first the activation phase that triggers the T cell recognition of the tumor antigen, then the affective phase, which the tumor killing takes place by the tumor-primed T cells. However, either phase of the treatment cycle can be blocked by naturally occurring phenomena. PD1/PDL1 is a pair of cell surface receptor that can present as check point road block, inhibiting immune cells from recognize the tumor target. By inhibiting the interaction of this pair "check point block" would allow T cells to work better in recognizing the tumor target (Mariathasan et al. 2018). In addition, tumor vaccine could be utilized to target the specific genetic mutation of tumors.

At the same PMWC 2018 meeting, Andrew Allen, co-founder of Gritstone Oncology in Emeryville, California, gave his brief discussion (excerpt by this author) on "Delivering Personalized Neoantigen Immunotherapy" (PMWC 2018):

Although small molecule target led the way for precision medicine some 20 years ago, immuno-therapy offers the possibility of cure. Relied on the action of T cells and their recognition on cancer surface markers, dosage control of immune-therapy is important. As some of those markers are also present in normal cells, high doses of drug can target normal cells as well, thus can cause patient fatality.

Since blocking PDL1 can induce T cell response to tumor neo-antigens, identifying tumor neo-antigens accurately could now be the key to immune-therapy (Balachandran et al. 2017). Big Data could facilitate this identification process. In addition, the novel use of mass spectrometry could help generating fresh tumor tissue-derived data on tumor peptide sequences. Assisted by artificial intelligence, particularly Deep Learning program, target neo-antigen peptide could be better predicted. Once tumor neo-antigen peptides are generated, effective vaccine could be generated to treat cancer in conjunction with antibody against check point blocker (like PDL1).

Dr. Ugur Sahin, CEO of BioNTech in Mainz, Germany, also discussed the progress of "Cancer Vaccines to Boost the Immune System" at this meeting (PMWC 2018) and his discussion is excerpted below.

By asking the question why cancer patients respond to immune-therapy, we know that MHC class II molecules on a given patient's cancer cells are the key to provide strong T cell-target antigen recognition. What we have improved now is the speed this precision medicine process for making cancer vaccine, based on mRNA vaccine technology (Sahin et al. 2017). From sequence patients' tumor mRNA on MHC class I and class II molecules to the vaccine delivery, the entire tumor targeting vaccine production time has been cut down to about 6 weeks.

CAR T-Cell Therapy

Dr. Crystal L. Mackall, Professor of Pediatrics at Stanford University, briefly discussed in this PMWC 2018 meeting "Cell-based Immunotherapies" and the excerpt is described below (PMWC 2018):

In addition to the fact that Immunotherapy is only effective treatment for some childhood cancers, immunotherapy seems to be overall more effective than some other forms of cancer therapies (MacKall et al. 2014). One of the outstanding examples of immune-oncology in pediatric patient population is the clinical evidence that autologous CD19 CAR T cells could completely eliminate specific childhood B cells cancer in some patients (Kochenderfer et al. 2010, Fry et al. 2018, NCI 2018b, NIH 2018). CAR stands for chimeric antigen receptor and more discussions are delineated in the session below. Nevertheless, there are challenges. Target issue remains the key challenge especially for solid tumors and target antigen needs to get more specific so as to provide better clinical results. In addition, there are microenvironment issues, such as checkpoint inhibition, microbiome, etc. Length of therapy varies from one disease to another. Moreover, engineering in immunotherapies such as PD1-knockout induces T cell malignancy as an unexpected adverse effect and thus needs to be cautious in manipulation of immune system.

CAR-T cell therapy, also a T cell-based immunotherapy, merits special discussion here. This fast emerging approach on immunotherapy is an "adoptive cell transfer" or ACT, consisting three steps:

- collecting patient's white blood cells and isolating their T cells
- modifying and/or activating them in the laboratory
- reintroducing these cells back to patient's blood.

In fact, this modified/activated T cells is like a "living drug" as they are indeed alive and serving as a drug. The most advanced ACT method is the "CAR T cell therapy", where CAR is the abbreviation for "chimeric antigen receptor" (NIH 2018). The first successful report of using this technique, through which patient's T cells are engineered to form a novel T cell receptor recognizing a tumor antigen, was by a NIH research group led by Dr. Rosenberg, resulting in complete eliminating a form of childhood leukemia (Kochenderfer et al. 2010). Due to such remarkable clinical results, two CAR T cell therapies were approved by FDA in 2017, one for childhood acute lymphoblastic leukemia and one for adult advanced lymphoma (NCI 2018b). To monitor long-term success and adverse effects, particularly due to the use of a non-pathogenic viral vector, lentivirus or retrovirus vector, in the engineering of the CAR, NIH has initiated a long-term follow-up study, starting on November 2, 2015, with estimated complete date on September 5, 2036 and 620 participating patients (NIH 2018). A Schematic representation of 3rd generation of engineered CAR (chimeric antigen receptor) of T cell is depicted in Fig. 2. CAR commonly is consisted of an ectodomain of single chain variable fragment of an antibody, a hinge and spacer segment, a transmembrane domain, and an internal signal domain (such as CD3). Costimulatory domains (such as CD28 and 4-1BB) have been added to the later generations of CARs to enhance the transgenic T cell

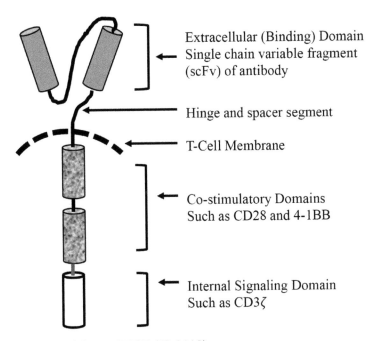

Extracellular (Binding) Domain
Single chain variable fragment
(scFv) of antibody

Hinge and spacer segment

T-Cell Membrane

Co-stimulatory Domains
Such as CD28 and 4-1BB

Internal Signaling Domain
Such as CD3ζ

Derived from: (BIOLAB 2018)

Fig. 2. Engineered CAR (Chimeric Antigen Receptor) on T Cell.

expansion and longevity. The construction of the binding ectodomain with single chain variable fragment (scFv) of antibody has the advantage of bypassing the MHC restriction of T cells, thus allowing the direct activation of effector cells for killing cancer cells.

One-stop CAR T-cell therapy development service is now commercially available for each of the CAR T-cell construction and validation step (BIOLAB 2018):

- *Identification and Selection of Cancer Biomarker*: Using multiple gene and protein analyses to discover biomarkers and select most suitable CAR T-cell target.
- *Generation of High Affinity scFv*: With thousands of antibody products, antibody sequence data, human antibody library, and hybridomas, generate and measure scFv affinity.
- *CAR Design and Construction*: Design CAR of various generations and build lentiviral or retroviral CAR construct.
- *CAR Gene Delivery*: Quality-controlled viral transfection tailored to specific cell types.
- *CAR-T in vitro Assay*: Validate CAR expression and T cell proliferation, screen multiplex cytokines, test cytotoxicity against target cancer cells in culture systems.
- *CAR-T in vivo Assay (Pre-clinical)*: Examination of efficacy of CAR-T cells on cytotoxic killing of target cancer cells that were present as xenografts on laboratory animals such as immunodeficient mice.

Disease Prevention

Because once a given disease is occurred, the expenses and man power needed for the acute treatment, maintenance, and recovery, are substantial, and it is always better to prevent the disease to occur the first place. Thus it will be ideal if precision medicine could help us in achieving disease prevention. Now if we could collect lots of data regarding the initial symptoms and detectable biomarkers before the onset of a given disease and such data could then become a "warning sign" for healthcare providers to take action, either to find ways to stop the disease on its track or to treat the disease at its earliest stage. Some academic centers are gradually turning the attention to "Precision Prevention", by utilizing the tool of precision medicine for preventative works (UCSF 2018). Dr. Aric Prather, a faculty member of the Psychiatry Department of UCSF, has studied the effect of insufficient sleep on human immune system and has demonstrated that individuals who have less than 5 hours of sleep were several times more likely to get sick than those who have 7 or more hours of sleep, when they are voluntarily exposed to a common cold rhinovirus through nasal administration (Prather et al. 2015). "Specifically, those sleeping < 5 h (odds ratio [OR] = 4.50, 95% confidence interval [CI], 1.08–18.69) or sleeping between 5 to 6 h (OR = 4.24, 95% CI, 1.08–16.71) were at greater risk of developing the cold compared to those sleeping > 7 h per night; those sleeping 6.01 to 7 h were at no greater risk (OR = 1.66; 95% CI 0.40–6.95). This association was independent of prechallenge antibody levels, demographics, season of the year, body mass index, psychological variables, and health practices" (Prather et al. 2015). This data supports the clinical survey result, that insufficient sleep is linked to higher incidence of respiratory infection (Prather and Leung 2016).

In another study, Dr. Prather found that individuals who have shorter sleep duration have lower secondary antibody response than those who have longer sleep duration, suggesting sufficient sleep is essential in developing immune protection against hepatitis B immunization (UCSF 2018, Prather et al. 2012).

The results from these studies by Dr. Prather are very consistent with other studies that sleep deprivation is linked to other lower immune functions such as reduction of T cell proliferation, shifting of T cell cytokine responses, decrease of natural killer cell cytotoxicity, and increase of pro-inflammatory pathway activation (Irwin et al. 1996, 2006, Sakami et al. 2002, Bollinger et al. 2009, Prather et al. 2015).

Cancer prevention through precision medicine is obviously worthy the effort, since most cancers became incurable at the time of advanced stage. According to an expert in the field, cancer prevention can be implemented at several levels (Rebbeck 2014):

- *Primary level*: cancer risk assessment, carcinogen avoidance, chemoprevention, risk-reduction surgeries, vaccination.
- *Secondary level*: cancer screening, early cancer detection, precancerous lesions treatment.
- *Tertiary level*: minimize undesirable outcomes.

Different from precision medicine that has typically focused on molecular and cellular data to target individual tumors, the implementation of precision prevention needs a broader data frameworks including behaviors, epidemiology, and socioeconomy, in addition to biological data (Pebbeck 2014). On the biology perspective, precision medicine would provide molecular information for cancer risk and cancer biomarkers so healthcare providers could assess those risks in patients they care for, provide counseling for their patients identified with cancer risk, and perform risk-reducing surgeries. On the epidemiology perspective, precision medicine could help indicating environmental carcinogens information so government policy could be enacted to remove or minimize the exposure. From the behavioral perspective, precision medicine, like the above publications mentioned above on sleep deprivation and health, would provide solid data regarding behaviors that endanger our health to the public square, in the hope that people would modify their behaviors accordingly. In addition, government policy can be enacted to ensure all occupations would guarantee a minimum sleep duration. On the socioeconomic perspective, precision medicine could help pinpointing the health issues affected by socioeconomy in certain needy communities for potential government policy intervention.

Next-generation of Functional Cancer Diagnostics

In the field of cancer therapy, there is a general tendency to equate precision medicine with genomic data. On one hand, the NGS has made substantial progress in making large scale of tumor genetic data and tens of thousands of mutations available for the healthcare providers to understand the origin of cancer. On the other hand, the recent clinical outcomes of cancer therapeutics, however, have not yet matched the outstanding (> 90%) success of chronic BCR-ABL-mutated chronic myeloid leukemia treated by imatinib, which led the initial launch of the genomic cancer treatment era (Friedman et al. 2015). Furthermore, we still have a limited functional understanding on thousands of tumor genetic mutations. Here precision medicine will

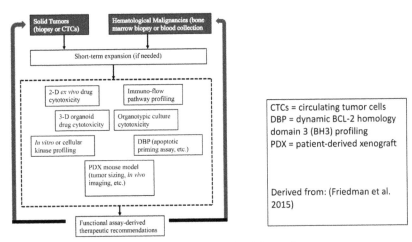

Fig. 3. Schematic illustration of the next generation of cancer functional diagnostics.

help fine-tuning our treatment strategy based on the next and higher level of data, the functional data. In the final analysis, it is the function that really counts in the living biological system. Thus, a dynamic test which assess the vulnerability of the tumor in addition to the appropriate target on the tumor, will serve the patients better. The first generation of functional cancer diagnostic mostly tested with chemotherapeutic agents on either homogeneous cell culture grown out of tumors or biopsy specimens of tumor, but very few of those tests were conducted in prospective, controlled, and randomized manner that are robust enough to address the central question of if the tests improve the clinical outcomes. Consequently, none of these first generation of assays became standard of care. The recently developed next-generation of functional diagnostic methodologies such as novel method of tumor manipulation, molecular level of precision assay of tumor responses, and device-based in-situ detection, aim to address the limitation of older generation of chemical sensitivity-based tests (Friedman et al. 2015). Figure 3 is a schematic illustration of the next generation of functional diagnostics for cancers. Essentially, small quantity of cancer cells will be collected from the patients and the cells will then be expanded in the laboratory to achieve a sufficient quantity. The cancer cells will be subjected to *in vitro, ex vivo,* or *in vivo* functional tests against the potential medications. Depending on the responses of cancer cells to the testing medications, healthcare providers will have better idea for deciding the treatment options. In order for functional diagnostic assays to be practical, the entire process needs to be completed in a short time frame, so as not to delay the initiation of cancer treatment. Therefore, to rapidly obtain sufficient amounts of cancer cells from solid tumors for functional assays, researchers sometimes have used a method called "conditional reprograming" which utilized an irradiated fibroblast feeder cell layer, growth factor-enriched culture media, and an inhibitor for RHO-associated kinase. There are different types of functional assays that can be employed. For the *in vitro* cellular kinase profiling, kinase substrate activities are usually measured. In the immune-flow pathway profiling, phosphor-antibody-based flow cytometry (fluorescence-activated cell sorting, FACS) method is commonly used. The measurements for 2-D and 3-D drug cytotoxicity and organotypic culture cytotoxicity assays are usually ATP levels, MTT (metabolic tetrazolium dye), cellular proliferation, and cell number counting. To examine the response by DBP (dynamic BCL2 homology domain 3 profiling), apoptotic priming assay is typically used. For the *in vivo* assay with PDX (patient-derived xenograft) mouse model, tumor size measurement and *in vivo* imaging are commonly employed (Friedman et al. 2015).

Next-generation of Structure-based Vaccine Design

Immunologists are very familiar with and extremely proud of the significant contribution of vaccination to human disease prevention and are keenly aware that antibody development is the principle immune mechanism that protect us from viral infection (Crowe et al. 2001, Crowe 2017). While the anti-viral protection depends on the development of antibody, the effectiveness of the anti-viral antibody in turn relies on its effectiveness of binding to virus and of neutralizing virus. After the method of generating hybridoma cell lines in conjunction with murine antibody revolutionized antibody production in the 1970s, immunologists slowly began to recognize the limitation of murine antibody system as animal germline gene segments lack certain specific sequences needed to develop particular antigen binding capacity. Consequently, there is a need for more precise understanding how human immunity against virus work on a molecular level and here precision medicine could help (Crowe et al. 2001, Almagro et al. 2014, Yu et al. 2012, Sela-Culang et al. 2015, Crowe 2017). Indeed, it should be abundantly clear that for the purpose of human vaccination, understanding of human antibodies is superior and more precise than understanding of other animals' antibodies. Thus in the last decade, immunologists have started to seek better antibody generation by looking for answers in human monoclonal antibodies (mAbs). Methods have been developed to isolate and to study the structures and functions of human mAbs, and to correlate the mAbs' *in vitro* neutralization abilities with their corresponding *in vivo* (real-life) immune protection. The recent studies have concluded that humans could make a very big variety and effective antibodies against almost any pathogens. Thorough sequencing of isolated human mAbs enables investigators to identify fine structure of neutralizing antibodies. Detailed studies utilizing electron microscopic and crystallographic methods have provided investigators the abilities in delineating fine structure foundation for viral molecular recognition and determined modes of antibody-virus interaction. In addition, utilizing artificial intelligence, particularly machine learning, researchers have found ways to improve their abilities to design the complementarity determining region (CDR) sequences in the antibody-antigen recognition interface that is the very essence of antibody binding (Yu et al. 2012, Sela-Culang et al. 2015).

Having delineated the molecular level of what contribute to the protective antibody formation, the "rules of engagement" if you will, the investigators can now derive a "reverse vaccination" design, based on the structural data and the rationale that antigenic sites and epitopes revealed by naturally occurring human antibodies should be the best choices for vaccine (Crowe 2017). In fact, few proof-of-concept studies have recently been initiated (Crowe 2017, Correia et al. 2014, Briney et al. 2016, Escolano et al. 2016, Steichen et al. 2016, Tian et al. 2016). Figure 4 depicts a general schematic diagram of how a structure-based vaccine design can be achieved. This scheme provides a general framework, rather than a very specific protocol, as new breakthrough methods for better and faster reverse vaccine designs are projected to occur in the not-too-distant future, the students will be greatly benefited from focusing at learning the general principles.

Having detailed the potentials of this structure-based next-generation of vaccine design in helping immunologists in defense of pathogens, it is equally important to point out that the benefit of this new antibody building strategy is by no means limited to immunological defense against pathogens. It should certainly be applicable to therapeutic antibody generation for treatments of cancers, autoimmune diseases,

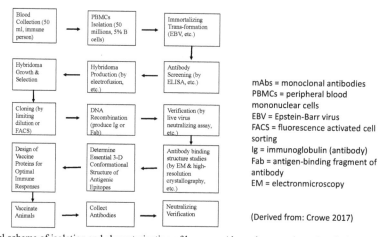

mAbs = monoclonal antibodies
PBMCs = peripheral blood mononuclear cells
EBV = Epstein-Barr virus
FACS = fluorescence activated cell sorting
Ig = immunoglobulin (antibody)
Fab = antigen-binding fragment of antibody
EM = electronmicroscopy

(Derived from: Crowe 2017)

Fig. 4. General scheme of isolation and characterization of human mAbs and reversed vaccine design.

and inflammatory diseases, as antibody medicines (biologics) have been widely used for these diseases in the recent years (Almagro et al. 2014, Carvalho et al. 2016, Ellebrecht and Payne 2017).

One additional important development in the next generation of vaccine design is the research on promoting robust T cell response by improving the efficiency of antigen presenting cells (dendritic cells), since effective T cell priming depends on dendritic cells' efficiency. Recent studies are undergone to examine the selection of adjuvants, the design of multifaceted vehicles, and the choice of surface molecules to target dendritic cells, which is now classified into 6 different subsets (Gornati et al. 2018).

Genome Editing

The epitome of precision medicine is reserved for the technique of genome editing, the tool that can actually change the genome itself. The recent development of utilizing a technique named "Clustered Regularly Interspaced Short Palindromic Repeats (CRISPR)/CRISPR-associated (Cas) Protein 9 System" (CRISPR/Cas9), signifies the progress, implications and challenges for the future gene therapy (Zhang et al. 2014, Hartenian and Doench 2015). The discovery of this unique molecular system dated back in 1987 when a study of the *Escherichia coli* genome revealed gene loci with a characteristics of repeat sequences that contain unknown functions (Ishino et al. 1987) and the subsequent investigation that identified similar repeated sequences in other prokaryotes (Jansen et al. 2002). Further studies of CRISPR systems determined that these prokaryotic loci possess the ability to genetically modify and adapt to booster immune defense and subsequently classified into three types (Barrangou et al. 2007, Makarova et al. 2011). Additional studies subsequently determined the ability of CRISPR/Cas9 to locate and cleave double-stranded genomic DNA. In fact, this bacterial system is like human's immune system and is used by bacteria to defend against viral invasion, in this case by deleting the invading viral genome inserted into the bacterial genome. While the type I and type III CRISPR systems require a complex of multiple proteins in order to locate and cut DNA, type II CRISPR systems only need single protein for such functions and thus became the most studied system thus far. The type II CRISPR systems are commonly derived from *Streptococcus thermophiles* and *Streptococcus pyogenes*, and Cas9 from the later actually became the widest used due to it requires just minimal additional sequence to perform its functions (Hartenian and Doench 2015).

Mechanism

Mechanistically, CRISPR/Cas9 system works through 2 essential molecular steps: spacers acquisition and double-stranded DNA cutting (Zhang et al. 2014, Hartenian and Doench 2015). Figure 5 illustrates this mechanism in a schematically manner.

Spacer Acquisition. This step is critical in lining up Cas9 enzyme with the specific cutting sites. The CRISPR/Cas9 system contains a guide RNA (gRNA, also termed crRNA), which is a stretch of about 20

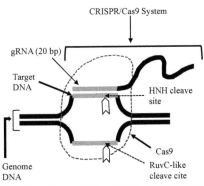

(Derived from: Zhang et al. 2014 and Hartenian and Doench 2015)

Fig. 5. Schematic diagram of CRISPR/Cas9 mechanism.

RNA bases and is complementary to the DNA target. In the natural bacteria defense, this gRNA will be complementary to the invading viral genome. In the medically indicated genome editing, this gRNA will be engineered to be complementary to the targeted DNA, be that a mutated gene in a genetic disease or a critical gene in a cancer cell. Situated at the end of a long RNA scaffold, gRNA locates and binds the target DNA, thus ensuring the correct DNA cutting site for Cas9 enzyme.

Double-stranded DNA Cutting. Guided by gRNA, the double-stranded DNA cleavage by Cas9 enzyme is conducted by two nuclease domains, namely the HNH domain and the RuvC-like domain. Whereas the HNH domain nuclease cleaves the DNA strand directly bound by gRNA, the RuvC-like domain nuclease cuts the opposite side of the double-stranded DNA.

Once the double-stranded DNA is cut, the damage is recognized by the cell and a repair process will be initiated. In a natural bacterial defense situation, the repair will be a simple recombination once the viral genome is removed. On the other hand, for a medically indicated genome editing scenario such as in gene therapy, the scientists will provide an exogenous DNA segment with correct sequence for the repair once the error sequence is removed.

Applications

Thus far, many studies have been conducted to examine the genome-editing capabilities of CRISPR/Cas9 system in the following areas of science and medicine:

- Genetic screening (Hartenian and Doench 2015).
- Cancer therapy (Albayrak et al. 2018, Belvedere et al. 2018, Zhen and Li 2018).
- Functional delineation (Albayrak et al. 2018, Chen et al. 2018a, Chen et al. 2018b, Miao et al. 2018).
- Gene therapy (Cromer et al. 2018, Gupta et al. 2018, Shariati et al. 2018, Vetchinova et al. 2018).
- Genetic reprograming (Melo et al. 2018).
- Gene knockout animal modeling (Martin et al. 2018, Sato et al. 2018).
- Infection therapeutics (Deng et al. 2018, Mahas and Mahfouz 2018).
- Transcriptional regulation (Zhang et al. 2014).
- Plant gene editing (Hirohata et al. 2018, Lee et al. 2018a, Lee et al. 2018b, Naim et al. 2018, Sugano and Nishihama 2018).

Clinical Trials of Precision Medicine

Another way to examine the trend of precision medicine for the future healthcare application is to survey the outlook of clinical trials involved precision medicine (Roper et al. 2015). A study published in 2015, the first of this kind, analyzed all adult interventional cancer trials registered on ClinicalTrials.gov between September 2005 and May 2013 was conducted to fulfill that purpose (Roper et al. 2015). Of the initial 18,797 clinical trials that required a genomic alteration (mutation) as the inclusion criterion, only 8% (684 trials) were subsequently classified as "precision cancer medicine" trials, whereas the remaining and large portion (92%, 8410 trials) were determined to be non-precision cancer medicine trials. The 684 identified precision cancer medicine trials involved a total 38 unique genomic alterations for enrollment and the top 10 genomic mutations of these trials were listed in Table 1. Overall, the proportion of "precision cancer medicine" trials among all cancer treatment trials increased from 3% in 2006 to 11% in 2011 and to 16% in 2013, reflecting a 5-fold increase from 2006 to 2013. Although this substantial increase in percentage is a step forward for achieving the goal of precision medicine, i.e., to provide the right treatment to the right patient, precision cancer medicine trial is still a small percentage (< 20%) of all cancer treatment trials (Roper et al. 2015). Thus, we have a long way to go before we could provide "precision medicine" for all patients suffering from cancers. Furthermore, these "precision cancer medicine" trials lack the true broad

Table 1. The top 10 genomic alterations targeted in precision cancer medicine trials between 2006 and 2015.

Genomic alteration identified	Number of trials (%)
HER2	287 (33)
BRAF	94 (11)
KRAS	77 (9)
EGFR	60 (7)
ABL1	50 (6)
BRCA1/2	42 (5)
FLT3	32 (4)
ALK	31 (4)
PIK3CA	20 (2)
MLL	17 (2)

Derived from Roper, N. et al. 2015. The landscape of precision cancer medicine clinical trials in the United States. Cancer Treatment Review 41: 385–390.

applicability since about 50% of these trials focused on cancers on three major organs: breast, lung, and skin (Roper et al. 2015). Diversification of clinical trials to cover more types of cancers should be considered.

Precision Medicine: Future Directions

At the 2018 Precision Medicine World Conference Silico Valley held in Mountain View California (January 2018), Sir. John Bell, Professor of Medicine at Oxford University and the Keynote Speaker for the Immunotherapy Section, delivered the following short message on the landscape of precision medicine in UK (excerpt, by this author):

- *On Accuracy of Precision Medicine*: Precision medicine will be more precise in the future; Genomic medicine has led the initial momentum; Correlations of genotype and phenotype are now needed to be implemented; Big Data of genotype and phenotype are important for precision medicine accuracy; Large scale clinical data and analytics are essential.

- *On Priority of Precision Medicine*: Make precision medicine applications earlier in the disease process; It is better to prevent disease or treat disease earlier; Early diagnostic will be based on data of risk assessment; Big Data set will facilitate the implementation of precision medicine on risk assessment.

- *On Driving Forces of Precision Medicine*: Big Data and AI are driving the precision medicine in the future.

Since United States is closely related to UK in general culture and in healthcare, the precision medicine landscape in the US will likely to be similar to that of UK. In this chapter, we have discussed these precision medicine landscapes in the US, namely the correlation of genotype and phenotype and the disease prevention. As for the driving forces, two independent chapters in this book deals exclusively with Big Data (Chapter 13) and AI (Chapter 14), and the students will have the opportunity to learn them in greater details.

Precision Medicine: Challenges in Fulfilling Its Missions

Looking into the future, there are several challenges that can be identified (Wagle et al. 2012, Gray et al. 2014, Hall 2014, Roper et al. 2015).

Defining "Actionable Targets"

Along with the fast pace of advanced genomic sequencing, came the phrase of "actionable" genomic target. Experts in the field, however, have recently called for a better definition for the meaning of actionable target (Roper et al. 2015). Such unclear terminology may cause confusion among healthcare providers.

Changing Healthcare Providers' Mindset

The very practice of precision medicine requires a change on the mindset of healthcare providers. From discussing the anatomical pathology results with the patients to the discussion of genomic testing reports with the patients; from the makeup of traditional tumor board (usually consisted of medical oncologist, pathologist, radiologist, and oncological surgeon) to that of a molecular tumor board (instead consisted of medical oncologist, molecular biologist, geneticist, and oncological surgeon); from medical decision relying solely on human judgement to decision assisted by artificial intelligence; there are big departures from the care tradition in terms of healthcare providers' comfort level on genetic knowledge and healthcare providers' feeling of autonomy. Thus, providers will be benefited from appropriate education and training.

Enhancing Healthcare Providers' Confidence

Recent studies have revealed a big variability of how willing healthcare providers are to utilize the advanced technology of multiplex genomic testing and their attitudes towards disclosing uncertain significance of genomic information to their patients (Gray et al. 2014, Roper et al. 2015). Since providers' confidence towards the treatment of a particular kind of cancer will be based on the strong evidence in the clinical setting, the most definitive answer to this challenge may be the statistically significant and positive outcomes of prospective clinical trials that determine the efficacy of treatment towards an identified genomic target. That leads to the follow-up questions of whether it is possible to obtain such "actionable target" determination in all cancer cases and how best to obtain these targets.

Building Precision Medicine "Ecosystem"

Experts in the field also urge the establishing of precision medicine "ecosystem" containing building blocks that could optimally connect healthcare providers, patients, researchers, and clinical laboratories to each other, so as to effectively accelerate the practice of precision medicine by leveraging genomic advancement. Constructing this "ecosystem", according to the field experts, will require fundamental changes in the infrastructures and mechanisms for data collection, storage, and sharing (Aronson and Rehm 2015). Figure 6 exhibits a proposed precision medicine ecosystem for optimizing the practice of precision medicine by Aronson and Rehm (2015).

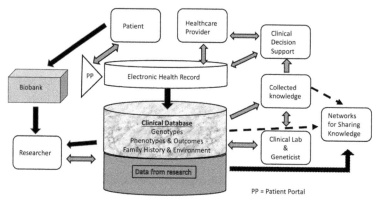

Fig. 6. A proposed precision medicine "ecosytem' with optimal connections of all stakeholders. (Derived from Aroson and Rehm 2015)

Multiplex Cancer Genomic Testing for Prospective Clinical Trials

With the rapid proliferation of multiplex cancer genomic testing and the corresponding exponential increase of potential genomic targets, it is not very difficult to imagine that the speed of testing has vastly outpaced the speed of determination of "actionable targets" by prospective clinical trials, as trials takes substantial financial and human resources to conduct (Roper et al. 2015). One possible way to handle this challenge could be to shift the resources from multiplex genomic testing to prospective clinical trials. This may make good sense, as without the back up of strong evidence from prospective clinical trials, more data from multiplex genomic testing may not ultimately help our patients to achieve the goal of precision medicine after all.

Overcoming Gene Therapy Obstacles

With the CRISPR/Cas9's big potential in correcting genetic diseases and other genome editing benefits came with its greatest ethical consideration of how to avoid potential abuse of this technology. Very recently, Dr. Francis Collins, the Director of NIH along with FDA, have proposed to revise the human gene-therapy oversight by moving the oversight role solely to the FDA, so as to "encourage further advances in this rapidly evolving field" (Collins and Gottlieb 2018). Furthermore, one of the major technical concerns is off-target mutations, where unintended mutations could be created by CRISPR/Cas9 if genome contains loci similar or identical to the intended target loci (Zhang et al. 2014). Moreover, the most studied Cas9 is derived from *Streptococcus pyogenes* and its size is too large to package for certain applications and therefore available small-sized and well-studied Cas9 would be needed (Hartenian and Doench 2015). In addition, the efficiency remains very low for site-specific incorporation of exogenous DNA sequence, after the removal of targeted DNA by Cas9, thus more studies in recombination/repair strategies are necessary (Hartenian and Doench 2015). The improvement of repair process after double-stranded cutting is essential for the future gene therapy, where the exogenously prepared correct genes must replace the removed error genes in the process.

The Cost Equation

Along with the promise of precision medicine for curing certain diseases came the big price tag. One such example is the CAR-T therapy, in which the two FDA-approved treatments cost about $400,000 each, not counting the ancillary costs. Recently, United Healthcare Insurance have asked the Centers for Medicare and Medicaid Services (CMS) to come up with a national coverage analysis, with the purpose of leveling a financial playing field for competing plans and ensuring equal access (Bach 2018).

Knowledgeable AI Operators

With the upcoming role played by AI in precision medicine comes with the challenge of availability of able personnel in handling AI-assisted diagnostic analysis. Experts in the field have pointed out the need to educate and train such able operators, if AI's advantages in assisting the mission of precision medicine would be fully utilized (Bughin et al. 2017). Realizing this problem, some Tech giants are working hard to find way to democratize the utilization of AI, particularly machine learning (ML). Dr. Fernanda Viegas, a graduate of MIT, became a leader of Google's "People + AI Initiative" with the goal to make ML accessible and user friendly. The initiative has created a program termed "deeplearn.js", allowing users to conduct ML operation locally on the browser. The "deeplearn.js" provides pre-trained models or untrained neural networks for the users to train themselves to analyze data or "learn" from data (Igoe 2018). This kind of initiative will be important to solve the AI man power issue and help popularizing the utilization of AI for precision medicine.

Summary

Precision medicine, with its goal of delivering the "right medicine" to the "right patients", is propelled by the rapid advance of computer and biotechnology, chief of them is the development of next generation of genomic sequencing. Although there are substantial contributions of genomic sequencing to more targeted cancer treatments, it is now clear that genomic data alone may not be sufficient to achieve the mission precision medicine seeks and that proteomic and functional assays may be a better pathway to obtain a more precise treatment option, at least in certain cancer cases. Thus, proteomics and functional diagnostics are now pursued by investigators to fine tune the precision in medicine. Recently, immunotherapy has risen to a top approach to some cancer treatment. Precision medicine is also actively engaged in making better and more precise human vaccine design, aiming to deliver superior outcomes for immunization and for therapeutics in autoimmune diseases, inflammatory diseases, and cancers. Furthermore, AI is playing an important role in areas of precision medicine. In the future, precision, as applied to medicine, will take on more roles in disease prevention. Nevertheless, several major challenges need to be overcome if the goals of precision medicine are to be fulfilled.

References

Albayrak, G., E. Konac, A. Ugras Dikmen and C.Y. Bilen. 2018. FOXA1 knock-out via CRISPR/Cas9 altered Casp-9, Bax, CCND1, CDK4, and fibronectin expression in LNCaP cells. Exp Biol Med (Maywood) July 25. Doi: 10.1177/1535370218791797.

Almagro, J.C., G.L. Gilliland, F. Breden, J.K. Scott, D. Sok, M. Pauthner et al. 2014. Antibody engineering and therapeutics, the annual meeting of the antibody society. MAbs 6: 577–618.

Aronson, S.J. and H.L. Rehm. 2015. Building the foundation for genomics in precision medicine. Nature 526: 336–342.

[ASPE] About ASPE. 2016. American Society of Precision Engineering. [aspe.net/about-aspe] accessed July 20, 2016.

[AZ QUOTES] AZ Quotes. 2016. [www.azquotes.com/quote/254231?refs=precision] accessed July 20, 2016.

Bach, P.B. 2018. National coverage analysis of CAR-T therapies – policy, evidence, and payment. N Engl J Med August 15. Doi: 10.1056/NEJMp1807382.

Bahl, M., R. Barzilay, A.B. Yedidia, N.J. Lorcascio, L. Yu and C.D. Lehman. 2018. High-risk breast lesions: A machine learning model to predict pathologic upgrade and reduce unnecessary surgical excision. Radiology 286: 810–818.

Balachandran, V.P., M. Kuksza, J.N. Zhao, V. Makarov, J.A. Moral, R. Remark et al. 2017. Identification of unique neoantigen qualifies in long-term survivors of pancreatic cancer. Nature 551: 512–516.

Barrangou, R., C. Fremaux, H. Deveau, M. Richards, P. Boyaval, S. Moineau et al. 2007. CRISPR provides acquired resistance against viruses in prokaryotes. Science 315: 1709–1712.

Belvedere, R., P. Saggese, E. Pessolano, D. Memoli, V. Bizzarro, F. Rizzo et al. 2018. MiR-196a is able to restore the aggressive phenotype of annexin A1 knock-out in pancreatic cancer cells by CRISPR/Cas9 genome editing. Int J Mol Sci Jly 6; 19(7). Pil. E1967. Doi: 10.3390/ijms19071967.

[BIOLAB] 2018. One-stop CAR-T Therapy Development Services. Creative Biolabs. [https://www.creative-biolabs.com/car-t/one-stop-car-t-therapy-development-services.htm] accessed July 7, 2018.

Bollinger, T., A. Bollinger, L. Skrum, S. Dimitrov, T. Lange and W. Solbach. 2009. Sleep-dependent activity of T cells and regulatory T cells. Clin Exp Immunol 155: 231–238.

[BRAINY] Brainyquote. 2016. Quote of precision. [http://www.brainyquote.com/quotes/keywords/precision.html] accessed July 20, 2016.

Briney, B., D. Sok, J.G. Jardine, D.W. Kulp, P. Skog, S. Menis et al. 2016. Tailored immunogens direct affinity maturation toward HIV neutralizing antibodies. Cell 166: 1459–1470.

Budnitz, D.S., N. Shehab, S.R. Kegler and C.L. Richards. 2007. Medication use leading to emergency department visits for adverse drug events in older adults. Ann Intern Med 147: 755–765.

Bughin, J., E. Hazan, S. Ramaswamy, M. Chui, T. Allas, P. Dahlstrom et al. 2017. Artificial intelligence: the next digital frontier? McKinsey Global Institute. June 2017.

Burroughs, V.J., R.W. Maxey and R.A. Levy. 2002. Racial and ethnic differences in response to medicines: towards individualized pharmaceutical treatment. J National Med Assoc 94: 1–24.

Butler, M.O. and N. Hirano. 2014. Human cell-based artificial antigen-presenting cells for cancer immunotherapy. Immuol Rev 257(1): Doi: 110.1111/o,r/12129.

Carvalho, S., F. Levi-Schaffer, M. Sela and Y. Yarden. 2016. Immunotherapy of cancer: from monoclonal to oligoclonal cocktails of anti-cancer antibodies. IUPHAR Review 18. Br J Pharmacol 173: 1407–1424.

[CDC] Center for Disease Control and Prevention. 2017. Prostate Cancer Statistics. [https://www.cdc.gov/cancer/prostate/statistics/index.htm] accessed December 1, 2017.

[CGARN] Cancer Genome Atlas Research Network. 2015. The molecular taxonomy of primary prostate cancer. Cell 163: 1011–1025.

Chen, L., Y. Ye, H. Dai, H. Zhang, X. Zhang, Q. Wu et al. 2018a. User-friendly genetic conditional knockout strategies by CRISPR/Cas9. Stem Cells Int Jun 14. Doi: 10.1155/2018/9576959.

Chen, X., L. Kozhaya, C. Tastan, L. Placek, M. Dogan, M. Home et al. 2018b. Functional interrogation of primary human T cells via CRISPR genetic editing. J Immunol July 18. Pii: ji1701616. Doi: 10.4049/jimmunol.1701616.

Collins, F.S. and S. Gottlieb. 2018. The next phase of human gene-therapy oversight. N Engl J Med August 15. Doi: 10.1056/NEJMp1810628.

Conner-Simons, A. 2017. Using artificial intelligence to improve early breast cancer detection: Model developed at MIT's computer science and artificial intelligence laboratory could reduce false positives and unnecessary surgeries. MIT News. October 16, 2017. [https://news.mit.edu/2017/artificial-intelligence-early-breast-cancer-detection-1017] accessed February 6, 2018.

Correia, B.E., J.T. Bates, R.J. Loomis, G. Baneyx, C. Carrico, J.G. Jardine et al. 2014. Proof of principles for epitope-focused vaccine design. Nature 507: 201–206.

Cromer, M.K., S. Valdyanathan, D.E. Ryan, B. Curry, A.B. Lucas, J. Camarena et al. 2018. Global transcriptional responses to CRISPR/Cas9-AAV6-based genome editing in CD34+-hematopoietic stem and progenitor cells. Mol Ther July 10. PlI: S1525-0016(18)30261-2. Doi: 10.1016/j.ymthe.2018.06.002.

Crowe, J.E., R.O. Suara, S. Brock, N. Kallewaard, F. house and J.-H. Weitkamp. 2001. Genetic and structural determinants of virus neutralizing antibodies. Immunologic Res 23: 135–145.

Crowe, J.E. 2017. Principles of broad and potent antiviral human antibodies: Insights for vaccine design. Cell Host & Microbe 22: 193–206.

Cullis, P. 2015. The Personalized Medicine Revolution: How diagnosing and treating disease are about to change forever. Greystone Books. Vancouver, Canada.

[DCBIOSCIENCES] 2018. Mass Sepctrometry based proteomics. DC Biosciences. [https://www.dcbiosciences.com/proteomics/proteomics-services/] accessed July 8, 2018.

Deng, Q., Z. Chen, L. Shi and H. Lin. 2018. Developmental progress of CRISPR/Cas9 and its therapeutic applications for HIV-1 infection. Rev Med Virol July 19. E1998. Doi: 10.1002/rmv.1998.

Ellebrecht, C.T. and A.S. Payne. 2017. Setting the target for pemphigus vulgaris therapy. JCI Insight 2: e92021 doi. 10.1172jci.insight.92021.

Escolano, A., J.M. Steichen, P. Dosenovic, D.W. Kulp, J. Golijamin, D. Sok et al. 2016. Sequential immunization elicits broadly neutralizing anti-HIV1 antibodies in Ig knockin mice. Cell 166: 1445–1458.

[FDA] Food and Drug Administration. 2016. "Table of Pharmacogenomic Biomarkers in Drug Listing." [http://www.fda.gov/drugs/scienceresearch/researchareas/pharmacogenetics/] accessed August 22, 2016.

[FDA] Food and Drug Administration. 2017. News. [http://www.gotoper.com/news/breakthrough-designated-companion-diagnostic-approved-for-over-300-genes-5-tumor-types] accessed December 1, 2017.

Flores-Morales, A. and D. Iglesias-Gato. 2017. Quantitative mass spectrometry-based proteomic profiling for precision medicine in prostate cancer. Frontiers Oncology 7: 1–8.

[FOUNDATION] Foundation Medicine. 2017. Genomic Testing. [https://www.foundationmedicine.com/genomic-testing] accessed December 1, 2017.

Fraser, M., V.Y. Sabelnykova, T.N. Yamaguchi, L.E. Heisler, J. Livingstone, V. Huang et al. 2017. Gernomic hallmarks of localized, non-indolent prostate cancer. Nature 54: 359–364.

Friedman, A.A., A. Letai, D.E. Fisher and K.T. Flaherty. 2015. Precision medicine for cancer with next-generation functional diagnostics. Nat Rev Cancer 15: 747–756.

Fry, T.J., N.N. Shah, R.J. Orentas, M. Stetler-Stevenson, C.M. Yuan, S. Ramakrishna et al. 2018. CD22-targeted CAR T cells induce remission in B-ALL that is naïve or resistant to CD19-targeted CAR immunotherapy. Nat Med 24: 20–28.

[GENOS] Genos beta. 2016. [https://genosresearch.com] accessed August 18, 2016.

[GENOTEK] 2018. DNA Genotek. Academic Research. Microbiome. [https://www.dnagenotek.com/us/industries/academic-research-microbiome.html] accessed July 7, 2018.

Gopalakrishnan, V., B.A. Helmink, C.N. Spencer, A. Reuben and J.A. Wargo. 2018. The Influence of the Gut **Microbiome** on Cancer, Immunity, and Cancer Immunotherapy. Cancer Cell 334: 570–580.

Gornati, L., I. Zanoni and F. Granucci. 2018. Dendritic cells in the cross hair for the generation of tailored vaccines. Front Immunol June 27. Doi: 10.3389/fimmu.2018.01484.

Grasso, C.S., Y.M. Wu, D.R. Robinson, X. Gao, S.M. Dhanasekaran, A.P. Khan et al. 2012. The mutational landscape of lethal castration-resistant prostate cancer. Nature 487: 239–243.

Gray, S.W., K. Hicks-Gourant, A. Cronin, B.J. Rollings and J.C. Weeks. 2014. Physicians' attitudes about multiplex tumor genomic testing [internet]. J Clin Oncol 32: 1317–1323.

Gupta, N., K. Susa, Y. Yoda, J.V. Bonventre, M.T. Valerius and R. Morizane. 2018. CRISPR/Cas9-based targeted genome editing for the development of monogenic disease models with human pluripotent stem cells. Curr Protoc Stem Cell Biol 45(1):e50. Doi: 10.1002/cpsc.50.

Hall, M.J. 2014. Conflicted confidence: academic oncologists' views on multiplex pharmacogenomics testing [internet]. J Clin Oncol 32: 1290–1292.

Hartenian, E. and J.G. Doench. 2015. Genetic screens and functional genomics using CRISPR/Cas9 technology. FEBS J 282: 1383–1393.

Hillis, D.M., D. Sadava, R.W. Hill and M.V. Price. 2014. Principles of Life. 2nd Ed. Sinauer/MacMillan, Sunderland, MA, USA.

Hirohata, A., I. Sato, K. Kaino, Y. Iwata, N. Koizumi and K.I. Mishiba. 2018. CRISPR/Cas9-mediated homologous recombination in tobacco. Plant Cell Rep July 13. Doi: 10.1007/s00299-018-2320-7.

Igoe, K.J. 2018. Alumna-and Google-make machine learning easy. Slice of MIT. [https://alum.mit.edu/slice/alumna-and-google-make-machine-learning-easy] accessed June 30, 2018.

Irwin, M., J. McClintick, C. Costlow, M. Fortner, J. White and J.C. Gillin. 1996. Partial night sleep deprivation reduces natural killer and cellular immune responses in humans. FASEB J 10: 643–653.

Irwin, M., M. Wang, C.O. Campomayor, A. Collado-Hidalgo and S. Cole. 2006. Sleep deprivation and activation of morning levels of cellular and genomic markers of inflammation. Arch Intern Med 166: 1756–1762.

Ishino, Y., H. Shinagawa, K. Makino, M. Amemura and A. Nakata. 1987. Nucleotide sequence of the iap gene, responsible for alkaline phosphatase isozyme conversion in *Escherichia coli*, and identification of the gene product. J Bacteriol 169: 5429–5433.

Jain, K.K. 2016. Role of proteomics in the development of personalized medicine. Adv Protein Chem Struct Biol 102: 41–52.

Jansen, R., J.D.A.V. Embden, W. Gaastra and L.M. Schouls. 2002. Identification of genes that are associated with DNA repeats in prokaryotes. Mol Microbiol 43: 1565–1575.

Kochenderfer, J.N., W.H. Wilson, J.E. Janik, M.E. Dudley, M. Stetler-Stevenson, S.A. Feldman et al. 2010. Eradication of B-lineage cells and regression of lymphoma in a patient treated with autologous T cells genetically engineered to recognize CD19. Blood 116: 4099–102.

Lee, K., Y. Zhang, B.P. Kleinstiver, J.A. Guo, M.J. Aryee, J. Miller et al. 2018a. Activities and specificities of CRISPR/Cas9 and Cas12a nucleases for targeted mutagenesis in maize. Plant Biotechnol J July 4. Doi: 10.1111/pbj.

Lee, Z.H., N. Yamaguchi and T. Ito. 2018b. Using CRISPR/Cas9 systems to introduce targeted mutation in Arabidopsis. Methods Mol Biol 1830: 93–108.

MacKall, C.L., M.S. Merchant and T.J. Fry. 2014. Immune based therapies for childhood cancer. Nature Reviews Clinical Oncology 11: 693–703.

Mahas, A. and M. Mahfouz. 2018. Engineering virus resistance via CRISPR-Cas systems. Curr Opin Virol 32: 1–8.

Makarova, K.S., D.H. Haft, R. Barrangou, S.J.J. Brouns, E. Charpentier, P. Horvath et al. 2011. Evolution and classification of the CRISPR-Cas systems. Nat Rev Microbiol 9: 467–477.

Mariathasan, S., S.J. Turley, D. Nickles, A. Castiglioni, K. Yuen, Y. Wang et al. 2018. TGFβ attenuates tumor response to PD-L1 blockade by contributing to exclusion of T cells. Nature 554: 544–548.

Martin, K., B. Begaud, P. Latry, G. Miremont-Salame, A. Fourrier and N. Moore. 2004. Differences between clinical trials and postmarketing use. Br J Clin Pharmacol 57: 86–92.

Martin, G.J., A. Baudet, S. Abelechian, K. Bonderup, T. d'Altri, B. Porse et al. 2018. A new genetic tool to improve immune-compromised mouse models: Derivation and CRISPR/Cas9-mediated targeting of NRG embryonic stem cell lines. Genesis July 16. Doi: 10.1002/dvg.23238.

Melo, U.S., L.F. de Souza, S. Costa, C. Rosenberg and M. Zatz. 2018. A fast method to reprogram and CRISPR/Cas9 gene editing from erythroblasts. Stem Cell Res 31: 52–54.

Miao, J., Y. Chi, D. Lin, B. Tyler and X.L. Liu. 2018. Mutations in ORP1 conferring oxathiapiprolin resistance confirmed by genome editing using CRISPR/Cas9 in Phytophthora capsici and P. sojae. Phytopathology July 6. Doi: 10.1094/PHYTO-01-18-0010-R.

Naim, F., B. Dugdale, J. Kleidon, A. Brinin, K. Shand, P. Waterhouse et al. 2018. Gene editing the phytoene desaturase alleles of Cavendish banana using CRISPR/Cas9. Transgenic Res July 9. Doi: 10.1007/s11248-018-0083-0.

[NATIONAL ACADEMIES] 2011. Report calls for creation of a biomedical research and patient data network for more accurate classification of disease, move toward 'precision medicine'. Nov. 2, 2011. [http://www8.nationalacademies.org/onpinews/newsitem.aspx?recordid=13284] accessed July 5, 2018.

[NCI] National Cancer Institute. NIH. 2018a. Precision medicine in cancer treatment. [https://www.cancer.gov/about-cancer/treatment/types/precision-medicine] accessed July 5, 2018.

[NCI] National Cancer Institute. NIH. 2018b. CART Cells: Engineering Patients' Immune Cells to Treat Their Cancers. [https://www.cancer.gov/about-cancer/treatment/research/car-t-cells] accessed June 24, 2018.

[NHGRI] National Human Genome Research Institute. 2016. NIH. DNA sequencing costs: data from the NHGRI Genome Sequencing Program (GSP). [https://www.genome.gov] accessed August 18, 2016.

[NIH] 2018. CD19 CART Long Term Follow UP (LTFU) Study. ClinicalTrials.gov. [https://clinicaltrials.gov/ct2/show/NCT02445222] accessed June 24, 2018.

Perica, K., A. De Leon Medero, M. Durai, Y.L. Chiu, J.G. Bieler, L. Sibener et al. 2014. Nanoscale artificial antigen presenting cells for T cell immunotherapy. Nanomedicine 10:119–129.

[PMWC] 2018. Precision Medicine World Conference Silicon Valley. January 22–24, 2018. [https://www.pmwcintl.com/2018sv-info/] accessed July 4, 2018.

Prather, A.A., M. Hall, J.M. Fury, D.C. Ross, M.F. Muldoon, S. Cohen et al. 2012. Sleep and antibody response to hepatitis B vaccination. Sleep 35: 1063–1069.

Prather, A.A., D. Janicki-deverts, M.H. Hall and S. Cohen. 2015. Behaviorally assessed sleep and susceptibility to the common cold. Sleep 38: 1353–1359.

Prather, A.A. and C.W. Leung. 2016. Association of insufficient sleep with respiratory infection among adults in the United States. JAMA Internal Med. 176: 850–852.

Rebbeck, T.R. 2014. Precision prevention of cancer. AACR Journal Doi: 10.1158/1055-9965.EPI-14-1058.

Reif, L.R. 2017. Transforming automation is coming: The impact is up to us. Boston Globe. Op-ed, November 10, 2017. [https://www.bostonglobe.com/opinion/2017/11/10/transformative-automation-coming-the-impact/az0qppTvsUu5VUKJyQvoSN/story.html] accessed February 2, 2018.

Reitz, C. 2016. Toward precision medicine in Alzheimer's disease. Ann Transl Med 4: 107. Doi:10.21037/atm.2016.03.05.

Robinson, D., E.N. Van Allen, Y.M. Wu, N. Schultz, R.J. Lonigro, J.M. Mosquera et al. 2015. Integrative clinical genomics of advanced prostate cancer. Cell 161: 1215–1228.

Roper, N., K.D. Stensland, R. Hendricks and M.D. Galsky. 2015. The landscape of precision cancer medicine clinical trials in the United States. Cancer Treatment Rev 41: 385–390.

Sahin, U., E. Derhovanessian, M. Miller, B.P. Kloke, P. Simon, M. Lower et al. 2017. Personalized RNA mutanome vaccines mobilize poly-specific therapeutic immunity against cancer. Nature 547: 222–226.

Sakami, S., T. Ishikawa, N. Kawakami, T. Haratani, A. Fukui, F. Kobayashi et al. 2002. Coemergence of insomnia and a shift in the Th1/Th2 balance toward Th2 dominance. Neuroimmunomodulation 10: 337–343.

Sato, M., M. Ohtsuka, S. Nakamura, T. Sakurai, S. Wantanabe and C.B. Gurumurthy. 2018. *In vivo* genome editing targeted towards the female reproductive. Arch Pharma Res July 4. Doi: 10.1007/s12272-018-0153z.

Schroder, F.H., J. Hugosson, M.J. Roobol, T.L. Tammela, S. Ciatto, V. Nelen et al. 2009. Screening and prostate-cancer mortality in a randomized European study. N Eng J Med 360: 1320–1328.

Sela-Culang, I., S. Ashkenazi, B. Peters and Y. Ofran. 2015. PEASE: Predicting B-cell epitopes utilizing antibody sequence. Bioinformatics 31: 1313–1315.

Shariati, L., F. Rohani, H.N. Heidari, S. Kouhpayeh, M. Boshtam, M. Mirian et al. 2018. Disruption of SOX6 gnee using CRISPR/Cas9 technology for gamma-globin reactivation: an approach towards gene therapy of β-thalassemia. J Cell Biochem July 16. Doi: 10.1002/jcb.27253.

Steichen, J.M., D.W. Kulp, T. Tokatilan, A. Escolano, P. Dosenovic, R.L. Stanfield et al. 2016. HIV vaccine design to target germline precursors of glycan-dependent broadly neutralizing antibodies. Immunity 45: 483–496.

Strait, J.E. 2017. The Father of the Microbiome. Washington University in St. Louis. March 3, 2017. [https://source.wustl.edu/2017/03/the-father-of-the-microbiome/] accessed July 8, 2018.

Sugano, S.S. and R. Nishihama. 2018. CRISPR/Cas9-based genome editing of transcription factor genes in Marchantia polymorpha. Methods Mol Biol 1830: 109–126.

Sultana, J., P. Cutroneo and G. Trifirò. 2013. Clinical and economic burden of adverse drug reactions. J Pharmacol Pharmacother 4(Suppl1): S73–S77.

Szolovits P. 1982. Artificial intelligence and medicine. Westview Press, Colorado, USA.

[THERMO] 2018. Protemoics and Protein Mass Spectrometry. Thermo Fisher Scientific. [https://www.thermofisher.com/us/en/home/industrial/mass-spectrometry/proteomics-protein-mass-spectrometry.html] accessed July 13, 2018.

Tian, M., C. Cheng, X. Chen, H. Duan, H.L. Cheng, M. Dao et al. 2016. Induction of HIV neutralizing antibody lineages in mice with diverse precursor repertoires. Cell 166: 1471–1484.

[UCSF] 2018. Precision Medicine at UCSF. Spotlight: Turning precision medicine into precision prevention. [https://precsionmedicine.ucsf.edu/content/spotlight-turning-precsion-medicine-precsion-prevention] accessed July 2, 2018.

[USNLM] U.S. National Library of Medicine. 2018. Genetic Home Reference. NIH. [https://ghr.nlm.nih.gov/primer/precisonmedicine/definition] accessed July 5, 2018.

Vetchinova, A.S., V.V. Simonova, E.V. Novosadova, E.S. Manuilova, V.V. Nenasheva, V.Z. Tarantul et al. 2018. Cytogenetic analysis of the results of genome editing on the cell model of Parkinson's disease. Bull Exp Biol Med July 13. Doi: 10.1007/s10517-018-4174-y.

Wagle, N., M.F. Berger, M.J. Davis, B. Blumenstiel, M. Defelice, P. Pochanard et al. 2012. High-throughput detection of actionable genomic alterations in clinical tumor samples by targeted, massively parallel sequencing [internet]. Cancer Discov 2: 82–93.

Whelan, B. and J. Taylor. 2013. Precision Agriculture for Grain Production Systems. CSIRO Publishing, Coilingwood, Victoria, Australia.

Wiese, S., K.A. Reidegeld, H.E. Meyer and B. Warscheid. 2007. Protein labeling by iTRAQ: a new tool for quantitative mass spectrometry in proteome research. Proteomics 7: 340–350.

Yu, C.M., H.P. Peng, I.C. Chen, Y.C. Lee, J.B. Chen, K.C. Tsai et al. 2012. Rationalization and design of the complementarity determining region sequences in an antibody-antigen recognition interface. PLoS One 7: e33340.

Zhang, F., Y. Wen and X. Guo. 2014. CRISPR/Cas9 for genome editing: progress, implications and challenges. Hu Mol Genet 23:R40-R46. Doi. 10.1093/hmg/ddu125.

Zhang, Q. [ed.]. 2016. Precision Agriculture Technology for Crop Farming. CRC Press, Boca Raton, Florida, USA.

Zhen, S. and X. Li. 2018. Application of CRISPR/Cas9 for long non-coding RNA genes in cancer research. Hum Gene Ther Jul 26. Doi: 10.1089/hum.2018.063.

QUESTIONS

1. Why is precision important in medicine as it is important for engineering?

2. What is the main rationale for initial launching of genomic medicine?

3. What are the critical elements of science and engineering that enable the development of precision medicine?

4. What are the biggest advantage of precision medicine vs conventionally delivered medical care?

5. Why is treatment based on next generation of functional diagnostic better than that simply based on genomic data alone?

6. What propels the recent development of cancer immunotherapy?

7. What is the rationale behind the structure-based human antibody design?

8. What are the cautions when considering gene therapy?
9. What are the future directions of precision medicine as you can see?

PROBLEMS

Each student will brainstorm and provide a real-life medical/healthcare problem that you think it needs the beneficial contribution from precision medicine. With this problem in mind, please develop in principle a strategy using genomic, engineering, or other functional assays to help resolve this problem. Student will present the idea and the precision medicine strategy in the class room as an independent project.

13

Big Data Analytics

Lawrence S. Chan

QUOTABLE QUOTES

"Information is the oil of the 21st century, and analytics is the combustion engine."

Peter Sondergaard, Senior Vice President, Gartner Research (WORLD ECONOMIC FORUM 2018)

"Hiding within those mounds of data is knowledge that could change the life of a patient, or change the world."

Atul Butte, MD, PhD, Professor, Stanford University (WORLD ECONOMIC FORUM 2018)

"How you gather, manage, and use information will determine whether you win or lose."

Bill Gates, Entrepreneur and Founder of Microsoft (AZ QUOTES 2018)

"Big data has brought about radical innovations in the methods used to capture, transfer, store and analyze the vast quantities of data generated every minute of every day."

S.M. Bagshaw et al. (Badshaw et al. 2016)

"The potential of "big data" for improving health is enormous but, the same time, we face a wide range of challenges to overcome urgently."

C. Auffray et al. (Auffray et al. 2016)

"At the core of the healthcare crisis is fundamental lack of actionable data."

Bruce R. Schatz (Schatz 2015)

Learning Objectives

The learning objectives of this chapter are to help the students to acquire the basic knowledge about Big Data, its information sources, requirements for proper function, potential clinical usefulness, and challenges. After finishing this chapter, students are expected to

- Understand the definition of Big Data.
- Understand the key elements of Big Data.
- Understand the current source and status of healthcare Big Data.
- Understand the potential benefits of healthcare Big Data.

University of Illinois College of Medicine, 808 S Wood Street, R380, Chicago, IL 60612 Tel. (312) 996-6966; larrycha@uic.edu

- Understand the challenge facing the optimum utilization of healthcare Big Data.
- Able to utilize the potential of Big Data to design problem-solving methods for real-life clinical challenges.

Healthcare Big Data Defined

Big Data, a loosely defined term, is generally understood to indicate the gathered data sets that are so large and complex that they are awkward to work with standard-capacity statistical software (Snijders et al. 2012). This original scope of Big Data has been significantly expanded over the years to not only refers to the data itself but also a set of technologies that capture, store, manage and analyze large and variable collections of data for the purpose of solve complex problems (SNS 2017). Big Data of healthcare, in addition to being large in size, possesses some unique characteristics, with its two major characteristics being its high energy and its limited lifespan, as described by some experts (Dinov 2016b). In terms of high energy, due to its large size and its purely empirical observation data that is free from specific assumptions it carries a "holistic" characteristic that its impact is greater than the sum of each component of integrated data. At the same time, since healthcare data decay rapidly, its limited lifespan indicates that a rapid acquisition, processing, and sharing of the collected data is critical for its full benefits to be realized (Dinov 2016b). Other experts often defined Big Data by the three "Vs": Volume, Variety, and Velocity. Volume refers to its large size, with most data set contains at least 1 petabyte (10^{15} bytes). Variety denotes the multiple different data sources that comprise the data set. Velocity indicates the speed of combining and analyzing large data sets to result in timely information (Rumsfeld et al. 2016). The primary sources of today's healthcare Big Data are derived from electronic medical records (EMR), insurance claims, and registry data from healthcare providers and payers. Not surprisingly, genetic, imaging, and other disease data have been incorporated into the EMR and integrated to the Big Data as well. In the future, data generated by patients during active monitoring, passive monitoring through internet, smart phone activities and medical sensors could be integrated into the EMR data base. Data from outside of medical field, such as government financial data, could be linked to the Big Data as well (Dinov 2016b, Monteith et al. 2016). The engineering-medicine aspects of Big Data lie in its innovative approach to medical data gathering and utilization and its potential to improve healthcare efficiency and for integrated healthcare. In this chapter, we will discuss three major areas of Big Data in the context of healthcare: Its key components (Dinov 2016b, Monteith et al. 2016, Toga and Dinov 2015); its potential clinical applications (Bagshaw et al. 2016, Lebo et al. 2016, Austin and Kusumoto 2016, Jellis and Griffin 2016, Perry 2016, Frakt and Pizer 2016, Waldman and Terzic 2016, Rodriguez et al. 2016, Iqbal et al. 2016, Suresh 2016, Broughman and Chen 2016, Baillie 2016, Murphy et al. 2016, Souliotis et al. 2016, Toga et al. 2015, Viceconti et al. 2015, Costa 2014, Mathias, et al. 2016, Janke et al. 2016, Issa et al. 2014, Rumsfeld et al. 2016), and its challenges (Auffray et al. 2016, Dinov 2016a, Dinov 2016b, Toga and Dinov 2015, Lebo et al. 2016, Frakt and Pizer 2016, Mathias et al. 2016, Rumsfeld et al. 2016,).

Healthcare Big Data Essentials

Key Components of Healthcare Big Data

- *Data collection*: In order to establish a set of data, these data need to be entered to a data site.
- *Data storage*: Once these data are entered into the data site, they need to be stored and maintained in a secure manner.
- *Data analysis and interpretation*: For the stored data to be useful for specific purpose, a data analysis needs to be performed and the resulting analysis needs to be interpreted as specific to the questions raised.
- *Data sharing*: Big Data could be useful for the general population only if they can be shared among the needed end users, i.e., the healthcare providers and healthcare decision makers.

Potential Benefits of Healthcare Big Data

In its recent and comprehensive report, SNS Telecom has presented an in-depth assessment of Big Data in the healthcare and pharmaceutical industry which includes application areas, user case studies, strategies, and future roadmap, as well as challenges, technologies, key market drivers, vendor directory, and forecasts for Big Data hardware, software and professional services investments from 2017 through 2030 (SNS 2017). The key findings from this report include the followings:

- In 2017, Big Data vendors will generate $4 Billion revenue from hardware, software and professional services in the healthcare and pharmaceutical industry and the future growth will be estimated to be 15% per year, eventually reaches $5.8 Billions by 2020.
- With the utilization of Big Data technologies, healthcare facilities have been able to obtain 10% cost reduction, variable degrees of outcome improvements (up to 20% in certain cases), 30% revenue growth, and 35% increase in patient access.
- Big Data technologies have been playing a vital role in accelerating the transition to accountable and value-based care models.

We will focus our discussions on several key application areas including healthcare delivery, diagnostics, therapeutics, drug discovery, prevention, and public health.

Healthcare Delivery

Big Data can provide accurate and comprehensive data for healthcare decision makers to base on, in the design and construction of innovative, high efficiency, high quality healthcare facilities (Baillie 2016). Such innovative facilities would help reducing operation downtime, duplicated administrative works, and provider manpower usage: reducing medical error, reducing waste; and reducing the overall cost of care (Baillie 2016, Mehta and Pandit 2018). Taking the advantage of the large data size, Big Data system can help investigators to extract information to make analysis regarding whether diagnostic procedures such as inpatient echocardiogram have been over utilized. Such operation can provide an optimum method to monitor and reduce healthcare wastes, thus making healthcare delivery more efficient with lower cost (Jellie and Griffin 2016). In one study, investigators reviewed the application of a pediatric risk score, a Big Data system that was integrated into a local hospital's EMR with the capability of delivering early warning signs for clinical deterioration of pediatric patients in real time (Suresh 2016). Such Big Data system could potentially assist the healthcare providers to deliver a higher quality of care, since earlier medical interventions would be better in controlling the disease progression and preventing unnecessary use of dramatic and expensive life saving measures. It would also help reducing variations in care delivery (Suresh 2016).

As one author pointed out, "At the core of the healthcare crisis is fundamental lack of actionable data", and proposed that this kind of "actionable data" can be gathered for useful purpose through future technology with mobile devices (Schatz 2015). He proposed an improved predictive modeling termed "3 M" strategy for population health that consists the three "M" steps of "Monitor", "Measure", and "Manage". His logical arguments are that "To manage populations to decrease cost and increase quality, it is necessary to measure populations across all the rings to determine health status. To measure population health deeply, it is necessary to monitor individuals on a continuous basis" (Schatz 2015). He further suggested that mobile phones could be a strategic adoption for monitoring purpose since it is commonly utilized, it has no additional requirement for behavior change, and it remains a constant device as it also functions as a needed communication tool (Schatz 2015). One example this "3 M" strategy can improve healthcare is a case of diabetes management. Utilizing implanted diabetes monitor and Big Data system, the diabetes management system can send alerts through smartphones to individual patients, warning them low blood sugar or need for insulin injection, thereby improve the healthcare through speedy normalization of sugar levels (Schatz 2015). Big Data is believed to be the key in correcting the current healthcare system inefficiencies characterized by poorly managed scientific insights, poorly utilized medical evidence, and poorly captured patient care experience (Rumsfeld et al. 2016). It has been estimated that Big Data system has the potential to improve healthcare outcomes and to reduce medically related wastes in applications such as predictive models, population management, and drug and medical device surveillance that many billions of dollars could be saved annually in the US alone (Rumsfeld et al. 2016).

Diagnostics

Because of the inflow of large amount of complex data from diverse domains of genomics, proteomics, metabolomics, and phenomics (imaging, biometrics, clinical data), the interplay among these domains could reveal patterns and relationships that eventually would permit investigators to locate early biomarkers for complex diseases such as Parkinson's and Alzheimer's (Toga et al. 2015). These biomarkers could then be used to enhance early detection of disease and may ultimately lead to prevention or reversal of disease. By incorporating powers of nanotechnology and Big Data system, investigators are now in the process of building computer-aided diagnostic systems that are capable of precise medical diagnostics leveraged the abilities to integrate and interpret data from different sources and format. It is hope that in the near future such Big Data diagnostic system will be available routinely in daily clinical practice (Rodrigues et al. 2016). Another potential benefit of Big Data is in the hospitals' emergency rooms, where the settings are characteristically high-flow and acute encountering that demand speedy and decisive decisions, predictive value is hugely important and is where Big Data could lend a helping hand (Janke et al. 2016). Armed with the predictive value of Big Data, emergency physicians would be empowered, when encountering patients, to make better decisions in ordering diagnostic testing and medications, thus would ultimately reduce the risk of adverse events such as serious drug reactions, incidental findings that are not relevant to the central acute care, and unnecessary costs from avoidable diagnostic procedures. Instead the emergency providers can zoom in the critical issues, with resulting cares that are higher in efficiency and lower in cost (Janke et al. 2016).

Therapeutics

One of the examples is the Caner-Associations Map-Animation (CAMA) Big Data system established to chart the association of cancers with other diseases over time (Iqbal et al. 2016). This CAMA system can now provide a dynamic visualization tool to view cancer-disease associations across different age groups and gender, so it could detect cancer co-morbidities earlier than manual inspection. Thus it could assist physicians and researchers to efficiently explore early stage disease detection, develop new cancer treatment strategies, and identify new cancer risk factors and modifiers (Iqbal et al. 2016). Paring therapeutic regimens with therapeutic outcomes in the context of large data set, Big Data has the potential not only to retrospectively assess the quality of cancer patient care, but also help healthcare providers to deliver high-quality care in real time (Broughman and Chen 2016). Aided by our current advanced biological discovery era of "omics", where individual patients' detailed genomics, transcriptomics, proteomics and metabolomics can be delineated, the Big Data informatics thus could enable network-driven systems to integrated with the scientific data to develop personalized biochemical fingerprints useful for individual-specific treatments (personalized medicine) and to discover new biological associations and drug targets (Viceconti et al. 2015, Costa 2014, Issa et al. 2014). In the specialty of surgery, there is also a call for integrating Big Data into surgical practice (Mathias et al. 2016). One example of such integration is the American College of Surgeons' Clinical Risk Calculator, a Big Data set aims to use the predictive value of Big Data to improve cost-effectiveness of surgical practices in a "personalized surgery" fashion (Mathias et al. 2016).

Drug Discovery

New drug discovery is essential to healthcare. Optimal care can be delivered to the patients so long as we have the proper medications to offer. Some of the enemies of human, cancer cells and infectious microorganisms, for example, can mutat-genetically in response to treatment and become drug resistant (Choi et al. 2018, Wang et al. 2018). Thus, we are in constant need for new medication in responding to these resistances. Big Data, when properly acquired and interpreted, could enhance project timelines and reduce clinical attrition by improving early decision making (Brown et al. 2018). Additionally, Big Data, in conjunction with artificial intelligence, can potentially help accelerating medicinal chemistry in the drug discovery process (Griffen et al. 2018). A good example is Alzheimers disease, for which a large number of data is present. New informatics analysis platform, if developed, can be used to organize and extract Big Data resources into actionable mechanism to assist effective drug repurposing or *de novo* drug design (Maudslev et al. 2018).

Prevention

In the past decades, medical education has been focused primarily on the diagnosis and treatment of diseases after the diseases have revealed through symptoms or physical signs and confirmed by various diagnostic measures such as serology, radiography or histopathology. Recently many leaders of medical practice and education have come to recognize the importance of disease prevention. By preventing diseases to occur in the first place, not only that lives are saved and suffering are eliminated, but also the healthcare system as a whole save huge sum of money that is required to perform therapeutic operations, either by medical or surgical means. In fact, Big Data could be leveraged to do just that and do it very well. By navigating through its large data base of complex information, Big Data could help healthcare providers to spot patterns of markers prior to disease onset, thus giving physicians the very tool they need to enter into the battle against the disease onset in preventative care (Iqbal et al. 2016, Toga et al. 2015).

Public Health

Another great potential that Big Data can benefit healthcare is in the area of precision public health (Dolley 2018). As the experts in the field defined, "precision public health is a new field driven by technological advances that enable more precise descriptions and analyzes of individuals and population groups, with a view to improving the overall health of populations" (Baynam et al. 2017). Although there may be some inherent risk involved its use for precision public health analytics, Big Data can bring value into the areas of precision disease surveillance and signal detection, public health risk prediction, identifying targets for treatment or intervention, and improving disease understanding (Dolley 2018). In the area of precision disease surveillance and signal detection, for example, Big Data has been shown to be effective in several health-related conditions or diseases including electromagnetic field exposure, pollution, medication safety, drowning, bacterial resistance to antibiotics, cholera, influenza, Lyme disease, dengue, and whooping cough (Dolley 2018). In the area of risk prediction, Big Data has been utilized for many health issues and infectious diseases including diabetes mellitus, lead contamination, air pollution, bacterial antibiotic resistance, and infectious diseases caused by parasite Plasmodium, Zika virus, West Nile virus, Human Immunodeficiency virus, Ebola virus, and bird-transmitted influenza virus (Dolley 2018). In the area of treatment and intervention targeting, Big Data has been utilized in health-related problems and infectious diseases such as airway allergic diseases, childhood obesity, opioid drug misuse, non-smoking tobacco use, and infections due to parasite Plasmodium, hepatitis C virus, Zika virus, and Human Immunodeficiency virus (Dolley 2018). In addition, Big Data analytics has increased understanding of many health conditions and diseases like opioid epidermis, premature birth, chronic diarrhea, diabetes mellitus, and infections by bacteria Vibrio cholera, influenza virus, Zika virus, and chikungunya virus (Dolley 2018).

The Big Data Challenges

Accuracy in Data Entry

This seems to be obvious. The quality of data coming out of the Big Data sets are as good as the data that enter in. If the data entered are inaccurate, inconsistent, or otherwise compromised, the bigness of the data will make it worse than the small data. The founder of Microsoft said it very well "The first rule of any technology…is that automation applied to an efficient operation will magnify the efficiency. The second is that automation applied to inefficient operation will magnify inefficiency" (Mathias et al. 2016). Thus, the entry guidance into a given Big Data set need to be clarified and data watchdogs are also needed to maintain the accuracy of a given data set. The bottom line is that it is essential to eliminate the possibility of the "garbage in and garbage out" kind of undesirable scenario.

Inherent Weakness

The limitation of observational information inherent in the Big Data, such as bias of treatment selection, raises the possibilities that Big Data could misinform healthcare providers and healthcare administrators for the patient care information they seek to obtain (Rumsfeld et al. 2016). Another intrinsic limitation of Big Data is due to the lack of control, uniformity and technological reliability for interrogating huge sample size (Dinov 2016b).

Data Noise

Noise in data can potentially hinder the usefulness of Big Data and these noise include incomplete and disparate data set, non-uniform data collection approaches, non-uniform data resolution and determination, non-uniform data formats, and the like. If the data from variety of sources are collected in different format or platform, it will end up as a big hurdle when the information is extracted for actionable data. It will even prevent the proper scope of required information to be extracted (Auffray et al. 2016). The other issue is Noise reduction through filtering. A good solution maybe to stride for a balance between hard filtering resulting in loss of some useful data vs. soft filtering resulting in some noise retention. The adaptive filtering with artificial intelligence may help resolving this problem, at least partially.

Correct Human Judgment

Human judgment is inevitably involved in any kind of data gathering and extraction, and Big Data is no exception. Thus, people with the skills and knowledge in both the informatics field and medical field are needed to be in charge of the maintenance and the utilization of Big Data, if Big Data is to be of great benefit to the patients (Auffray et al. 2016).

Technology Limitation

As of today, the data size of health-related subjects is approaching zetabyte (10^{21} bytes of data) just in the US alone (Rumsfeld et al. 2016). To maintain, secure, and share this huge size of data requires substantial commitment of technology infrastructure and technical man powers. Obviously new innovative technologies also need to be developed to enhance and optimize the management and processing of these large scale of complex and heterogeneous data sets (Dinov 2016b). Thus technical limitation could become a key implementation barrier. Increasingly, the tasks of understanding and interpretation of Big Data have become so big that human technicians found them difficult to handle and researchers now start to turn to artificial intelligence for help (Beam and Kohane 2018, Brown et al. 2018, Cao et al. 2018, Hueso et al. 2018, Lee et al. 2018, Low et al. 2018). The following chapter (Chapter 14) in this book deals with artificial intelligence in medicine with greater details.

Financial Constraints

As indicated in the above technical issue, the substantial commitment of technical infrastructure and technical man powers has a big price tag. Currently, federal and foundation funding is insufficient. Collectively US government and industry are spending about $200 billion dollars annually for Cloud resources to support storage, computation, and social networking functions (Dinov 2016b). Thus the actual utilization may be limited to institutions with the financial means to pursue the application of Big Data in their healthcare activities (Toga and Dinvo 2015, Dinov 2016b).

Healthcare Inequity

As discussed above, to take advantage of the Big Data benefit, medical centers would need a large investment of fund to build internet infrastructure, hire technically competent employees to extract useful information, and to train medical expert to utilize the data. The advantage of Big Data in medicine may in some way amplify social inequality as medical centers with sufficient fund may be able to take advantage of this advancement, whereas those with insufficient fund may lag behind (Auffray et al. 2016, Toga and Dinov 2015, Mathias et al. 2016).

Evidence of Real-life Use of Big Data Analytics

A recent systematic review of Big Data analytics in healthcare found that currently there is only a small number of published information providing evidence of real-world use of Big Data analytics in healthcare. In particular, quantitative data are commonly not available to give healthcare providers the confidence on its usefulness or benefits. Researchers also found that there is a lack of consensus on the operational definition of Big Data in healthcare (Mehta and Pandit 2018).

Administrative Policies

As we have discussed, Big Data gathering and usage involved a very large communities of professionals of a diverse backgrounds and experiences. Each of these communities may have different expectation and standard. Therefore, comprehensive and coherent policies, guidelines, and procedures for Big Data collaboration, access, and sharing are also very important to ensure data security, proper community governance, data maintenance, stakeholder trust, and wide distribution of the Big Data informatics (Auffray et al. 2016, Toga and Dinov 2015). In its recent and comprehensive examination on Big Data in the healthcare and pharmaceutical industry, SNS Telecom reports a key finding that addressing privacy and security is an essential necessity for fully leverage the benefits of Big Data (SNS 2017). Thus, investments in data encryption, cyber security, and defensive de-identification techniques, and user security trainings are needed. In addition, policies need to be established to integrate Big Data tool into the daily clinical practices, as mere existence of Big Data does not influence the very healthcare outcome we seek to obtain (Rumsfeld et al. 2016).

Summary

In this chapter, we define briefly the meaning and key elements of Big Data. We then discuss the potential benefits of Big Data in medicine, including potential use in more effective and efficient healthcare delivery, more accurate diagnoses and therapeutics, facilitating new drug discovery, as well benefits for disease prevention and public health. We also point out some of the challenges facing the implementation of Big Data and the hurdles on receiving the full benefit from Big Data in medicine.

References

Auffray, C., R. Bailing, I. Barroso, L. Bencze, M. Benson, J. Bergeron et al. 2016. Making sense of big data in health research: Towards an EU action plan. Genomic Med 8:71. Doi.10.1186/s13073-016-0323-y.

Austin, C. and F. Kusumoto. 2016. The application of Big Data in medicine: current implications and future directions. J Interv Card Electrophysiol 47: 51–59.

[AZ QUOTES] 2018. Bill Gates. AZ Quotes. [www.azquotes.com/author/5382-Bill-Gates] accessed May 11, 2018.

Bagshaw, S.M., S.L. Goldstein, C. Ronco and J.A. Kellum. 2016. ADQI 15 Consensus Group. Acute kidney injury in the area of big data: the 15(th) consensus conference of the acute dialysis quality initiative (ADQI). Can J Kidney Health Dis 2016; 3: 5. Doi:10:1186/s40697-016-0103-z. eCollection.

Baillie, J. 2016. How 'Big data' will drive future innovation. Health Estate 70: 59–64.

Baynam, G., A. Bauskis, N. Pachter, L. Schofield, H. Verhoef, R.L. Palmer et al. 2017. 3-dimensional facial analysis-facing precision public health. Front Public Health 5: 31 Doi: 10. 3389/fpubh.2017.00031.

Beam, A.L. and I.S. Kohane. 2018. Big Data and machine learning in health care. JAMA 319: 1317–1318.

Broughman, J.R. and R.C. Chen. 2016. Using big data for quality assessment in oncology. J Comp Eff Res 5: 309–319.

Brown, N., J. Cambruzzi, P.J. Cox, M. Davies, J. Dunbar, D. Plumbley et al. 2018. Big Data in drug discovery. Prog Med Chem 57: 277–356.

Cao, C., F. Liu, H. Tan, D. Song, W. Shu, W. Li et al. 2018. Deep learning and its applications in biomedicine. Genomics Proteomics Bioinformatics 16: 17–32.

Choi, Y.M., S.Y. Lee and B.J. Kim. 2018. Naturally occurring hepatitis B virus reverse transcriptase mutations related to potential antiviral drug resistance and liver disease progression. World J Gastroenterol. 24: 1708–1724.

Costa, F.F. 2014. Big data in biomedicine. Drug Discover Today 19: 433–40.

Dinov, I.D. 2016a. Volume and value of big healthcare data. J Med Stat Inform 4. Pii: 3.

Dinov, I.D. 2016b. Methodological challenges and analytic opportunities for modeling and interpreting big healthcare data. Gigascience 5: 12. Doi: 10.1186/s13742-016-0117-6.

Dolley, S. 2018. Big Data's role in precision public health. Front in Public Health 6: 68 Doi: 10.3389/fpubh.2018.00068.

Frakt, A.B. and S.D. Pizer. 2016. The promise and perils of big data in healthcare. Am J Manag Care 22: 98–9.

Griffen, E.J., A.G. Dossetter, A.G. Leach and S. Montague. 2018. Can we accelerate medicinal chemistry by augmenting the chemist with Big Data and artificial intelligence? Drug Discov Today. Mar 22. Doi: 10.1016/j.durdis.2018.03.011 [Epub ahead of print].

Hueso, M., A. Vellido, N. Montero, C. Barbieri, R. Ramos, M. Angoso et al. 2018. Artificial intelligence for the artificial kidney: Pointers to the future of a personalized hemodialysis therapy. Kidney Dis (Basel). 4: 1–9.

Iqbal, U., C.K. Hsu, P.A. Nguyen, D.L. Clinciu, R. Lu, S. Syed-Abdul et al. 2016. Cancer-disease associations: A visualization and animation through medical big data. Comput Methods Programs Biomed 127: 44–51.

Issa, N.T., S.W. Byers and S. Dakshanamurthy. 2014. Big data: the next frontier for innovation in therapeutics and healthcare. Expert Rev Clin Pharmacol 7: 293–8.

Janke, A.T., D.L. Overbeek, K.E. Kocher and P.D. Levy. 2016. Exploring the potential of predictive analytics and big data in emergency care. Ann Emerg Med 67: 227–36.

Jellis, C.L. and B.P. Griffin. 2016. Are we doing too many inpatient echocardiogram? The answer from Big Data may surprise you! J Am Coll Cardiol 67: 512–4.

Lebo, M.S., S. Sutti and R.C. Green. 2016. "Big Data" gets personal. Sci Transl Med 8:322fs3-3fs3. Doi: 10.1126/scitranslmed.aad9460.

Lee, S.I., S. Celik, B.A. Logsdon, S.M. Lundberg, T.J. Martins, V.G. Oehler et al. 2018. A machine learning approach to integrate big data for precision medicine in acute myeloid leukemia. Nat Commun. 9:42. Doi: 10.1038/s41467-017-02465-5.

Low, S.K., H. Zembutsu and Y. Nakamura. 2018. Breast cancer: The translation of big genomic data to cancer precision medicine. Cancer Sci 109: 497–506.

Mathias, B., G. Lipori, L.L. Moldawer and P.A. Efron. 2016. Integrating "big data" into surgical practice. Surgery 159: 371–4.

Maudslev, S., V. Devanarayan, B. Martin and H. Geerts. 2018. Intelligent and effective informatics deconvolution of "Big Data" and its future impact on the quantitative nature of neurodegenerative disease therapy. Alzheimers Dement March 15. Doi: 10.1016/j.jalz.2018.01.014. [Epub ahead of print].

Mehta, N. and A. Pandit. 2018. Concurrence of big data analytics and healthcare: A systematic review. Int J Med Inform 114: 57–65.

Monteith, S., T. Glenn, J. Geddes, P.C. Whybrow and M. Bauer. 2016. Big data for bipolar disorder. Int J Bipolar Disord 4: 10. Doi:10.1186/s40345-016-0051-7.

Murphy, D.R., A.N. Meyer, E. Bhise, E. Russo, D.F. Sittig, L. Wei et al. 2016. Computerized triggers of Big Data to detect delays in follow-up of chest imaging results. Chest 150: 613–620.

Perry, P.M. 2016. Harnessing the power of big data and data analysis to improve healthcare entities. Healthc Financ Manage 70: 74–5.

Rodrigues, J.F., Jr., F.V. Paulovich, M.C. de Oliveira and O.N. de Oliveira, Jr. 2016. On the convergence of nanotechnology and Big Data analysis for computer-aided diagnosis. Nanomedicine (London) 11: 959–82.

Rumsfeld, J.S., K.E. Joynt and T.M. Maddox. 2016. Big data analytics to improve cardiovascular care: promise and challenges. Nature Rev 13: 350–9.

Schatz, B.R. 2015. National survey of population health: Big Data analytics for mobile health monitors. Big Data 3: 219–229.

Snijders, C., U. Matzat and U.-D. Reips. 2012. Big Data: big gaps of knowledge in the field of internet science. Int J Internet Sci 7: 1–5.

[SNS] SNS Telecom. 2017. Big Data in the healthcare & pharmaceutical industry: 2017–2030—opportunities, challenges, strategies & forecasts. [www.snstelecom.com/bigdatahealthcare] accessed January 3, 2018.

Souliotis, K., C. Kani, M. Papageorgiou, D. Lionis and K. Gourgoulianis. 2016. Using Big Data to assess prescribing patterns in Greece: the case of chronic obstructive pulmonary disease. PLoS One 11: e0154960. Doi: 10.1371/journal.pone.0154960.

Suresh, S. 2016. Big Data and predictive analysis: Applications in the care of children. Pediatr Clin North Am 63: 357–66.

Toga, A.W. and I.D. Dinov. 2015. Sharing big biomedical data. J Big Data 2: 7.

Toga, A.W., I. Foster, C. Kesseman, R. Madduri, K. Chard, E.W. Deutsch et al. 2015. Big biomedical data as the key resource for discovery science. J Am Med Inform Assoc 22: 1126–31.

Viceconti, M., P. Hunter and R. Hose. 2015. Big data, big knowledge: Big data for personalized healthcare. IEEE J Biomed Health Inform 19: 1209–15.

Waldman, S.A. and A. Terzic. 2016. Big Data transforms discovery-utilization therapeutics continuum. Clin Pharmacol Ther 99: 250–4.

Wang, L., L. Ma, F. Xu, W. Zhai, S. Dong, L. Yin et al. 2018. Role of long non-coding RNA in drug resistance in non-small cell lung cancer. Thorac Cancer. May 3. doi: 10.1111/1759-7714.12652. [Epub ahead of print].

[WORLD ECONOMIC FORUM] 2018. The most revealing big data quotes. World Economic Forum. [https://www.weforum.org/agenda/2015/01/the-most-revealing-big-data-quotes/] accessed May 6, 2018.

QUESTIONS

1. What is your definition of Big Data? Please describe in your own word.
2. What are the key elements of Big Data that are needed to function properly?
3. What are the potential benefits that Big Data can provide to medicine?
4. Which is the greatest potential benefit of Big Data to medicine and why?
5. What are the challenges in implementing Big Data in medicine?
6. Which is the biggest challenge in implementing Big Data and why?

PROBLEM

Students will be formed into groups of 3–5 people. Within each group, students will first brainstorm for a real-life medical problem to solve that can be approached with Big Data analytics. Then each group will provide detailed step-by-step methods to solve their selected problem, using the engineering solution method depicted in Chapter 9.

14

Artificial Intelligence

Buyun Zhang **and** *James P. Brody**

QUOTABLE QUOTES

"Nobody phrases it this way, but I think that artificial intelligence is almost a humanities discipline. It's really an attempt to understand human intelligence and human cognition."

Sebastian Thrun, German Computer Scientist, and CEO of Kitty Hawk Corporation (Marr 2017)

"Artificial intelligence will reach human levels by around 2029. Follow that out further to, say, 2045, we will have multiplied the intelligence, the human biological machine intelligence of our civilization a billion-fold."

Ray Kurzweil, American Computer Scientist, and Inventor (Marr 2017)

"Some people worry that artificial intelligence will make us feel inferior, but then, anybody in his right mind should have an inferiority complex every time he looks at a flower."

Alan Kay, American Computer Scientist, and Fellow of American Academy of Arts and Sciences (Marr 2017)

"Anything that could give rise to smarter-than-human intelligence—in the form of artificial intelligence, brain-computer interfaces, or neuroscience-based human intelligence enhancement—wins hands down beyond contest as doing the most to change the world. Nothing else is even in the same league."

Eliezer Yudkowsky, American Researcher of Artificial Intelligence (Marr 2017)

"A year spent in artificial intelligence is enough to make one believe in God."

Alan Perlis, American Computer Scientist and Professor (Marr 2017)

Patient: "I'm having trouble breathing. I want my nurse."

Hospital Administrator: "We have something better than nurses, algorithms!"

Patient: "That sounds like a disease"

Hospital Administrator: "Algorithms are simple mathematical formulas that nobody understands. They tell us what disease you should have, based on what other patients have had."

Patient: "Okay that makes no sense. I'm not other patients, I'm me."

University of California, Irvine, Department of Biomedical Engineering, 3120 Natural Sciences II, Zot 2715, Irvine, CA 92697-2715.
* Corresponding author: jpbrody@uci.edu

Hospital Administrator: "Look, it's not all about you. We've spent millions on algorithms, software, computers . . ."

Patient: "I need my nurse!"

National Nurses United Radio Ad. (National Nurses United, 2014)

"When you're fundraising, it's artificial intelligence. When you're hiring, it's machine learning. When you're implementing, it's linear regression. When you're debugging, it's printf()"

Baron Schwartz, Chief Technical Officer at VividCortex (B. Schwartz, 2017)

Learning Objectives

After completing this chapter, the reader should be able to:

- Understand the general definition of artificial intelligence (AI).
- Understand the relationship between Big Data and AI.
- Describe how classification algorithms are related to diagnosis of medical conditions.
- Read and understand a receiver operator characteristic curve.
- Understand the benefits and limits to characterizing a receiver operating characteristic (ROC) curve by its area under the curve (AUC).
- Understand the distinction between supervised and unsupervised learning.
- Give different examples of machine learning algorithms.
- Understand how ethical and regulatory issues effect the application of machine learning to medicine.

Introduction

Machine Learning vs. Artificial Intelligence

Artificial intelligence is a broad term that includes many different technologies and has a long history in medicine (Schwartz et al. 1987). In its broadest sense, artificial intelligence is defined as a machine that has thinking, reasoning, problem solving, and learning capabilities similar to humans.

Examples of artificial intelligence include knowledge bases and expert systems. Table 1 gives an example of an expert system for diagnosing congenital heart disease from 1968 (Gorry and Barnett 1968).

Table 1 demonstrates the early importance placed on natural language processing, the ability to extract information from written language. Natural language processing is still an area of active research with some success. However, today's most successful artificial intelligence applications in medicine use machine learning.

The recent popularity of machine learning can be attributed to two major innovations, first the availability of extremely large digitized datasets and second, the development of computational techniques to control and reduce overfitting known as *regularization*.

These large digital datasets are often called Big Data. Big Data has led to a number of innovations. Large collections of text that have been translated into two or more languages led to machine translations that are almost as good as humans (Monroe 2017). Thousands of hours of television programs that include closed captions for the hearing impaired led to better voice recognition systems. Enormous sets of tagged images have vastly improved automated image recognition (Halevy et al. 2009).

Artificial intelligence advancements in a field follow the existence of Big Data in that field. A big impetus for the adoption of electronic health records in the United States was the passage of the Health Information Technology for Economic and Clinical Health (HITECH) Act, part of the American Recovery and Reinvestment Act of 2009. We are beginning to see the applications of machine learning to medicine now.

Table 1. An artificial intelligence package for diagnosis, 1968.

```
PLEASE TELL ME ABOUT THE PATIENT
The patient here with me is 23 years old
PLEASE CONTINUE
He shows no evidence of cyanosis
OK PLEASE GO ON
There is no indication of the squatting symptom
ALL RIGHT.  GO ON PLEASE.
I'm afraid that's all I know about him at present.
CHECK THE HILAR VESSELS.  ARE THEY INCREASED OR DECREASED
They are increased
I SEE
LOOK FOR A LEFT SECOND SYSTOLIC MURMUR
First tell me what the three most probable diseases are at this time
2 WITH PROBABILITY 74
18 WITH PROBABILITY 7
30 WITH PROBABILITY 3
PLEASE CONTINUE
LOOK FOR A LEFT SECOND SYSTOLIC MURMUR
And the name of disease 18 is
PATENT DUCTUS ARTERIOSUS
ALL RIGHT. GO ON PLEASE
LOOK FOR A LEFT SECOND SYSTOLIC MURMUR
He has a left second midsystolic murmur
ALL RIGHT
IS THE AXIS IN THE ECG RIGHT, LEFT, OR NORMAL
It is the right axis
OK
I RECOMMEND A DIAGNOSIS OF DISEASE 2
What is your current view of the problem
MY CURRENT VIEW IS
        DISEASE      PROBABILITY
           2             93
           4              2
Name disease 4 for me
ATRIAL SEPTIC DEFECT WITH PULMONARY HYPERTENSION
I'm finished with this case. Thank you for your help
I HOPE THAT I HAVE BEEN OF SERVICE.  GOODBYE
```

Diagnosis, Prognosis in Medicine vs. Classification in Machine Learning

Diagnosis is a key part of medicine. A patient presents with several symptoms and laboratory tests. Based on these symptoms and tests, the physician is called on to classify the patient's condition into a disease. Once diagnosis is complete, treatment can begin.

Classification is a key part of machine learning. A dataset is characterized by a number of variables. Based upon these variables, the machine learning algorithm is called on to classify the dataset into a particular class. An example is optical character recognition. An image containing a single numeral can be digitized into a 20×20 array of pixels. The algorithm can classify these 400 pixel values into one of ten possible digits, 0–9.

Thus diagnosis in medicine is a natural target for the application of machine learning. Both medical diagnoses and machine learning classification share a set of terminology that describes the performance of diagnosis and classification tests.

Diagnosis or classification tests are judged by their ability to correctly predict both the correct answer and avoid the incorrect answer. For the simplest case, a binary classification, four rates are relevant: the true positive, false positive, true negative and false negative rates. The true positive and true negative rates are the correct answers, while the false negative and false positive errors are the incorrect answers.

Sensitivity versus Specificity

These four numbers can be combined into two: sensitivity and specificity. The sensitivity is defined as the probability of a positive test given that the sample is known to be positive. While the specificity is the probability of a negative test given that the sample is known to be negative.

Sensitivity and specificity depend on cutoff values. Different cutoff values give different sensitivity and specificity values. Ideally, a test will have both high sensitivity and high specificity, but a tradeoff exists between the two. Often a test will produce a numerical value. All values above a threshold have a positive test result and all values below are negative. The exact values of the sensitivity and specificity will depend on the threshold value. If one chooses a low threshold value, one gets a high true positive rate (high sensitivity), and a high false positive rate (low specificity). On the other hand, if one chooses a high threshold value, one gets a low true positive rate (low sensitivity) and a low false positive rate (high specificity).

Characterizing such a test is a very general problem first addressed in the field of signal detection theory (Peterson et al. 1954). The problem was posed this way: "Suppose an observer is given a voltage varying with time during a prescribed observation interval and is asked to decide whether the source is noise or is signal plus noise. What method should the observer use to make this decision and what receiver is a realization of that method?"

Petersen, Birdsall and Fox answered the question they posed. The best method to decide whether you have signal, or just noise is to set a threshold. If the voltage exceeds the threshold, then one can claim to have detected the signal. Of course, this alters the question to "how does one set a threshold". One sets the threshold based on the acceptable true positive rate and false positive rate. In the electrical engineers' formulation of the problem, the "test" was an electronic receiver that detected the voltage. Thus, the main task was to characterize their receiver's operating condition. They formulated a graphical expression of their receiver's performance that they named the receiver operating characteristic (ROC) curve. In our case, a better name might be the test characteristic curve, but the ROC nomenclature is firmly embedded in science.

ROC Curves and Area Under Curve to Quantify Quality of Test

The receiver operating characteristic curve expresses the full capabilities of a test. One often needs to answer, "How good is the test?" The naïve would expect an answer like "80% accurate." Where the naïve might define accuracy as "probability of a positive test given that the sample is known to be positive," which we have defined as sensitivity. Instead, we can first define two extreme answers to the question, "how good is the test?" We can have the answer, "it's a perfect test", and "it's a completely random test".

A perfect test is one with 100% specificity and 100% sensitivity. It would have an ROC curve that looks like Fig. 1.

If the predictions are made at random, the specificity will be equal to the sensitivity for all thresholds, it looks close to Fig. 2.

The full range of possible threshold values and the associated true positive rate and false positive rate can only be expressed by a receiver operating characteristic curve (ROC). However, a summary of the test's receiver operating characteristic curve can be computed by taking the integral under the receiver operating characteristic curve. This quantity is widely known as the area under the receiver operating characteristic curve or area under the curve or AUC.

A perfect test has an AUC = 1.0. A completely random test has an AUC = 0.5. One can now answer the question, "how good is the test?" with a number. The test has an AUC = 0.7. A useful shorthand is to think of the AUC as a grade one might get in a class.

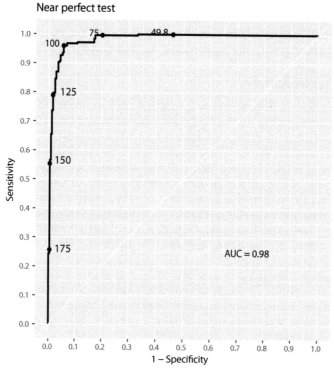

Fig. 1. A near perfect test has an AUC close to 1.0. In this figure, the threshold values are indicated on the curve.

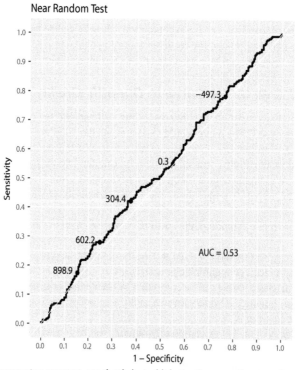

Fig. 2. This ROC curve represents a poor test, one that is just a bit better than guessing at random.

Table 2. AUC's can be interpreted as grades.

AUC	Grade
> 0.9	A
0.80–0.89	B
0.70–0.79	C
0.60–0.69	D
0.50–0.59	F

Near random test that works perfectly on small subpopulation

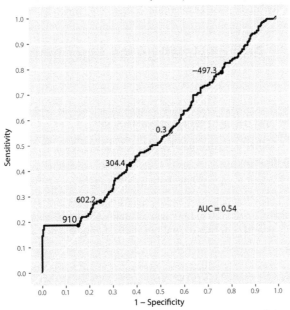

Fig. 3. This test has almost the same AUC as the one in Fig. 2, but it performs very well for a small subpopulation. This subpopulation performance is shown by the steep slope of the curve in the lower left. This figure demonstrates the importance of examining the entire AUC curve, and not just the summary statistic, AUC.

The AUC is only a summary of the test, and a test with a low AUC can be useful in a small subset of cases. Figure 3 shows an example from a BRCA-like test, where the test predicts outcome well for a small fraction of the population.

The AUC is widely used to quantify tests in machine learning, medicine, psychology and many other fields. In some fields, the AUC has other names including c-statistic, concordance statistic, and c-index.

One appealing feature of the ROC/AUC is that it is insensitive to class imbalance. Suppose a test set contains 90% of normal patients and 10% diseased patients. The machine learning task is to classify whether a particular patient is normal or diseased. An algorithm that simply guesses "normal" for all unknown patients will have an accuracy of 90%. The AUC, however, will be 0.50. The AUC is a better measure of the algorithms performance than the accuracy.

In cases of extreme class imbalance, screening for a rare cancer for instance, one often wants to identify the small number of patients most likely to be diagnosed with the rare cancer. In these cases, it is better practice to use a lift chart, which identifies what percentage of the target population (those with the rare cancer) can be identified with the smallest possible group. As an example, suppose the rare cancer occurs at a rate of 1 in 100,000. If we had an algorithm that could narrow identify a subset of the population in which the rare cancer occurs at a higher rate, say 1 in 10,000, that algorithm would have a lift of 10. The lift is computed as the ratio of the rate after prediction to the rate before prediction. An algorithm that provides no information (random) has a lift of 1.

Table 3. Several common medical tests and their published AUC values.

Test	AUC	Reference
PSA to detect prostate cancer	0.68	(Thompson et al. 2005)
Cardiac troponin to diagnose acute myocardial infarction	0.96	(Reichlin et al. 2009)
Cell free DNA test for Down's syndrome	0.999	(Norton et al. 2015)
HbA1c for diagnosing diabetes	0.958	(Tankova et al. 2012)
HEART score to predict major adverse cardiac events in 6 weeks from patients presenting with chest pain	0.86	(Poldervaart et al. 2017)
Circulating tumor cells to diagnose lung cancer	0.6	(Tanaka et al. 2009)

Artificial Intelligence Algorithms

Overfitting

In many cases, the primary objective of a machine learning algorithm is to predict a value or diagnose a condition based upon an input. Biomedical examples abound:

- Can one diagnose whether a patient has lung cancer based upon an immunofluorescent studies of cells captured from the blood?

- Can one diagnose whether a patient has diabetes by measuring glycated hemoglobin levels in the blood?

- Can one predict whether a patient will have a heart attack in the next month based upon a combination of EKG, age, body mass index, tobacco smoking status, family history, and measurements of troponin in the blood?

In the simplest case, one has an example set of results y and input values x. The goal is to identify the best function that will predict the value of y from the input values, x: $y = f(x)$ as shown in Fig. 4. This

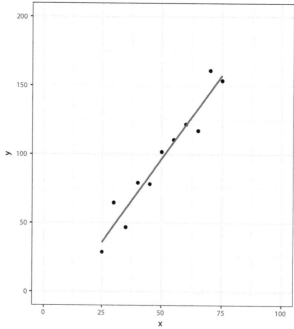

Fig. 4. The line represents the best prediction of y, given x. The points indicate observations of y, and the corresponding x value. The line is represented by two parameters a slope and an intercept.

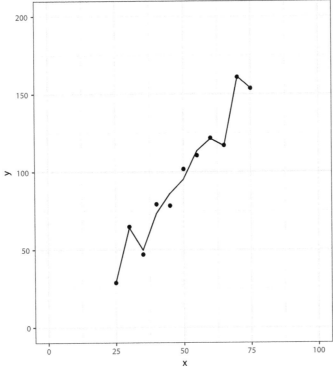

Fig. 5. This black curve represents a clear case of overfitting. The same data as shown in Fig. 4 are presented. In this case, the datapoints are fit with an 8th degree polynomial.

process is referred to as supervised learning. The challenge is to derive this function f without overfitting. An example of overfitting is shown in Fig. 5.

Every real set of data contains both signal and noise. The goal of machine learning is to build a model of the signal, while ignoring the noise. The problem is that the noise is often indistinguishable from the signal.

Many approaches to reducing overfitting exist. The first, and simplest, is to have some fundamental understanding of the relationship between y and x. For instance, if there is good reason to believe that y is linearly related to x, then the set of functions, f, should be limited to those in which x and y are linearly related: $y = \beta x + c$. If the relationship between y and x is complex and not well understood, then methods that are more complex are needed.

The main approaches to reduce overfitting in complex functions are known as regularization and drop out. To understand regularization better, we first need to state the cost function. A typical linear least square fit minimizes the cost function

$$C = \frac{1}{2n}\sum_{i=1}^{n}(y_i - \beta x_i)^2,$$

This cost function is called the mean squared error. It takes the squared difference between the actual value, y_i and the predicted value β x. The problem can be posed as a minimization problem, where the goal is to minimize the cost function by adjusting β.

If multiple input variables, x_j, exist, instead of a single input variable, x, then a set of coefficients, β j is also needed. The cost function now looks like this, when there are a total of p input variables and n independent observations:

$$C = \frac{1}{2n}\sum_{i=1}^{n}\left(y_i - \sum_{j=1}^{p}x_{ij}\beta_j\right)^2,$$

Minimizing this function will often lead to overfitting.

Overfitting can be reduced by adding a new term to the cost function that penalizes functions that are more complex. This addition to the cost function is known as regularization. The idea is that a simpler model is one that uses fewer of the x_j variables. The penalty is implemented by adding a term to the cost function that is proportional to the absolute value of the coefficient:

$$C = \frac{1}{2n} \sum_{i=1}^{n} \left(y_i - \sum_{j=1}^{p} x_{ij} \beta_j \right)^2 + \lambda \sum_{j=1}^{p} | \beta_j |$$

This addition to the cost equation is known as L1 or Lasso Regression. If, instead, the cost function includes a factor proportional to the square of the coefficient:

$$C = \frac{1}{2n} \sum_{i=1}^{n} \left(y_i - \sum_{j=1}^{p} x_{ij} \beta_j \right)^2 + \lambda \sum_{j=1}^{p} \beta_j^2,$$

The process is known as L2 regularization.

In each case, the cost function is increased by the final term, which is proportional to the parameter λ.

Regularization Revolution

Dropout

Dropout is a revolutionary method to prevent overfitting (Srivastava et al. 2014). Dropout has primarily been applied to deep learning algorithms, but similar techniques are also used in other algorithms. The principle underlying the dropout mechanism is that by randomly dropping different connections within a network during the training process, the network becomes more robust and less susceptible to overfitting. The advent of dropout led to an immediate improvement in the performance of most deep learning algorithms.

Machine learning algorithms fall into two categories: supervised and unsupervised. Supervised algorithms require a batch of training data: a set of chest x-ray images from patients with pneumonia, for instance. The supervised algorithm uses the training data to build a model that can be applied to similar data (a chest x-ray) with an unknown diagnosis. An unsupervised algorithm is applied to a set of data to discover sub classifications. For instance, an analysis based on six variables from patients with adult onset diabetes found that these patients could be grouped into five different types. The different types had different risk of complications (Ahlqvist et al. 2018).

Supervised

The goal of supervised algorithms is to predict either a classification or a numerical result, called regression. Classification is ubiquitous in medicine. The patient presents with symptoms, the physician attempts to classify the patient, based upon symptoms into one of several possible diagnoses. Machine learning algorithms usually provide a measure of probability that a set of data belongs to one class or another. Regression is less common, but still useful. Regression can answer questions like, how many days is the patient expected to be in the hospital based on a set of data like age, initial diagnosis, vital signs, height, weight, days since last hospital stay, etc.

Most supervised algorithms are based upon one of two types of algorithms: decision trees and neural networks.

Decision Trees

A basic decision tree is shown in the figure. A patient has seven binary variables: A, B, C, D, E, F and G. Each variable has one of two values, indicated by the corresponding lower or upper case letter. In this particular decision tree, the results are independent of C, D, and E. The patient can be classified as either "cancer" or "normal" based upon the values of A, B, F, and G.

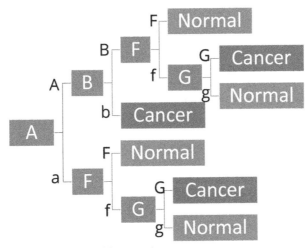

Fig. 6. This decision tree classifies a dataset as either normal or cancer.

This basic algorithm can be extended to cases that are more general. Instead of a definitive diagnosis for "cancer" or "normal" a probability is given. Continuous variables rather than binary variables can be used. The algorithm would find the optimal point at which the continuous variable be split into less than or greater than. Multiple decision trees can be built and an ensemble probability is computed by averaging the result of the many different decision trees. One can vary the number of different trees and the maximum depth of each tree. Specific popular algorithms based upon decision trees are XGBoost (Chen and Guestrin 2016), GBM (Friedman 2001, 2002), and LightGBM (Ke et al. 2017).

Neural Networks

Artificial neural networks are based upon biological neural networks found in the brain. An input layer, one or more hidden layers, and an output layer characterize an artificial neural network. Each layer is composed of multiple artificial neurons, the basic building block of the artificial neural network. An artificial neuron has multiple inputs and a single output. The artificial neuron takes the numbers provided on its input and applies some function to compute an output number. The output number of each neuron in a layer is then fed to neurons in the next layer. This process continues until the output layer.

An artificial neural network might diagnose pneumonia from chest x-rays. First, a network topography is identified. The input layer needs one neuron for each input variable. The output layer needs one neuron for each possible output. The geometry of the hidden layer varies, and identifying the best hidden layer geometry is one of the major challenges of building an effective neural network.

Next, all x-rays images are converted to a uniform image shape, say 1024×768. Each pixel of those images, 786,432 in total, corresponds to one neuron in the input layer. (Thus, this neural network would have an input layer with 786,432 input neurons, one or more hidden layers each with multiple neurons, and two neurons in the output layer.) The two neurons in the output layer correspond to the two states: "patient with pneumonia" and "patient without pneumonia".

The neural network is then "trained". A batch of chest x-rays previously identified as either from patients with pneumonia or from patients without pneumonia are identified. Training consists of applying the values of each pixel in the x-ray to the corresponding input neuron. Then the properties of each neuron in the hidden layer (known as weights) are adjusted to produce the appropriate output, a value of 1 in the appropriate output neuron and a value of 0 in the other neuron.

Training a large neural network can take significant computer time. If one had 10,000 chest x-rays from each class, each with 786,432 pixels, training could easily take weeks on a typical 2018 computer. Various methods are used to speed up training. The most common is to employ a graphical processing unit (GPU) in the computer. Another is specialized hardware: Google developed a Tensor Processing Unit to speed up training neural networks (Jouppi et al. 2017). Sometimes, the data is down sampled to reduce the training time.

Hyperparameters

Training a neural network requires not only determining the parameters (weights of the neurons), but also the hyperparameters. The hyperparameters are the properties of the network that are determined before the determination of the weights of the individual neurons. Hyperparameters include the structure of the network (how many hidden layers, how many neurons in each layer) and information on how the weights are to be determined during the training process (factors known as the learning rate, batch size, and momentum are examples).

Unsupervised

k means Clustering

The most common unsupervised machine learning algorithm is known as *k*-means clustering. The algorithm is rather old; it dates from at least the 1980's (Lloyd, 1982). The goal of this algorithm is to identify subgroups in the dataset. The number of subgroups must be pre-specified, and is known as *k*. With a complex dataset, the algorithm will find a close to ideal partition of the data into *k* different clusters. Usually, subgroups have some important characteristic that makes them useful with a common diagnosis into subgroups. A typical process is to divide a group of patients into sub-groups, where each sub-group has a different survival time for that disease.

One of the critiques of this process is that it will identify clusters even when none exists. Given a dataset with 1000 samples from a single subject, where the only difference between samples is noise from the measurement method, *k*-means will happily identify *k* different clusters in the data.

Hierarchical Clustering

Hierarchical clustering follows a different approach. This algorithm does not require that one specify several clusters beforehand. Instead, it computes a similarity score between different samples of the dataset. Pairs of samples that are most similar are then organized adjacent to one another in a dendrogram, see Fig. 7.

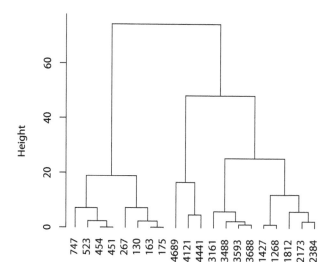

Fig. 7. This dendogram presents a hierarchical clustering of 20 patients. Each patient has a number of clinical variables associated with them. The patients are arranged such that the height of the line connecting them represents how difference in the patient's clinical variables.

Terminology: Machine Learning and Clinical

Supervised learning with hyperparameters requires that one be very careful to not over fit the data. Ideally, one has access to an unlimited dataset, or more data than one can easily handle. Often, however, only a limited set of data is available.

Given a limited dataset, best practices are to split the data into three parts known as the training data, the validation data, and the test data. A typical split is 70% training, 15% validation, and 15% test data.

The training data is used to determine the parameters of the initial model. The initial model is evaluated by applying it to the validation data. The results of the model application to validation data can be used to determine when the training data is being over fit and also to determine the hyperparameters of the model. The test data is only used to perform a final evaluation of the model.

One common pitfall is modifying the model after looking at how it performs on the test data. Any such modification is a form of fitting to the test data. Improvements gained through such modifications inevitably will not translate to the next batch of fresh data. A second pitfall is when the data is preprocessed/sub selected before the test data is extracted. This preprocessing can lead to information derived from the test data leaking into the training process.

The terminology for these sub-datasets: train, validate, test is commonly used by the machine learning field, but it is not universal. Clinical laboratory tests are developed on one dataset, and then validated on an independent dataset. Thus, clinical papers often refer to the final independent dataset as the validation step.

Applications

Applications of machine learning to medical data have grown rapidly in the past few years. We can organize these applications by different types of medical data. Examples of different data structures found in medical data include tabular data, medical images, time series, and natural language.

Tabular data refers to the unstructured data collected for each patient: sex, visit date, diagnosis, height weight. Although medical test results could be considered unstructured and included in tabular data, most medical tests might also vary over time and could also be considered time series: systolic and diastolic blood pressure, temperature, and many different blood tests (electrolytes, glucose, cholesterols) can vary on timescales as short as minutes or hours.

More traditional time series medical data include pulse rate, respiration, and electrocardiogram. The characteristic of time series data is that each data point includes both a time and a value. Different data points in the dataset are related to one another by the time. In contrast, tabular data refers to data in which the order is irrelevant and each data point is unrelated to another. One could switch the order of the data in tabular data with no effect, but not with a time series dataset.

Medical image data can be found in x-rays, ultrasound, magnetic resonance imaging, and computed tomography. Even plain old photography provides valid medical images and is often used by pathologists to record tissue samples.

Natural language is an important source of data, primarily from unstructured physician notes. These notes are still the most common and complete form of documentation in many electronic health records.

Supervised Learning

Deep Learning Image Analysis: Classification of Skin Cancer with a Neural Network

Melanoma, or skin cancer, starts as a discolored patch of skin. At this stage, it is easily removed and once completely removed will not recur at that location. Occasionally cells from this discolored patch of skin can travel, or metastasize, to other locations in the body. Common landing sites are the brain, liver, or lungs. The five-year survival rate of metastatic melanoma is less than 10%, while the survival 5 year survival rate for localized melanoma is over 90% (Jemal et al. 2017).

Diagnosing and removing melanoma in its earliest stages should lead to a significant reduction of metastatic melanoma and deaths due to melanoma. Early stage melanoma superficially appears similar to a mole, which is a general term for a harmless discoloration on the skin. Thus, differentiating early stage melanoma from a harmless mole, which is ubiquitous, is a regular task that dermatologists perform.

A group led by Sebastian Thrun developed a neural network to classify skin lesions as either benign or malignant (Esteva et al. 2017).

They started with a large dataset. Their training/validation dataset consisted of 129,450 images of skin lesions. Each image had been previously classified by a trained dermatologist. Some of the skin lesions depicted in the images had biopsies performed on them, providing an absolute truth to the classification problem.

They used a pre-trained convolutional neural network to build their classifier. Pre-trained neural networks use transfer learning (Quattoni et al. 2008). The disadvantage is that the supplied images must match the geometry of the pre-trained network expects. In this case, they started with a network called the Inception-V3 (Szegedy et al. 2015) network, which uses 299 × 299 pixel images. They pre-trained this network with 1.28 million images that contained 1000 different object categories, these categories are common descriptors and unrelated to skin or melanoma. Pre-training like this can substantially improve performance on a new unrelated dataset.

The pre-trained network was then trained with specific dermatology data. They split the 129,450 images into a train/validate dataset containing 127,463 images and a test dataset containing 1,942 test images. Each of the test images had a biopsy-confirmed classification.

The model, trained on the 127,463 images, was tasked with classifying the 1,942 test images. A subset of the 1,942 test images were also graded by at least 20 experienced dermatologists. The results of the model could then be compared to the dermatologists, as shown in Fig. 8.

In the specific case shown in the figure, for diagnosing melanoma, the machine learning algorithm had an AUC of 0.94.

Recall that the ROC curve is determined by varying the threshold and plotting the resultant sensitivity/specificity that one would achieve with the threshold. Actual dermatologists do not have a specified threshold. Instead, they have an internal, hard to quantify, threshold. By measuring a particular dermatologist's true positive and false positive rate, one can put a single point on the ROC plot. Including several dozen dermatologists, results in an approximate curve on the ROC plot.

The conclusion of the paper is that the trained algorithm performed at level similar to trained dermatologists.

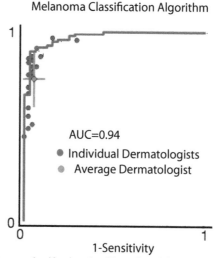

Fig. 8. The ROC curve for the melanoma classification algorithm, adapted from [21].

Deep Learning Image Analysis: Diagnosis of Diabetic Retinopathy from Fundus Photography

A leading cause of blindness in adults is diabetic retinopathy. Retinopathy occurs in untreated type 1 diabetes and can be reduced or delayed with tight control of glucose levels (Nathan 2014). Retinopathy is the degradation of small blood vessels in the retina. Using a special type of photography, called fundus photography, images of these blood vessels can be acquired, see Fig. 9. The degree of retinopathy can be quantified and tracked. At the earliest stages, retinopathy is easily diagnosed with fundus photography while it has no discernable effect on the patient's vision. At later stages, retinopathy can cause blurred vision, blank areas in the vision field and ultimately complete loss of vision.

Ophthalmologists typically grade the degree of retinopathy in fundus photographs. These ophthalmologists will have slightly different grades for identical photographs, known as inter-observer variability. The same grader shown the same fundus photograph at two different times will provide a slightly different grade, known as intra-observer variability (Abràmoff et al. 2010). Inter-observer and intra-observer variability hamper the measuring of small changes in retinopathy in patients. Thus, an ongoing goal is to have a consistent reliable method of grading fundus photography for the degree of retinopathy present.

This problem, the automated detection of micro vessel damage in fundus photographs, is one of ongoing interest to the biomedical imaging community. A challenge, known as the Retinopathy Online Challenge, was organized in 2009 to provide a uniform dataset and grading criteria (Niemeijer et al. 2010). The challenge provided 50 training images with the reference standard and another 50 test images where the reference standard was withheld. Competitors provided "answers" for the 50 test images and were graded based upon those answers. The best competitor algorithms did not reach the level of human experts.

In 2016, a team lead by Google employees developed a predictive model to analyze fundus photographs that does compare to the level of human experts (Gulshan et al. 2016). This work was noteworthy not only for its development of a deep learning algorithm, but also for its accumulation of a very large dataset of graded fundus photographs.

The Google group first built a dataset of 128,175 retinal photographs. Each was graded from 3 to 7 different experts. The experts consisted of a paid panel of 54 ophthalmologists who were licensed in the US or were senior residents. All grading occurred within an 8-month period in 2015.

The model building started with the Inception-v3 architecture neural network (Szegedy et al. 2015). Recall, this architecture requires that all input images are 299 × 299 pixels. This neural network was pre-trained with the ImageNet database, using transfer learning. The first step was to scale the fundus photographs such that the diameter of the dark circle was 299 pixels across. They split the data into 80% for training and 20% for validation (tuning hyperparameters). The training optimized the values of the 22 million parameters in the neural network

They had two separate independent test datasets consisting of just over 10,000 images. Seven or eight trained ophthalmologists graded each of these images for referable diabetic retinopathy. The grading resulted in a binary decision: the patient should be either "referred" or "not referred" for further evaluation. The trained neural network performed comparably to trained ophthalmologists.

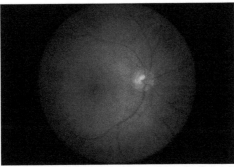

Fig. 9. A fundus photograph captures the microvasculature in the eye. Image by Danny Hope, CC by 2.0. (Image available at https://commons.wikimedia.org/wiki/File:Righ_eye_retina.jpg).

The results for a few other interesting experiments were also presented. One showed that the 128,175 initial retinal photographs in the training set might be too many. The results indicate that they could achieve similar results with half as many images. If they had used only 20,000 images, their results would be about 20% worse. A second experiment measured the effect of changing the number of experts who grade the training images. They found that increasing the average number of grades per image led to increases in the results right up to the limits of the data they had. This result implies that their neural network could be improved with more graders on the initial dataset, where the algorithm learns.

Deep Learning Image Analysis: Cardiovascular Risk from Retinal Images

Fundus photographs offer one of the easiest ways to image the microvasculature system. Many cardiovascular diseases stem from problems that occur with the blood vessels. Fundus photographs may offer a window into cardiovascular disease.

Several cardiovascular risk calculators exist. One of the first was the Framingham risk score (Lloyd-Jones et al. 2004, Pencina et al. 2009). The Framingham risk score provides a number that indicates the likelihood of the patient developing cardiovascular disease within the next ten years. The factors considered are age, sex, total blood cholesterol, history of cigarette smoking, HDL cholesterol levels in blood, and systolic blood pressure. Taken together, these factors can predict cardiovascular disease with an area under the receiver operator characteristic curve AUC of ROC of about 0.8 (Günaydın et al. 2016).

A group of researchers at Google built a cardiac risk predictor that uses only fundus photographs as input (Poplin et al. 2018). They reasoned that since much of cardiovascular disease manifests itself in microscopic blood vessels and fundus photographs offer the clearest, easily obtainable images of the microvasculature, fundus photographs might be able to predict cardiovascular disease.

They first acquired the training data, which they refer to as the development dataset. They used 48,101 patients from the UK Biobank project, which has retinal fundus images along with other health information. They combined the UK Biobank data with an additional 236,234 patients from EyePACS. The UK Biobank health information included sex, ethnicity, BMI, blood pressure, and tobacco smoking status, while the EyePACS patient information included age, sex, ethnicity, and HbA1c blood levels.

They also constructed a test dataset (which they refer to as the validation dataset, consistent with clinical terminology). The test dataset consisted of 12,026 patients from the UK Biobank dataset and 999 from the EyePACS data.

They trained a deep neural network model to predict clinical factors from the fundus images. The model is again based on the Inception-v3 architecture and was pre-trained with objects from the ImageNet dataset. In this case, the initial images were scaled to be 587 pixels square.

Separate models were trained to predict discrete factors (sex and smoking status) and continuous factors (age, BMI, blood pressure, and HbA1c blood test results).

Overall results were good. They found that they could predict age, blood pressure, and BMI from fundus photographs significantly better than just guessing. However, they could not predict glycosylated hemoglobin (HbA1c) blood levels better than guessing. Glycosylated hemoglobin measures long-term average glucose levels in the blood. High levels of glucose are an indication of diabetes, which is a major risk factor for cardiovascular disease.

Finally, they were able to predict major adverse cardiac events in the following five years with an AUC of 0.70. This AUC is comparable to state of the art risk prediction using clinical factors like age, sex, blood pressure, and BMI. However, it does not represent much of an advance in medicine, since those clinical factors are easily measured. While this application of machine learning is useful, it will not change the practice of medicine.

Image Analysis: Human Level Performance is Often the Goal, the Limitation is using Large Medical Image Sized Images

Images of interest to most consumers contain substantial information at large scales, but much less information on small scales. A face, for instance, is recognizable from its overall structure and shape. The information at smaller scales: wrinkles or blemishes for instance, do not contribute much to its recognizability. Thus many of these images can be downsized substantially without losing important information. Medical images, on the other hand, often contain the relevant features, a crack in a bone or a small lump in the breast, at the small scale.

When applying deep learning to medical images, the goal is often to meet the level of performance that the best-trained humans can provide. Several of the applications mentioned above meet that level, but do so with downsized images. Most medical images have resolution in the 1000s by 1000s pixel range, but this many pixels is often too large computationally for machine learning algorithms to handle in reasonable amounts of time. To make the problem tractable, images are down sampled to 299×299 pixels, or 587×587 pixels. One question is how does this down sampling effect the ability of the machine learning algorithm to diagnose as compared to human readers who diagnose using the full resolution of the original medical image.

Image Analysis: Breast Cancer Screening

Screening for breast cancer presents a good opportunity to test the effect of down sampling. First, breast cancer screening often involves multiple images. Each image is an x-ray taken at a different angle of the breast.

Currently, a radiologist will examine these different views and assign the patient into one of seven classes. These classes are defined by the American College of Radiology and known as the Breast Imaging Reporting and Data System (BI-RADS) (American College of Radiology. BI-RADS Committee 2013). BI-RADS has seven levels:

Table 4. The seven levels of BI RADS grading.

0	Incomplete –Need additional imaging evaluation
1	Negative
2	Benign
3	Probably Benign
4	Suspicious
5	Highly suggestive of malignancy
6	Known biopsy-proven malignancy

In a typical screening population about 10–15% of patients receive a score of 0, necessitating a second mammogram (Geras et al. 2017). Of these, less than 1% will ultimately lead to a diagnosis of cancer. Nevertheless, this second mammogram induces anxiety and significant medical costs (Tosteson et al. 2014). Reducing these incomplete mammograms is an important goal of breast screening research.

A group at New York University examined how well a deep learning network could assign mammography images to one of the BI-RADS categories (Geras et al. 2017). They specifically asked the question of how well the machine learning algorithm performed as a function of the image size.

They started by building a dataset of 201,698 breast screening exams from 129,208 different patients. All the patients were from the New York City metropolitan area. This dataset consisted of 886,437 different images. Each exam typically consisted of two images of each breast. One image provides the craniocaudally view, while the second provides the mediolateral oblique view. Each image was taken at a resolution of 2600×2000 pixels. The entire dataset consisted of about four terabytes. All the images had been previously assigned BI-RADS scores of 0, 1, or 2 from a radiologist.

They split the dataset into three parts: training, validation, and test datasets. Their dataset had about 22,000 exams with a BI-RADS score of 0, 75,000 exams with a BI-RADs score of 1 and 67,000 exams with a BI-RADS score of 2. Then they constructed a deep learning network to train a model on classifying the exams into one of the BI-RAD classes.

To evaluate the model, the treated the problem as three separate binary classification problems: BI-RADS 0 vs. BI-RADS 1 and BI-RADS 2, BI-RADS 1 vs. BI-RADS 0 and BI-RADS 2, and BI-RADS 2 vs. BI-RADS 0 and BI-RADS 1. They computed the area under the curve for the receiver operator characteristic curve (AUC) for each of the three problems, and then take the average of the three AUCs to arrive at a macro average AUC (macAUC). This value, macAUC, is the primary performance metric of their model.

The results show that downscaling the datasets images by a factor of 8 (325×250 images rather than the 2600×2000 images) reduces the macAUC from 0.73 to 0.68. This downscaling corresponds to a reduction of the total size of the images by a factor of 64. They measured the effect of decreasing the resolution at several intermediate values and found a monotonic increase in macAUC with the resolution of the images.

They also tested the effect of changing the size of the training data. As expected, the increasing the amount of training data leads to an increase in the macAUC. For instance, when using 100% of the training data, the macAUC was 0.73. If, however, one only used 10% of the training data, the macAUC dropped to 0.68.

One can compare the two effects to see that decreasing the resolution of the images had a larger effect on the model than decreasing the amount of training data. If one is constrained by the total size in bytes, it is better to reduce the number of training cases than to decrease the resolution of the individual images.

Unsupervised Clustering

A common observation in medicine is that patients with identical diagnoses can have vastly different outcomes. Perhaps some respond to a particular therapy, while others do not. Some survive for years, while others die within a short period. This observation suggests that subcategories exist for the given diagnosis.

Type 2 Diabetes

Diabetes is a good example of a diagnosis with heterogeneous outcomes. Diabetes is classified into two forms: type 1 and type 2. The outcome for patients with type 2 diabetes is particularly heterogeneous. Some patients develop severe forms of kidney disease, but have no vision problems. Other patients, who share the same type 2 diabetes diagnosis, develop vision problems, but have no kidney problems.

Identifying subgroups is a branch of machine learning known as unsupervised clustering. The most widely used algorithm for unsupervised clustering is known as k-means (Kanungo et al. 2002). The k means algorithm takes a list of patients, each of whom has several clinical variables related to the diagnosis, and groups them into k subgroups. The grouping is done in a way such that the patients in each group have similar clinical variables. This algorithm is always successful; it will find subgroups. The key question is whether the subgroups are useful: do the subgroups predict outcomes better than any other method?

A group from Sweden and Finland recently applied k-means clustering to type 2 diabetes (Ahlqvist et al. 2018). They suggested that a better classification of type 2 diabetes patients could identify individual patients with increased risk of specific complications and then allow customized treatments to prevent those complications.

The first step was constructing the dataset. The group used data on about 30,000 diabetes patients from five Scandinavian studies that had been running for several years. The largest study, called the ANDIS project ("All New Diabetics In Scania—ANDIS | Swedish National Data Service" n.d.), was responsible for over half the patients. The ANDIS project is a study based upon the National Diabetes Register in Sweden (Gudbjörnsdottir et al. 2003).

The second step was selecting the relevant clinical variables. Since they were using patient samples that had been previously collected, they were constrained to clinical variables that these studies had collected. They selected model variables that should be related to insulin production and demand. They ended up with body mass index, age of onset, calculated estimates of beta cell function and insulin resistance (based on c-peptide concentrations) and the presence or absence of GADA, an antibody indicative of autoimmune conditions.

The clusters were probably not just later stages of diabetes. They found similar clustering in both recently diagnosed patients and in patients who were diagnosed years before.

It is not clear if the clusters represent different forms of diabetes. Understanding how diabetes originates and progresses is of fundamental interest, and a better understanding of this process could lead to both treatments and preventative measures. This clustering does not necessarily point to different causes, but it does indicate different outcomes for each cluster.

Unsupervised Classification of Brain Tumors

Brain tumors are another example of a heterogeneous disease. Brain tumors are classified using histology. Major divisions of brain tumors like gliomas, astrocytomas, medulloblastomas, each have many well defined subtypes (Louis et al. 2016). However, diagnosis within subtypes is variable. For instance, different pathologists will classify gliomas differently. These different sub classifications result in a different course of treatment for the patients. This variability has an effect on both clinical trials, and on applying the results of clinical trials to other patients (van den Bent 2010). An accurate diagnosis of the specific type of brain cancer a patient has should result in better treatment of the patient.

One potential use for classification is then to accurately classify brain tumors. A group led by Stefan Pfister tackled this important problem (Capper et al. 2018). The group started with the WHO classification of central nervous system tumors (Louis et al. 2016). They collected a total of 2,801 samples, with at least eight cases for each group, and then analyzed each patient sample for a genome wide methylation profile (Capper et al. 2018). The methylation profile provides about 450,000 measurements of methylation at different locations across the genome.

Once they collected this data, they performed an unsupervised clustering. They ultimately ended up with 82 different central nervous system tumor classes. Each class was characterized by a distinct DNA methylation profile. Of these 82 classes of tumors, 29 mapped directly to a single entity in the WHO classification scheme (for instance, diffuse midline glioma), another 29 mapped to a subcategory of an entity in the WHO classification (for instance glioblastoma G34). The remaining 24 classes were more complicated: 19 of them could be associated with several WHO classes (a one to one mapping did not exist) and the remaining five were classes that simply had been defined by WHO.

Their unsupervised clustering used the random forest algorithm. This algorithm gave probability estimates that the tumor sample belongs to each class. They first used cross validation to establish the consistency of the algorithm. The cross validation established that the about 95% of the tumors were consistently classified. Most of the inconsistently classified tumors occurred within a small group of closely related classes that have no clinical difference.

Having established their classifier, they applied it to a new set of 1,155 central nervous system tumors. This test set consisted tumors extracted from both adult (71%) and children (29%) patients with 64 different histopathological classifications. This test set was enriched with rare cases and does not represent a typical population. First, about 4% of the samples had to be discarded because they could not obtain a methylation profile from the sample. Of the remaining samples, 88% of the samples matched to one of the classes they established with the training set. Of these, about three quarters agreed with the histopathology evaluation, but one quarter did not. These one quarter underwent a second histopathology examination. In 90% of these mismatches, the revised classification matched with the methylation profile.

The authors demonstrated that they could consistently classify methylation profiles from different brain tumors. This classification system is useful clinically. The authors set up a web site where a pathologist can upload the methylation profile of a tumor sample, and the web site will provide a report on the classification of the tumor ("MolecularNeuropathology" n.d.).

Natural Language: Reading Doctors Notes

Natural language processing extracts computer readable data from free flowing text. A perfectly working natural language processing system would find many uses in biomedicine. For instance, a project dubbed "Literome" attempts to extract molecular interactions that might be involved in complex disease by analyzing the scientific literature and extracting gene networks and genotype-phenotype associations (Poon et al. 2014).

With the conversion of health records to electronic form, significant medical information is present in computer readable free form text. Physician notes and pathology reports are two particular areas that often contain text. Physician notes can contain everything from the patient's complaints, physician's observations, and patient's family history to the comments from the patient's primary caregiver. Physician notes can be inaccurate, incomplete, or poorly worded. These conditions make the natural language processing of physician notes challenging.

Pathology reports are more structured and focused than physician notes. These properties should make them more amenable to natural language processing. However, even these present challenges. One study of 76,000 breast pathology reports illustrates the problem (Buckley et al. 2012). They found 124 different ways of saying "invasive ductal carcinoma", even more troubling they found over 4000 different ways of saying "invasive ductal carcinoma was not present". Even with these challenges, a computerized system had 99.1% sensitivity at a level of 96.5% specificity compared to expert human coders.

One study of 500 colonoscopy reports provides a good test for a natural language processing system. The study, led by Thomas F. Imperiale, tested how well a natural language processing system could extract the highest level of pathology, and then the location of the most advanced lesion and the largest size and number of adenomas removed. This study found that the natural language processing system could identify the highest level of pathology correctly 98% of the time, but only could report the number of adenomas removed with 84% accuracy (Imler et al. 2013).

This 2013 study used the most popular open source natural language processing system, known as cTAKES (Savova et al. 2010). This software package was originally developed at the Mayo Clinic. Its name derives from the phrase "clinical Text Analysis and Knowledge Extraction System". It is now part of the Apache Software Foundation, which generally means that it is under active development and one can expect significant improvements over time. Performing the same test today should give substantially better results than the 2013 test.

Regulatory Issues

As the use of artificial intelligence and machine learning in medicine comes into widespread use, regulation should soon follow. The regulatory agencies are still creating a framework for dealing with these new medical tools. The International Medical Device Regulation Forum (IMDRF) is working on general principles that should be applicable to regulatory agencies around the world. IMDRF has representatives from Australia, Brazil, Canada, China, European Union, Japan, and the United States. In 2017, the IMDRF issued a working paper, "Software as a Medical Device (SaMD): Clinical Evaluation" (International Medical Device Regulation Forum 2017).

The US Food and Drug Agency (FDA) issued draft guidance in late 2017 on their approach for what they call "Clinical and Patient Decision Support Software" (Federal Drug Agency 2017). This guidance clarified what types of machine learning would and would not be subject to regulatory approval. The general principle that they follow is that software that makes a recommendation, but allows a physician to review the recommendation and the basis for the recommendation is exempt from FDA regulation. The FDA commissioner gave an example of software that knows current clinical guidelines and drug labelling that suggests ordering liver function tests first, when a physician attempts to prescribe statins for a patient (Gottlieb 2017).

However, the FDA would continue to regulate software that analyzes medical images, data from in vitro diagnostic tests, or signals from medical instrumentation like EKGs to make treatment recommendations. They view these types of software as an integral part of the medical device, which they already regulate. The FDA's justification is that if the software provides inaccurate information, significant patient damage could occur

Ethical Issues

The use of machine learning in medicine will bring new ethical issues to the forefront. Several ethical issues are predictable, have long been known (Schwartz et al. 1987), and should be addressed early on. Others will be more subtle.

Biases can be carried over from training. A machine learning algorithm is only as good as its training data. Since the majority of people in the United States are white, most medical data in the United States is collected from white people. However, the majority of genetic diversity in the United States comes from African Americans (Campbell and Tishkoff 2008, Gibbs et al. 2015). The dilemma then is how one determines the ethnic distribution of the training set. Should one aim to obtain a training set that benefits all people fairly well, or one that benefits one subgroup over another?

An example that we have worked on illustrates the problem. We have a machine learning algorithm that predicts a future diagnosis of ovarian cancer based on the patient's germ line DNA (Toh and Brody 2018). This algorithm was developed and trained with data from the Cancer Genome Atlas (TCGA) dataset (Bell et al. 2011). To test how the racial makeup of the training dataset effects predictions, we trained the model using a subset of the TCGA data solely from white patients. Then we applied the model to three different test sets each containing members labelled as "white", "Asian", or "black/African American". We found the AUC for predicting a future diagnosis of ovarian cancer to be 0.93 for the white test group, 0.98 for the Asian test group, and 0.70 for the black/African American test group. This general trend, where the AUC for the black/African American group is substantially lower, was also true for the other cancers tested: a form of brain cancer (glioblastoma multiforme), breast cancer, and colon cancer.

A second ethical issue is whether the machine learning algorithm should be designed to do what is best for the individual patient or for the medical system (Char et al. 2018). In our current system, physicians make medical decisions. This physician has an ethical obligation to treat patients to the best of their abilities. Medical systems, on the other hand, have no such obligation. A natural tension exists between physicians and medical systems, where the physician often takes the role of protecting the patient's best interest and the medical system tries to accommodate the physician, but must also deal with economic realities.

These issues might come up in a clinical decision support system. Suppose a private designer develops and markets a software package to review medical histories and recommend the next step in care. Should the private designer consider the effect of recommendations on the medical system's profit and quality metrics? Neither of these factors effect the patient's health, but both would be key selling points to the decision makers in the medical system who will authorize the purchase of such a software package (Char et al. 2018).

Conclusion

Today's machine learning applications in medicine are much more advanced than those from 50 years ago. Machine learning has made substantial progress on medical image analysis and the identification of subclasses of disease. The widespread adoption of electronic health records should lead to more innovations using this source of data. Regulatory agencies are developing new frameworks to ensure patient safety.

References

Abràmoff, M.D., J.M. Reinhardt, S.R. Russell, J.C. Folk, V.B. Mahajan, M. Niemeijer et al. 2010. Automated early detection of diabetic retinopathy. Ophthalmology 117(6): 1147–1154. doi:10.1016/j.ophtha.2010.03.046.

Ahlqvist, E., P. Storm, A. Käräjämäki, M. Martinell, A. Dorkhan, A. Carlsson, P. Vikman et al. 2018. Novel subgroups of adult-onset diabetes and their association with outcomes: a data-driven cluster analysis of six variables. The lancet. Diabetes & Endocrinology 6(5): 361–369. Elsevier. doi:10.1016/S2213-8587(18)30051-2.

All New Diabetics In Scania - ANDIS | Swedish National Data Service. n.d. Retrieved May 26, 2018, from https://snd.gu.se/en/catalogue/study/EXT0057.

American College of Radiology. BI-RADS Committee. 2013. ACR BI-RADS atlas: breast imaging reporting and data system. American College of Radiology.

Bell, D., A. Berchuck, M. Birrer, J. Chien, D.W. Cramer, F. Dao et al. 2011. Integrated genomic analyses of ovarian carcinoma. Nature 474(7353): 609–615. Nature Publishing Group. doi:10.1038/nature10166.

Bent, M.J. van den. 2010. Interobserver variation of the histopathological diagnosis in clinical trials on glioma: A clinician's perspective. Acta Neuropathol. 120(3): 297–304. doi:10.1007/s00401-010-0725-7.

Buckley, J.M., S.B. Coopey, J. Sharko, F. Polubriaginof, B. Drohan, A.K. Belli et al. 2012. The feasibility of using natural language processing to extract clinical information from breast pathology reports. J. Pathol. Inform. 3(1): 23. Medknow Publications and Media Pvt. Ltd. doi:10.4103/2153-3539.97788.

Campbell, M.C. and S.A. Tishkoff. 2008. African genetic diversity: implications for human demographic history, modern human origins, and complex disease mapping. Annu. Rev. Genomics Hum. Genet 9: 403–33. NIH Public Access. doi:10.1146/annurev.genom.9.081307.164258.

Capper, D., D.T.W. Jones, M. Sill, V. Hovestadt, D. Schrimpf, D. Sturm et al. 2018. DNA methylation-based classification of central nervous system tumours. Nature 555(7697): 469–474. Nature Publishing Group. doi:10.1038/nature26000.

Char, D.S., N.H. Shah and D. Magnus. 2018. Implementing Machine Learning in Health Care—Addressing Ethical Challenges. N. Engl. J. Med. 378(11): 981–983. Massachusetts Medical Society. doi:10.1056/NEJMp1714229.

Chen, T. and C. Guestrin. 2016. XGBoost. Proc. 22nd ACM SIGKDD Int. Conf. Knowl. Discov. Data Min. - KDD '16, 785–794. New York, New York, USA: ACM Press. doi:10.1145/2939672.2939785.

Esteva, A., B. Kuprel, R.A. Novoa, J. Ko, S.M. Swetter, H.M. Blau et al. 2017. Dermatologist-level classification of skin cancer with deep neural networks. Nature 542(7639): 115–118. Nature Publishing Group. doi:10.1038/nature21056.

Federal Drug Agency. 2017. Clinical and Patient Decision Support Draft Guidance for Industry and Food.

Friedman, J.H. 2001. Greedy function aproximation: A gradient boosting machine. Ann. Stat. 29(5): 1189–1232. doi:10.1214/aos/1013203451.

Friedman, J.H. and H., J. 2002. Stochastic gradient boosting. Comput. Stat. Data Anal. 38(4): 367–378. Elsevier Science Publishers B. V. doi:10.1016/S0167-9473(01)00065-2.

Geras, K.J., S. Wolfson, Y. Shen, S.G. Kim, L. Moy and K. Cho. 2017. High-Resolution Breast Cancer Screening with Multi-View Deep Convolutional Neural Networks.

Gibbs, R.A., E. Boerwinkle, H. Doddapaneni, Y. Han, V. Korchina, C. Kovar, S. Lee et al. 2015. A global reference for human genetic variation. Nature 526(7571): 68–74. Nature Publishing Group. doi:10.1038/nature15393.

Gorry, G.A. and G.O. Barnett. 1968. Sequential Diagnosis by Computer. JAMA J. Am. Med. Assoc. 205(12): 849. American Medical Association. doi:10.1001/jama.1968.03140380053012.

Gottlieb, S. 2017. On advancing new digital health policies to encourage innovation, bring efficiency and modernization to regulation. FDA Statement.

Gudbjörnsdottir, S., J. Cederholm, P.M. Nilsson, B. Eliasson and Steering Committee of the Swedish National Diabetes Register. 2003. The National Diabetes Register in Sweden: an implementation of the St. Vincent Declaration for Quality Improvement in Diabetes Care. 26(4): 1270–6. American Diabetes Association. doi:10.2337/DIACARE.26.4.1270.

Gulshan, V., L. Peng, M. Coram, M.C. Stumpe, D. Wu, A. Narayanaswamy et al. 2016. Development and Validation of a Deep Learning Algorithm for Detection of Diabetic Retinopathy in Retinal Fundus Photographs. JAMA 316(22): 2402. American Medical Association. doi:10.1001/jama.2016.17216.

Günaydın, Z.Y., A. Karagöz, O. Bektaş, A. Kaya, T. Kırış, G. Erdoğan et al. 2016. Comparison of the Framingham risk and SCORE models in predicting the presence and severity of coronary artery disease considering SYNTAX score. Anatol. J. Cardiol. 16(6): 412–8. Turkish Society of Cardiology. doi:10.5152/AnatolJCardiol.2015.6317.

Halevy, A., P. Norvig and F. Pereira. 2009. The Unreasonable Effectiveness of Data. IEEE Intell. Syst. 24(2): 8–12. doi:10.1109/MIS.2009.36.

Imler, T.D., J. Morea, C. Kahi and T.F. Imperiale. 2013. Natural language processing accurately categorizes findings from colonoscopy and pathology reports. Clin. Gastroenterol. Hepatol. 11(6): 689–694. Elsevier. doi:10.1016/j.cgh.2012.11.035.

International Medical Device Regulation Forum. 2017. Software as a Medical Device (SaMD): Clinical Evaluation.

Jemal, A., E.M. Ward, C.J. Johnson, K.A. Cronin, J. Ma, A.B. Ryerson et al. 2017. Annual Report to the Nation on the Status of Cancer, 1975–2014, Featuring Survival. JNCI J. Natl. Cancer Inst. 109(9). Oxford University Press. doi:10.1093/jnci/djx030.

Jouppi, N.P., A. Borchers, R. Boyle, P. Cantin, C. Chao, C. Clark and et al. 2017. In-Datacenter Performance Analysis of a Tensor Processing Unit. Proc. 44th Annu. Int. Symp. Comput. Archit. - ISCA '17, Vol. 45, 1–12. New York, New York, USA: ACM Press. doi:10.1145/3079856.3080246.

Kanungo, T., D.M. Mount, N.S. Netanyahu, C.D. Piatko, R. Silverman and A.Y. Wu. 2002. An efficient k-means clustering algorithm: analysis and implementation. IEEE Trans. Pattern Anal. Mach. Intell. 24(7): 881–892. doi:10.1109/TPAMI.2002.1017616.

Ke, G., Q. Meng, T. Finley, T. Wang, W. Chen, W. Ma et al. 2017. LightGBM: A Highly Efficient Gradient Boosting Decision Tree.

Lloyd-Jones, D.M., P.W. Wilson, M.G. Larson, A. Beiser, E.P. Leip, D'Agostino et al. 2004. Framingham risk score and prediction of lifetime risk for coronary heart disease. Am. J. Cardiol. 94(1): 20–24. Excerpta Medica. doi:10.1016/J.AMJCARD.2004.03.023.

Lloyd, S. 1982. Least squares quantization in PCM. IEEE Trans. Inf. Theory 28(2): 129–137. doi:10.1109/TIT.1982.1056489.

Louis, D.N., A. Perry, G. Reifenberger, A. von Deimling, D. Figarella-Branger, W.K. Cavenee et al. 2016. The 2016 World Health Organization Classification of Tumors of the Central Nervous System: a summary. Acta Neuropathol. 131(6): 803–820. doi:10.1007/s00401-016-1545-1.

Marr, B. 2017. 28 Best Quotes About Artificial Intelligence. Forbes. Retrieved August 30, 2018, from https://www.forbes.com/sites/bernardmarr/2017/07/25/28-best-quotes-about-artificial-intelligence.

MolecularNeuropathology. (n.d.). Retrieved May 31, 2018, from https://www.molecularneuropathology.org/mnp.

Monroe, D. 2017. Deep learning takes on translation. Commun. ACM 60(6): 12–14. doi:10.1145/3077229.

Nathan, D.M. 2014. The diabetes control and complications trial/epidemiology of diabetes interventions and complications study at 30 years: Overview. Diabetes Care 37(1): 9–16. American Diabetes Association. doi:10.2337/dc13-2112.

National Nurses United. 2014. Insist on a Registered Nurse. Retrieved from https://soundcloud.com/national-nurses-united/radio-ad-algorithms.

Niemeijer, M., B. van Ginneken, M.J. Cree, A. Mizutani, G. Quellec, C.I. Sanchez et al. 2010. Retinopathy online challenge: automatic detection of microaneurysms in digital color fundus photographs. IEEE Trans. Med. Imaging 29(1): 185–195. doi:10.1109/TMI.2009.2033909.

Norton, M.E., B. Jacobsson, G.K. Swamy, L.C. Laurent, A.C. Ranzini, H. Brar et al. 2015. Cell-free DNA Analysis for Noninvasive Examination of Trisomy. N. Engl. J. Med. 372(17): 1589–1597. Massachusetts Medical Society. doi:10.1056/NEJMoa1407349.

Pencina, M.J., R.B. D'Agostino, M.G. Larson, J.M. Massaro and R.S. Vasan. 2009. Predicting the 30-year risk of cardiovascular disease: the framingham heart study. Circulation 119(24): 3078–84. doi:10.1161/CIRCULATIONAHA.108.816694.

Peterson, W., T. Birdsall and W. Fox. 1954. The theory of signal detectability. Trans. IRE Prof. Gr. Inf. Theory 4(4): 171–212. doi:10.1109/TIT.1954.1057460.

Poldervaart, J.M., M. Langedijk, B.E. Backus, I.M.C. Dekker, A.J. Six, P.A. Doevendans et al. 2017. Comparison of the GRACE, HEART and TIMI score to predict major adverse cardiac events in chest pain patients at the emergency department. Int. J. Cardiol. 227: 656–661. Elsevier. doi:10.1016/J.IJCARD.2016.10.080.

Poon, H., C. Quirk, C. DeZiel and D. Heckerman. 2014. Literome: PubMed-scale genomic knowledge base in the cloud. Bioinformatics 30(19): 2840–2842. doi:10.1093/bioinformatics/btu383.

Poplin, R., A.V. Varadarajan, K. Blumer, Y. Liu, M.V. McConnell, G.S. Corrado et al. 2018. Prediction of cardiovascular risk factors from retinal fundus photographs via deep learning. Nat. Biomed. Eng. 2(3): 158–164. Nature Publishing Group. doi:10.1038/s41551-018-0195-0.

Quattoni, A., M. Collins and T. Darrell. 2008. Transfer learning for image classification with sparse prototype representations. 2008 IEEE Conf. Comput. Vis. Pattern Recognit., 1–8. IEEE. doi:10.1109/CVPR.2008.4587637.

Reichlin, T., W. Hochholzer, S. Bassetti, S. Steuer, C. Stelzig, S. Hartwiger et al. 2009. Early Diagnosis of Myocardial Infarction with Sensitive Cardiac Troponin Assays. N. Engl. J. Med. 361(9): 858–867. Massachusetts Medical Society. doi:10.1056/NEJMoa0900428.

Savova, G.K., J.J. Masanz, P.V. Ogren, J. Zheng, S. Sohn, K.C. Kipper-Schuler et al. 2010. Mayo clinical Text Analysis and Knowledge Extraction System (cTAKES): architecture, component evaluation and applications. J. Am. Med. Informatics Assoc. 17(5): 507–513. Oxford University Press. doi:10.1136/jamia.2009.001560.

Schwartz, B. 2017. Tweet. Retrieved August 30, 2018, from https://twitter.com/xaprb/status/930674776317849600.

Schwartz, W.B., R.S. Patil and P. Szolovits. 1987. Artificial Intelligence in Medicine. N. Engl. J. Med. 316(11): 685–688. Massachusetts Medical Society. doi:10.1056/NEJM198703123161109.

Srivastava, N., G. Hinton, A. Krizhevsky, I. Sutskever and R. Salakhutdinov. 2014. Dropout: a simple way to prevent neural networks from overfitting. J. Mach. Learn. Res. 15: 1929–1958.

Szegedy, C., V. Vanhoucke, S. Ioffe, J. Shlens and Z. Wojna. 2015. Rethinking the Inception Architecture for Computer Vision. IEEE Conf. Comput. Vis. Pattern Recognit. 2818–2826.

Tanaka, F., K. Yoneda, N. Kondo, M. Hashimoto, T. Takuwa, S. Matsumoto et al. 2009. Circulating tumor cell as a diagnostic marker in primary lung cancer. Clin. Cancer Res. 15(22): 6980–6. American Association for Cancer Research. doi:10.1158/1078-0432.CCR-09-1095.

Tankova, T., N. Chakarova, L. Dakovska and I. Atanassova. 2012. Assessment of HbA1c as a diagnostic tool in diabetes and prediabetes. Acta Diabetol. 49(5): 371–378. doi:10.1007/s00592-011-0334-5.

Thompson, I.M., D.P. Ankerst, C. Chi, M.S. Lucia, P.J. Goodman, J.J. Crowley et al. 2005. Operating Characteristics of Prostate-Specific Antigen in Men With an Initial PSA Level of 3.0 ng/mL or Lower. JAMA 294(1): 66. American Medical Association. doi:10.1001/jama.294.1.66.

Toh, C. and J.P. Brody. 2018. Analysis of copy number variation from germline DNA can predict individual cancer risk. bioRxiv 303339. Cold Spring Harbor Laboratory. doi:10.1101/303339.

Tosteson, A.N.A., D.G. Fryback, C.S. Hammond, L.G. Hanna, M.R. Grove, M. Brown et al. 2014. Consequences of False-Positive Screening Mammograms. JAMA Intern. Med. 174(6): 954. doi:10.1001/jamainternmed.2014.981.

QUESTIONS

1. What types of medical problems can be tackled by machine learning? What types of problems cannot?

2. When is the area under the curve (AUC) an inappropriate metric for quantifying the performance of a classification algorithm?

3. What types of biases will a machine-learning algorithm contain?

4. When is a machine learning algorithm subject to FDA regulatory approval?

PROBLEMS

Each student will suggest a real life dataset that machine learning could be applied to. The dataset should represent a single medical condition or disease. The dataset is easiest to imagine as a table. Each row represents a different patient. The first column should indicate whether the patient is positive/negative for the medical condition. The other columns are variables specific to the patient. Examples of variables are blood pressure, resting pulse, height, weight, results from any blood test, etc.

15

Quality Management

Lawrence S. Chan

QUOTABLE QUOTES

"Quality is never an accident; it is always the result of high intention, sincere effort, intelligent direction and skillful execution; it represents the wise choice of many alternatives, the cumulative experience of many masters of craftsmanship; and it also marks the quest of an ideal after necessity has been satisfied and usefulness achieved."

Will A. Foster, American Business Executive (FORBES 2017)

"People forget how fast you did a job-but they remember how well you did it."

Howard W. Newton, American Advertising Executive (FORBES 2017)

"It's the quality of the ordinary, the straight, the square, that accounts for the great stability and success of our nation. It's a quality to be proud of."

Gerald R. Ford, American 38th President (WOQ 2017)

"Quality is the degree to which a specific product conforms to a design or specification."

H.L. Gilmore, Author and Product Quality Control Expert (Render et al. 2016)

"In its broadest sense, quality is an attribute of a product or service. The perspective of the person evaluating the product or service influences her judgment of the attribute."

Patrice L. Spath, Healthcare quality expert (Spath 2018)

Learning Objectives

The learning objectives are to familiarize the students with the engineering principle of quality management and the importance of quality in health care, and to guide the students in their group projects of conducting quality control and improvement in real-life medical encounters.

After completing this chapter, the students should:

- Understand the meaning of quality.
- Understand the engineering concept and function of quality control.

University of Illinois College of Medicine, 808 S. Wood Street, R380, Chicago, IL 60612; larrycha@uic.edu

- Understand the importance of quality management in medicine.
- Understand the principle of healthcare quality management.
- Understand the principle and application of Lean method.
- Understand the principle and application of Six Sigma method.
- Understand the principle and application of SMART goal setting method.
- Understand the principle and application of statistical control chart.
- Understand the principle and application of Fishbone diagram method.
- Understand the principle and application of PDSA cycle for continuous improvement.
- Understand the principle and application of Failure Mode and Effects Analysis (FMEA) method.
- Understand the principle and application of Root Cause Analysis (RCA) method.
- Recognize the challenge of implementing healthcare quality management.
- Has the ability of utilizing engineering quality management in real-life medical practice.

What is Quality

There is no universally accepted definition of quality and it is indeed difficult to define and quantify. According to one expert in the area of quality management, quality share three common and essential elements (Spath 2018):

- Quality should meet or exceed customer expectations.
- Quality is dynamic, as today's quality may not meet tomorrow's quality expectation.
- Quality is improvable.

Engineering Principles of Quality Management

Quality control has long been a key character engrained in engineering as applied to industry. In fact, it is unimaginable nowadays that a finished product of engineering does not go through a vigorous process of quality control, which we sometimes tend to take it for granted. If we buy a computer from a tech store, we certainly expect that it will function as its specifications state. When we purchase a 20-megapixel digital camera from a photographic store, we would expect the camera will produce the 20-megapixel quality of photos as manufacturer promises. As we will delineate in the following section, it was not always the case that we have the product quality we enjoy today. The engineering concept of quality control can be defined as "a system for verifying and maintaining a desired level of quality in an existing product or service by careful planning, use of proper equipment, continued inspection, and corrective action as required" and it is now commonly categorized as a part of job description for industrial engineers (QC 2017, BLS 2017). In fact, a specific career track of "Quality Control Engineer" is currently offered with a set of very specific job descriptions (CP 2017, PURDUE 2017).

To understand the significance of quality control, it is proper to trace back to the past to look at how the quality control was developed and to examine the forces behind this development that is so critical for the industry as whole globally and in particularly the effects in the United States. Historically, the concept of quality control has begun in the ancient times but the practice was not standardized or widely applied until recent times, particularly after World War II. Of particular importance, the practice of quality control moved through a strong change period led by the Japanese quality revolution in the last few decades (Juran 1995a, 1995b, 1995c). In the United States, quality management went through a long-period of growing pain (Juran 1995b, 1995c). During the colonial America time, goods were primarily sold in nearby locality, the "manufacturers" of that time, i.e., the craftsmen, had personal stake in ensuring the quality of their products, so as to retain their clients. Thus, they personal inspected the products before selling. The industrial revolution initiated in Europe, did much of the changes in terms of quality control, since mass production necessitated the creation of "inspectors", whose sole function was to ensure the product quality. During the late 19th century, a new form of industrial management, known as Frederick W. Taylor's System of Scientific Management, was adopted in the US, with the goal of increasing productivity without simultaneous increase of skilled workers. This new system, however, resulted in unfortunate reduction of quality and a subsequent response was the creation of inspection department with the task to keep defected products from reaching the market place. In 1926, US saw a new wave of quality control methodology known as "statistical quality control" initiated by the Bell System. This quality control system utilized a new Shewhart control chart and theory of probability to sampling product scientifically. During the World War II, a second wave of statistical quality control re-emerged and the War Production Board was created in part to help dealing with the quality problems. However, the focus was on tool instead of product outcome, thus did not make much of improvement as a result. When World War II ended in 1945, the US manufacturers, who had been devoted much of their efforts to produce war-related products, mobilized to speed up civilian merchant production to meet the market demand. But as a result of putting priority in production volume, quality suffered substantially. Predictably, during the second half of 20th century, US witnessed several emerging forces concerning the quality issues, including the growth of consumerism, increase of quality-based litigation, development of governmental regulation, and oversea-based quality revolution led by Japan. The growth of consumerism and litigation was for most parts a response to several factors relating to misleading advertisement, product failure during usage, inadequate after-sell services, and insufficient redress for consumer complaints. To respond to these forces, there had been establishment of consumer test services, product certification and certification marking/listing services, standards for materials and finishing products, data bank on consumer complaints (like Better Business Bureau), complaint settlements (using mediation or arbitration), consumer organizations, and government intervention. The Federal Trade Commission, which was in charge of the oversight of "unfair or deceptive practice in commerce", had two major tasks to ensure that "The advertising, labeling, and other product information must be clear and unequivocal as to what is meant by seller's representation" and that "The product must comply with the representation" (Juran 1995b). In addition, specific federal statue provided authorities to government agencies to investigate consumer's complaints and product failure, inspect manufacturer's process, test product at any stage, recall sold products, inform product deficiency to consumers, revoke manufacturer's right, and issue cease-and-desist orders (Juran 1995b). One of the major forces for the 20th century quality improvement was the Japanese quality revolution. The loss of the war in Japan indirectly provided an impetus for such movement as Japanese recognized the need to improve their product quality through learning from oversea experts. As a revolution, the Japanese sent their teams abroad, invited quality experts like W. Edward Deming and J.M. Juran to give lectures on statistical methods and quality management, and provided extensive training for their engineers and work force for quality control and improvement (Juran 1995c). As the Japanese products' quality improved, their market share increased in the US, forcing the US companies to deal with the quality crisis of their products and their shrinking market shares. In 1980s, few US companies made strong initiatives and significant efforts for quality improvement and were able to attain world-class quality. The name of "quality control and improvement" was gradually replaced by the term "quality management". In 1990s, the quality management methods of these US model companies were widely learned by other US companies (Juran 1995c). After reviewing the entire history of quality management in the US, J.M. Juran summarized following key elements essential for the success of quality management (Juran 1995c):

- Focusing on customers
- Leadership in upper managers
- Strategic quality planning
- Expanding quality to services and business
- Training in quality management
- Quantifying quality
- Benchmarking
- Empowerment
- Motivating through recognition and reward

Obviously, quality management benefits consumers greatly by providing them with predictably good products. Not only good quality benefits customers, it would also benefit the producers directly. It has been argued by business leaders that product quality could positively impact business in 5 different ways: (1) Building trust with customers; (2) Fueling word-of-mouth & social media recommendation; (3) Reducing customer complaints and product returns; (4) Increasing customers' satisfaction with good aesthetics; and (5) Generating a higher return of investment (BUSINESS 2017). Empirical studies in several industries have demonstrated the importance of quality management practices for the industrial branding, productivity and profitability (Agus et al. 2009, BUSINESS 2017). In addition, a recent study conducted by National Institute of Standards and Technology, a division of U.S. Department of Commerce, tracking a hypothetical stock investment in Malcolm Baldrige National Quality Award winners showed that these quality-winning companies soundly outperformed the Standard & Poor's 500 for the eighth year in a row from 1995–2002 (NIST 2002). Thus, it is clear to all concerned that the results of good engineering quality management benefits not only the consumers, but also the producers. On a long term basis, it is a win-win situation. So why then we cannot apply the principles of quality management in industry to medicine?

Quality Management in Engineering-Medicine

Having stated the importance of quality management in industrial practice, we now turn our attention to the quality management in medicine. If quality management is so essential for the well beings of industries, why it is not so emphasized in medical practice? And if customers expect certain quality from other commercial products, why do they not also demand the similar degree of quality when they encounter medical care services? Are there some substantial differences that set the patient care apart from other commercial production? When we do the thoughtful analysis, we will soon recognize that there are some significant differences between other commercial products and patient care. First, the difference in "materials". In commercial products, the raw materials are essentially identical between one final product and another final product. When it comes to patient care, one patient's basic health condition or predisposition to certain disease, due to genetic or environmental factor, may differ from that of the other, thus may result in the varied medical care outcomes. Second, the difference in timing. In commercial products, the raw materials are utilized to produce the same batch of final products at essentially the same time frame. In patient care, however, the time frame from disease onset to initial physician visit, due to differential factors such as patients' education level, healthcare access, and transportation availability, could be varied substantially between patients. This disparity in time frame of diagnosis will in turn lead to difference in severity of disease, resulting in difference of medical care outcomes. Third, the difference in knowledge. For commercial products, the same knowledge (through the operator or robot) would conduct the manufacturing of multiple products, thereby lead to products of identical or nearly identical quality. On the other hand, physicians' knowledge (which could be a function of training level or diagnostic threshold) about the signals or warning signs of a given disease varied. Therefore, this knowledge variability might then lead to the delay of disease diagnosis and management, resulting in disparity of medical care outcomes. Fourth, the expectation difference. In purchasing commercial products, customers expect, and rightly so, one product's quality should be identical to that of another product of the same kind. On the other hand, patients in general do not expect the clinical outcome of one patient to be identical to that of the other, with the understanding that medicine is not an exact science. Furthermore, direct comparison of one patient's outcome to that of other is not easily obtained. Fifth, the guarantee difference. In most commercial products, the manufacturers commonly offer a sort of warranty, so that the customers can get a refund or an exchange of product if the purchased product is found defective or otherwise not fully functional as stated during the warranty period of time. Again, there is no guarantee for any medical treatment or procedure a patient obtains from a medical facility. Sixth, the standard difference. In commercial production, the setting of standard is universally practiced. Whereas in medical care, such standard is not universally and consistently practiced to the same extent as commercial productions. Even if the standards are set, there is no easy way the patients could determine if the quality of care in a particular physician or healthcare facility matches that of the standard, since a complete transparency is not currently existed in informing public the medical outcomes (Makary 2012). These differences, particularly the last one, could be targeted by quality management. From an ethical perspective, quality management in healthcare is imperative, as some of the health quality statistics are quite alarming (Makary 2012). Furthermore, healthcare quality is recognized as the biggest concern of the patients when choosing a healthcare plan, way above low cost, benefit range, and physician choice (Makary 2012, KAISER 1996).

History of Healthcare Quality Management

Described succinctly by a recent publication, healthcare quality improvement "began with an acknowledgment of the role of quality in healthcare, and gradually evolved to encompass the prioritization of quality improvement and development of systems to monitor, quantify, and incentivize quality improvement in healthcare" (Marjoua and Bozic 2012). Some traced the early healthcare quality control and improvement movement back to 19th century when obstetrician Ignaz Semmelweis who urged the importance of hand washing for medical care, the English nurse Florence Nightingale who identified the link between high mortality rate of army hospital-treated soldiers and poor living conditions, and the surgeon Ernest Codman who pioneered the hospital standards (Marjoua and Bozic 2012). The contemporary healthcare quality improvement movement has started in the late 1960s and early 1970s, when a series of academic publications have begun to informing the deficiencies in healthcare quality along with suggestion of quantifying healthcare quality (Marjoua and Bozic 2012, Dunabedian 1968, 1988, Williamson 1971, Brook and Stevenson 1970, Brook and Appel 1973, Brook 2010). Efforts of various dimensions have been promoted to improve healthcare quality, including healthcare delivery re-engineering, peer-review, competition incentives, good performance reward, poor performance reprimand, quality data collection and publication, and professional education (Marjoua and Bozic 2012, Dunabedian 1968, 1988, Williamson 1971, Brook and Stevenson 1970, Brook and Appel 1973, Brook 2010, McIntyre et al. 2001, Loeb 2004). In 2001, a publication by the Institute of Medicine, now renamed as National Academy of Medicine, has alerted the medical community on the seriousness of quality control and improvement for healthcare and has provided a definition for healthcare quality as "the degree to which health services for individuals and populations increase the likelihood of desired health outcomes and are consistent with current professional knowledge" (Marjoua and Bozic 2012, IOM 2001). In a more recent statement, the National Academy of Medicine has depicted six dimensions of healthcare quality (Spath 2018):

- *Safe*: Do no harm
- *Effective*: Scientifically based to help patients
- *Patient-centered*: Respectful and responsive to patients' preferences, needs and values
- *Timely*: Provide prompt care
- *Efficient*: Eliminate waste
- *Equitable*: Best care to everyone

Ultimately, we also need to have a clear concept that the healthcare quality can be significantly influenced by external factors and that healthcare organizations may not be able to successfully achieve the desirable outcomes due to certain external factors uncontrollable by the healthcare organizations. These external factors include government regulation (like regulatory mandates by federal, state, and local governments), accreditation agency's policies (such as guidelines and requirements set by Joint Commission), and insurance agency's rules (like directives set by Medicare, Blue Cross Blue Shield) (Spath 2018).

Principles of Healthcare Quality Management

Regardless the changes and evolutions of healthcare management, the general principles remain essentially the same and are recognized by the field experts on healthcare management as three essential steps in sequence: measurement, assessment, and improvement (Spath 2018).

Measurement

As one expert in the field point out, "Measurement is the first step that leads to control and eventually to improvement", healthcare quality management starts with a very critical step of measurement (QUOTEMASTER 2018). Measurement, in essence, is to determine the current state of the quality. According to the field expert, measurement is the very key for the quality management and it must fulfill four essential characteristics (Spath 2018):

- *Correct*: Data must be accurate, true, and valid.
- *Beneficial*: Data must be useful for evaluating performance.
- *User Friendly*: Data must be easily interpreted.
- *Consistent*: Data must be collected in identical method throughout.

In healthcare quality management, the data is usually collected in one of the four formats: absolute number, percentage, average, and ratio (Spath 2018). By comparison, percentage form of presentation is preferred over absolute number form, since it will better communicate with the users in terms of comparative performance. In preparation for quality measurement, the organization that conducts the survey should set the target or goal for the performance, as this set target or goal will be needed in the next step of assessment. In addition, four different of categories of measurements are generally used (Spath 2018):

- *Structure*: measuring the physical and organizational environment and resources for the healthcare delivery and services.
- *Process*: measuring the various activities of healthcare delivery and services.
- *Outcome*: measuring the results of the healthcare delivery and services by objective and quantitative means, essentially measuring the effects of structure and process from organization's perspectives.
- *Patient experience*: measuring the subjective feedback of patients regarding the healthcare delivery and services (clinical and administrative), essentially measuring the effects of structure and process from patients' perspectives.

Assessment

Once quality item is measured accurately and consistently and the data are present in a useful and user-friendly manner, the next step is to analyze them and compare with the quality the organization has set as the goal or target (Spath 2018).

Improvement

If the quality measurement meets the goal or target of the quality set by the organization, no immediate improvement action is needed at that point. But the quality management does not stop here even if no immediate improvement action is required. The ultimate objective of quality management is to generate a sustainable quality by continuous monitor the quality. Thus, the step of quality management will cycle back to measurement (Fig. 1, Spath 2018).

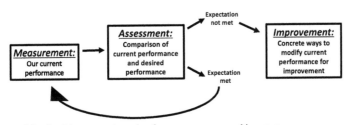

Fig. 1. Essential Steps of Quality Management: cycle of measure, assess, and improve.

On the other hand, If the quality measurement does not meet the goal or target of the quality set by the organization, immediate improvement action is needed at that point (Fig. 1, Spath 2018). In this respect, we are reminded that improvement should not be viewed as a one-time event but should be regarded as a continuous process, as our ultimate objective is to achieve a sustainable healthcare quality through this quality management process (Fig. 1. Spath 2018).

Specific Techniques of Healthcare Quality Management

Having defined the general principles governing healthcare quality management, we will now discuss in the following several technology-based methodologies that can be utilized to achieve better healthcare quality. Many more methods can be used for quality management and the entire scope cannot be covered by this book. The students are encouraged to review more methods in this area from other textbook (Spath 2018).

SMART Goal Setting Method

When organization set its goal or target, it is recommended that they define their target with the following characteristics such that their first alphabets form an acronym of SMART. This target forming method has been referred as SMART goal setting method (Spath 2018):

- **S**pecific
- **M**easurable
- **A**chievable
- **R**ealistic
- **T**imely

Lean Method

Originally developed by Japanese automobile companies, particularly Toyota, the Lean method is a special method of improvement with a narrow aim of eliminating waste and inefficiency. The Lean method steps are all value-based and the end result will be a waste-terminator (Womack and Jones 2003, Spath 2018). The Lean method has also been applied in healthcare for the purpose of improve efficiency and decrease the occurrence of medical errors and healthcare organizations commonly utilize the following 5 principles to reduce waste and increase efficiency (Spath 2018):

- *Value Identification*: determine and focus on what customer treasures.
- *Value streaming*: ensure all activities are necessary and value-adding.
- *Flowing process*: strike for continuous process through value streaming.
- *Demand-based*: use demand to drive production.
- *Perfect production*: prevent defect and rework.

Six Sigma Method

Originated at Motorola and subsequently refined by General Electric, Six Sigma method is a systematic and data-driven improvement technique that is based in part on the statistical process control philosophies initiated by Shewhart and a statistic field known as Process Capability Studies (Spath 2018). The Sigma letter, σ, a Creek alphabet, means variability. In the Six Sigma method, the higher the sigma level, the higher amount of defect-free (or error-free) result. Sigma means standard deviation, as one sigma indicates one standard deviation from the mean and Six Sigma (6 σ) represents six standard deviations from the mean (+/– 6 standard deviations). It is therefore important to indicate that Six Sigma quality means that the quality of a product or service has reached a very high level such that the defect occurrence rate is less than 3.4 per 1 million opportunities (3.4 DPMO). In other word, the product or service in question is 99.99966 percent defect-free. Most healthcare processes operate at or below 3 σ level (or 93.3% error free or worse). By comparison, FedEx is ranked at or above 6 σ level. The calculation of sigma is by a standard deviation formula if the calculation is for the entire population (Eq. 1) or with a representative sample (Eq. 2). Standard deviation is the measure of average distance from the mean value. Most Six Sigma quality improvement processes commonly follow a 5-step methodology with an acronym of DMAIC (Barry et al. 2017, Spath 2018):

- **D**efine the problem
- **M**easure the key process aspects
- **A**nalyze the data
- **I**mprove the system
- **C**ontrol and sustain the improvement

$$S = \sqrt{\Sigma_{i=1}^{N}[(xi - \overline{x}) \wedge 2]/(N)} \qquad \text{(Eq. 1)}$$

$$S = \sqrt{\Sigma_{i=1}^{N}[(xi - \overline{x}) \wedge 2]/(N-1)} \qquad \text{(Eq. 2)}$$

$\overline{x}$ = mean, N = number of samples

Statistical Control Chart

Statistical control chart, also known as statistical process control, is an excellent tool that can be used to evaluating performance (Render et al. 2016, Spath 2018). Statistical control chart contains a line graph, with the value points at its y-axis and the time points at its x-axis. Horizontally paralleled with a central line denotes the mean value are the upper and lower limits of the normal range which also termed control limits (Render et al. 2016, Spath 2018, Fig. 2). Using this control chart, the organization can plot the data collected over a period of time on it and monitor if there is a significant shift or trend of variation. A significant shift in performance has occurred when there are 7 and 8 consecutive points above or below the mean value in a line graph with less than 20 time points and 20 or more time points, respectively (Fig. 2). A performance trend is when there are 7 or 8 consecutive points incline or decline regardless if points are below or above the mean value. If there is a significant shift or trend of performance, the organization could then initiate effort to find the causes and solve the problems.

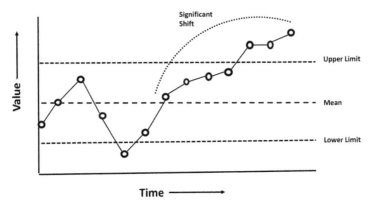

Fig. 2. Statistical Control Chart: an example.

Fishbone diagram method

One of the useful techniques for performance improvement is brainstorming method aiming to determine all possible causes of an effect. This method is named "Cause and Effect Diagrams" and is also termed "Fishbone Diagram" as the diagram utilized many lines that denote the major causes resembling that of a fish backbone (Spath 2018, Fig. 3). Fishbone method is also known as Ishikawa method, recognizing Kaoru Ishikawa as the pioneer user of this method to improve quality (Spath 2018). Fishbone method, of course, can also be used to set objectives. The diagram is intuitively constructed in this manner: The "main fishbone" is a left-to-right pointing arrow to the final effect. This effect could be a problem (for improvement purpose) or an objective (for goal setting purpose). The main categories of potential causes will be indicated as "side bones" pointing to the "main bone" in a tangential fashion toward the right direction. And the potential causes resulted from brainstorming will be placed under each of these main categories of potential causes. For most uses, the starting point of generating categories of potential causes are policies, procedures, people, and plant, or 4 P's (Spath 2018). Figure 3 illustrates a blank diagram of Fishbone method utilized to brainstorm the potential 4 main categories of causes (environment, procedures, equipment, and people) and sub-causes, leading to the effect (on the far right center). Once the Fishbone diagram is constructed, the organization will need to collect data to determine which of these potential causes are the real culprit. Once the true causes are isolated and validated, the organization can then find ways to solve the problem and improve the quality in question.

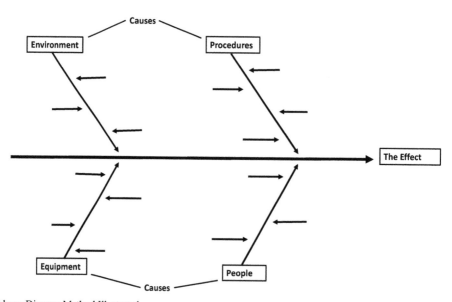

Fig. 3. Fishbone Diagram Method Illustrated.

PDSA Cycle Method

Another performance improvement technical tool is one that was developed by Walter Shewhart and is a technique utilizing statistical process control (Spath 2018), Walter Shewhart named it as the Plan-Do-Check-Act Cycle (abbreviated as PDCA Cycle) (Best and Neuhauser 2006). Subsequently, this cycle method was modified and renamed by W. Edward Deming as Plan-Do-Study-Act Cycle or PDSA Cycle (Best and Neuhauser 2005). PDSA Cycle is generally considered as the most known improvement process today (Spath 2018, Fig. 4). PDSA Cycle emphasizes a key characteristics of continuous improvement and thus the cycle steps cycle perpetually. The essential contents pertaining to the individual step of this PDSA Cycle are summarized below (Spath 2018):

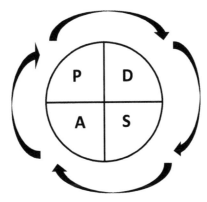

Fig. 4. Plan, Do, Study, Act (PDSA) Cycle Method.

Plan (P):
- Depict the improvement objective.
- Decide on the improvement needs.
- Design process change for the objective.
- Develop plan to implement the change: when, where, who, and what.
- Determine data collection for desirable outcome.

Do (D):
- Execute changes on small scale.
- Examine problem and unexpected results.
- Evaluate collected data on impact of change.

Study (S):
- Determine effectiveness of change by data analysis.
- Determine if outcomes meet expectation.
- Determine lessons learned from change implementation.

Act (A):
- Change successful (fully or partially): larger scale implementation or modification.
- Change unsuccessful: repeat the PDSA Cycle.
- Change success not yet determined: predict outcomes.

Failure Mode and Effects Analysis (FMEA)

This method is a commonly used for proactive risk assessment, primarily utilized for patient safety in healthcare operations (Spath 2018) and it is sometimes termed "Failure Mode Effects and Criticality Analysis" to reflect issue's critical importance (Powell et al. 2014). Since it emphasizes in risk assessment, FMEA technique could be an excellent way to prevent potential medical errors. Risk is a potentially occurring event in just about any activity in our life. The aim of FMEA is to identify these risks, to find ways to prevent them from happening, and to design method to minimize the effects should these risky events occur (Spath 2018). Actually we probably all use this kind of technique for our human activities. This summer, my family planned to go for a special cruise trip in Northern Europe, including a three-day visit to St. Petersburg, Russia, my favor city. So we asked ourselves what could go wrong in terms of our chance for this trip. Well, what could go wrong would be that the cabins could be sold out or that we may not be able to go due to unforeseen issues. So if either of these events does occur, we could face one of these two negative effects respectively: we would be disappointed for cabin sold out (in first event) or for the combination of inability to go and loss of cruise fare (in second event). Thus, we made two preventative moves to counter these two potential failures, at least partially: we purchased the cruise cabins more than 6 months in advance (to prevent cabin sold out disappointment) and the travel insurance (to prevent loss of cruise fare in the event we could not go after purchasing the cruise cabin). Using the same logics, FMEA aims at preventing failure by analyzing potential risks, risk prevention, and risk effect minimization before risky event occurs. The FMEA method is composed of six steps and can be arranged in the PDSA cycle, with the first three steps included in the P (plan) cycle step (Spath 2018):

- Organize information for the process
- Carry out hazard analysis   ----- Plan (P) step
- Determine change for the process
- Conduct pilot test the change for the process   ----- Do (D) step
- Evaluate if the achieved result meet the goal   ----- Study (S) step
- Revise and retest the change of process (if not goal not met),   ----- Act (A) step
  or establish permanent change for the process (if goal met)

Root Cause Analysis (RCA)

This method follows the same logical steps as the FMEA technique, but is not a proactive one. Rather, it analyzes the event after it occured, for the sake of preventing future event of the same or similar one to happen (Spath 2018). Healthcare organizations that are accredited by the Joint Commission have the requirement to conduct a RCA, after a sentinel event occurred in that organization. A sentinel event is defined as a health incident where a patient was seriously harmed or died. Also consisted of 6 steps, RCA is arranged in parallel with the PDSA cycle, with the first 3 steps in the Plan (P) of PDSA cycle step (Spath 2018):

- Understand what happened in the sentinel event
- Determine root cause   ----- Plan (P) step
- Establish risk-reducing future strategies
- Pilot test risk-reducing strategies   ----- Do (D) step
- Evaluate if goal of risk-reducing strategies met   ----- Study (S) step
- Revise and retest risk-reducing strategies (if goal unmet) or   ----- Act (A) step
  permanently implement the risk-reducing strategies (if goal met)

Challenges in Implementing Quality Control in Medicine

Despite many efforts made and few short-term successes obtained in implementing healthcare quality control and improvement in the US, it is generally recognized that a sustainable quality improvement for the complex system of healthcare has not yet been achieved thus far (Marjoua and Bozic 2012). Some have delineated the following challenges this movement currently faces (Marjoua and Bozic 2012, Gonzalo et al. 2017, Nicklin et al. 2017, Reid 2017, Powis et al. 2017, Tappen et al. 2017). Most of these challenges are of non-technical nature. Overcoming these challenges will therefore go a long way to achieve a sustainable improvement of the healthcare quality in the future:

- Physician leadership
- Infrastructural support and resources
- Medical culture for the prioritization of quality improvement
- Integration of healthcare quality improvement into medical education curriculum
- Scientism
- Achievable benchmarking
- Complexity of healthcare
- Ability for aligning quality with accreditation

Summary

Through learning the lessons of quality control and improvement in engineering, we also come to the understanding of the importance of quality control and improvement in healthcare. One important historic lesson we learned from the engineering quality improvement is that whenever the priority is placed solely on production volume, the product quality would suffer a sure decline. It is equally clear from the engineering perspectives that quality improvement provides great benefits not only to consumers, but also to the manufacturers that offer their products in better quality. High quality products boost the branding, attract and retain consumers, promote sells, reduce unnecessary product return and refund, and increase overall return of investment. Having delineated the beneficial aspects of quality control and improvement in engineering, we also realize that the translation of this engineering concept to healthcare has encountered many challenges, as healthcare is a much more complex system with a large group of stakeholders. This chapter offers the general principles and some technical methodologies and depicts some of the major challenges for the potential benefits to healthcare quality improvement. The difficulties of implementing healthcare quality improvement notwithstanding, this task is indeed the right thing to do, as our patients demand it.

References

Agus, A., M.S. Ahmad and J. Muhammad. 2009. An empirical investigation on the impact of quality management on productivity and profitability: Associations and mediating effect. Contemporary Management Res 5: 77–92.

Barry, R., A.C. Smith and C.E. Brubaker. 2017. High-reliability healthcare: Improving patient safety and outcome with Six Sigma. 2nd Ed. Health Administration Press, Chicago, USA.

Best, M. and D. Neuhauser. 2005. W. Edwards Deming: father of quality management, patient, and composer. Quality and Safety in Healthcare 14: 137–145.

Best, M. and D. Neuhauser. 2006. Walter A. Shewhart, 1924, and the Hawthorne Factory. Quality and Safety in Healthcare. 15: 142–143.

[BLS] Bureau of Labor Statistics. 2017. Occupational Employment Statistics: 17-2112 Industrial Engineers. [https://www.bls.gov/oes/current/oes172112.htm] accessed September 21, 2017.

Brook, R.H. and R.L. Stevenson, Jr. 1970. Effectiveness of patient care in an emergency room. N Eng J Med 283: 904–907.

Brook, R.H. and F.A. Appel. 1973. Quality-of-care assessment: choosing a method for peer review. N Engl J Med 288: 1323–1329.

Brook, R.H. 2010. The end of the quality improvement: long live improving value. JAMA 304: 1831–1832.

[BUSINESS] Business.com. 2017. Elevating expectations: 5 ways product quality impacts your brand. [https://www.business.com/articles/5-reasons-why-product-quality-matters/] accessed September 23, 2017.

[CP] Career Planner.com. 2017. "Quality Control Engineer": Job description and jobs. [https://dot-job-descriptions.careerplanner.com/QUALITY-CONTROL-ENGINEER.cfm] accessed September 21, 2017.

Dunabedian, A. 1968. The evaluation of medical care programs. Bull N Y Acad Med 14: 117–124.

Dunabedian, A. 1988. The quality of care. How can it be assessed? JAMA 260: 1743–1748.

[FORBES] Forbes Quotes. 2017. Thoughts on the Business of Life. [https://www.forbes.com/quotes/9698] accessed September 22, 2017.

Gonzalo, J.D., K.J. Caverzagie, R.E. Hawkins, L. Lawson, D.R. Wolpaw and A. Chang. 2017. Concerns and responses for integrating health systems science into medical education. Acad Med Oct. 24. Doi: 10.1097/ACM.0000000000001960.

[IOM] Institute of Medicine. 2001. Crossing the quality chasm: a new health system for the 21st Century Committee on Quality of Health Care in America, ed. National Academy Press. Washington, DC USA.

Juran, J.M. [ed.]. 1995a. A history of managing for quality: The evolution, trends, and future directions of managing for quality. ASQC Quality Press. Milwaukee, USA.

Juran, J.M. 1995b. A history of managing for quality in the United States-Part 1. Quality Digest. [https://www.qualitydigest.com/nov95/html/histmang.html] accessed September 26, 2017.

Juran, J.M. 1995c. A history of managing for quality in the United States-Part 2. Quality Digest. [https://www.qualitydigest.com/dec/juran-2.html] accessed September 26, 2017.

[KAISER] Kaiser family foundation. 1996. New national survey: are patients ready to be health care consumers? October 28, 1996. [http://www.kff.org/insurance/1203-qualrel.cfm] accessed November 6, 2017.

Loeb, J.M. 2004. The current state of performance measurement in health care. Int J Qual Health Care 16: Suppl 1: i5–9.

Makary, M. 2012. Unaccountable: What hospitals won't tell you and how transparency can revolutionize healthcare. Bloomsbury Press, New York, USA.

Marjoua, Y. and K.J. Bozic. 2012. Brief history of quality movement in US healthcare. Curr Rev Musculoskelet Med 5: 265–273.

McIntyre, D., L. Rogers and E. Heier. 2001. Overview, history, and objectives of performance measurement. Health Care Financ Rev 22: 7–21.

Nicklin, W., T. Fortune, P. van Ostenberg, E. O'Conner and N. McCauley. 2017. Leveraging the full value ad impact of accreditation. Int J Qual Health Care 29: 310–312.

[NIST] National Institute of Standards and Technology. 2002. Baldrige award winners Beat the S&P 500 for eighth year. March 7, 2002. [https://www.nist.gov/news-events/news/2002/03/Baldrige-award-winners-beat-sp-500-eighth-year] accessed September 23, 2017.

Powell, E.S., L.M. O'Connor, A.P. Nannicelli, L.T. Barker, R.K. Khare, N.P. Selvert et al. 2014. Failure mode effects and criticality analysis: Innovative risk assessment to identify critical areas for improvement in emergency department sepsis resuscitation. Diagnosis 1: 173–181.

Powis, M., R. Sutradhar, A. Gonzalez, K.A. Enright, N.A. Taback, C.M. Booth et al. 2017. Establishing achievable benchmarks for quality improvement in systemic therapy for early-stage breast cancer. Cancer 123: 3772–3778.

[PURDUE] Purdue Engineering. 2017. Quality Engineering. [https://engineering.purdue.edu/ProEd/programs/masters-degrees/interdisciplinary-engineering/quality-engineering] accessed September 21, 2017.

[QC] Quality Control. 2017. Dictionary.com. [www.dictionary.com/browse/quality-control] accessed September 21, 2017.

[QUOTEMASTER] Quote Master 2018. H. James Harrington. [www.quotemaster.org] accessed December 5, 2018.

Reid, L. 2017. Scientism in medical education and the improvement of medical care: opioids, competencies, and social accountability. Health Care Anal Oct 6. Doi. 10.1007/s10728-017-0351-9.

Render, B., R.M. Stair, Jr., M.E. Hanna, T.S. Hale and T.N. Badri. 2016. Quantitative Analysis for Management. Pearson India Education, Uttar Pradesh, India.

Spath, P.L. 2018. Introduction to healthcare quality management. 3rd Ed. Health Administration Press, Chicago, IL.

Tappen, R.M., D.G. Wolf, Z. Rahemi, G. Enstrom, C. Rojido, J.M. Shutes et al. 2017. Barriers and facilitators in implementing a change initiative in long-term care usig the INTERACT quality improvement program. Health Care Manag (Frederick) 36: 219–230.

Williamson, J.W. 1971. Evaluating quality of patient care. A strategy relating outcomes and process assessment. JAMA 218: 564–569.

Womack, J. and D. Jones. 2003. Lean Thinking: Banish waste and create wealth in your corporation. 2nd Ed. Free Press. New York, USA.

[WOQ] World of Quotes. 2017. [http://www.worldofquotes.com/topic/quality/1/index.html accessed September 20, 2017.

QUESTIONS

1. Why is quality management important in healthcare?

2. What are the differences between general commercial products and healthcare in terms of the characteristics of quality management?

3. What are the factors that can hinder the proper utilization of quality management in healthcare delivery and services?

4. Explain the key difference between the Failure Mode and Effects Analysis (FMEA) and Root Cause Analysis (RCA).

5. What role does the government play in healthcare quality management?

6. What is physician's role in healthcare quality management?

PROBLEM

As a final project of this course, the students will be asked to brainstorm and propose a healthcare quality problem they encountered during their training that can be improved. Students will then be asked to define what is the nature of the problem and the specific reason they select that particular problem. Having decided on the particular problem as their target, the students will present to the class on what specific technical method(s) (SMART goal setting method, Lean method, Six Sigma method, Fishbone diagram method, PDSA cycle method, FMEA, RCA, or statistical control chart) they will utilize to achieve better quality management and also the rationale behind the method selection.

16

Cellular and Molecular Basis of Human Biology

Lawrence S. Chan

QUOTABLE QUOTES

"I could prove God statistically. Take the human body alone—the chances that all the functions of an individual would just happen is a statistical monstrosity."

George Gallup, American Pioneer on Survey Technique and Inventor of Gallup Poll
(BRAINYQUOTE 2018a)

"A human body is a conversation going on, both within the cells and between the cells, and they're telling each other to grow and to die, when you're sick, something's gone wrong with that conversation."

W. Daniel Hillis, American Inventor, Entrepreneur, Scientist, and Writer (BRAINYQUOTE 2018a)

"There's not a pill or an injection that's going to give me, going to give any player the hand-eye coordination to hit a baseball."

Mark McGwire, American Professional Baseball Player (BRAINYQUOTE 2018b)

"The moment I saw the model and heard about the complementing base pairs I realized that it was the key to understanding all the problems in biology we had found intractable—it was the birth of molecular biology."

Sydney Brenner, South African Biologist and Nobel Laureate in Physiology or Medicine 2002
(BRAINYQUOTE 2018c)

"Human Genome Project has given us a genetic parts list."

Leroy Hood, American Biologist and National Medal of Science Awardee (BRAINYQUOTE 2018d)

"The brain is the most complicated organ in the universe. We have learned a lot about other human organs. We know how the heart pumps and how the kidney does what it does. To a certain degree, we have read the letters of the human genome. But the brain has 100 billion neurons. Each one of those has about 10,000 connections."

Francis Collins, American Physician, Geneticist, and Director of US National Institutes of Health
(BRAINYQUOTE 2018d)

University of Illinois College of Medicine, 808 S. Wood Street, R380, Chicago, IL 60612; larrycha@uic.edu

Learning Objectives

The learning objectives of this chapter are to familiarize students the general understanding of human biology, particularly focusing on cellular and molecular levels.
After completing this chapter, the students should:

- Understand the overview of all systems of human biology.
- Understand the major human cell types and their functions.
- Understand the essential make up of human cells and their corresponding functions, including cell membrane, cell organelles, cellular cytokines, enzymes and the antibodies they produce.
- Understand the molecular biological aspects of human biology, including genome, proteome, and their major functions.
- Understand the operation principles of molecular biology technology, including polymerase chain reaction, real-time PCR, and cDNA microarray.

Introduction to Human Biology

Biomedical engineering students and medical students acquire knowledge on human biology. For the biomedical engineering students, this area of knowledge enables them to have a general understanding of the special structure of human body and how human body function in a harmonious way, so as to be better equipped to conduct logical engineering designs. For the medical students, learning human biology will help them appreciate how normal human body functions are, how diseases would alter normal body functions, and how appropriate treatment would restore the normal body functions. The traditional teaching of human body functions has been focused on system, organ, and tissue levels. Nowadays, much of the cutting edge works in biomedical engineering have been at the cellular and molecular levels (BOSTONU 2018). This chapter, therefore, will place more emphasis on the cellular and molecular levels of human biology. Nevertheless, an overview of the human body systems will be first discussed, so the students could have bird's eye view of the entire make up of human biology. The biology teaching arrangement of biomedical engineering programs are likely vary, while some programs integrate biology materials into each biomedical engineering course, other programs may conduct the teaching differently (UCI 2018). Regardless the variation of methods, this chapter serve the purpose of simply providing the needed materials for whatever method a given program chooses to educate their students.

Overview of Human Body Systems

Traditionally, major human body systems are divided into nervous, cardiovascular, musculoskeletal, gastrointestinal, respiratory, immunological, endocrine, urinary, and reproductive. In some of the newer biology textbooks, this traditional system arrangement is no longer held exactly but still quite similar (Reece et al. 2014). In conducting the overview of major human body systems, an emphasis on the connectivity and coordination between these systems is the main goal, so the students would come away with the understanding that all these systems are working together, rather than working independently, to keep the human body in a perfect harmony.

Nervous System

As the central command of the human body, nervous system is perhaps the most complicated and least understood system, although we have learned a lots about it already. As Dr. Francis Collins, the Director of the US National Institutes of Health, pointed out, the complexity of nervous system starts with the facts that there are 100 billion neurons in our brain. Each one of those individual neuron has about 10,000 connections (BRAINYQUOTE 2018d). The nervous system provides electrical signaling that guide the muscle movement, guide our thinking, memory, language, and emotion, sensing temperature, pain, taste, smell, touch, sound, and vision, and help us for the interpretation of our encounter in environmental conditions. Again, the nervous system cannot fulfill all of its functions without help from other systems either. For example, it requires the musculoskeletal system to provide a steady structural frame of support for the central part of system. In addition, the help from cardiovascular system is needed to sustain the nervous system by providing the needed oxygen and nutrient. Without the blood floating to the brain for just a few minutes, brain tissue death from hypoxia will start to occur, leading quickly to the death of the entire brain, and for this very reason CPR was developed to be the first medical response in case of cardiac arrest. For the year of 2016, there were more than 350,000 incidence of outside hospital cardiac arrest and in 46% cases CPR actions were performed by bystanders, resulting in a survival rate of 12% (AHA 2018). The nervous system is commonly divided into the central and peripheral nervous system.

Central Nervous System

This includes the brain and spinal cord. It has in many components, including the brain itself (cerebral cortex, cerebellum, diencephalon), brain stem, and spinal cord. The brain cerebral cortex is further divided into several lobes: frontal, parietal, occipital, and temporal, each of which has special controls of body functions (Reece et al. 2014).

- *Frontal lobe*: Controlling speech, decision making, and skeletal muscle motion.
- *Parietal lobe*: Controlling sensory functions.
- *Occipital lobe*: Controlling vision (image and object recognition).
- *Temporal lobe*: Controlling auditory functions (hearing).

The cerebellum, located posterior and inferior to the occipital lobe, controls the balance and coordinates movements, such as hand-eye coordination.

The diencephalon, located deep in the brain and inferior to the middle cerebral cortex, has several components, each with distinct sensing and endocrine functions:

- *Thalamus*: Controlling the input of sensing from afferent (ascending) nerves.
- *Hypothalamus*: Regulating body temperature, regulating hunger, thirst, sexual behavior via pituitary gland, also initiating "fight-or-flight" response.
- *Pineal gland*: Regulating biorhythm.
- *Pituitary gland*: Regulating reproductive system in organ maturation.

The brain stem functions as to integrate sensing signals, coordinate visual reflex, communicating between the peripheral and central nervous system, and controlling several automatic homeostatic functions: breathing, activities of heart and blood vessels, digestion, vomiting, and swallowing.

Peripheral Nervous System

This included cranial peripheral nerves (of the face and head), ganglia, spinal nerves, and nerve endings. The afferent (ascending) arm of the peripheral nervous system is consisted of senor receptors and afferent nerves that responsible for sending sensory signals to the central nervous system. The efferent arm of the peripheral nervous system is composed of efferent neurons, which in turn divided to motor system that controls skeletal muscles and autonomic nervous system that controls smooth muscles, cardiac muscles, and glands. The autonomous nervous system is further divided into three divisions: sympathetic, parasympathetic, and enteric. While sympathetic division activates "fight-or-flight" kind of response including increase heart rate, dilate pupils, inhibit digestion, convert glycogen to glucose and secrete epinephrine (adrenaline), parasympathetic division does the opposite. The enteric division controls some smooth muscle, pacemaker cells, blood vessels, mucosal glands, and epithelia (Reece et al. 2014). Whereas the human central nervous system cannot be regenerated naturally, peripheral nervous system can go through natural regeneration, albeit slowly, to repair injury (Kyritsis et al. 2014, Bergmeister et al. 2018). Peripheral nerve system also has important roles in many special organs such as visual (eye, vision-related), auditory (hearing-related), olfactory (smell-related), larynx (voice-related), skin (temperature, pain, and touch senses-related), tongue (taste-related) organs. Peripheral nervous system is responsible for sending the sensing signals to the brain, which make a decision how to respond, and the peripheral nervous system will then carry out the response command from the central nervous system. Details of these controls are beyond the scope of this chapter.

Musculoskeletal System

The act of human being can stand on two legs is in reality the true demonstration of the functions of the musculoskeletal system. This system provides a structural framework of human body, together with strength and flexibility, so that human not only can stand, but walk, run, throw, kick, swim, and all other physical activities. Watching a sport event is in fact a living witness of this wonderful human system. This system is composed of bone, joints, muscles, and tendons. Again, the functions of this musculoskeletal system also need the support from other systems, such as the nervous and the cardiovascular systems. The intended movements of human body are directed by the signals from our brain, through peripheral nerves, guide our muscles to perform the intended motions, simple or complex, coarse or smooth, crude or delicate. When this neurological pathway is malfunction, such as in Parkinson's disease, patient develops tremor when they conduct intentional motion, significantly interfering their ability to carry proper social interactions and daily activities (Heusinkveld et al. 2018). Furthermore, cardiovascular system will be required to deliver the needed nutritional materials for generating energy, adenosine triphosphate (ATP), the very power source for muscle movement (Sahlin 2014). The obtaining of the nutritional materials, in turn, depends on the gastrointestinal system we will now discuss. In a later chapter of this book, the subject of utilizing robotic technology and artificial intelligence in helping patients of musculoskeletal system disabilities is discussed (Chapter 24).

Gastrointestinal (GI) System

To obtain proper nutrition to nurture the body is likely the main function of this system, which is consisted of mouth, tongue, teeth, esophagus, stomach, small and large intestines, rectum, pancreas, gall bladder, and liver. When we consume food, our teeth inside of our mouth will reduce the food size and mix them by our tongues with saliva that contain digestive enzymes. The food will then be transport through the constriction movement of esophagus down to the stomach, where acid and enzymes are added and further digestion is performed. The digested food will be continuously moved down to the small intestine and nutrients are absorbed into the body. Here the full function of the GI system requires the coordinative functions of the cardiovascular and endocrine systems. The absorbed nutrients will need the blood circulation to carry them for distribution to other needed body areas. In addition, the GI system will need the proper secretion of insulin from the endocrine system to store the glucose absorbed through the blood stream into the liver. When this endocrine assistance is loss in disease, the proper GI function is inhibited. One such example is type I diabetes mellitus, where the secretion of insulin is insufficient, results in high concentration of glucose in the blood stream, leading to multiple organ diseases (Mancini et al. 2018). Furthermore, the GI function would also need the help from the nervous system, which provides the autonomous nerve to perform alternating constriction and relaxation motions for propelling the food down the track. Not to be forgotten, GI system also participates in waste disposal function by removing waste through colon and out of the body by way of rectum.

Respiratory System

The main function of this system is to provide human body with sufficient oxygen, which is required for cellular respiration, which turns nutrients into energy. This system is comprised of nose, trachea, bronchus, lung. Obviously, the respiratory system would not be able to conduct its oxygen-acquiring functions without the help from another system: the red blood cells that perform the real oxygen capturing activities. The red blood cells are born from the bone marrow, which is a part of musculoskeletal system. Moreover, the respiratory system can never accomplish its function of delivering oxygen to the needy body parts without the coordinative work of the cardiovascular system. The deoxygenated red blood cells, returned from major vein will be actively pumped by the cardiovascular system and travel from right chamber of the heart to the lung where the oxygen capture occurs. The oxygenated red blood cells then travel back from the lung to the left chamber of the heart, where they can be pumped out to different areas of the body, again through the cardiovascular system. The coordination is nearly perfect in order to accomplish its optimal oxygen delivery function. When the coordination between cardiovascular and respiratory system is interrupted in some ways, the proper respiratory functions would be hindered. One such example is a heart condition called "patent foramen ovale", where a tunnel between and left and right upper heart chambers (atria) existed during fetal development, remains open at adulthood. Such condition causes part of oxygenated blood from upper left heart chamber to be pushed back to the upper right heart chamber during heart contraction and allows part of the deoxygenated blood from the upper right heart chamber to leak into the upper left heart chamber, thus reduces the delivery of oxygenated blood to targeted organs. As a result, many undesirable health conditions are observed, including stroke, migraines headache, sleep apnea, and high altitude pulmonary edema (Mojadidi et al. 2015). A similar, but potentially more serious condition, "patent ductus arteriosus", is due to a similar opening between the lower left and right heart chambers (ventricle), also resulting in reduction of oxygenated blood being delivered to the needed body areas (Schneider 2012).

Cardiovascular System

As mentioned above, the cardiovascular system is working parallel and in perfect coordination with the respiratory system for the delivery of oxygen. One cannot do without the other. The system is composed of the heart, arteries, veins, and capillaries. Cardiovascular system also has other important functions as well, as it will deliver nutrients to needed body areas, in addition to oxygen. Moreover, it delivers the immune components, like antibodies, white blood cells, cytokines, to needed body areas quickly and efficiently. Besides the need of respiratory system, this system also needed nervous and endocrine systems to fulfill its functions. For example, thyroid hormone from the endocrine system is instrumental in affecting the heart beat rate and nervous system controls the arterial tone, which in turn affects the blood pressure. Therefore, when endocrine functions are abnormal, the cardiac functions would be inevitably affected. Hyperthyroidism, with excessive thyroid hormone secreted, leads to increase in resting heart rate, blood volume, stroke volume, myocardial contractility, ejection fraction of the heart, and ultimately to high-output heart failure. Hypothyroidism, with insufficient thyroid hormone, on the other hand, results in lower heart rate and weakened myocardial contraction, and could lead to increased risk of coronary heart disease (Vargas-Uricoechea et al. 2014).

Immunological System

The primary function of the immune system is to defend the host, in this case human, from attacks by foreign organisms, be it bacteria, viruses, fungi, or parasites. The many components of this system include white blood cells, antibodies, cytokines, other immune-related chemicals, and several immune organs: thymus, spleen, lymph nodes, and lymphatic vessels. Immunologists commonly categorize the system into innate immune system and adaptive immune system.

Innate System

The innate immune system, which does not require prior "immune training" to function properly, including some cytokines and many types of white blood cells, including neutrophils, eosinophils, basophils, mast cells, natural killer cells, and some epithelial cells. In addition, innate immune system also uses several cell surface recognition systems to recognize "enemies", like the toll-like receptors, C-type lectin receptors, NOD-like receptors, and RIG-I-like receptors (Kawai and Akira 2011). Innate system acts instinctively against invaders, without the need for prior exposure or training. It does not have immune "memory" either, therefore, there is no accentuated immune response to previously exposed "enemies".

Adaptive System

The adaptive immune system, on the other hand, requires a previous exposure to the invader or part of the invader (as in vaccination) as an immune "training" to build a long-term memory of immune defense and it is composed of B-lymphocytes, T-lymphocytes, and antigen presenting cells (APCs), which can be monocytes/macrophages, Langerhans' cells (in the skin), and other dendritic cells. Once trained by exposing to the "invaders", the subsequent exposure to the same "invaders" will induce an enhanced immune response from this system, and this is exactly what the immunologists count on to protect the human hosts through vaccination.

As important as it is, immune system also depends on other system to function fully. For a start, most of the immune system components are generated in the bone, specifically in the bone marrows, which is a part of musculoskeletal system. Also, the immune system depends on the cardiovascular system to transport their immune "soldiers" to the sites of defensive actions. When the cardiovascular system is malfunction in some ways, the immune system's function suffers as well. One good example is illustrated in patients with diabetes mellitus, where the disease causes partial blood vessel blockage, which combines with other factors, leading to reduction of immune function effectiveness (Lipsky et al. 2006). More details on immune system will be described in the Cellular Biology section below.

Endocrine System

Though not easily visualized or recognized by its small physical footprint, the endocrine system has many important regulation functions for the human and its major components include components of the nervous system (pineal gland and pituitary gland), thyroid gland, parathyroid gland, adrenal gland, pancreas (which is also part of GI system), ovary and testes (also part of reproductive system) (Reece et al. 2014).

The overlapping of elements between endocrine and GI systems and between endocrine and reproductive systems underscore the interconnectivity and interdependence of human biological systems in human body. Moreover, the human body functions depend heavily on endocrine system to perform its regulatory functions. Major regulations and their affected systems are (Reece et al. 2014):

- Biorhythm by melatonin of the pineal gland.
- Milk production by prolactin of the pituitary gland—affecting reproductive system by vasopressin of the pituitary gland.
- Reproduction maturation by of follicular-stimulating hormone and luteinizing hormone of the pituitary gland—affecting reproductive system.
- Metabolic regulation by thyroid hormones of the thyroid gland—affecting reproductive and cardiovascular systems.
- Reduce blood calcium level by calcitonin of the thyroid gland—affecting musculoskeletal system.
- Increase blood calcium level by parathyroid hormone of the parathyroid gland—affecting musculoskeletal system.
- Constriction of peripheral blood vessels by epinephrine of adrenal gland—affecting cardiovascular system.
- Raise blood glucose level by glucocorticoids, epinephrine, and norepinephrine of adrenal gland.
- Uptake of Na^+ and excrete K^+ in kidney by mineralocorticoids of adrenal gland—affecting urinary system.
- Reduce and increase blood glucose levels respectively by insulin and glucagon produced by pancreas.
- Promotion of male sexual maturation and sperm production by androgens of testis.
- Promotion of female reproductive organ maturation by estrogens and progesterone of ovary.

When nervous system functions are interrupted, endocrine functions will suffer with it. As an example, traumatic brain injury could lead to pituitary dysfunction, resulting in long-term physical, cognitive, and psychological disability, including hypogonadism (underdevelopment of sexual organs), hypothyroidism, prolactin deficiency, and excessive prolactin (Bondanelli et al. 2004).

Urinary System

Even though it is generally recognized as a system of waste disposal, the urinary system also has important regulatory functions for the human body. The system is consisted of kidney, urinary duct (ureter), urinary bladder, and urinary outlet (urethra). Besides the function of removing metabolites and breakdown products out of the human body, the kidney also performs its osmoregulation by titrating the salt concentration through selectively close or open the "salt gate" or "water gate" to maintain a normal level of salt concentration in the human body. The urinary system is also dependent on the cardiovascular system to deliver the "waste" to it for filtering and disposal. In addition, the endocrine system also influences the urinary system in that mineralocorticoids of the adrenal gland promotes reabsorption of Na^+ and excretion of K^+ (Reece et al. 2014). The proper function of urinary system also requires the intact nervous function. Neurological diseases could lead to urinary dysfunction, such as neurogenic bladder, which could result in renal failure, upper urinary tract dilatation and infections (Amarenco et al. 2017).

Reproductive System

Essential for the propagation of next generation, this system's sole function is to produce important reproductive elements and provide the nurturing environment for the reproduction of offspring. In the male reproductive system, it is primarily composed of sperm factory (testis), sperm delivery ducts (vas deferens and ejaculation duct), prostate gland (provides sperm nutrient), and penis (containing the sperm outlet, urethra). The counter part of female reproductive system is consisted mainly of ovary (egg factory), oviduct (egg transport duct) uterus (fetus' home at pregnancy), vagina, and mammary gland (for baby feeding) (Reece et al. 2014). The reproductive system, like any other system in the human body, also required the collaboration of other systems to function properly. The cardiovascular system is needed to equip the sexual organs to proper function during sexual intercourse and it is also needed to develop a mother-to-child oxygen and nutrient transportation during pregnancy. The development and maturation of the reproductive organs require the endocrine system to provide the proper hormonal stimulation. The very desire of sexual act is also dependent on hormones from the endocrine system. Angiogenesis, the ability of generating new blood vessels, in the uterus, is absolutely essential for the implantation of the fertilized egg and the subsequent development of placenta that sustains the growth of the fetus and it is a function of the cardiovascular system (Reynolds et al. 1992).

Cellular Biology

Cellular biology examines the biological structures and functions at the cell level. Cell is the smallest functional unit of human body. As the basic unit of the body, understanding cell and its functions pave an excellent way for our understanding of the human body as a whole.

Cell Membrane

In order to function as an inclusive unit, cell has to be self-contained, so to speak. To accomplish this role, the human cells have a cell membrane to govern what goes in and what comes out, sort of like a gate-keeper for the cell. Figure 1 illustrates a schematic diagram of a typical cell membrane. Composed primarily with phospholipid bilayers, the cell membrane arranged its hydrophilic heads facing the aqueous environments inside and outside the cell, with hydrophobic tails facing inside of the cell membrane. Since the middle of cell membrane is hydrophobic, water and proteins, both are hydrophilic in nature, cannot pass through and can only through the cell membrane by way of protein channel. The outer surface of cell membrane also has carbohydrate, glycoprotein, and glycolipid embedded, some of them serve as cell receptors. These cell receptors are particularly important for immune cells, as they need the receptors to recognize danger and to activate immune response against enemies, real or perceived. We should see one essential example in the interaction between T cell and antigen presenting cells in dendritic cell section below (Fig. 2). In a later chapter of this book, the students will learn more details on the importance of cell surface receptors. In the chapter of Precision, we will illustrate how investigators are finding ways to modify T cell surface receptors to trigger immune response to kill cancer cells (Chapter 12).

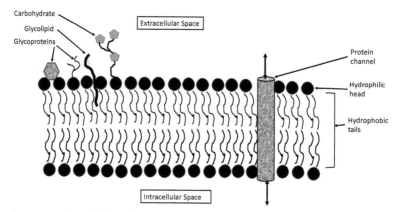

Fig. 1. Schematic representation of cell membrane composed of phospholipid bilayer.

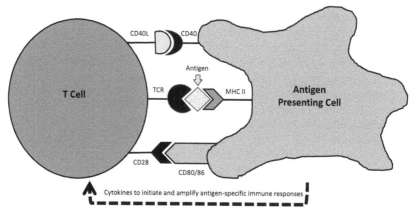

Fig. 2. Schematic diagram depicting interaction between T cell and antigen presenting cell.

Organelles

Within the cell membrane, human cells contain many small machineries that function to maintain the health of the cells, including the following major organelles with specific and essential functions (Hillis et al. 2014, Reece et al. 2014).

Mitochondria

Organelle that are responsible for cellular respiration and cellular energy adenosine triphosphate (ATP) generation, by converting carbohydrate and oxygen into ATP.

Endoplasmic Reticulum

Organelles that participate in membrane synthesis and metabolism, calcium storage, glycogen degradation, and lipid and steroid synthesis.

Lysosome

Organelles that function to digest macromolecules.

Nucleus

Enclosed by a nucleus envelope, it contains the genetic materials DNAs that are responsible for replication or transcribe to mRNA.

Golgi Apparatus

Organelles that synthesize, modify, sort, and secrete cell products.

Ribosomes

These are free small complexes that synthesizes proteins (or translation).

Cytoskeleton

These meshwork of protein filament-formed polymers provides strength and movement for the cells. It supports and maintains the cell shape, holds the organelles in proper positions, and helps anchoring the cells by interacting with extracellular structures.

Stationary Cells

Many cells in the human body are stationary in nature, that means that they stay in certain body location all their lives. Major stationary cells include the followings:

- Epithelial cells (of the skin, mucous membranes, hair follicles, cornea, retina, and lining of esophagus, intestines, and bladders).
- Fibroblasts (cells that produce many extracellular matrices like collagens).
- Melanocytes (pigment-producing cells of skin, mucous membrane, retina, and iris).
- Endothelial cells (of the lining of blood or lymphatic vessels).
- Neurons (of the nerve and brain).
- Muscle cells (cells that constitute the muscle mass).
- Bone-building cells (osteoblasts).
- Bone-breakdown cells (osteoclasts), and many more.

Some stationary cells, when become malignant, however, tend to become mobile and pick up the ability to invade distant organs by a process of metastasis. One example is melanoma cells, the malignant skin pigmented cells, could migrate to invade brain and were able to survive in a neuronal environment by adopting to a neurological phenotype (Redmer 2018).

Mobile Cells

Most of the mobile cells are bone-marrow derived cells, those cells are generated in the bone marrow and their various functions include oxygen delivery (by red blood cells) and immune protection (by white blood cells and through their cellular products). Mobile cells, by definition, are able to move, or to be moved, to different parts of the body during their life time.

Red Blood Cells

These cells, contained oxygen-binding and oxygen-release abilities, function to capture oxygen when they travel by way of blood vessels to lung and release oxygen when they move into tissue level. Red blood cells do not have nucleus and therefore cannot replicate.

White Blood Cells

The white blood cells are immune cells with the central goals to perform immune protection by surveying and destroying invading microorganisms and viruses. There are several different types of white blood cells in human body. In general, they can be categorized as innate immune cells (neutrophils, eosinophils, basophils, mast cells, natural killer cells, monocytes/macrophages) or adaptive immune cells (T- and B-lymphocytes, macrophages, dendritic cells).

Neutrophils. Also known as polymorph, neutrophils has an appearance of multiple lobes of nuclei under microscope lens. It is an early bird of immune reaction particularly in infection defense.

Eosinophils. Identified by their eosinophilic cytoplasm, eosinophils are usually present when there is a parasitic infection. They are also commonly observed in tissues where allergic reactions occurred.

Basophils. With a basophilic cytoplasm, basophils are a small percent of white blood cells that have the ability to release histamines, thus they are observed in allergic reaction sites.

Mast Cells. Clearly identified by its basophilic cytoplasmic granules, mast cells are local residents on the skin. They possess a special antibody receptor on their surface that accept immunoglobulin E (IgE). The binding of IgE to IgE receptor on mast cells will trigger release of histamines and other inflammatory cytokines. The presence of large amount of mast cells indicate an allergic type of reaction.

Monocytes/Macrophages. Monocytes develop into macrophages. These cells are capable of perform phagocytosis (ingestion) and killing of invaders. They also participate in the adaptive immune functions as antigen-presenting cells.

Natural Killer Cells. As a member of lymphocytes, natural killer cells can distinguish cancer cells and virus-infected cells and perform killing on these "bad" cells by lysis.

T-lymphocytes. T-lymphocytes or T-cells, with a cell surface marker of CD3, are responsible for adaptive immune response. Possessed on the surface a T cell receptor (TCR) that can interact with MHC molecules on the surface of APCs, T cells are instrumental for long-term memory of immune defense, such as in the case of vaccine-initiated immune defense. The recent development of utilizing molecularly engineered T-cells for immunotherapy against cancer truly illustrates the important role of T cells in the health affairs of humans (Butler and Hirano 2014, Perica et al. 2014). Two major T cell types are important for human immune defense, the CD4+ helper T cells and the CD8+ cytotoxic T cells. More details on T-cells' role in human medicine are discussed in Precision Chapter (Chapter 12).

B-lymphocytes. Also a major contributor of the adaptive immune response, B-lymphocytes or B-cells are the precursors of makers of antibodies. Through interaction with T-cells, B-cells mature and became committed antibody-producing cells, plasma cells. Antibodies are responsible for many immune defense against viral infection and can also become a destructive force when they take on the role of auto-antibodies against the human host. In a later chapter of this book, students will learn how investigators are utilizing precision medicine method to construct a superior way of making vaccination by harvesting and analyzing human monoclonal antibodies against infections (Chapter 12).

Dendritic Cells. These are the "professional" APCs, responsible for "presenting" part of the antigen they digested and formatted on their cell surface through a MHC class I or II receptor, leading to proliferation and the T cells that would subsequently attack the antigen-barring invaders. Dendritic cells that possess MHC Class II receptor would present antigen to CD4+ T cells (or helper T cells), whereas those possess MHC Class I receptor would only present antigen to CD8+ T cells (or cytotoxic T cells). Figure 2 illustrate schematically the interaction of a CD4+ T-cell with a MHC Class II-barring APC. Dendritic cells first engulf and digest the antigen inside their cells before they present a processed antigen on their cell surface through their surface MHC class II molecule. The T cell interacts through their surface T cell receptor (TCR) with MHC molecule of dendritic cell. This interaction is followed by co-stimulatory molecules by the dendritic cells to T cells, by ways of surface receptors CD28:CD80/CD86 and CD40L:CD40 interaction and cytokine release, leading to T cell proliferation and development into long-term memory immune responses (Fig. 2). The recent medical advancements in utilizing artificially engineered APCs to stimulate T cells for immunotherapeutic and for generation of tailored vaccine purposes underscore the significant role of dendritic cells in human medicine (Butler and Hirano 2014, Perica et al. 2014, Gornati et al. 2018). There are now 6 classified subsets of dendritic cells, each with slightly different functional characteristics (Gornati et al. 2018, Table 1). More discussions of the APCs will be included in the Precision chapter (Chapter 12).

Table 1. Subset of Dendritic Cell (DC) and their characteristic features.

Subset	Surface Markers	Characteristic Features
DC1	CLEC9A+, CD141+	Cross-presentation
DC2	DC1c+, MHC II^{high}, FCGR28/CD32B+	Naïve T cell proliferation stimulators
DC3	DC1c+, CD14+, S100A9+, S100A8+, CD163+, CD36+	Naïve T cell proliferation stimulators
DC4	FCGR3A+/CD16+	Anti-viral responses
DC5	AXL+, SIGLEC6+, SIGLEC1+	Newly defined subset, intermediate between cDCs and pDCs
DC6	IL3RA/CD123+, CLEC4C/DC303+	Classic pDCs

CLEC = C-type lectin-like receptor, a transmembrane protein receptor involving adhesion and other immune functions.
CD = Cluster designation, or cluster differentiation.
FCGR = Fc gamma receptor, capable of binding Fc portion of IgG antibody, cell surface receptors that mediate immune responses like phagocytosis and antibody-dependent cell-mediated cytotoxicity.
AXL = a family member of receptor tyrosine kinase, a mediator of cellular growth and survival SIGLEC = Sialic acid-binding Ig-like lectins, cell surface receptors involving adhesion, phagocytosis, and other immune functions.
cDCs = Conventional dendritic cells that prime naïve T cells and initiate antigen-specific adaptive immune responses.
pDCs = Plasmacytoid dendritic cells that produce type I interferon response during viral infections.
Derived from: Gornati et al. 2018

Cytokines

These hormone-like small, secretory, cellular protein molecules are important in many immunological functions. While some of them are pro-inflammatory in functional nature, others are anti-inflammatory in characters. Since the initial discover of interleukin-1 in the 1970s, more than 60 cytokines have been reported and designed as interleukins, plus many cytokines that are not classified as interleukins (Akdis et al. 2016). Cytokines are present in nearly all cells, whether they are primarily immune cells or non-immune cells. Table 2 lists major cytokines and their essential functions.

Enzymes

These small cellular chemicals are responsible for either building up or breaking down cellular components, like DNA, RNA, or proteins. Enzymes catalyze cellular and extracellular reactions either to construct a molecule or to degrade a molecule and they make the chemical reactions go quicker than without the enzymes. They have important role of maintain a balance between providing sufficient components and eliminating excessive components in human body functions in a homeostatic way. Enzymes are generally named according to their functions. For example, the function of RNA polymerase is to synthesize RNA polymer (or RNA strand). DNA ligase functions to joint two pieces of DNA together. The six major groups of human enzymes classified by an international enzyme committee and their functional mechanisms are depicted in Table 3.

Antibodies

These mobile proteins are the products of committed B-cells, plasma cells. In human, 5 classes of antibodies are produced: IgM, IgG, IgA, IgE, and IgD. These classes of antibodies are immunologically distinguished by the difference of their heavy chain sequences. As for light chain, only two classes are produced, lambda and kappa chains. In human, IgG has 4 subclasses, IgG1, IgG2, IgG3, and IgG4; whereas IgA has 2 subclasses, IgA1 and IgA2. IgD is a B cell surface receptor and is not usually released into the extracellular environment. Whereas IgM is the initially released antibody response, IgG comes later in the immune response but become the dominant class, constituting 80% of circulating antibodies. IgA is primarily responsible for mucosal immune defense. IgE involves in allergic reaction by binding on IgE receptors of mast cells, triggering mast cells to release histamines.

Table 2. Major human Cytokines and their known functions.

Cytokine Family	Name of Cytokine	Major Characteristics and Functions of Cytokine
IL-1 Family (11 members)	IL-1 alpha	Pro-inflammatory, Th17 cell differentiation, hematopoiesis
	IL-1 beta	Same as IL-1 alpha
	IL-18	Pro-inflammatory, IFN-gamma induction, promote Th1 and Th2 cells responses
	IL-33	Pro-inflammatory, induce mucosal Th2 response, repress transcription, DC maturation, ILCs induction
	IL-36 alpha	Pro-inflammatory, early response to injury/infection
	IL-36 beta	Same as IL-36 alpha
	IL-36 gamma	Same as IL-36 alpha
	IL-1 RA	Anti-inflammatory, antagonist to IL-1
	IL-36 RA	Anti-inflammatory, antagonist to IL-36
	IL-37	Anti-inflammatory, inhibit IL-18 activity
	IL-38	Anti-inflammatory, inhibit Th17 cytokine production, antagonist to IL-36
Common Gamma Chain Family	IL-2	Growth and proliferation factors for T & B cells, Treg development, stimulate antibody production, NKs proliferation and differentiation, ILCs cytokine and proliferation stimulation
	IL-4	Induce Th2 cell differentiation; upregulate B-cell MHC class II, IL-4R, and CD23; IgE class switching; survival factor for T and B cells
	IL-7	Pre-B cell and thymocyte proliferation, naïve T cell survival, ILCs development and maintenance, monocyte inflammatory mediator synthesis induction
	IL-9	T and mast cell growth factor, Th1 cytokine inhibition, CD8+ T and mast cell proliferation, IgE and chemokine production promotion
	IL-15	T cell activation and proliferation, NK cell activation, gamma-delta T cell differentiation, CD8+ memory, Th2 cell differentiation enhancement, anti-apoptosis of neutrophils and eosinophils
	IL-21	B cell proliferation, differentiation, and survival; T cell growth factor; NKT cell proliferation
IL-10 Family	IL-10	Immunosuppressive to APCs and T cell subsets, suppressive IgE, induction IgG

Table 2 contd. ...

...Table 2 contd.

Cytokine Family	Name of Cytokine	Major Characteristics and Functions of Cytokine
	IL-19	Induction of Th2 cytokines, enhancing IL-6, IL-10, and TNF-alpha in monocytes
	IL-20	Skin biology
	IL-22	Pathogen defense, wound healing, tissue reorganization
	IL-24	Suppression of tumor
	IL-26	Epithelial cell activation and regulation
	IL-28	Th2 downregulation, Th1 upregulation, tolerogenic DCs induction, Treg promotion and expansion
	IL-29	Same as IL-28
IL-12 Family	IL-12	Th1 cell development and maintenance, NK cell activation, DC maturation, cytotoxicity induction
	IL-23	T cell proliferation enhancing, memory T cell promotion, NK cells activation, antibody production regulation, IL-17 production stimulation
	IL-27	Th1 cell differentiation promotion, Th17 cell response inhibition, T-bet induction
	IL-35	T cell proliferation reduction, increase IL-10 production and Treg proliferation
Type 2 Immune Response Cytokines Family	IL-4	See IL-4 functions above
	IL-5	Myeloid cell differentiation, enhance eosinophil chemotactic activity and adhesion, wound healing
	IL-9	T cell and mast cell growth factor, Th1 cytokine inhibition, CD8+ cells and mast cells proliferation, IgE and chemokine production stimulation
	IL-13	IgG4 and IgE switching promotion; B cell MHC II and CD23 upregulation; monocyte CD11b, CD11c, CD18, CD23, CD29, and MHC II induction, eosinophil and mast cell activation; eosinophil recruitment and survival; parasitic infection defense
	IL-25 (IL-17E)	Th2 response induction; Th1 and Th17 responses inhibition; production of IgE, IgG1, IL-4, IL-5, IL-9 and IL-13 induction
	IL-31	Eosinophil IL-6, IL-8, and chemokines induction, epithelial cell chemokine and growth factor induction, epithelial cell proliferation and apoptosis inhibition
	IL-33	Repressing transcription activity, Th2 inflammation induction, bone marrow-derived DCs maturation factor, eosinophil and basophil integrin expression

Table 2 contd. ...

...Table 2 contd.

Cytokine Family	Name of Cytokine	Major Characteristics and Functions of Cytokine
		and inflammatory cytokine release enhancer, ILCs inducer
	TSLP	Promoting Th2 immune responses by actions on DCs, monocytes, CD4+ T cells, mast cells, and B cells; respond to viral, bacterial, and parasitic pathogens
Chemokine-cytokines Family	IL-8	Member of CXC chemokine family; Hematopoietic stem cell mobilization; neutrophil, T cell, NK cell, eosinophil, and basophil chemoattractant; angiogenesis
	IL-16	T cell response modulation, CD4+ T cell, CD8+ T cell, monocyte, mast cell, and eosinophil chemoattractant
IL-17 Family	IL-17A	Activate, recruit and induce pro-inflammatory cytokines, chemokines, and metalloproteinases of neutrophils
	IL-17B	Induction of anti-microbial peptides and pro-inflammatory cytokines, chemokines, and metalloproteinases; chondrogenesis, osteogenesis
	IL-17C	Induction of anti-microbial peptides and pro-inflammatory cytokines, chemokines, and metalloproteinases; intestinal barrier functions
	IL-17D	Induction of anti-microbial peptides and pro-inflammatory cytokines, chemokines, and metalloproteinases; myeloid progenitor cell proliferation suppression
	IL-17F	Induce pro-inflammatory cytokines, chemokines, and metalloproteinases; activate and recruit neutrophils
Other interleukins	IL-3	Hematopoietic growth, DCs and Langerhans cells differentiation, enhance antigen uptake and phagocytosis, basophil and eosinophil activation
	IL-6	Induce acute-phase proteins, leukocyte trafficking and activation, T-cell & B-cell differentiation, IgM, IgG, and IgA production, hematopoiesis, neoangiogenesis, bone resorption, cartilage degradation, fibroblast proliferation, induction of adrenocorticotropic hormone synthesis
	IL-11	Growth factors for myeloid, erythroid, and megakaryocyte progenitor cells, and plamacytoma cells, acute phase protein induction, monocyte and macrophage activity inhibition, neuronal

Table 2 contd. ...

...Table 2 contd.

Cytokine Family	Name of Cytokine	Major Characteristics and Functions of Cytokine
		development promotion, osteoblast inhibition, osteoclast stimulation
	IL-14	Activated B cell proliferation
	IL-32	Induction of epithelial cell apoptosis, IL-6, IL-8, and TNF-alpha
	IL-34	Myeloid lineage differentiation, proliferation, and survival regulator; microglial proliferation
Interferon Family	IFN-alpha	Orchestrate adaptive immune defense against viral infection, dendritic cell's antigen presentation enhancement, macrophage's antibody-dependent cytotoxicity stimulation, naïve T cell activation, tumor cell and virus-infected cell apoptosis triggering
	IFN-beta	Same as IFN-alpha
	IFN-gamma	Expressed by Th1 cells, activate microbe-killing macrophages, promote Th1 differentiation and cytotoxicity, skin and mucosal epithelial cell apoptosis induction, MHC I and II upregulation, cell growth inhibition.
TGF Family	TGF-beta	Coordinate cardiac and bone development, immune cell precursors growth reduction, Th cell subset differentiation regulator, Treg and immune tolerance induction
TNF Family	TNF-alpha	Double roles in pro-inflammatory (initiation of strong inflammatory responses) and anti-inflammatory (limiting extent and duration of inflammation); host defense; inhibition of autoimmunity, tumor generation, and epithelial cell apoptosis

APC = antigen presenting cell
CXC chemokine = chemoattractant with CXC motiff
DC = dendritic cell IL = interleukin
ILC = innate lymphoid cell
FN = interferon
MHC = major histocompatibility complex
NK cell = natural killer cell
NKT cell = natural killer T cell
TGF = transforming growth factor
Th1 = helper T cell-type 1
Th2 = helper T cell-type 2
Th17 = helper T cell-type 17
TNF = tumor necrosis factor
T-bet = T-box transcription factor (T cell associated)
Treg = regulatory T cell TSLP = thymic stromal lymphopoietin
Derived from: Akdis et al. 2016

Table 3. International Classification of Enzymes (Six Major Categories) by Enzyme Commission (EC) of the Nomenclature Committee of the International Union of Biochemistry and Molecular Biology (NC-IUBMB) – First Level Classification.

EC Group	Class	Equation of Chemical Reaction	Enzyme Examples
EC1	Oxidoreductase	$X + Y: \rightleftharpoons X: + Y$ (electron give-reductase; and electron take-oxidase)	Alcohol dehydrogenase, Lactose dehydrogenase
EC2	Transferase	$A + BX \rightarrow AX + B$ (transfer chemical group to other)	Peptidyl transferase, Hexokinase
EC3	Hydrolase	$A + H_2O \rightarrow B + C$ (breaking bond with H_2O, hydrolysis)	Sucrase, Serine hydrolase
EC4	Lyase	$A \rightarrow B + C$ (breaking bond without hydrolysis or oxidation)	Histidine decarboxylase, Argininosuccinate lyase
EC5	Isomerase	$A \rightarrow A'$ (convert to its isomer)	Phosphoglucose isomerase, Fumarase
EC6	Ligase	$X + Y \rightarrow XY$ (join two compounds together)	DNA ligase, Pyruvate carboxylase

Derived from: Cuesta, S.M., S.S. Rahman, N. Furnham and J.M. Thomton. 2015. The classification and evolution of enzyme function. Biophys J 109: 1082–1086.

Molecular Biology

Molecular biology studies the human biology at the molecular level. Commonly molecular biology focuses on the study of human genome structure and function. A brief description of the human genome functions is discussed below.

Human Genome

Perhaps the most important element of molecular biology is the biology of human genome, which defines the entire genetic makeup of the human body. Human genome takes on such critical role because it directs all other activities of human body. Our understanding of human genome has taken a fast turn when the entire human genome was successfully cloned and reported in 2001 (Lander et al. 2001). The technology advancement of genome sequence nowadays has the ability to sequence the entire genome of a person in just about one week with a price tag of just $1,000 dollars.

The human genome is consisted of 20,000 different genes and nucleic acids composed of 3 billion base pairs (Green et al. 2015). For the basic components of nucleic acid, the nucleotides, there are only 4 distinct nitrogenous bases: adenosine (short for A), thymine (T), guanine (G), and cytosine (C) and each nucleotide also contains a phosphate group and a sugar deoxyribose. One interesting and important fact is that virtually all differentiated cells have the identical and entire genome. Yet some proteins are produced by certain cell types and not by other cell types. The key factor is transcriptional factors, which function to bind to promoter region of DNA in initiating mRNA transcription. The binding of transcriptional factors enables the binding of RNA polymerase to DNA for the transcription. Activation and inactivation of certain transcriptional factors during human development determine the ability and inability to express certain proteins, and by extension their phenotype expressions, respectively (Hillis et al. 2014). Another interesting and also essential fact is that reactivation of cell-specific transcriptional factors could change a cell's phenotype. For example, transgenic introduction of neuron-specific transcription factors into fibroblasts turn these connective tissue protein-producing cells into functional neurons with characteristic neuronal synapses (Hillis et al. 2014).

Human DNA could undergo three major biological activities: replication, transcription and translation.

Replication

Replication takes place when cells are divided and one copy of DNA will become two identical copies, just before the actual cell division occurs (Hillis et al. 2014). The human cell reproduction or division is under precisely controlled and occurred infrequently. Growth factors have the role of stimulate cells for division when it is needed for the entire organism. The enzymes involved in the DNA replication are: helicase, primase, DNA polymerase, and ligase. The basic steps of DNA replication are (Hillis et al. 2014):

- A large protein complex (pre-replication complex) binds specific site (called *ori*, or origin of replication) on DNA molecule.
- The binding on *ori* is followed by unwinding of the double-stranded DNA through the action of an enzyme helicase, leads to formation of replication fork.
- An enzyme primase binds to DNA and synthesizes a short RNA fragment (a primer) complementary to the DNA, initiating the replication process. The primase then releases from the DNA.
- A large enzyme complex, DNA polymerases bind to the site of primers of the two single-stranded DNA, start to synthesize the new DNA strands by adding one nucleotide at a time in a 5' to 3' direction, such that these two new DNA strands are synthesized in the opposite directions. DNA polymerases add deoxyribonucleoside triphosphates (dATP, dTTP, dGTP, and dCTP) to the new strand to "match" the DNA templates in a "A-T" and "G-C" fashion.
- A different DNA polymerase hydrolyses the RNA primer and replaces with DNA.
- A ligase fills the gap in the new strand between short segments.

Transcription and Translation

Transcription and translation are two closely occurring sequence of events, for the purpose of protein synthesis. Whereas DNA transcription is the activity that a gene is transcribed into a message that can be used to make protein, i.e., synthesizing a messenger RNA (mRNA) by utilizing a DNA template, translation is the process of synthesizing protein using a mRNA template. While the transcription process occurs in the nucleus region of a cell, translation takes place in the ribosome region of a cell (Reece et al. 2014).

Transcription. This process can be viewed as three steps: initiation, elongation, and termination (Hillis et al. 2014).

- *Initiation*: RNA polymerase binds to the promoter of DNA, initiating the unwinding of the double-strand.
- *Elongation*: The bound RNA polymerase acts as a primer, to initiate the transcription on the template strand of DNA, which is complementary to the coding strand of DNA. RNA polymerase adds ribonucleoside triphosphates (ATP, UTP, CTP, GTP) to build the RNA strand along the template DNA from 3' to 5' of template direction).
- *Termination*: When RNA polymerase reaches the terminating site on the DNA template, the newly synthesized RNA is released. This RNA strand thus have same base sequence as the coding strand DNA, except TTP is replaced by UTP in the RNA.

Translation. Like transcription, this process also takes place in three steps: initiation, elongation, and termination (Hillis et al. 2014). This process takes place in ribosome, only after the pre-mRNA has been processed and introns (non-coding parts of pre-mRNA) are removed.

- *Initiation*: A small ribosomal subunit binds to the initiation site of mRNA, usually in the area surrounding initiation/start codon AUG. This is followed by the binding of anticodon (UAC) of methionine-charged tRNA (transfer RNA) to the start codon AUG. Then a large ribosomal subunit comes to jointly form a completed initiation complex.
- *Elongation*: tRNAs that are charged with amino acids will add one amino acid at a time to the growing peptide chain. Again, the binding is through the anticodon of the charged tRNA to the codon of the mRNA. Once the amino acid-charged tRNA binds next to the methionine-charged tRNA, methionine forms a bond with the newly incorporated amino acid and released from its tRNA and the uncharged tRNA is released. The cycle will repeat to add another new amino acid.

- *Termination*: The elongation process finally stops when it arrives at a termination codon, which may be UAA, UAG, or UGA. Within the initiation complex, the stop condon binds a protein release factor, leading to hydrolysis of bond between polypeptide chain and tRNA, separating the polypeptide from the ribosome.

More details on how delineating human genome and proteome can facilitate the medical diagnosis and treatment are discussed in Precision chapter (Chapter 12).

Proteome

Although molecular biology focuses on examining human genome structure and function, one cannot neglect the studies of products of human genome, proteome. Proteome defines the entire protein makeup of the human body. Whereas the human genome is the central command of the human body, proteins are actually the foot soldiers that carry out the real works, i.e., the biological structures and functions. Most human genes have to be first translated into functional proteins, which then exert the real biological effects on their targets. We now know, however, certain RNA can exert influence in biological functions without translational effort. For example, interfering RNA (iRNA) and micro RNA (miRNA) by themselves have inhibitory functions to protein synthesis (Wilson and Doudna 2013). From a gene to a protein, the transcriptional and translational processes require many critical steps and error can potentially occur in each of those steps, leading to a "loss in translation" so to speak, finally resulting in a non-functional protein. These potential transcriptional and translational errors and roadblocks are depicted below:

- *RNA interference*: iRNAs and miRNAs are double-stranded RNA molecules that regulate gene expression during development and in response to stress situation such as viral infection. These RNAs, with one strand preferentially selected to use in guiding sequence-specific silencing of complementary mRNA target by endonucleolytic cleavage or translational repression (Wilson and Doudna 2013).

- *Transcriptional errors*: RNA polymerase II, the critical enzyme responsible for the initial transcriptional effort by reading DNA template and synthesizing the pre-mRNA, can encounter potential errors of DNA lesions, which occurred in a frequency of one million per day per cell. Encountering such lesions may result in misincorporation in RNA, leading to non-functional proteins. Other threat to the fidelity of RNA synthesis include damaged ribonucleotides that can be incorporated into the RNA, leading to altered secondary or 3-dimensional structures. Three main fidelity checkpoints with regard to RNA polymerase II include insertion step (specific nucleotide selection and incorporation), extension step (from a matched vs. a mismatched 3'-RNA terminus), and proofreading step (removing misincorporated nucleotides from 3'-RNA terminus). Error occurred in any of these three steps could result in non-functional protein translation (Xu et al. 2015).

- *Post-transcriptional modification*: Initially transcribed RNA is a pre-mRNA and it requires modification before translation can occur properly. Since the initial copy of RNA contains non-coding region (intron) transcript as well as coding region (exon) transcript, alternative splicing is a needed post-transcriptional process typically occurs in multi-exon genes. Error can occur during this alternative splicing process and would lead to error in protein translation, even lead to diseases such as neurodegenerative diseases (Mills and Janitz 2012).

- *Translational error*: Although protein translational errors are known facts, there is a debate on whether this kind error is well tolerated by the organisms or may even have a purpose of protein diversification, since the error rate is also regulated (Ribas de Pouplana et al. 2014).

- *Post-translational modification*: After a protein is translated, it commonly goes through a post-translational modification process, where other biological materials, such as carbohydrate (glycosylation), acetyl group (acetylation), methyl group (methylation), are added into the protein chain. Abnormal post-translational modifications may affect the protein folding or aggregation in some ways that could lead to a disease process, such as Parkinson's disease (Barrett and Timothy Greenamyre 2015).

Structural Proteins

Structural proteins are responsible for creating a functional frameworks of human structure and function, like muscles, tendon, bone, cartilage, skin, hair, nail, etc. Although the major skeletal muscle proteins include common contractile proteins of slow type 1 and fast types 2A and 2X myosins, actins, tropomyosins, troponin complexes, and metabolic proteins, there is a high variability in terms of relative composition since muscle protein types vary with age, activity type and level, and gender (Gelfi et al. 2011). In bone, about 90% of protein is in the form of type I collagen (Lammi et al. 2006). In cartilage, type II collagen and aggrecans (large aggregating chondroitin sulfate proteoglycan) predominate (Lammi et al. 2006). In tendon, the major proteins are type I (most abundant), II, and III collagens (Buckley et al. 2013). For the skin, collagen is one of the major protein components. Hair and nail proteins are primarily keratins.

Soluble Proteins

Many types of mobile proteins are very important for human functions. They are soluble and usually circulate around the human body through the blood stream. When they are called upon to act, they will be ready to respond in carrying out their duties. The major human soluble proteins, cytokines, enzymes, and antibodies have been discussed in the Cellular Biology section above. Other soluble proteins include fibrinogen, a clotting protein in the blood that can form fibrin network in the process of blood clotting. Albumin is another blood-containing protein with a function of contributing to the circulation volume homeostasis (Garcia-Martinez et al. 2013).

Molecular Biology Techniques

Biologists utilize many advanced techniques to study biology at the molecular levels. Some of these techniques are discussed below.

Polymerase Chain Reaction

Polymerase chain reaction (PCR) is a technique developed to amplify a small quantity of DNA or mRNA, so that the researchers can identify and quantify these genes or gene products (Hillis). This technique makes molecular biology studies much easier and quicker, especially for those genes or gene products that are not abundant in natural conditions.

PCR is a novel and smart way to study DNA. The inventor of PCR method, Kary B. Mullis, an American Scientist, shared the 1993 Nobel Prize in Chemistry for this work, which was described by the Nobel Committee as "An organism's genome is stored inside DNA molecules, but analyzing this genetic information requires quite a large amount of DNA. In 1985, Kary Mullis invented the process known as polymerase chain reaction (PCR), in which a small amount of DNA can be copied in large quantities over a short period of time. By applying heat, the DNA molecule's two strands are separated and the DNA building blocks that have been added are bonded to each strand. With the help of the enzyme DNA polymerase, new DNA chains are formed and the process can then be repeated. PCR has been of major importance in both medical research and forensic science" (NOBEL 2018).

DNA PCR. To perform DNA PCR, researchers need to prepare several essential components: The DNA target template, a designed primer pair that is complementary to DNA target at each end of desired DNA segment to be amplified, heat-stable DNA polymerase commonly Taq polymerase that is derived from a hot-spring-grown bacteria *Thermus aquatica*, mixture of deoxynucleoside triphosphate, and a bivalent cation-containing buffer solution. Magnesium is the common cation used. Of note, the bacteria *Thermus aquatica* was found, isolated, and named by an American Scientist Dr. Thomas D. Brock and his student when they studied microorganisms at a large hot spring near the Great Fountain Geyser inside the Yellowstone National Park (Brock 1994).

Although there are many variations used by different researchers, PCR method is in general consisted of multiple repeats of a three-step process:

- *First step—Denaturing*: The target DNA molecule (or a mixture of DNA containing the target DNA molecule) is heated to about 93°C to denature the double strands of DNA, so that the target DNA is now open to two single strands.

- *Second step—Annealing*: Investigators cool down the mixture to 50–65°C, at which temperature the two DNA target strands remained separated. The annealing temperature is somewhat dependent on the primers used. A pair of primers (short stretch of single-strand DNA) designed to be complementary to the two ends of DNA of interest will be added to the mixture, allowing these two primers to anneal (bind) the open DNA target strands, with one primer binds to the 5'-end (upstream) and the other primer binds to the 3'-end (downstream) of DNA of interest.

- *Third step—Extending/Elongating*: Investigators would now add a heat-stable DNA polymerase (an enzyme that builds DNA strand) and a mixture of deoxynucleoside triphosphates (dATP, dTTP, dCTP, dGTP), so the two new strands of DNA will be synthesized from each end (5'-end and 3'-end) at temperature around 72°C. The process will be repeated 30 to 40 times (cycles) to achieve a quantity can be measured and identified. The size of the synthesized DNA by PCR will be restricted by the designed primers which determine the 5'-end and 3'-end of the PCR-generated DNA.

RNA PCR. To examine mRNA, which is unstable and vulnerable to RNase degradation, investigators would need to first convert mRNA to a stable DNA form, called complementary DNA, or cDNA. This step is achieved by a reverse transcriptase, which reverse the DNA to mRNA process of transcription. The reverse transcription technology was built upon the knowledge that retrovirus, a type of RNA virus, has the ability to reverse this DNA to mRNA process by reversely transcribing mRNA to cDNA for their replication process. After the cDNA is generated, the PCR process will basically follow the DNA PCR method as described above. This RNA PCR is commonly designated as RT-PCR (or reverse transcription-PCR).

With PCR method, researchers can now study many molecular biological subjects with great speed and accuracy. These subjects include study of ancient DNAs from ancestors, extinct animals, fossils, or dinosaurs. In addition, PCR can facilitate other molecular biological techniques for research, such as gene cloning (to isolate a previously unknown gene), microarrays, transgenesis (to artificially express a gene in a cell or an animal), gene knockouts (to delete a gene from a cell or an animal), DNA sequencing (to delineate the sequence of the base pair arrangement), and many others. Medically PCR is useful for rapid infectious disease diagnosis such as HIV and tuberculosis, is useful for cancer diagnostics in detection of mutated oncogenes, and is useful for matching organ transplant donor and recipient in tissue typing process. Legally, PCR is utilized to perform genetic fingerprinting to link forensic evidence to criminal.

Real-time PCR

Another biotechnology method that builds on the success of PCR is real-time PCR (Deepak et al. 2007). This technology enables the research to see the amplification results from every step (cycle) of the PCR process in real-time, so that the research can select result from area of the slope can best represent the result scientifically. This ability to monitor and select results of PCR process is important, as enzymatic reaction depends on the concentrations of enzyme and substrate and the best results are located at the linear slope areas and not in the plateau areas when either the enzyme or substrate is exhausted. One of advantages of real-time PCR is its ability to perform quantitative genotyping, genetic variation of inter and intra organisms, early diagnosis of cancer and infectious diseases, forensic medicine, and many others (Deepak et al. 2007).

cDNA Microarrays

cDNA microarray technology allows researchers to examine a wide arrange of gene expressions, i.e., mRNA) with a single experiment (Xiang and Chen 2000). The basic technique is the following:

- A tissue sample or cell sample is extracted for RNA.
- The extracted RNA is reversely transcribed to cDNA.

- The cDNA mixture is labeled with dye, commonly cyanine 3 (green color) or cyanine 5 (red color).
- The dye-labeled cDNA mixture of interested is placed on a microarray, which embedded with multiple different known cDNA probes in different slot of the microarray.
- The dye-labeled cDNA mixture placed on the slot of microarray is allowed to hybridize to the embedded cDNA probes.
- The microarray is washed to eliminate unbound the residues.
- The Hybridized microarray is scanned to detect the dye-labeled cDNA hybridized to the known probes and measure the intensity of hybridized dye-labeled cDNA.
- The scan reports the results both qualitatively (presence or absence of particular mRNA in the tissue or cell sample) and quantitatively (how much).
- The quantitation is usually a relative one, by comparing two different samples, for example, between a normal tissue and a disease tissue, or between cell sample untreated or treated with a specific cytokine.

The many applications of cDNA microarray include gene expression and discovery, biochemical pathways prediction, drug discovery and development, and many others (Xiang and Chen 2000).

Summary

In this chapter, an overview of major human body systems is discussed. The cellular and molecular basis of human biology was introduced. Since a single chapter cannot delineate this subject matter in great details, the goal of this chapter is a general introduction, with emphasis on the interconnectivity of different body systems in human biology, so the students will come away with a view of human body as a harmonized and integrated single system, thus paving the way for the introduction of next chapter, Systems Biology (Chapter 17).

References

[AHA] 2018. CPR & First Aid. American Heart Association. Statistical Update. [https://cpr.heart.org] accessed August 31, 2018.

Akdis, M., A. Aab, C. Altunbulakli, K. Azkur, R.A. Costa, R. Crameri et al. 2016. Interleukins (from IL-1 to IL-38), interferons, transforming growth factor b, and TNF-a: Receptors, functions, and roles in diseases. J Allergy Clin Immunol 138: 984–1010.

Amarenco, G., S. Sheikh Ismael, C. Chesnel, A. Charlanes and F. LE Breton. 2017. Diagnosis and clinical evaluation of neurogenic bladder. Eur J Phys Rebahil Med 53: 975–980.

Barrett, P.J. and J. Timothy Greenamyre. 2015. Post-translational modification of a-synuclein in Parkinson's disease. Brain Res 1628(Pt B): 247–253.

Bergmeister, K.D., S.C. Daeschler, P. Rhodius, P. Schoenie, A. Bocker, U. Kneser et al. 2018. Promoting axonal regeneration following nerve surgery: a perspective on ultrasound treatment for nerve injuries. Neural Regen Res 13: 1530–1533.

Bondanelli, M., L.De Marinis, M.R. Ambrosio, M. Monesi, D. Valle, M.C. Zatelli et al. 2004. Occurrence of pituitary dysfunction following traumatic brain injury. J Neurotrauma 21: 685–696.

[BOSTONU] 2018. The biomedical engineering teaching and innovation center. Boston University College of Engineering. [https://www.bu.edu/eng/alumni/bme-tic/] accessed August 15, 2018.

[BRAINYQUOTE] 2018a. Brainy Quote. [https://www.brainyquote.com/search_results?q=human+body] accessed August 15, 2018.

[BRAINYQUOTE] 2018b. Brainy Quote. Coordination Quotes. [https://www.brainyquote.com/topics/coordination] accessed August 15, 2018.

[BRAINYQUOTE] 2018c. Brainy Quote. [https://www.brainyquote.com/search_results?q=molecular+biology] accessed August 15, 2018.

[BRAINYQUOTE] 2018d. Brainy Quote. [https://www.brainyquote.com/search_results?q=human+genome] accessed August 15, 2018.

Brock, T.D. 1994. Life at High Temperatures. Yellowstone Association, USA.

Buckley, M.R., E.B. Evans, P.E. Matuszewski, Y.L. Chen, L.N. Satchel, D.M. Elliott et al. 2013. Distributions of types I, II and III collagen by region in the human supraspinatus tendon. Connect Tissue Res 54: 374–379.

Butler, M.O. and N. Hirano. 2014. Human cell-based artificial antigen-presenting cells for cancer immunotherapy. Immunol Rev 257(1): Doi: 10.1111/imr.12129.

Deepak, S.A., K.P. Kottapalli, R. Rakwal, G. Gros, K.S. Rangappa, H. Iwashashi et al. 2007. Real-time PCR: revolutionizing detection and expression analysis of genes. Curr Genomics 8: 234–251.

Garcia-Martinez, R., P. Caraceni, M. Bernardi, P. Gines, V. Arroyo and R. Jalan. 2013. Albumin: pathophysiologic basis of its role in the treatment of cirrhosis and its complications. Hepatology 58: 1836–1846.

Gelfi, C., M. Vasso and P. Cerretelli. 2011. Diversity of human skeletal muscle in health and disease: contribution of proteomics. J Proteomics 74: 774–795.

Gornati, L., I. Zanoni and F. Granucci. 2018. Dendritic cells in the cross hair for the generation of tailored vaccines. Front Immunol June 27. Doi: 10.3389.fimmu.2018.01484.

Green, E.D., J.D. Watson and F.S. Collins. 2015. Twenty-five years of big biology. Nature 526: 29–31.

Heusinkveld, L.E., M.L. Hacker, M. Turchan, T.L. Davis and D. Charles. 2018. Impact of tremor on patients with early stage Parkinson's disease. Front Neurol August 3; 9: 628. Doi: 10.3389/fneur.2018.00628.

Hillis, D.M., D. Sadava, R.W. Hill and M.V. Price. 2014. Principles of Life. 2nd Ed. Sinauer, Sunderland, MA, USA.

Kawai, T. and S. Akira. 2011. Toll-like receptors and their crosstalk with other innate receptors in infection and immunity. Immunity 34: 637–650.

Kyritsis, N., C. Kizil and M. Brand. 2014. Neuroinflammation and central nervous system regeneration in vertebrates. Trend Cell Biol 24: 128–135.

Lammi, M.J., J. Hayrinen and A. Mahonen. 2006. Proteomic analysis of cartilage- and bone-associated samples. Electrophoresis 27: 2687–2701.

Lander, E.S., L.M. Linton, B. Birren, C. Nusbaum, M.C. Zody, J. Baldwin et al. 2001. Initial sequencing and analysis of the human genome. Nature 412(6846): 565.

Lipsky, B.A., A.R. Berendt, H.G. Deery, J.M. Embil, W.S. Joseph, A.W. Karchmer et al. 2006. Diagnosis and treatment of diabetic foot infections. Plast Reconstr Surg 117: 212S–238S.

Mancini, G., M.G. Berioli, E. Santi, F. Togari, G. Toni, G. Tascini et al. 2018. Flash glucose monitoring: a review of the literature with a special focus on type I diabetes. Nutrients July 29; 10(8); pii: E992. Doi. 10.3390/nu10080992.

Mills, J.D. and M. Janitz. 2012. Alternative splicing of mRNA in the molecular pathology of neurodegenerative diseases. Neurobiol Aging 33(5): 1012.e11–24. Doi: 10.1016/j.neurobiolaging.2011.10.030.

Mojadidi, M.K., P. Christia, J. Salamon, J. Liebeit, T. Zaman, R. Gevorgyan et al. 2015. Patient foramen ovale: Unanswered questions. Eur J Intern Med 26: 743–751.

[NOBEL] Kary B. Mullis. The Nobel Prize. [https://www.nobelprize.org/prizes/chemistry/1993/mullis/facts/] accessed August 31, 2018.

Perica, K., A. De Leon Medero, M. Durai, Y.L. Chiu, J. Glick Bieler, L. Sibener et al. 2014. Nanoscale artificial antigen presenting cells for T cell immunotherapy. Nanomedicine 10: 119–129.

Redmer, T. 2018. Deciphering mechanisms of brain metastasis in melanoma—the gist of the matter. Mol Cancer July 27; 17(1): 106. Doi: 10.1186/s12943-018-0854-5.

Reece, J.B., L.A. Urry, M.L. Cain, S.A. Wasserman, P.V. Minorsky and R.B. Jackson. 2014. Campbell Biology. 10th ed. Pearson, Boston, USA.

Reynolds, L.P., S.D. Killilea and D.A. Redmer. 1992. Angiogenesis in the female reproductive system. FASEB J 6: 886–692.

Ribas de Pouplana, L., M.A. Santos, J.H. Zhu, P.J. Farabaugh and B. Javid. 2014. Protein mistranslation: friend or foe? Trends Biochem Sci 39: 355–362.

Sahlin, K. 2014. Muscle energetics during explosive activities and potential effects of nutrition and training. Sports Med 44: s167–173.

Schneider, D.J. 2012. The patent ductus arteriosus in term infants, children, and adults. Semin Perinatol 36: 146–153.

[UCI] 2018. Department of Biomedical Engineering. UCI. [http://engineering.uci.edu/dept/bme/undergraduate] accessed August 18, 2018.

Vargas-Uricoechea, H., A. Bonelo-Perdomo and C.H. Sierra-Torres. 2014. Effects of thyroid hormones on the heart. Clin Investig Arterioscler 26: 296–309.

Wilson, R.C. and J.A. Doudna. 2013. Molecular mechanisms of RNA interference. Annu Rev Biophys 42: 217–239.

Xiang, C.C. and Y. Chen. 2000. cDNA microarray technology and its applications. Biotechnology Advances 18: 35–46.

Xu, L., W. Wang, J. Chong, J.H. Shin, J. Xu and D. Wang. 2015. RNA polymerase II transcriptional fidelity control and its functional interplay with DNA modifications. Crit Rev Biochem Mol Biol 50: 503–519.

QUESTIONS

1. Why the focuses of current biomedical engineering are now on cellular and molecular levels of biology?

2. Please provide examples that human biological systems are interconnected and interdependent?

3. Why are antigen presenting cells so important for T cell memory?

4. What are the recent utilizations of engineered antigen presenting cells for medicine?

5. Why are engineered T-cells useful for immunotherapy against cancers?

6. What type of T-cells do MHC class I-barring antigen presenting cells react?

7. What type of T-cells do MHC class II-barring antigen presenting cells react?

8. What are cytokines and what are their functions in general?

9. Can you name few cytokines the possess "anti-inflammatory" property?

10. Are all cells in human possess the same genome or genetic information?

11. Why do some cells produce certain proteins and other cells do not?

12. What are some reasons that a cell possess the proper gene may not produce a functional protein from that specific gene?

13. Can you name the three major steps of polymerase chain reaction?

14. What is the most critical ingredient in carrying out polymerase chain reaction?

PROBLEM

The students will be grouped into 3–5 people per group. Within the groups, students will discuss and decide on a real-life cancer for which effective therapy is not yet available. The students will brainstorm and derive a theoretical basis for immunotherapy for that cancer and present their detailed works in the entire class.

17

Systems Biology
An Introduction

Lawrence S. Chan[1],* and *William C. Tang*[2]

QUOTABLE QUOTES

"Systems biology is based on the understanding that the whole is greater than the sum of the parts".

Institute for Systems Biology (SYSTEMS BIOLOGY 2016)

"Biological systems such as organisms, cells, or biomolecules are highly organized in their structure and function. They have developed during evolution and can only be fully understood in this context."

E. Klipp and Colleagues, Systems Biologists (Klipp et al. 2016)

"Some of today's scientists devoted their entire doctoral dissertations to trying to understand a single gene or protein, and for good reason: deconstructing systems into their constituent components has taught us much of what we know about biology today. But we now also have the capacity to start putting those pieces back together to understand how they make life work. That's systems biology."

James R. Valcourt, System Biology Researcher (Valcourt 2017)

"This 'modern activity' is characterized by experiments that reveal the behavior of entire molecular systems and so come to be called systems biology."

Brian P. Ingalls, MIT Professor (Ingalls 2013)

"A computer does not substitute for judgement any more than a pencil substitutes for literacy. But writing without a pencil is no particular advantage."

Robert McNamara (Ingalls 2013)

Learning Objectives

The learning objectives of this chapter are to introduce to the students with the concept of understand biology as a connected and coordinate systems to make life work.

---

[1] University of Illinois College of Medicine, 808 S. Wood Street, R380, Chicago, IL 60612.

[2] University of California, Irvine, Department of Biomedical Engineering, 3120 Natural Sciences II, Zot 2715, Irvine, CA 92697-2715; wctang@uci.edu

* Corresponding author: larrycha@uic.edu

After completing this chapter, the student should:

- Understand the meaning of systems biology.
- Capable of handling basic mathematical tools, calculus, linear algebra, and probability, necessary for computational modeling.
- Able to construct a simple mathematical model.
- Capable of thinking the human body as an interconnected and interactive system.

Systems Biology: History, Definition, and Rationale

Systems biology is analogous to ecology. This year's Spring break, the lead author of this chapter has a wonderful opportunity to visit Costa Rica with his daughter Angelina. During this oversea trip, they visited an area of tropical rain forest called Cano Negro Wildlife Refuge along a River Rio Frio by boat. In just one half mile stretch of river, they were able to observed 1 wood stock, 2 howler monkeys, 1 white-faced monkey, 2 mangrove warblers, 1 sungrebe bird, 5 cormorants, 1 sloth, 2 iguanas, 2 lizards, 3 turtles, and 3 fresh water crocodiles, plus hundreds of different plant species. There are many more they did not see with their own eyes. This amazing animal and plant diversities are special in tropical area. Such diversity provides an outstanding environment for ecologists to examine how do these animals live harmoniously with the surrounding plants and other animals in such sustainable way, for thousands of years, in this unique ecosystem. Likewise, systems biology is examining how the many components work harmoniously within a biological system. Systems biology is a new and rapidly building field in biology and biomedical engineering. As defined by the National Institute of Health, the very institution that provides the largest amount of funding for biomedical research, "Systems biology is an approach in biomedical research to understanding the larger picture—be in at the level of the organism, tissue, or cell—by putting its pieces together. It's in stark contrast to decades of reductionist biology, which involves taking the pieces apart" (Wanjek 2011). As one of the researchers in the field pointed out, the top-ranked university like Harvard, has just recently matriculated its first class of Systems Biology PhD Programs by 2005 (Valcourt 2017). Furthermore, who should be included in the field of Systems Biology are still being debated among academicians, as some viewed it as someone whose study of biology involving large quantity of data while others considered it as researcher who performs mathematical modeling for biology (Valcourt 2017). A careful examination of the subjects of systems biology will inevitably lead one to point out that biologists have always been studying these "biological systems" before it is categorized as "systems biology". Biologists have studied, for example, the immune system, that is consisted of immune organs (like lymph nodes, spleen, bone marrow), innate immune components (like neutrophils, mast cells, natural killer cells) and acquired immune components (like B lymphocytes, T lymphocytes, and antibodies). And these organs and cells together protect the human body from being overcome by invading pathogens. Biological scientists for the most part have been studying biological system from the bottom up, by deconstructing the entire system into their basic constituent components and performing detailed examinations on each of them. While these biology studies have allowed us to learn a great deals of biology, a complete picture of how a whole system works needs not only the data on each of its parts, but also the knowledge of how the individual parts interact to make the entire system work (Valcourt 2017). The knowledge of how these individual parts interact to make the system work in life, a bodily teamwork if you will, is in fact the definition of systems biology. To understand those essential interactions, in turn, we need a huge computational power and that is where mathematical modeling comes in to complete the jig saw puzzle. Some of field experts who have contributed to and influenced on the development of systems biology include Uri Alon, Brian P. Ingalls, Eberhard Voit, among others (Alon 2006, Ingalls 2013, Voit 2018, Valcourt 2017),

As one textbook described it in this manner, "Systems biology is an exciting, young discipline with deep roots across biology, chemistry and physics, as well as engineering, mathematics, and computing. It analyzes biological phenomena within their natural context by addressing systemic questions" (Voit 2018). Some of these "systemic" questions include the followings, but many more will be added to this list in the future:

- How does a biological component interact with other components and its environment?
- How is the biological component affected by the interactions with other components and its environment?
- What and how the functions of the biological component are regulated?
- What actually regulate the regulator?
- What kind of responses emerged from these interactions?
- What roles does this biological component have within its integrated system?
- What changes would occur if features of this biological component are modified or one of this biological components is deleted?
- What causes the disease to occur during the interaction of this biological component and its environment?

Rather than independent of the traditional biology study, systems biology supplements it. It actually depends on the traditional biological data, lots of them. As a result of utilizing and integrating large qualitative and quantitative traditional biology data, and in combination with throughput measurements and computational modeling, systems biology brings in unique values of new insights, explanations, and hypothesis. As one textbook stated, it can be viewed as a new framework of thinking that visualizes and approaches life science as a "dynamic system of systems" (Voit 2018).

Similarly, another textbook describes it this way, "Systems biology is the scientific discipline that studies the systemic properties and dynamic interactions in a biological object, be it a cell, an organism, a virus, or an infected host, in a qualitative and quantitative manner and by combining experimental studies with mathematical modeling" (Klipp et al. 2016).

Yet, an even simpler but brilliant definition of systems biology offered by a young and bright researcher, a PhD candidate at Harvard University, stated in his recent book that systems biology is "the study of connected groups of biological parts that all work together" (Valcourt 2017). Another biomedical engineering textbook defined systems biology as "the study of interactions between the components of a biological system and how these interactions give rise to the function and behavior of that system" (Saltzman 2015).

As to the big question of why do we need to study biology through the systems approach, the key answer is in the understanding of how all these systems act together. It is analogous to examine how the entire forest sustains itself, rather than how each tree survives and thrives. In the field of traditional biology study, aided by the advanced technology of high-speed molecular biological techniques, refined biomedical imaging, and others, we now have obtained a very detailed knowledge of cellular machinery. However, in order to determine how all these systems naturally act together, or how cells could work as complex, robust system, it is insufficient by simply cataloging and understanding of single-cell components. Thus, there is a need to capture the "global dynamics" between these components and this is where mathematical/computational modeling comes into play, as stated so succinctly by these systems biologists: "Biological systems such as organisms, cells, or biomolecules are highly organized in their structure and function. They have developed during evolution and can only be fully understood in this context" (Klipp et al. 2016).

Systems biology, indeed is a rapidly developing field of scientific study, where the central aim is to understand complex processes in living organisms, with the tools that not only include the "good old" laboratory experimentation, but also "state-of-the-art" high-throughput instrumentation and computational modeling. The main objectives of this chapter, however, are not to provide a comprehensive survey on systems biology. In fact, a comprehensive review in systems biology cannot be achieved by any length of materials since the field is moving forward in a past pace and new information is being added continuously. Instead, this chapter aims to provide a general understanding on the framework of systems biology, so the students would be able to explore more details and new additions of the field in a future date.

Systems Biology in Engineering-Medicine

Having delineated the meaning and rationale of systems biology, we now focus on the application of systems biology in the engineering-medicine curriculum. Obviously, teaching the entire course of systems biology is not the goal of this textbook, nor it will be a practical approach from the engineering-medicine curriculum load perspective. Rather, the central features and essence of systems biology will be emphasized here for the engineering-medicine curriculum. Drawing from the discussions on the session above, two topics will be focused: mathematical modeling and human body as one interconnected and integrated system.

Mathematical Modeling

Advantages of Mathematical Modeling

Before we discuss the various mathematical modeling, it is prudent to first delineate what mathematical modeling is and what advantages will we have if we utilize such models. Mathematical modeling is the core essence of computational analysis of systems biology. A mathematical model of systems biology can be defined as "an artificial construct in the language of mathematics that represents a process or phenomenon in biology" and modeling is the "process of creating such construct and squeezing new insights out of it" (Voit 2018). Such a model can be simple or complicated in terms of complexity, small or large in terms of model size, and intuitive or abstract in terms of easiness to understand (voit 2018). Regardless of its complexity, size, or easiness to understand, the essential benefit of such modeling is the ability to generate insights into processes or systems that we would not otherwise be able to obtain. It will help us to grasp the essence of and the interconnectivity of large number of isolated and apparently unrelated data and observations, just as one researcher in the field beautifully stated "Some of today's scientists devoted their entire doctoral dissertations to trying to understand a single gene or protein, and for good reason: deconstructing systems into their constituent components has taught us much of what we know about biology today. But we now also have the capacity to start putting those pieces back together to understand how they make life work. That's systems biology" (Valcourt 2017). Furthermore, mathematical models will enable scientists to extrapolate and make sensible prediction on conditions that have not been experimentally investigated and situations that cannot be experimentally examined. In addition, the results of such modeling may also lead researchers to formulate new hypotheses, novel manipulative method of biology, or even breakthrough in medically diagnostic and therapeutic initiatives (Voit 2018). Since mathematical modeling is like a "working hypothesis", it allows the dynamic behaviors, all aspects and time points, to be unambiguously probed that cannot be possibly achieved in the laboratory setting. In addition, model simulation can be conducted much faster and cheaper than empirical biological experiments. Furthermore, the modeling process may also detect previously unknown biological knowledge gap, thus facilitating the discovery of novel scientific insights (Ingalls 2013).

Fundamental principles of modeling biological systems can be represented by input-system-output models:

Real life: input(s) → biological system → output(s).

Mathematical model: input(s) → mathematical model → approximated output(s).

Given identical input(s), a good and useful mathematical model will produce output(s) that closely approximate the output(s) of the biological system.

There is a tradeoff between closeness of the approximation and the complexity of the mathematical model given a fixed set of input(s). The closer we wish the output(s) of the mathematical model to emulate the biological system output(s) the more complicated and time-consuming the mathematical model will be. Another tradeoff is the range of input(s) that the mathematical model needs to accommodate and still generates close approximation output(s). The more number of inputs, the broader range of inputs, and the greater variety of input behaviors will require significantly more complicated and costly mathematical model. Usually the complexity and cost of such a model go up exponentially with incremental demand on generalizing the input(s) and the closeness of the approximating output(s). Therefore, a good practice in constructing a mathematical model is to first carefully limit the number, variability, and range of inputs that represent a meaningful real-life scenario, and to decide on the level of approximation in the output(s) that serve practical scientific and clinical purposes.

Limitations of Mathematical Modeling

Since mathematical modeling acts as a "working hypothesis", it can never replace laboratory experiments and cannot predict biological behaviors in a definitive manner. In fact, solid experimental data are essential for the building of Systems Biology as they served as input and to ensure reality check (Wanjek 2011). At time, mathematical modeling results in inconsistencies between the model prediction and the laboratory experimental data. Although identifying such inconsistencies suggest a "false hypothesis" of the model, it does pave the way to refinement of the model hypothesis with the hope for a refined model. Ultimately, mathematical modeling would aim to achieve a fully predictable simulation of the real biological behavior (Ingalls 2013). Consistent with this aim is the definition of Systems Biology given by Dr. Ron German, Chief of National Institute of Allergy and Infectious Diseases' Laboratory of Systems Biology, "a scientific approach that combines the principles of engineering, mathematics, physics, and computer science with extensive experimental data to develop a quantitative as well as a deep conceptual understanding of biological phenomena, permitting prediction and accurate simulation of complex (emergent) biological behaviors" (Wanjek 2011).

The Modeling Basic Tool I: Calculus

Rate of Change

Before we go into the detailed discussion of mathematical modeling, it is prudent that we will first review the technical tools necessary for the modeling process, i.e., mathematics, and calculus in particular. This review is especially important for the non-engineering students and is compiled primarily from three sources (Finney et al. 2003, Ingalls 2013, Voit 2018). But why we have to use calculus rather than the other types of mathematical tools? The key here is the rate of change we need for mathematical modeling. Let us consider a standard function $y = f(x)$, whereas x is the variable input (or called argument), y is the output (the function value), and $f()$ is the function itself. A single-output function is not limited to one input, for example, $y = g(x_1, x_2, x_3, x_4, x_5)$. The non-calculus method in calculating the rate of change is sufficient if the change is on a straight line (a constant/linear rate of change). In this incidence, the rate of change will be mathematically expressed as $\Delta y / \Delta x$ (or slope) or "rise over run", with Δy (the rise) denotes the change of matter in question and Δx (the run) denotes the change of time. The equation of the rate of change will be written and graphed (Fig. 1) as below:

$$\{\text{Rate of Change}\} = \frac{f(x_2) - f(x_1)}{x_2 - x_1} = \frac{y_2 - y_1}{x_2 - x_1} = \frac{\Delta y}{\Delta x}$$

However, non-calculus method will face a huge challenge if the change occurs in a curve (with non-constant rate of change), instead of a straight line. In overwhelming majority of biological conditions, the rate of change is non-linear and the simple slope equation $\Delta y / \Delta x$ would not be able to give us the answer

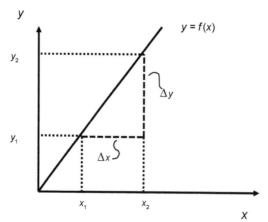

Fig. 1. Linear Rate of Change.

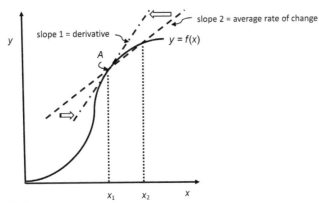

Fig. 2. Non-linear rate of change.

accurately. Here calculus has the ability to solve the instantaneous rate of change in any given point on a curve line. In fact, the central goal of calculus is to deal with the rate of change, under any and all conditions.

For example, we could consider the following condition where we need to find the rate of change at point A for non-linear function $y = f(x)$ (Fig. 2).

While non-calculus method cannot accurately determine the rate of change at point A (slope 1), we could estimate the rate of change by using other mathematical rule, for example, by drawing a line that touches point A and intersects the curve (secant line, slope 2). We now can estimate the rate of change (average rate of change) at point A by using the same formula:

$$\{\text{Rate of Change}\} = \frac{f(x_2) - f(x_1)}{x_2 - x_1} = \frac{\Delta y}{\Delta x}$$

Now let us consider a more accurate way to determine the rate of change at point A. Suppose we now move slope 2 towards (or approaching) slope 1 as close as possible yet not reaching slope 1 (as shown by the hallow arrows in the above graph) while still in touch with point A, we can get a slope that is very, very, very close to slope 1. Here is where calculus comes into play, by a process called limit.

In fact, the three major calculus tools that are important for mathematical computation are limit, derivative, and integral.

Limit

Limit deals with how a function does as it approaches a particular value. For example, $\lim_{x_2 \to x_1} f(x)$ is the limit of the function $f(x)$ as x_2 approaches x_1.

Derivative

Derivative describes a function's rate changes, the slope of tangent of instantaneous rate of change at a given point. Derivative actually derives from the concept of Limit. The concept of derivative, in fact, describes a closest approximation.

For example, derivative is the $\lim_{x_2 \to x_1} \left(\dfrac{f(x_2) - f(x_1)}{x_2 - x_1} \right)$ of $f(x)$ with respect to x at the point x_1 Derivative of a function is the rate of change of the function at a given point, i.e., at a particular input variable. And this above derivative is in fact the slope 1 at point A of the above graph, in the sense of closest approximation. It is important to note that the denominator of Lim cannot become zero, otherwise it will be meaningless. Thus x_2 can approach x_1 very, very, very closely, but cannot become x_1.

Several notations of derivative have been used:

$\dfrac{d}{dx} f(x)$ is denoted as the derivative of the function $f(x)$ (a Leibniz's notation).

$\dfrac{d}{dx} f(x)$ can be expressed also as $\dfrac{dy}{dx}$, since $y = f(x)$.

Derivative can also be expressed as f' (Lagrange's notation) and y' (Newton's notation) with the implicit understanding that the derivatives are with respect to an independent variable, x.

Integral

Integral calculates the area under a curve between two points.

For example, $\int f(x)dx$ denotes integral, the integration of $f(x)$ with respect to x, which is the inverse of derivative.

There are many rules of derivative, which are commonly used for convenient and speedier mathematical calculation (Table 1):

Table 1. Rules of Derivative (Rules of Differentiation).

Name of rule	Function $f(x)$ or f	Derivative $\dfrac{d}{dx}f(x)$ or f'
Constant	C	0
Identity	x	1
Square	x^2	$2x$
Square root	$\sqrt{x}$	$\dfrac{1}{2\sqrt{x}}$
Power	x^n	$nx^{(n-1)}$
Exponential	e^x	e^x
Natural Log	$\ln(x)$	$\dfrac{1}{x}$
Constant multiplier	$Cf(x)$	$C\dfrac{d}{dx}f(x)$
Sum	$g(x)+h(x)$	$\dfrac{d}{dx}g(x)+\dfrac{d}{dx}h(x)$
Difference	$g(x)-h(x)$	$\dfrac{d}{dx}g(x)-\dfrac{d}{dx}h(x)$
Product	$g(x)\cdot h(x)$	$g(x)\dfrac{d}{dx}h(x)+h(x)\dfrac{d}{dx}g(x)$
Quotient	$\dfrac{g(x)}{h(x)}$	$\dfrac{h(x)\dfrac{d}{dx}g(x)-g(x)\dfrac{d}{dx}h(x)}{[h(x)]^2}$
Reciprocal	$\dfrac{1}{g(x)}$	$\dfrac{-\dfrac{d}{dx}g(x)}{[g(x)]^2}$
Nested	$g[h(x)]$	$\dfrac{d}{dx}h(x)\cdot\left[\dfrac{d}{dw}g(w)\Big|_{w=h(x)}\right]$
Trigonometry	$\sin(x)$	$\cos(x)$
	$\cos(x)$	$-\sin(x)$
	$\tan(x)$	$\dfrac{1}{\cos^2(x)}=\sec^2(x)$

$\dfrac{d}{dw}g(w)\Big|_{w=h(x)}$ means to evaluate the derivative of $g(w)$ with respect to w at $w=h(x)$.

Differential Equation

This term simply means both sides of the equations are differentiated (or derivatized), so that the derivative can be solved. Differential equations can be explicit or implicit as illustrated below:

- *Explicit differentiation*: When the highest-order derivative is explicitly written as a function of the independent variable (x), dependent variable (y), and lower order derivatives. The general form can be expressed as

$$y^{(k)} = G(x, y, y', ..., y^{(k-1)})$$

A special case of explicit differential equation is the first-order ordinary differential equation:

$$y' = bx - ay$$

Where a and b are constants. Rearranging terms and writing out the differential explicitly gives:

$$\frac{dy(t)}{dt} + ay(t) = bx(t)$$

This first-order ordinary differential equation is encountered quite often in many physical systems. The solutions have been solved for many different $x(t)$. For the special case of $x(t) = u(t) = \begin{cases} 0 & \text{for } t < 0 \\ 1 & \text{for } t \geq 0 \end{cases}$ where $u(t)$ is called the unit function, the solution is given as

$$y(t) = \frac{b}{a} + \left[y(0) - \frac{b}{a} \right] e^{-at}$$ where $y(0)$ is the value of $y(t)$ when $t = 0$, and is simply called initial value. We will revisit this special case shortly.

Here is an example of solving an explicit differential equation:

$$3 f(x) = 3x + x^3$$

Differentiating both sides of the equation:

$$\frac{d}{dx} 3 f(x) = \frac{d}{dx} (3x + x^3)$$

$$3 \frac{d}{dx} f(x) = \frac{d}{dx} (3x) + \frac{d}{dx} (x^3)$$

$$3 \frac{d}{dx} f(x) = 3 \frac{d}{dx} (x) + 3x^2$$

$$3 \frac{d}{dx} f(x) = 3(1) + 3x^2 = 3(1 + x^2)$$

$$\frac{d}{dx} f(x) = 1 + x^2$$

- *Implicit differentiation*: When function values are implicitly defined by an equation involving input variable and the function value, we are dealing with a function where its value is difficult to evaluate. However, we can resolve its derivative by implicit differentiation as illustrated in the example below. First, we differentiate both sides of the equations, then we could go ahead to solve the derivative of the function $f(x)$:

Given the following implicit equation:

$$x^2 + f(x) = \sqrt{x} \cdot [1 + f(x)]^2$$

Differentiating both sides:

$$\frac{d}{dx}\left[x^2 + f(x)\right] = \frac{d}{dx}\left\{\sqrt{x} \cdot \left[1 + f(x)\right]^2\right\}$$

$$\frac{d}{dx}x^2 + \frac{d}{dx}f(x) = \sqrt{x}\frac{d}{dx}\left[1 + f(x)\right]^2 + \left[1 + f(x)\right]^2\frac{d}{dx}\sqrt{x}$$

$$2x + \frac{d}{dx}f(x) = \sqrt{x}\frac{d}{dx}f(x) \cdot 2\left[1 + f(x)\right] + \left[1 + f(x)\right]^2\frac{1}{2\sqrt{x}}$$

$$\frac{d}{dx}f(x) - \sqrt{x}\frac{d}{dx}f(x) \cdot 2\left[1 + f(x)\right] = \left[1 + f(x)\right]^2\frac{1}{2\sqrt{x}} - 2x$$

$$\left\{1 - 2\sqrt{x}\left[1 + f(x)\right]\right\}\frac{d}{dx}f(x) = \left[1 + f(x)\right]^2\frac{1}{2\sqrt{x}} - 2x$$

$$\frac{d}{dx}f(x) = \frac{\left[1 + f(x)\right]^2\dfrac{1}{2\sqrt{x}} - 2x}{1 - 2\sqrt{x}\left[1 + f(x)\right]}$$

Partial Differential Equation

If a function is dependent on multiple input variables or arguments, the rate of change of the function with respect to each of these argument will be unique and separate from all other variables that the function is also dependent on. These rates of change are termed the partial derivatives of the function. Partial derivatives are generated by fixing every variable but one single input and by treating the function dependent on a single input (Ingalls 2014).

Consider this function containing multiple input variables: $h(x_1, x_2, x_3, x_4, x_5)$

Partial derivative with respect to the x_1 variable is written as: $\dfrac{\partial}{\partial x_1}h\left(x_1, x_2, x_3, x_4, x_5\right)$

The "curly d" notation "∂" is used for indicating a partial derivative.

Boundary Conditions

These conditions basically specify the two ends of the interval so that partial differential equations can be properly solved (Ingalls 2013).

Delay Differential Equations

If it is desirable, one may incorporate an explicit delay for the effect of one variable on another by including a specific term of $(t - \tau)$ as the independent variable when describing the rate of change (Ingalls 2013). For example, we have the following equation delayed, such that the effect of S_2 is only felt after a time lag of τ units:

Ordinary equation: $S_1(t)$

Delay equation: $S_2(t) = S_1(t - \tau)$

Having established the essential aspects of calculus, we should now consider an example that calculus can be used to analyze and understand a real biomedical application. Intravenous (IV) drug delivery is perhaps one of the most common approaches in delivering medications to a hospitalized patient, by which medicine is directly and continuously administered into the circulation system. There are several methods to access the circulation system. The easiest and most common way is with the peripheral intravenous (PIV) line, in which a hypodermic needle is passed through the skin directly into the vein in either the back of the hand or the median cubital vein at the elbow. The other end is usually connected to an IV bag hung on a pole through a line with a drip control, which then relies on gravity to deliver the medication. Sometimes, the line is connected to an infusion pump for programmable control of dosage over time. Suppose the total volume of blood in the patient is V_C, which can be assumed to be constant over time. The drug is introduced at a constant rate starting at time 0, $R_D u(t)$, where R_D is a constant and $u(t) = \begin{cases} 0 \text{ for } t < 0 \\ 1 \text{ for } t \geq 0 \end{cases}$ is the unit function described previously. The drug is constantly being removed by the liver at a rate proportional to its concentration [D], with K_L as the constant of proportionality. Therefore, the rate of change in [D] is given by the following first-order ordinary differential equation:

$$\frac{d[D](t)}{dt} = \underbrace{-K_L[D](t)}_{\text{Liver}} + \underbrace{\frac{R_D}{V_C}u(t)}_{\text{IV Drip}}$$

Comparing the above with the standard form $\frac{dy(t)}{dt} + ay(t) = bx(t)$, we have

$[D](t) = y(t)$, $K_L = a$, and $\frac{R_D}{V_C} = b$.

Therefore, the solution is

$$[D](t) = \frac{R_D}{K_L V_C}\left(1 - e^{-K_L t}\right) \text{ with the initial condition of } [D](0) = 0.$$

In this example, the mathematical analysis tells us several important medically relevant information: (1) Because there is an exponential term $e^{-K_L t}$ in the solution, at the start of the drip, the concentration of the drug in the body rises from 0 quickly initially, and then tapers off and approach a constant level asymptotically; (2) given long enough time, i.e., as $t \to \infty$, the drug concentration settles to a constant given by $[D] = \frac{R_D}{K_L V_C}$. This constant is proportional to how fast the drug is infused (R_D) and inversely proportional to how fast the drug is removed by the liver (K_L); and (3) when the drip is shut off, say, at time $t = t_0$, the drug concentration drops quickly initially, and then drops more slowly as time progresses. These are consistent with experience and show how applying a simplified differential equation can model a real-life situation.

The Modeling Basic Tool II: Linear Algebra

Several terminologies of linear algebra useful for mathematical modeling, including vector, matrix, the identity matrix, the matrix inverse, the matrix transpose, and the nullspace (Ingalls 2013).

Vector

A vector is a set of numbers arranged into either a column (column vector, vertically arranged) or a row (row vector, horizontally arranged). For example,
a row vector is represented below:

Vr = [$x1, x2, x3, x4, x5$]

And a column vector is depicted below:

$$\mathbf{Vc} = \begin{bmatrix} x1 \\ x2 \\ x3 \end{bmatrix}$$

Matrix

A matrix is basically combination of several row vectors stacked into a rectangle, with rows and columns of numbers. For example, a matrix can be arranged:

$$\mathbf{M} = \begin{bmatrix} x1a & x2a & x3a \\ x1b & x2b & x3b \\ x1c & x2c & x3c \end{bmatrix}$$

By convention, the above matrix, **M**, is called 3-by-3 matrix, whereas the first number indicates the number of rows and the second number tells the number of columns. The above **M** is also termed square matrix since it has the same numbers of rows and columns, thus it is a "square", rather than a "rectangle".

Matrix Multiplication Matrices, as well as vectors, can be multiplied to become a new product, by the following fashion, provided that they have the same length, i.e., the number of a row on right multiplier equal the number of columns on the left multiplier. First, we look at matrix multiplying column vector. The following example works if the number of the rows ($a, b, c, ..., z$) in the second multiplier, the vector, equal to the number of columns ($A1, A2, A3, ..., Az$) in the first multiplier, the matrix:

$$\begin{bmatrix} A1 & A2 & A3 & \cdots & A_z \\ \vdots & \ddots & & & \vdots \\ M1 & M2 & M3 & \cdots & M_z \end{bmatrix} \bullet \begin{bmatrix} a \\ b \\ c \\ \cdot \\ \cdot \\ z \end{bmatrix} = \begin{bmatrix} A1a + A2b + A3c + \cdots A_z z \\ \vdots & \vdots \\ M1a + M2b + M3c + \cdots + M_z z \end{bmatrix}$$

For row vector multiplying column vector, the above will simply have only the first row and the product of this multiplication is called scalar, distinct from a vector.

The Identity Matrix This matrix, a square matrix, is defined by having the number "1" (the identity number) along the diagonal from top-left to bottom-right and zeros elsewhere, like the 3-by-3 identity matrix, denoted as $\mathbf{I}_3$, below:

$$\mathbf{I}_3 = \begin{bmatrix} 1 & 0 & 0 \\ 0 & 1 & 0 \\ 0 & 0 & 1 \end{bmatrix}$$

The Matrix Inverse This matrix, also a square matrix, is denoted as "**M**⁻¹". The defining characteristic of **M**⁻¹ is such that: $\mathbf{M} \bullet \mathbf{M}^{-1} = \mathbf{I}_n$, for example:

$$\mathbf{M} = \begin{bmatrix} 2 & 0 \\ 1 & 1/2 \end{bmatrix}, \qquad \mathbf{M}^{-1} = \begin{bmatrix} 1/2 & 0 \\ -1 & 2 \end{bmatrix}$$

Verifying: $\mathbf{M} \bullet \mathbf{M}^{-1} = \begin{bmatrix} 2 & 0 \\ 1 & 1/2 \end{bmatrix} \bullet \begin{bmatrix} 1/2 & 0 \\ -1 & 2 \end{bmatrix} = \begin{bmatrix} 1 & 0 \\ 0 & 1 \end{bmatrix}$

It is also important to realize that not every square matrix can have inverse.

The Matrix Transpose By definition, this transpose, denoted as "**M**ᵀ", is basically conducted by "flipping" the matrix, i.e., the rows become columns, and vice versa. Simply put, the transpose of a n-by-m matrix will become a m-by-n matrix:

If $\mathbf{M} = \begin{bmatrix} A & B \\ a & b \end{bmatrix}$, $\qquad$ then the $\mathbf{M}^{\mathrm{T}} = \begin{bmatrix} A & a \\ B & b \end{bmatrix}$

The Nullspace This occurs when the products of non-zeros vectors of matrices become zero, the vector is said to be in the nullspace of the matrix. For example:

If $\mathbf{M} \bullet \mathbf{Vc} = \begin{bmatrix} 0 & -1 & -2 & -1 \\ 1 & -1 & 0 & 0 \end{bmatrix} \bullet \begin{bmatrix} 1 \\ 1 \\ 0 \\ -1 \end{bmatrix} = \begin{bmatrix} 0 + (-1) + 0 + 1 \\ 1 + (-1) + 0 + 0 \end{bmatrix} = \begin{bmatrix} 0 \\ 0 \end{bmatrix}$

Therefore, **Vc** is in the nullspace of **M**.

The Modeling Basic Tool III: Probability

Probability deals with the phenomena of chance. Obviously it is concerned with random events and uncertainty, which is needed when we engage in stochastic computational modeling. We will now review some basic elements important for this aspect.

Probability and Sample Space

The easiest way to explain probability is by using a commonly encountered chance example, like flipping a coin during game time. A new US coin has one side with president's head figure and one side with a state-specific scene. We should call Head and Scene, respectively. When flipping the coin, the coin lands in either Head (H) or Scene (S) in a random fashion and of course by chance. Now suppose we will flip two such coins (in one event) and our multiple event experiment will end up with a sample space (*SaSp*), which is defined as the set of all possible outcomes of the coin flipping:

$SaSp = \{(H, S), (H, H), (S, H), (S, S)\}$

As we think logically, the probability (P) of any event occurred as defined in *SaSp* is between zero and one. For example, $P(S, S) = \frac{1}{4}$. On the other hand, the probability of the entire *SaSp* is one, or $P(SaSp) = 1$.

Discrete Random Variables and Related Functions

Discrete Random Variables Using the above coin flipping example, if we assign specific values of 3 and 1 to H and S, respectively, we define a random variable to be the sum of the results by flipping the two coins. The random variable is termed "discrete" because it takes on a specific value. Thus, the following 3 equations specify the "discrete probability distribution" for random variable "X":

$P(X = 6) = P\{(H, H)\} = \frac{1}{4}$

$P(X = 4) = P\{(H, S)\} + P\{(S, H)\} = \frac{1}{4} + \frac{1}{4} = \frac{1}{2}$

$P(X = 2) = P\{(S, S)\} = \frac{1}{4}$

Cumulative Distribution Function The above "discrete probability distribution" can be utilized to determine the probability of random variable "X" in taking a specific value less or equal to a particular value, like "b", denoted as "$F(b)$", for "cumulative distribution function":

$$F(b) = P(X \le b) = \begin{cases} 0 & \text{for } b < 2 \\ \frac{1}{4} & \text{for } 2 \le b < 4 \\ \frac{3}{4} & \text{for } 4 \le b < 6 \\ 1 & \text{for } 6 \le b \end{cases}$$

Expected Value Denoted as "$E[X]$", expect value, is defined as the weighted average over all X values (average of a random variable over many experiments). Using the same example above, the value can be calculated as below:

$$E[X] = \sum_{xi=2,4,6} Xi \cdot P(X = Xi) = 2 \cdot \frac{1}{4} + 4 \cdot \frac{1}{2} + 6 \cdot \frac{1}{4} = \frac{1}{2} + 2 + 1\frac{1}{2} = 4$$

Continuous Random Variables and Related Functions

Continuous Random Variables These are the random variables that take on values over a continuum. For most of such variables, the probability of having any specific value becomes extremely small and assigning a probability distribution is not possible. The alternative is that we will define "probability density function", denoted as "$f()$", to depict the probability of the variable taking up values at specific intervals. If we let X to denote a continuous ransom variable, then at the specific interval between a and b,

$P(a \le X \le b) = \int_a^b f(x)\, dx$

Then the cumulative distribution function, denoted as "$F()$" will be defined as:

$$F(b) = P(X \le b) = \int_{-\infty}^{b} f(x)\, dx$$

Also, the expected value, denoted as "$E[X]$", defined as weighted sum of the probabilities of all values, is assigned as:

$$E[X] = \int_{-\infty}^{\infty} x \cdot f(x)\, dx$$

Uniform Random Variable This is one standard example of continuous random variables. And on the interval between zero and one [0, 1], has the characteristic equal probability at every point in the interval range, so the probability density function "$f(x)$" is:

$$f(x) = \begin{cases} 1 & \text{for} \quad 0 \le x \le 1 \\ 0 & \text{for} \quad \text{all others} \end{cases}$$

So, the cumulative distribution function, $F(b)$, will be:

$$F(b) = P(X \le b) = \begin{cases} 0 & \text{for} \quad b < 0 \\ b & \text{for} \quad 0 \le b < 1 \\ 1 & \text{for} \quad 1 \le b \end{cases}$$

And the expected value, $E[X]$, will be:

$$E[X] = \int_{-\infty}^{\infty} x \cdot f(x)\, dx = \int_{0}^{1} x \cdot 1 \, dx = \frac{1}{2}$$

Exponential Random Variable This is another standard continuous random variable and depicts the time elapses between events occurring at a constant rate and continuously. For the probability density function $f(x)$ depending on a parameter "θ", it is assigned as:

$$f(x) = \begin{cases} 0 & \text{for} \quad x < 0 \\ \theta e^{-\theta x} & \text{for} \quad x \ge 0 \end{cases}$$

And the cumulative distribution function $F(b)$ will be:

$$F(b) = P(X \le b) = \begin{cases} 0 & \text{for} \quad b < 0 \\ 1 - e^{-\theta b} & \text{for} \quad b \ge 0 \end{cases}$$

The Modeling Basic Tool IV: Laplace Transform and Fourier Transform

Laplace Transform

As we have discussed in the previous section, differential equations are very useful to describe and model real physical and biological systems. First-order differential equation alone can be used to shed light on a range of systems that at first may not be understood intuitively. Second-order differential equations can be used to describe an even broader range of biological behaviors. However, higher-order differential equations quickly become intractable to be analyzed by hand. Laplace transform is a tool to help simplify the evaluation of differential equations in a way similar to how log and anti-log simplify multiplication and division in arithmetic to addition and subtraction.

Laplace transform of a time-domain function $f(t)$ is defined mathematically as

$$\mathcal{L}\left\{f(t)\right\} = F(s) \triangleq \int_{t=0}^{\infty} f(t) e^{-st} dt$$

The Laplace transform of the first-order derivative of $f(t)$ is then evaluated to be

$$\mathcal{L}\left\{\frac{df(t)}{dt}\right\} = \int_{t=0}^{\infty} \frac{df(t)}{dt} e^{-st} dt = \ldots = sF(s) - f(0)$$

Table 2. Common Laplace Transform Pairs.

$f(t),\ t \geq 0$	$F(s)$	$f(t),\ t \geq 0$	$F(s)$
$\delta(t)$	1	$\sin \omega_0 t$	$\dfrac{\omega_0}{s^2 + \omega_0^2}$
$u(t)$	$\dfrac{1}{s}$	$\cos \omega_0 t$	$\dfrac{s}{s^2 + \omega_0^2}$
t	$\dfrac{1}{s^2}$	$e^{-at} \sin \omega_0 t$	$\dfrac{\omega_0}{(s+a)^2 + \omega_0^2}$
t^n	$\dfrac{n!}{s^{n+1}}$	$e^{-at} \cos \omega_0 t$	$\dfrac{s+a}{(s+a)^2 + \omega_0^2}$
e^{-at}	$\dfrac{1}{s+a}$	$t \sin \omega_0 t$	$\dfrac{2\omega_0 s}{\left(s^2 + \omega_0^2\right)^2}$
te^{-at}	$\dfrac{1}{(s+a)^2}$	$t \cos \omega_0 t$	$\dfrac{s^2 - \omega_0^2}{\left(s^2 + \omega_0^2\right)^2}$
$t^n e^{-at}$	$\dfrac{n!}{(s+a)^{n+1}}$		

Similarly for the second-order derivative of $f(t)$:

$$\mathcal{L}\left\{\frac{d^2 f(t)}{dt^2}\right\} = \ldots = s^2 F(s) - s f(0) - f'(0)$$

Note that $f'(0)$ is the short form of $\left.\dfrac{d}{dt} f(t)\right|_{t=0}$. One can immediately see the pattern that the Laplace transform of higher-order derivatives is simply adding more "s" in front of the basic transform $F(s)$, plus taking care of the initial values $f(0), f'(0)$, etc. Table 2 has a list of common Laplace transform.

A large variety of functions can be structured with a combination of any of the above transform pairs shown in Table 2. After the operation in the s-domain, the time-domain function is obtained by performing inverse Laplace transform, which can most easily be done by table look-up.

An example of a system that can be described with a second order differential equation is

$$\frac{d^2}{dt^2} y(t) + 4\frac{d}{dt} y(t) + 3y(t) = x(t),$$ in which $y(t)$ is the output of the system in response to the input $x(t)$.

Note that the short form of this equation is $\ddot{y} + 4\dot{y} + 3y = x$ with the understanding that the independent variable is t.

Performing Laplace transform yields:

$$\mathcal{L}\{\ddot{y} + 4\dot{y} + 3y\} = \left[s^2 Y(s) - sy(0) - \dot{y}(0)\right] + \left[4sY(s) - 4y(0)\right] + 3Y(s) = X(s)$$

Rearranging and collecting terms:

$$\left(s^2 + 4s + 3\right)Y(s) - y(0)s - \left[4y(0) + \dot{y}(0)\right] = X(s)$$

$$\Rightarrow Y(s) = \frac{X(s) + y(0)s + 4y(0) + \dot{y}(0)}{s^2 + 4s + 3} = \frac{X(s) + y(0)s + 4y(0) + \dot{y}(0)}{(s+1)(s+3)}$$

Suppose in this system, the input is zero, and the initial conditions are $y(0) = 2$ and $\dot{y}(0) = 4$. Then,

$$Y(s) = \frac{X(s) + y(0)s + 4y(0) + \dot{y}(0)}{(s+1)(s+3)} = \frac{2s + 4(2) + 4}{(s+1)(s+3)} = \frac{2s + 12}{(s+1)(s+3)}$$

Further simplifying in order to perform inverse Laplace transform, we can use "Partial Fraction Expansion" as illustrated below:

$$Y(s) = \frac{2s+12}{(s+1)(s+3)} = \frac{A}{s+1} + \frac{B}{s+3}$$ where A and B are constants. By direct substitution:

$A(s+3) + B(s+1) = (A+B)s + (3A+B) = 2s+12 \Rightarrow A+B = 2$ and $3A + B = 12$. Solving the two unknowns in two equations yields $A = 5$ and $B = -3$. So

$$Y(s) = \frac{5}{s+1} - \frac{3}{s+3}$$

From Laplace transform table, we have $\mathcal{L}\left\{e^{-at}\right\} = \frac{1}{a+s}$. Therefore, the final solution is

$$y(t) = 5e^{-t} - 3e^{-3t}$$

Since the input $x(t)$ is 0, we call the output $y(t)$ "Natural Response" to the given set of initial conditions. On the other hand, if the input is not zero but the initial conditions of $y(t)$ and $\frac{d}{dt}y(t)$ are, then the response of the system is call "Forced Response", with the input $x(t)$ being the "force" driving the system. For example, if the input is an exponentially decaying force given by $x(t) = 4e^{-2t}$, then $X(s) = \mathcal{L}\left\{4e^{-2t}\right\} = \frac{4}{2+s}$, and so $Y(s) = \dfrac{X(s) + y(0)s + 4y(0) + \dot{y}(0)}{(s+1)(s+3)} = \dfrac{4}{(s+2)(s+1)(s+3)}$

Using Partial Fraction Expansion again to simplify, we have

$$Y(s) = \frac{A}{(s+2)} + \frac{B}{(s+1)} + \frac{C}{(s+3)}.$$ Solving for the constants A, B, and C, we have

$$Y(s) = \frac{-4}{(s+2)} + \frac{2}{(s+1)} + \frac{2}{(s+3)}.$$ Performing inverse Laplace transform by looking up the table, we have

$$y(t) = -4e^{-2t} + 2e^{-t} + 2e^{-3t}.$$

Fourier Transform

Most physiological activities are cyclical, following a certain regular rhythm. In breathing, our inhaling and exhaling form the two halves of a cycle. Our blood pressure continuously repeats the alternating phases of systole and diastole. On a longer time scale, we have our circadian cycle of wake and sleep. The regularity or the lack thereof of these rhythms are strong indicators of our physical wellbeing. The ability of physicians to identify irregularities of these rhythms are usually acquired, developed, and fine-tuned with bedside experience. It is highly desirable to develop an instrument to help physicians to detect and interpret even slight irregularities of these rhythms to enhance early diagnosis of diseases.

In order to do so, a good grasp of biomedical signal processing and analysis is important. The raw data of any signal over time consist of pairs of numbers, one is the time t, the other is the amplitude of the signal, y, taken at time t. (t_1, y_1) means that at one particular time point t_1 the signal is measured to be y_1. The raw data of a time series signal is simply $(t_1, y_1), (t_2, y_2), (t_3, y_3)\ldots$ which can be graphically represented by plotting all the y values along the t axis. An experienced physician can identify potential issues by looking at these kinds of plots such as electrocardiograms (ECG) to determine if there are any issues with the patient's heart. For a machine to develop the same "intuition" as a physician, or even offer suggestions based on minute irregularities, some sophisticated signal processing and interpretation are needed.

The most important mathematical tool for analyzing rhythmic signals is Fourier transform. Simply put, Fourier transform is a way to describe a cyclical signal by identifying the frequency contents of the signal instead of the amplitude over time.

Mathematically, it can be proved that all periodic signals, no matter how complicated, can be represented by a summation of a series of cosine and sine waves at different frequencies with different amplitudes. Suppose $x(t)$ is a periodic function with a period T, i.e., the value of $x(t)$ repeats after the duration of T. Mathematically, $x(t - nT) = x(t)$, $n = 1, 2, 3,...$ The period is related to the fundamental frequency ω_0 as $T = \dfrac{2\pi}{\omega_0}$. The Fourier series of $x(t)$ is given by:

$$x(t) = \frac{A_0}{2} + \sum_{k=1}^{\infty} A_k \cos\left(k\omega_0 t\right) + \sum_{k=1}^{\infty} B_k \sin\left(k\omega_0 t\right) \text{ with } k = 1, 2, 3,...$$

where the coefficients of all the terms, $A_0, A_1, A_2,..., A_k,...$ and $B_1, B_2, B_3,... B_k,...$ are constants. Note in particular the first constant term, $\dfrac{A_0}{2}$, is sometimes called the "DC" term. It accounts for any possible non-zero mean value of the signal. The coefficients can be evaluated with:

$$A_k = \frac{2}{T} \int_0^T x(t) \cos(k\omega_0 t) dt \text{ and } B_k = \frac{2}{T} \int_0^T x(t) \sin(k\omega_0 t) dt$$

And the DC term is evaluated with:

$$A_0 = \frac{2}{T} \int_0^T x(t) dt$$

With the Fourier series, one can then plot the coefficients against the discrete frequencies $n\omega_0$ with $n = 0,1,2,3,...$ This plot is called the "frequency spectrum" of the signal $x(t)$. The frequency spectrum is extremely useful in showing the "frequency contents" of the signal $x(t)$. Irregularities of the signal $x(t)$ when plotted in the time domain of $x(t)$ against t that are not immediately apparent may show up in the frequency spectrum.

In biomedical signal processing, the first two functions that always need to be performed after acquiring the raw signal are amplifying and filtering. Amplifying is to boost the weak signals for subsequent analysis. Filtering is to remove artifacts and noise from the signals. Filtering is done by identifying the useful signals from the frequency spectrum and suppressing all the other unwanted signals based on their frequencies. A low-pass filter, for example, is to suppress frequencies higher than a certain threshold, below which the signals are passed through to the next stage, i.e., allowing the low-frequency contents to pass. In addition to low-pass filters, there are high-pass and band-pass filters. Frequency spectra of various filters are needed to show what and how much frequencies can pass through.

Now, consider the scenario that the periodic signal $x(t)$ has a very long period T. In other words, $\omega_0 = \dfrac{2\pi}{T}$ is very small. The frequency spectrum will show very dense data points, since each data point

Table 3. Common Fourier Transform Pairs.

$x(t)$	$X(\omega)$	$x(t)$	$X(\omega)$
$\delta(t)$	1	$\sin \omega_0 t$	$j\pi\left[\delta(\omega + \omega_0) - \delta(\omega - \omega_0)\right]$
1	$2\pi\delta(\omega)$	$\cos \omega_0 t$	$\pi\left[\delta(\omega + \omega_0) + \delta(\omega - \omega_0)\right]$
$u(t)$	$\pi\delta(\omega) + \dfrac{1}{j\omega}$	$x(t)\sin \omega_0 t$	$\dfrac{j}{2}\left[X(\omega + \omega_0) - X(\omega - \omega_0)\right]$
$\mathrm{sgn}(t) = \begin{cases} 1 & t > 0 \\ -1 & t < 0 \end{cases}$	$\dfrac{2}{j\omega}$	$x(t)\cos \omega_0 t$	$\dfrac{1}{2}\left[X(\omega + \omega_0) + X(\omega - \omega_0)\right]$
$e^{-at}u(t)$	$\dfrac{1}{a + j\omega}$	$\mathrm{rect}\left(\dfrac{t}{\tau}\right) = \Pi\left(\dfrac{t}{\tau}\right) = \begin{cases} 0 & \text{if} & |t| > \dfrac{\tau}{2} \\ \dfrac{1}{2} & \text{if} & |t| = \dfrac{\tau}{2} \\ 1 & \text{if} & |t| < \dfrac{\tau}{2} \end{cases}$	$\tau\mathrm{sinc}\left(\dfrac{\tau\omega}{2\pi}\right)$
$t^n x(t)$	$j^n \dfrac{d^n}{d\omega^n} F(\omega)$	$\left(1 - 2\dfrac{|t|}{\tau}\right)\mathrm{rect}\left(\dfrac{t}{\tau}\right)$	$\dfrac{\tau}{2}\mathrm{sinc}^2\left(\dfrac{\tau\omega}{2\pi}\right)$

is separated by ω_0. Pushing this scenario to the extreme where T is infinitely long, ω_0 will then approach zero. The frequency spectrum will become a continuous line instead of discrete data points. This continuous line originates from the coefficients of the discrete frequencies when T is finite. It is denoted as a function of the continuous frequency ω as $X(\omega)$. To convert a time-domain signal $x(t)$ to a frequency-domain one $X(\omega)$ is called "Fourier Transform." The transform is evaluated with the following equation:

$$\mathcal{F}\{x(t)\} = X(\omega) = \int_{t=-\infty}^{\infty} x(t)e^{-j\omega t}dt$$

The Inverse Fourier Transform, going from $X(\omega)$ back to $x(t)$, is given by

$$\mathcal{F}^{-1}\{X(\omega)\} = x(t) = \frac{1}{2\pi}\int_{\omega=-\infty}^{\infty} X(\omega)e^{j\omega t}d\omega$$

Table 3 is a table of common Fourier Transform pairs.

It is important to note that pushing the period T of the signal $x(t)$ to infinity implies that the signal itself is no longer periodic. That means that any arbitrary signals can be evaluated for their frequency spectra by performing Fourier Transform, another reason why Fourier Transform is very central in biomedical signal analysis.

The Modeling Basic Tool V: Computational Software

The proper simulation and analysis of mathematical models of biology phenomena require the use of sophisticated computational software and the students of engineering-medicine are recommended to be familiar with the availability and the simple operational ability on these programs. The most commonly utilized programs are likely to be XPPAUT and MATLAB®. Whereas XPPAUT program is available free of charge for all researchers, MATLAB is a fee-for-service commercial program. For beginning users of these software, XPPAUT, which is designed for simulation and analysis of dynamic modeling, is recommended. One feature distinguish these two programs is that the interface of XPPAUT program is predominantly manu-based and the interface of MATLAB program is a command-based operation. Another distinct features between these two programs are that simulation and analysis in XPPAUT is directly carried out by a graphical user interface (GUI) and that these operations in MATLAB can be carried out by GUI with an additional add-on package of Systems Biology Toolbox. Other programs useful for computational modeling include Maple, Mathematica, Berkeley Madonna, and Copasi. A special language called Systems Biology Markup Language (SBML) has the ability to facilitate the model transfer between different programs (Ingalls 2013).

XPPAUT

Created and maintained by G. Bard Ermentrout, the software and documents of this program is available on the internet (XPPAUT). Two guidebooks are also available for beginning users (Ermentrout 2002, Fall et al. 2002, Ingalls 2013).

MATLAB

As a technology product of MathWorks, inc. MATLAB is available on the company's website (MATHWORKS 2018). The GUI needs an additional package of Systems Biology Toolbox, developed by Henning Schmidt (SBTOOLBOX 2018).

Types of Mathematical Modeling

There are two main types of mathematical models that have been employed for the purpose of understanding life through the systems biology approach: deterministic and stochastic (Ingalls 2013, Voit 2018). In this chapter, our focus will be on deterministic models and readers are encouraged to examine the stochastic models in other available textbooks (Ingalls 2013, Voit 2018).

Deterministic Model

This type of model is a mathematical model which the behavior is exactly reproducible.

Stochastic Model

On the contrary, this is a kind of mathematical model which the behavior is random and is influenced both by specific conditions and by unpredictable forces.

Example of Mathematical Model: Enzymatic Kinetics

Since the majority of cellular reactions are enzyme-catalyzed, it is therefore a practical exercise to examine how a mathematical model can be built for such a reaction (Ingalls 2013). From the outset, we should make it clear that the rate of enzyme-catalyzed reaction is a non-linear one, since as substrates increase to a degree where all enzymes are occupied the rate will be flattened (slow to increase and then stop to increase). For this reason, calculus is needed to describe the rate of change. Now let us consider a reaction in which a single substrate (S) is catalyzed by a single enzyme (E). According to the kinetic law called Michaelis-Menten Kinetics, the sequence of steps of such enzyme-catalyzed reaction is expressed as the following (Ingalls 2013):

$$S + E_f \rightleftarrows C_A (E_b + S) \rightleftarrows C_B (E_b + P) \rightleftarrows E_f + P \qquad [1]$$

S is the substrate

E_f is the free (unbound) enzyme

E_b is the bound enzyme

C_A and C_B represent the complexes between bound enzymes and their corresponding substrates and products, respectively

P is the product catalyzed by enzyme

The symbol of $\rightleftarrows$ denotes association and disassociation events, between enzyme and substrate or between enzyme or product.

Let us now to simplify the Eq. 1 with a couple of assumptions. Let us first assume there is a rapid equilibrium between C_A and C_B, that is the conversion between these two complexes are much speedier than the association and disassociation events, thus we could combine C_A and C_B to become C. The other assumption is that the product P will never bind to E_f, thus we will eliminate one left direction arrow. Now Eq. 1 becomes Eq. 2 (Ingalls 2013):

$$S + E_f \rightleftarrows C \rightarrow E_f + P \qquad [2]$$

Where the reaction $S + E_f \rightarrow C$ is by an association constant k_a, the reaction $C \rightarrow S + E_f$ is by disassociation constant k_d, and the reaction $C \rightarrow E_f + P$ is by catalytic constant k_c.

Using rule of mass action we can then write the following differential equations for the rates of changes of S, E_f, C, and P:

$d/dt\, S(t) = k_d C(t) - k_a S(t) \cdot E_f(t)$

$d/dt\, E_f(t) = k_d C(t) + k_c C(t) - k_a S(t) \cdot E_f(t)$

$d/dt\, C(t) = k_a S(t) \cdot E_f(t) - k_d C(t) - k_c C(t)$

$d/dt\, P(t) = k_c C(t)$

The difficulty here is that E_f is hard to determine at a given time. Since enzyme is not consumed per se, the total enzyme, denoted as E_T (a constant), we can rewrite it as $E_T = E_f + C$, so $E_f(t) = E_T - C(t)$, indicating that free enzyme at time t equals the total enzyme subtracting the enzyme bound in the complex at time t. Therefore, the differential equations can be rearranged as (Ingalls 2013):

$$d/dt\ S(t) = k_d\ C(t) - k_a\ S(t) \bullet [E_T - C(t)]$$

$$d/dt\ C(t) = k_a\ S(t) \bullet [E_T - C(t)] - k_d\ C(t) - k_c\ C(t)$$

$$d/dt\ P(t) = k_c\ C(t)$$

Now this mathematical model can be easily simulated by computer to get the rates of changes on S, C, and P at any time point.

Human Body as One Interconnected and Integrated System

The Interconnected Human Body

Having surveyed the major tools in system biology, we should now review one fundamental principle in human physiology that demands the best of these available tools to analyze: human body as one interconnected and integrated system. From an engineering perspective, the multiple subsystems in the body tightly interact with each other to adapt to the changing external environment while maintaining a relatively constant internal environment for the purpose to survive and thrive. This collection of multi-input-multi-output subsystems communicates via a vast and complex network of chemical, electrochemical, biomechanical, photochemical, and thermochemical signals. For any one part of the body to function properly, many other parts of the body must also function properly and in a coordinated fashion. For example (Stanfield 2017):

- Muscle requires oxygen provided by erythrocytes in blood that are manufactured in bone marrow.
- Erythrocyte synthesis requires erythropoietin, a hormone produced by the kidneys in response to decreased oxygen levels in the circulating blood.
- Oxygen is extracted from air breathed in by the lungs.
- Lung expansion is controlled by the nervous system and achieve by muscle contractions in the diaphragm and chest cavity.
- Blood is pumped by the heart, which depends on coordinated cardiac muscle contractions.
- Small intestine absorbs into blood stream sugar, which depends on the pancreas to secrete insulin for glucose storage in the liver.

Homeostasis

Despite its unfathomable complexity, however, there is one central organizing principle unifying all physiological activities: homeostasis (Stanfield 2017). Homeostasis is the ability to maintain a relatively constant internal environment even when the external environment changes. The internal environments that are regulated and maintained include temperature, volume, and composition. How homeostasis is maintained by checking and balancing ALL physiological aspects of the entire body requires organ systems integration. Disruption of homeostasis is the basis for disease and even death. In engineering terms, the aspects of the body that are being regulated are called "variables". The most common variables are body temperature, blood glucose concentration, pH level of blood, etc. The control method is called "negative feedback". When the regulated variable decreases below the "homeostatic level", the system responds to make it increase, and vice versa. The desired value of the variable, or "homeostatic level", is called "set point". Examples of set points are: body temperature set at 37°C; blood glucose concentration set at 100 mg·dL^{-1}; and blood pH set at 7.4. "Error signal" is the information needed for the negative feedback, and is the difference between the value of the set point and the value of the regulated variable. The "hardware" needed for the negative-feedback control are "sensors" that measure the current value of the variable, "integrating centers" that process the error signal and execute control, and "effectors" that actuate and change the regulated variable. The sensors are the receptors in the body such as thermoreceptors to sense the temperature they are exposed to, photoreceptors to sense light, baroreceptors to sense pressure, etc. They are specialized cells that transduce one form of energy into chemical or electrochemical energy, which serve as the signals for the integrating centers. The integrating centers then orchestrate appropriate responses. Many of these integrating centers are found in the brain and some in the spinal cord. The effectors are muscles and glands that carry out the commands from the integrating centers.

Disease: When Human Homeostasis Breakdown

A healthy person is usually unaware of how homeostasis works, or even if it exists, until something goes wrong, that is, when disease occurs.

One excellent example of homeostasis breaking down is diabetes mellitus, a metabolic disease manifests in elevated glucose levels in blood and in urine, elevated sensation of thirst, etc. All these "error signals" are not corrected in diabetes patients, and the results affect nearly every organ system. The high blood sugar lead to blood vessel damage and blood vessel leakage, resulting in retinopathy, where blood leaks out to retina and causes blindness. The high blood sugar also leads to cataract formation in the lenses of the eye, resulting in slow reduction of visual acuity. The high blood sugar in addition leads to damage of kidney, resulting in glucose and protein leakage to the urine. The list goes on and on.

Role of Engineering in Restoring Human Homeostasis

Biomedical engineers should be very proud that they play a significant role in restoring human homeostasis. To an engineer, the inner workings of the extremely complicated multi-nested network of negative-feedback controls that are central to homeostasis cannot begin to be approached without the latest advances in information science, big data, and artificial intelligence. At times, partial system analysis using the developed tools can still elucidate valuable insights that lead to the development of medical devices. Such devices can then assist patients in restoring partial or full functions or return the body to an acceptable level of homeostasis. In the case of diabetes, the automated insulin pump with glucose sensor and a negative-feedback control system is such an outstanding example. Although it does not permanently cure the disease, the insulin pump helps restoring the glucose homeostasis and alleviate some diabetic symptoms and reduce disease comorbidity. In addition, many other chapters in this book exemplify the role of engineering in restoring human homeostasis: Engineers help restoring the homeostasis of cancer-free state by designing more precise treatments for cancers (Chapter 12, Precision); Engineers help restoring the homeostasis of disease-free state by inventing and improving diagnostic instruments so that diseases can be more accurately and speedily determined (Chapter 7, invention and innovation; Chapters 19, magnetic resonance imaging; Chapter 20, molecular imaging; Chapter 21, optical coherence tomography; Chapter 22, photoacoustic imaging; Chapter 23, mass spectrometry); Engineers help restoring the homeostasis on body balance and mobility by designing robotic rehabilitation machines and devices (Chapter 24, rehabilitation). More sophisticated assistive devices in the future will be integrated more tightly with the natural physiological system for more extensive restoration of homeostasis. To achieve this goal requires concerted efforts in engineering-medicine, with engineers working closely with physicians. To be sure, advocating close collaboration between engineers and physicians is one of major goals of this text book.

Summary

In this chapter, we introduce the concept of systems biology, a new path of looking at biology from a systemic perspective. We have discussed the rationale behind the development of the field, some essential characteristics, and advantages and limitations of the field. We further reviewed the basic tools to build mathematical models for biological functions, including calculus, linear algebra, probability, Laplace and Fourier Transforms. A brief description of the common computational software was also included. Example of mathematical modeling was provided to introduce the students to this new knowledge element. The chapter concludes with a discussion on the human body as one interconnected and integrated system: homeostasis.

References Cited

Alon, U. 2006. An Introduction to Systems Biology: Design Principles of Biological Circuits. Chapman & Hall, London, UK.

Ermentrout, B. 2002. Simulating, Analyzing, and Animating Dynamical Systems: A guide to XPPAUT for researchers and students. SLAM. Philadelphia, USA.

Fall, C.P., E.S. Marland, J.M. Wagner and J.J. Tyson (eds.). 2002. Computational Cell Biology. Springer, New York, USA.

Finney, R.L., F.D. Demana, B.K. Waits and D. Kennedy. 2003. Calculus. Pearson Education, Inc. Upper Saddle River, New Jersey.

Ingalls, B.P. 2013. Mathematical Modeling in Systems Biology: An Introduction. MIT Press, Cambridge, MA.

Klipp, E., W. Liebermeister, C. Wierling and A. Kowald. 2016. Systems Biology: A textbook, 2nd ed. Wiley-VCH, Weinheim, Germany.

[MATHWORKS] 2018. [www,mathworks.com/matlab] accessed June 19, 2018.

Saltzman, W.M. 2015. Biomedical Engineering: Bridging Medicine and Technology, 2nd ed. Cambridge University Press, Cambridge, UK.

[SBTOOLBOX] 2018. [www.sbtoolbox.org] accessed June 19, 2018.

Stanfield, C.L. 2017. Principles of Human Physiology, 6th Ed. Pearson Education, San Francisco, Inc.

[SYSTEMS BIOLOGY] 2016. What is systems biology. Institute of Systems Biology. [https://www.systemsbiology.org] accessed July 5, 2016.

Valcourt, J.R. 2017. Systematic: How Systems Biology is Transforming Modern Medicine. Bloomsbury Publishing, London, UK.

Voit, E.O. 2018. A First Course in Systems Biology. 2nd ed. Garland Science, New York, USA.

Wanjek, C. 2011. Systems Biology as Defined by NIH: An intellectual resource for integrative biology. The NIH Catalyst. Intramural Research Program. NIH. November-December 2011. [https://irp.nih.gov/catalyst/v19i6/systems-biology-as-defined-by-nih] accessed July 9, 2018.

[XPPAUT] 2018. [www.math.pitt.edu/-bard/XPP/XPP.html] accessed June 19, 2018.

QUESTIONS

1. What is the rationale for the development of systems biology?
2. What is the chief objective of systems biology?
3. What is the key difference between conventional biology and systems biology?
4. What are the advantages and limitations of mathematical modeling?
5. Why is calculus important for mathematical modeling?
6. What is the key difference between deterministic model and stochastic model?
7. Why deterministic model is the most commonly used model?
8. Why do we need Laplace Transform?
9. What is the key purpose of Fourier Transform in the context biosignals?
10. How does the concept of homeostasis help engineers understand human physiology?

PROBLEMS

Each student will consider and bring several biological problems for which a mathematical model would be helpful to build. The student will then discuss these problems with his or her mentors before deciding on one problem to build a mathematical model. Under the guidance of the mentor, the student will complete the mathematical model for computer simulation.

18

Advanced Biotechnology

William C. Tang

QUOTABLE QUOTES

"Other kids went out and beat each other up or played baseball, and I built electronics."

Robert Moog, American Engineer and Pioneer of Electronic Music (BRAINYQUOTE 2018a).

"With electronics, they just get smaller and smaller."

Amy Heckerling, American Film Director (BRAINYQUOTE 2018b).

"The science of today is the technology of tomorrow."

Edward Teller, Hungarian-born American Theoretical Physicist (BRAINYQUOTE 2018c)

"Technology is a gift of God. After the gift of life it is perhaps the greatest of God's gift. It is the mother of civilizations, of arts and of science.

Freeman Dyson, England-born American Theoretical Physicist and Mathematician (BRAINYQUOTE 2018c)

"Good, bad or indifferent, if you are not investing in new technology, you are going to be left behind."

Philip Green, British Businessman (BRAINYQUOTE 2018c)

"Engineering or technology is all about using the power of science to make life better for people, to reduce cost, to improve comfort, to improve productivity, etc."

N.R. Narayana Murthy, Indian IT Industrialist (BRAINYQUOTE 2018c)

"Technology fives us power, but it does not and cannot tell us how to use that power. Thanks to technology, we can instantly communicate across the world, but it still doesn't help us know what to say."

Jonathan Sacks, British Orthodox Rabbi and Theologian (BRAINYQUOTE 2018c)

"We must not be afraid to push boundaries, instead, we should leverage our science and our technology, together with our creativity and our curiosity, to solve the world's problems."

Jason Silva, Venezuelan American Filmmaker (BRAINYQUOTE 2018c)

University of California, Irvine, Department of Biomedical Engineering, 3120 Natural Sciences II, Zot 2715, Irvine, CA 92697-2715; wctang@uci.edu

Learning Objectives

The learning objectives of this chapter are to help the students to acquire the overall perspectives on the major advances and trends in technology and engineering for the medical field. Specifically, the students are introduced to the key miniaturization technology, Microelectromechanical Systems (MEMS), and how it advances biomedical instrumentation.

After completing this chapter, students are expected to:

- Understand the essence of MEMS technology.
- Understand the broad view on medical applications of MEMS.
- Understand the two main applications of MEMS for medicine: microimplants and lab-on-a-chip.
- Able to utilize the knowledge of MEMS to understand other biomedical uses.

Technology and Engineering for Medicine

What are the latest advances in technology and engineering for medicine? One of the most authoritative sources to answer this question is the Institute of Electrical and Electronics Engineers (IEEE), the world's largest technical professional organization dedicated to advancing technology for the benefit of humanity (IEEE.org). By the end of 2017, the IEEE has more than 417,000 members in more than 160 countries. The 39 technical Societies published more than 4 million journal articles and conference papers. Among these 39 Societies, the Engineering in Medicine and Biology Society (EMBS) is itself the world's largest international society of biomedical engineers, with 11,000 members from 97 countries around the world (EMBS.org). EMBS provides its members with access to the people, practices, information, ideas and opinions that are shaping one of the fastest growing fields in science. This multidisciplinary field encompasses (EMBS.org):

Diagnostic Systems

- Conventional systems
- Point-of-care diagnostics
- Imaging and other tests

Therapeutic Systems

- Neuromuscular devices
- Cardiovascular devices
- Cancer treatment
- Drug delivery
- Artificial tissues and organs

Healthcare and Bioinformation Systems

- MIS
- E-medicine
- D2H2 (Distributed Diagnosis and Home Healthcare)
- Genomics, proteomics, and physiome
- Tools in drug discoveries

Technologies & Methodologies

- Biosignal processing
- Biomedical imaging
- Medical instrumentation and sensors
- MEMS and nanotechnology
- Neural engineering
- Rehabilitation engineering
- Biorobotics
- Biosystems modeling
- Computational bioengineering and bioinformatics

The last category of *Technologies & Methodologies* includes some of the most rapid infusion of technological and engineering advances into the medical field. Biomedical imaging, rehabilitation engineering, biorobotics, and biosystems modeling are covered in other chapters in this book. In this chapter, we will focus on Microelectromechanical Systems (MEMS) specifically. The reason is that MEMS, being the foundational and enabling technology that offer significant miniaturization of medical devices and tools, are being applied to other categories, particularly point-of-care diagnostics, drug discoveries and deliveries, medical sensors, neural engineering, and the like. Therefore, a good grasp of MEMS would be essential to understanding one of the fastest growing technological advances in medicine. Two fields of application of MEMS will be described to illustrate the power of miniaturization, integration, and cost-reduction of MEMS: microimplants and Lab-on-a-chip.

MEMS

In the past several decades, the application of microelectronics technology to the fabrication of micromechanical devices greatly stimulated research in micro-sensors and micro-actuators. MEMS, a term first coined in the late 1980s, has now become a ubiquitous acronym referring to the interdisciplinary field characterized by micromachining of mechanical devices and integration with electronics. Micromachining is a large collection of micro-fabrication techniques that has an origin in the integrated circuit (IC) industry. The foundation of various micromachining techniques is the planar lithographic process. The planar processes in the IC industry are inherently batch-fabrication techniques that enable parallel production of large number of highly precise electronic circuits through a single pass of processing sequences (Fig. 1). Micromachining benefits from the same precision batch-fabrication processes in the creation of micromechanical devices.

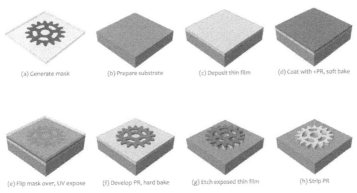

(a) Generate mask   (b) Prepare substrate   (c) Deposit thin film   (d) Coat with +PR, soft bake

(e) Flip mask over, UV expose   (f) Develop PR, hard bake   (g) Etch exposed thin film   (h) Strip PR

Fig. 1. Basic steps in photolithography.

As the variety of micromechanical devices grows, so does the variation of micromachining approaches. However, there are three common classes of micromachining techniques: wet bulk micromachining, surface micromachining, and soft-lithographic micromachining. Bulk-micromachined devices are primarily made by the accurate etching of a silicon substrate anisotropically along crystal planes. Surface micromachining is a construction process with sequential stacking and structuring of thin films. Both technologies use materials and processes from the IC industry. Soft-lithographic micromachining, on the other hand, uses soft materials to create flexible and transparent structures specifically for processing and manipulating fluids at the microscale, a distinct advantage for biomedical uses. In addition, wafer bonding and integration techniques are frequently used to further enhance the versatility of MEMS fabrication processes. This chapter provides a review of the three main classes of micromachining, wafer bonding processes, and integration techniques.

Bulk Micromachining

The advantages of mono-crystalline silicon as a mechanical material are detailed in (Petersen 1982). The most popular method in machining silicon wafers is with wet chemistry. There are two types of wet bulk micromachining: anisotropic and isotropic. These two techniques serve complementary roles in structuring mechanical devices from the bulk of a silicon wafer. A third method is based on reactive ion, which etch silicon without liquid-phase chemistry.

Anisotropic Wet Etching

Anisotropic etching of mono-crystalline silicon with strong alkaline solutions at elevated temperatures provides smooth etched surfaces and tight dimensional control. Since the early 1960s, strong alkaline solutions have been used to etch silicon selectively according to crystallographic orientations (Holms 1962). Generally, the etch rate is slowest in the ⟨111⟩ directions of the silicon cubic diamond crystalline structure, and fastest in the ⟨100⟩ and ⟨110⟩ directions. Selectivity is defined as the ratio of the etch rates of the desired direction to those of the undesired one. The higher the selectivity, the better defined the finished geometry. Some silicon etchants display a reduced etch rate in regions that are heavily doped with boron, adding flexibility in defining the finished structures. An externally applied electrical potential on one side of a *p–n* junction within the bulk silicon can also be used to influence the etch rate in the direction perpendicular to the junction (electrochemical etching) (Kloeck et al. 1989). Most of the silicon etchants have very low etch rates on silicon dioxide (SiO_2) or silicon nitride (Si_3N_4), making the latter suitable masking films for defining the desired geometry. Figure 2 illustrates how three different structures can be created from a (100)-silicon wafer in a single processing sequence: a square opening through the thickness of the wafer, a V-groove with a rectangular top-view, and a square diaphragm on the bottom of the

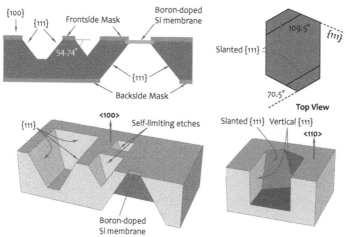

Fig. 2. Anisotropic wet etching of silicon.

wafer. The exposed {111} sidewalls all form a 54.74° with the ⟨100⟩ oriented surface. With (110)-Silicon wafer as the starting material, vertical sidewalls and shallow V-grooves can be created (Kendall 1975). Different alkaline solutions display different etching characteristics and selectivities, depending on the chemical compositions and temperature (Seidel et al. 1990). A commonly used alkaline etchant is an aqueous solution of potassium hydroxide, with concentration ranging from 10 to 50%. Isopropyl alcohol is frequently added to improve selectivity.

Isotropic Wet Etching

Acidic etchants, such as mixtures of hydrofluoric, nitric, and acetic acids (HF, HNO_3, and CH_3COOH) do not show crystal-orientation dependency (Schwartz and Robbins 1976). However, in a certain composition, it etches heavily doped regions ($p+$ or $n+$) much faster than lightly doped ones. With a 1:3:8 volume ratio (HF:HNO_3:CH_3COOH) at room temperature, the etch rate in the heavily doped region ($> 5 \times 10^{18}$ cm^{-3}) is between 50 and 200 μm·h^{-1} with a selectivity over lightly doped region ($< 10^{17}$ cm^{-3}) being 150 (Petersen 1982). Since its dopant-dependent selectivity is the opposite of that of strong alkaline systems, this acidic system can be used as a complementary etchant to enhance the flexibility of creating chemically etched structures. Si_3N_4 or Au are good masking materials.

Deep Reactive Ion Etching (DRIE)

In the mid-1990s, a method to create high-aspect-ratio silicon microstructures by vertically etching into the bulk of the silicon substrate was introduced (Lärmer and Schilp 1996). Unlike conventional reactive-ion assisted plasma etching, the process described in Lärmer and Schilp 1996, commonly known as Deep Reactive Ion Etching (DRIE), actually alternates frequently between plasma etching and plasma deposition of etch inhibitor within the same chamber. An inductively coupled RF magnetic field is also added to enrich the density of ions and neutrals within the plasma to increase the etch rate (inductively coupled plasma, or ICP). During the etch cycles, SF_6 is introduced into the chamber long enough to etch silicon materials for up to one or a few μm deep. Then during the deposition steps, C_4F_8 is introduced instead, which results in the deposition of about 50 nm-thick of fluorocarbon polymer, coating both the sidewalls and the bottom of the etch pits. In the following etch step with SF_6, the polymer is removed far more readily from the bottom of the etch pits than the sidewalls due to the directionality of the accelerating ions bombardment, allowing silicon etch to proceed downward while the sidewalls are protected by the fluorocarbon polymer. Common etch masks for this process are photoresist, with an etch selectivity of 100:1 or more, and SiO_2, with selectivity up to 300:1. Since its invention, ICP-DRIE has gained widespread popularity and has become one of the most used micromachining processes.

Surface Micromachining

In surface micromachining, thin films of materials are sequentially deposited and patterned on the surface of a wafer, which serves only as a carrier. Surface-micromachined devices in general are substantially smaller than bulk-micromachined ones, and promise order-of-magnitude improvements on device density. Figure 3 illustrates the basic surface-micromachining process. One or more of the intermediate layers serve only as spacers for the depositions of subsequent structural layers, and are removed in the final devices. This final release process is often called sacrificial etching, and can be performed with either wet or dry chemistries. The two most crucial aspects in surface micromachining are the choice of thin-film materials and the sacrificial etching process. The sacrificial etch must remove the spacer layers efficiently without attacking the structural materials. In many applications, there are only very narrow access channels or holes for the etchants to reach the sacrificial layers, requiring an extended etch time. Therefore, the etch selectivity between the sacrificial and the structural materials must be sufficiently high. This requirement drives the selection of the materials and the choice of the sacrificial etch process.

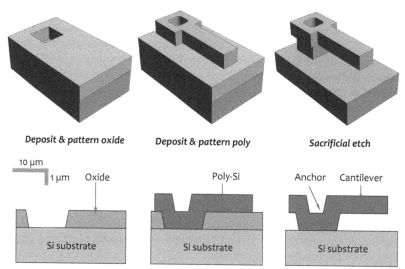

Fig. 3. Basic steps in surface micromachining.

Thin-Film Materials

The most common thin-film materials for surface micromachining are polycrystalline silicon (polysilicon) as the structural material and silicon dioxide for the sacrificial layer. The sacrificial etch is performed with hydrofluoric acid (HF), which remove the oxide material but not the silicon. Polysilicon films can be deposited by a variety of means. Good quality polysilicon can be deposited by Low-Pressure Chemical Vapor Deposition (LPCVD) using the silane pyrolysis at pressures of 300–500 mT and temperatures of 500–700°C (Tang et al. 1989).

Another structural material is single-crystal silicon film, which has a well-defined structure, and hence its mechanical properties are very reproducible. Single-crystal silicon films can be obtained directly from the top layer of a commercial silicon-on-insulator (SOI) wafer (Diem et al. 1993) or by epitaxial growth (Bartek et al. 1995). The buried oxide layer conveniently serves as the sacrificial layer. Commercial SOI wafers can be purchased with different top-layer thickness, and can be as much as 100 μm or more. Since it is monocrystalline, CMOS circuits can be fabricated on the top layer. Because of the superior mechanical properties, the available thickness, and the CMOS integration possibility, SOI wafers are popular in surface micromachined MEMS devices.

Silicon oxide is the sacrificial material of choice for polysilicon or SOI surface micromachining. It can be grown or deposited by a number of means. Exposure of silicon surfaces at high temperatures (1000°C and higher) in oxygen grows high-quality silicon dioxide films. This thermal oxidation process is common in the IC industry, and thus is well characterized and highly reproducible. However, the oxide film expands in volume by about 45% while consuming the adjacent silicon material, thus it creates a compressive stress in the silicon parts. Alternatively, oxide films can be deposited by LPCVD using the pyrolytic reaction of silane and oxygen at a pressure of 300–500 mT and a temperature of 450°C. Because of the substantially lower temperature of deposition compared to oxidation, LPCVD oxides are commonly called low-temperature oxides (LTO).

Sacrificial Etching

The sacrificial layer must be exposed somehow to allow access of etchants. Etch holes or channels are intentionally designed into the final devices. The opening should be large enough to allow the etch to proceed without requiring an unduly-long etch process. In long and narrow channels, diffusion effects will limit the etch rate considerably. Multiple openings cut the etch time dramatically and should be used whenever possible. Sealing may be performed to plug the holes or channels after the sacrificial etch.

Isotropic etching of SiO_2 is usually done in wet chemistry with diluted HF or buffered HF (BHF). BHF consists of an aqueous solution of HF with the addition of NH_4F to achieve a more uniform and consistent etch rate. Anisotropic etching of SiO_2 is accomplished by ion-assisted plasma etching. With a 1:1 combination of C_2F_6 and CHF_3, the resulting sidewalls are acceptably vertical.

Sealing

Sacrificial-etching techniques form cavities that are open to the surface through a small opening. The cavity can be hermetically sealed by plugging the opening. Sealed cavities are used in surface-micromachined pressure sensors, resonators, and infrared emitters. There are a few sealing techniques available. The earliest sealing technique (Burns 1988) used the thermal reactive growth of silicon oxide on the interior of a polysilicon cavity. The enlarged volume of the oxide on the cavity walls eventually seals the opening. The remaining oxygen in the cavity is consumed by further oxidation, creating a vacuum. The cavity can also be sealed by plugging the hole with a deposited material, such as LPCVD nitride (Mastrangelo and Muller 1989). In both reactive and deposited sealing techniques, the interior of the cavity is coated with the plug material, which may not be desirable. A sealing technique that does not coat the cavity interior is shadow plugging. In this method, the thin film is deposited from a point source that has poor step coverage. Sputtered, evaporated, and PECVD films (Sugiyama et al. 1986) are adequate for this purpose. The plugging occurs when enough material is deposited to block the entrance of the opening. If these materials have pinholes and cracks, it is necessary to coat the sealed areas with a better film on top of the deposited films, such as silicon nitride.

Wafer Bonding

Wafer-bonding techniques are commonly used for sealing microsensors and in the construction of composite bulk-micromachined sensors. Anodic bonding, fusion bonding, and bonding with intermediate layers are the most common wafer-bonding techniques (Knecht 1987, Ko et al. 1985).

Anodic Bonding

The technique of anodic or electrostatic bonding between a conductive substrate and a sodium-rich glass substrate was discovered in 1969 (Wallis 1969), which was later adapted to bond silicon and glass wafers. The two wafers are aligned and pressed together between two electrodes inside a chamber. The temperature is raised to between 350–450°C, which is sufficient to mobilize the sodium ions within the glass. When a voltage of 400–700 V is applied between the two electrodes with the glass substrate as the anode, the sodium ions are driven from the glass-substrate interface, creating a shallow ion-depletion region about 1 μm thick and a high electric field on the order of 7×10^6 V·m^{-1} (Albaugh 1988). The high field induces a large electrostatic pressure of several atmospheres that brings the silicon and glass into intimate contact. A good bond is established in a matter of minutes, forming a thin layer of SiO_2 at the interface, which is responsible for the high strength of the bond. The resulting bond is hermetic with bond strength exceeding that of the substrates. To avoid bimetallic warping effect, the glass and silicon wafers should have matched thermal expansion coefficients. Corning glass 7740 offers the closest match to silicon (3.2 ppm·°C^{-1}). Two silicon wafers can be anodically bonded using an intermediate layer of sputtered glass. Conversely, two glass wafers can be bonded anodically with an intermediate sputtered amorphous silicon film (Lee 2000).

Fusion Bonding

In fusion bonding, two silicon wafers are thermally "fused" when in contact at high temperatures. This technique is commonly used for the fabrication of silicon-on-insulator devices and pressure sensors (Shimbo et al. 1986). The flat surfaces of the two wafers are thoroughly cleaned and placed in contact, which are then immediately held together by weak Van der Waal forces. After a high temperature treatment at over 1100°C, the bond becomes permanent with no visible interface even in the presence of native oxide layers. Fusion bonding is an attractive method for applications where intermediate layers cause undesirable stresses. However, this method requires very high bonding temperatures, which precludes the presence of electronics at the time of bonding.

Bonding with Intermediate Layers

One of the intermediate layers for bonding is low-temperature glass, including borosilicate glasses, sputtered doped glass layers, spin-on glass slurries, and frits (Knecht 1987, Legtenberg et al. 1991). After the wafers are coated with the glass materials, they are pressed together and heated to establish the bond. Glass frits are paste-like mixtures of metal oxides in solvent. Under pressure, the frit forms a planarizing film that fills gaps between rough surfaces. Glass-frit bonding is popular because of the low bonding temperature requirements (~ 300–600°C) and its ability to bond non-flat surfaces. However, thermal expansion coefficients of commercially available frits range from 2 to 5 times larger than that of silicon.

Some metals can also be used as the intermediate bonding layers, which react with the silicon surfaces when heated, forming an alloy bond. Gold is a commonly used material for bonding silicon, with an alloying temperature of 363°C. Another metal is aluminum, which can be used on silicon to bond to glass (Cheng et al. 2000) for creating a hermetic vacuum protection of microstructures. The aluminum can be heated by rapid thermal annealing (RTA).

Organic compounds are also studied as materials for the intermediate bonding layers mainly for their low bonding temperatures. Wafers have been bonded using polyimides and epoxies with moderate strengths (den Besten et al. 1992). Reasonable-quality bonds are made at temperatures as low as 130°C.

Figure 4 illustrates a miniature pressure sensor made from wafer bonding a glass substrate with a through hole to a bulk-micromachined silicon structure with surface-micromachined features. Using MEMS technology, conventional pressure sensors sized tens of cubic centimeters in volume have been shrunk to a few cubic millimeters, a reduction of almost four orders of magnitude. Miniaturized pressure sensors are advantageous for monitoring a range of vital signs including pulse rates, blood pressures, breathing, intracranial pressure, etc., that demand extremely small form factors to achieve minimal invasiveness.

Integration

It is highly desirable to integrate micromachined devices with signal processing circuits to eliminate interconnects between chips, improve signal-to-noise ratios, and minimize the final package. One method is based on modular approaches, in which the CMOS fabrication procedures are left intact while adding processing steps to integrate micromechanical devices onto the same substrate. Also, integration of various functional components such as optical, chemical, fluidic, and mechanical devices onto the same substrate can broaden the versatility and utility of micro systems. However, these components are often fabricated with incompatible processes. Precision pick-and-place hybrid assembly can be used for the final integration, but is both costly and inefficient. As a result, research has been done in achieving precision parallel assembly of pre-fabricated components.

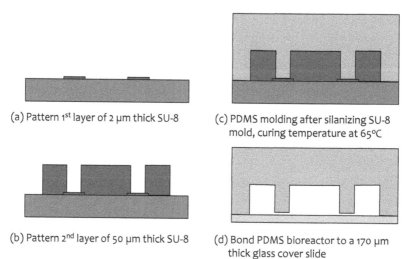

(a) Pattern 1st layer of 2 μm thick SU-8

(b) Pattern 2nd layer of 50 μm thick SU-8

(c) PDMS molding after silanizing SU-8 mold, curing temperature at 65°C

(d) Bond PDMS bioreactor to a 170 μm thick glass cover slide

Fig. 4. Miniature pressure sensor.

Process Integration with Complementary Metal-Oxide Semiconductors (CMOS)

Almost all electronics fabricated today is based on CMOS circuits. There are several alternatives in CMOS process integration with micromachining. Demonstrated approaches include customized processing sequence in which CMOS and surface micromachining fabrication steps are interleaved, micromachining performed on CMOS circuits or post-CMOS micromachining, and fabricating CMOS circuits on substrates with micromachined devices or pre-CMOS micromachining. In the custom approach, fabrication steps can be simultaneously optimized for both circuits and microdevices. However, the tradeoff is the loss of flexibility in accommodating other designs. More activities in the field have been on modular approaches such as the post- and pre-CMOS processes.

One key advantage of post-CMOS integration process is that the highly developed and optimized CMOS process is not disturbed. Commercially available CMOS chips with sophisticated electronics can be obtained from vendors. However, one of the challenges of fabricating mechanical parts on an CMOS chip is that the highly delicate circuits cannot be subjected to extreme processing steps. For example, polysilicon surface micromachining that includes process steps at elevated temperatures (600°C for polysilicon deposition, 850°C for nitride, and 1050°C for polysilicon stress anneal) cannot be used. An alternative is to use LPCVD silicon germanium as surface micromachining structural material and LPCVD germanium as the sacrificial material shows promising results, since both of which can be deposited at a low enough temperature (450°C) that regular CMOS circuits are not affected (Franke et al. 2000). Another additive post- CMOS integration involves depositing parylene-C multilayers to create fluidic channels onto a silicon substrate with photodiodes for fluorescence detection and analysis of DNA in an electrophoresis system (Webster et al. 2000).

In a subtractive post-CMOS integration process, instead of adding surface micro mechanical devices on CMOS circuits, microstructures are etched directly from either the surface or the bulk of a CMOS wafer. DRIE has been used to etch microstructures from the multilayered surface of a commercial CMOS (Xie et al. 2000). Similar approach with wet anisotropic etching has been demonstrated in creating bulk-micromachined accelerometers with 0.8-μm commercial CMOS circuit on the seismic mass (Takao et al. 2000). The etch was performed with heated TMAH (90°C) from the backside, with PECVD oxide as the etch mask. This process is combined with anodically bonded SW-3 glass cap on the bottom and silicon-aluminum-glass bonded cap on the top to form the final hermetically sealed device.

In pre-CMOS micromachining, the machining process, typically surface micromachining, is highly optimized to achieve the best mechanical performance. The challenge is that the surface topology of the wafer becomes too severe (as much as 5 μm for surface micromachining) for subsequent CMOS process. In order to realize MEMS-first with CMOS integration, the surface of the wafer must be returned to the planar state. In one approach, a shallow well is created on a starting silicon wafer, in which subsequent surface micromachining is performed (Gianchandani et al. 1998). The well is then filled with oxide followed by chemo-mechanical polishing (CMP). The result is the creation of embedded micromechanical devices below a planarized surface, on which conventional CMOS processes are carried out. The final step is the release of the embedded structure by sacrificial removal of the oxide materials that filled the well.

There are other approaches in integration and final assembly. Wafer-level batch transfer (Bannon III et al. 2000, Gianchandani and Najafi 1992, Nguyen et al. 2000, Singh et al. 1997) is the process by which micro components fabricated on one wafer is transferred to another. Micro self-assembly is a process that position and bind individual micro components onto predefined sites on a substrate in a fluid medium. One of the mechanisms to promote the positioning of micro components relies on gravity and shear forces to move the micro components with a trapezoidally shaped outline onto complementarily shaped holes on the substrate (Yeh and Smith 1994). This technique has demonstrated rapid assembly to 1-μm precision of GaAs light-emitting diodes onto silicon substrates. Other examples of positioning and binding mechanisms include the use of bridging flocculation in a dilute polymer solution (Nakakubo 2000) and electrostatic traps and ultrasonic vibration in a vacuum (Böhringer et al. 1998).

Soft Lithography

A relatively new class of micromachining is based on printing, molding, and embossing elastomeric materials and is collectively referred to as soft lithography (Qin et al. 2010). The most common soft lithography process is illustrated in Fig. 5. A thick layer of photoresist, up to several hundred microns, is spin coated on a silicon wafer and exposed through a photomask. After development, the thick features form the master mold for creating multiple molded parts. Poly(dimethylsiloxane) (PDMS) is a common elastomer for that purpose. The two precursors are mixed thoroughly and then poured onto the wafer with the thick features and cured at 60° to 80°C for several hours. It is then peeled off the master with negative features imprinted on the underside, forming fluidic chambers and channels of different sizes and shapes. Openings for fluidic injection and extraction are punched into the PDMS piece. The finishing step is to bond the piece onto a glass substrate. The advantages of forming a platform with PDMS is that it is cost effective, easy to mold, flexible, permeable to gas but not to liquid, biocompatible, and optically transparent to allow examination of the fluids and materials inside the device. Microfluidic devices fabricated with this technique have been used to perform cell sorting, capturing, tagging, and studied under various chemical or physical stimuli (Whitesides et al. 2001). It greatly enhances our ability to probe the physiological responses of single cells for the purpose of understanding both the natural and pathological implications of cell development.

A second common use of soft lithography is to create the PDMS piece with fine features molded from high-precision photoresist layers. The PDMS piece is then used as a stamp to transfer a thin layer of materials from the source onto a substrate by inking and stamping. This microcontact printing technique allows precise coating of thin layers of chemicals suitable for biomedical applications. For example, alkanethiols and proteins can be printed onto substrates coated with gold, silver, copper, palladium, or platinum. This offers an ability to engineer the properties of the metallic surfaces with molecular-level detail using self-assembled monolayers (SAMs) of alkanethiols. The patterned SAMs have been used for studying the role of spatial signaling in cell biology by controlling the molecular structure of a surface in contact with cells (Chen et al. 1997). Microcontact printing has also been used to print precise patterns of axon guidance molecules, which acted as templates for growing retinal ganglion cell axons (von Philipsborn et al. 2006). Further, by combining various biosensors on the chip with fluidic and cell manipulation capabilities, lab-on-a-chip can be realized to perform various functions that are traditionally done in a clinical or biological laboratory, greatly enhancing the broad access to advanced medical technology.

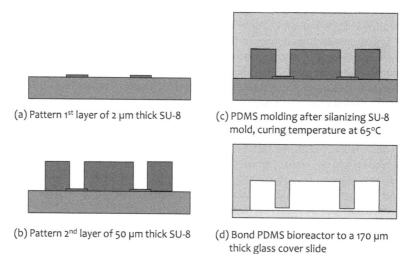

(a) Pattern 1st layer of 2 μm thick SU-8

(b) Pattern 2nd layer of 50 μm thick SU-8

(c) PDMS molding after silanizing SU-8 mold, curing temperature at 65°C

(d) Bond PDMS bioreactor to a 170 μm thick glass cover slide

Fig. 5. Soft lithographic process to create microfluidic device.

MEMS Application: Micro-implantable Devices

According to the U.S. Food & Drug Administration (USFDA), medical implants are devices or tissues that are placed inside or on the surface of the body. Many implants are prosthetics, intended to replace missing body parts. Other implants deliver medication, monitor body functions, or provide support to organs and tissues (FDA.gov).

Pacemaker

The first totally implanted device is the pacemaker in 1958, which, through many generations of innovative improvements, is still being used today (Joung 2013). There are several obvious advantages of implanting an assistive device like the pacemaker that helps regulate cardiac rhythms. First, such devices are highly targeted, avoiding any side effects common with pharmacological approaches. Second, their effectiveness is immediate, in contrast to drugs that take effect gradually. Third, they require almost no user involvement, unlike the need to take medicine on a regular routine. It is estimated that 8% to 10% of the population in America and 5% to 6% of people in industrialized countries have experienced an implantable medical device for rebuilding body functions, achieving a better quality of life, or expanding longevity (Jiang and Zhou 2010). The success of implantable devices coincides with the advances in engineering and technology, including the rapid growth of electronics, biocompatible materials, battery technology, etc. In particular, the advent of MEMS technology allows significant size shrinkage, a distinct advantage for developing miniaturized medical implants.

Cochlear Implants

One of the most successful miniaturized implantable medical devices is the cochlear implant. It is a small, complex electronic device that can help provide a sense of sound to persons with profound deafness. These individuals suffer from sensorineural deafness due to damage to the hair cells of the inner ear (Johnson 2014). Without these hair cells, the cochlea is unable to convert the mechanical vibrations of sound transmitted from the eardrum through the middle ear to the cochlea into nerve impulses needed by the auditory sensory system to hear sound. According to the National Institute on Deafness and Other Communication Disorders (NIDCD), approximately 324,200 registered devices have been implanted worldwide as of December 2012 (NIDCD.gov). Between June 2012 and June 2013, it is estimated that 50,000 cochlear implants were sold (MEDEL.com). With a conservative estimate of 10% annual growth rate, there should be more than 700,000 cochlear implants deployed worldwide by the end of 2018. Figure 6 illustrates the different components of a cochlear implant (NIDCD.nih.gov). The microphone picks up sound from the environment, which is

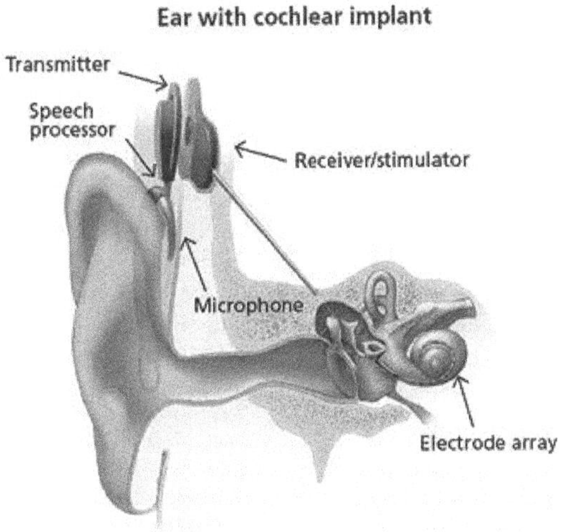

Ear with cochlear implant

Fig. 6. Cochlear implant (NIDCD.nih.gov).

then send to the speech processor that analyzes the sound and divides it into discrete frequency channels. Only sound with frequency contents between about 1,000 Hz to 10,000 Hz are processed and sent to the transmitter. The receiver that has been implanted beneath the skin picks up the signal and power wirelessly, and send the stimulation to the microelectrode array, which has been surgically introduced into the scalar tympani of the cochlea. The array is the key component of the cochlear implant consisting of a group of electrodes that collects the impulses from the stimulator and sends them to different regions along the depth of the cochlea. The cochlea is tonotopically arranged, with the region closest to the oval window being sensitive to high frequency at 20,000 Hz, while the innermost apex of the cochlea is responsive to low frequency at 20 Hz. Thus, the stimulation site on the microelectrode array closest to the tip will deliver stimulations representing the amplitude of high-frequency sound, while those closer to the base will deliver stimulations for low frequencies. Traditional manufacturing approach involves soldering individual wires (as small as 25 μm) under a microscope by hand and injection molding a silicone carrier around them (Johnson 2014). The resulting size of such hand-assembled arrays is too big to accommodate more than 20 channels. The labor-intensive manufacturing approach is time-consuming, expensive, and difficult to control uniform quality. The use of MEMS technology to manufacture these microelectrode arrays greatly reduces or eliminates these problems. Thin-film arrays can offer significant advantages by increasing the number of electrodes while at the same time reducing array size, allowing better frequency representation and deeper insertion into the scalar tympani (reaching lower frequency range), and reducing cost via automated batch fabrication. One of the latest development of MEMS demonstrated a robust and flexible 32-site prototype cochlear electrode array for a 128-site human prosthesis (Johnson 2014). The electrode array is designed to be self-curling to fit the spiral shape of the cochlea. These advances are possible only with the novel uses of MEMS technology.

Retinal Implants

Following the successful introduction of cochlear implants, researchers were exploring similar approaches to restore vision by implanting stimulating electrode arrays onto the retina (epiretinal implants) or inserting the array in-between the retinal tissue and the choroid (subretinal implants). The development of retinal implants not only benefit from but also are enabled by the powerful miniaturization capabilities offered by MEMS technology.

The World Health Organization estimates that, globally, 36 million people live with total vision loss and an additional 217 million people suffer from moderate to severe vision impairment (WHO.int). Retinitis pigmentosa (RP) is the leading cause of inherited blindness with 1.75 million patients worldwide (NEI.NIH. gov). It is characterized by progressive disintegration of the photoreceptors within the retina, starting with loss of peripheral vision and poor night vision. Severe cases of RP may progress to tunnel vision and even total blindness. Another vision disorder that compromises the retina is age-related macular degeneration (AMD), which is the leading cause of visual loss among adults over 65 years of age. The disease starts with blurry vision near the center of the visual field, the macula. Patients may develop blank spots that can severely interfere with daily activities. Because of population aging, AMD is projected to more than double from 2.07 million patients in 2010 to 5.44 million by 2050 in the US alone (NEI.NIH.gov).

Since these two diseases affect mainly the photoreceptors, the remaining neurons in the retina, including the bipolar cells and ganglion cells, can be electrically activated to elicit the sensation of sight (Weiland and Humayun 2014). The early versions of retinal implants adopted most of the system architecture from the successful cochlear implant systems, replacing the microphone with a miniaturized camera, the speech processing circuits with vision processor, and the linear electrode array with a 2D electrode matrix. The electronics and the camera are situated outside the body. The power and stimulation signals are wirelessly transferred to the stimulation electrodes inside the eye. In the early prototypes developed just before the turn of the century, there were only 16 electrodes arranged into a 4 × 4 array. Since then, the progressive adoption of MEMS technology, advances in electronics, uses of biocompatible materials, hermetic packaging techniques and better understanding of the retinal physiology and pathology have all contributed to FDA-approved retinal prosthetic devices. Argus® II Retinal Prosthesis System (Second Sight Medical Products, Inc), which is FDA-approved for blind individuals with severe to profound Retinitis Pigmentosa, consists of a 60-electrode epiretinal implant (SECONDSIGHT.com). The electronics are protected with a hermetic enclosure and implant longevity is estimated to be greater than ten years (Weiland and Humayun 2014). This system has allowed subjects to identify any letter of the alphabet, which restores a very important daily task, reading. Subjects using the implants are also able to navigate inside the house without bumping into walls or large objects.

Alternatively, the subretinal implant approach is to implant an electrode array behind the retina. This type of electrode is fabricated with microphotodiodes which collect incident light and transform it into electrical current to stimulate the retinal ganglion cells (Weiland and Humayun 2014). The idea is to directly replace the degenerated photoreceptors with engineered photodiodes, while retaining the use of the rest of the eye including the cornea, iris, and lens. The obvious advantage of the subretinal approach is the elimination of the need for an external camera, visual processor, and wireless transmission of signals and power, greatly simplifying the system. For epiretinal implants, the power consumption is directly proportional to the number of signal-processing channels, where one independent channel is needed per stimulation electrode. Due to power budget, the maximum available pixels (electrodes) to the user are currently 60. On the other hand, there is no signal processing required for subretinal implants, and thus the chip can be fabricated with significantly more electrodes, allowing the patient to see many more pixels. Finally, and perhaps most importantly, the fewer engineered substitutes of the natural biological system the easier it is for the user to adopt to. Users of the subretinal implants can move the eye, adjust the amount of light input with the iris, and accommodate viewing distance with the lens, which are not possible with epiretinal implants. However, the major drawback of subretinal implants is that engineered photodiodes are far less efficient than photoreceptors in converting light into electricity. The current subretinal implants include circuitry to amplify the photocurrent to an adequate level, which then requires an external power source. Nevertheless, a subretinal implant system with 1500 electrodes were demonstrated with the Alpha IMS® produced by Retina Implant AG, Reutlingen, Germany (Stingl et al. 2015). This device won the European CE Mark in 2013, and is the second commercially available retinal prosthetic system as of 2018. Initial trials seem to indicate that this subretinal implants tend to degrade faster than the epiretinal counterparts. A possible reason is that the subretinal electrode chip must be made very thin and small in order to fit behind the retina, and thus packaging that protect the chip from the biological environment is not as robust as epiretinal implants. Both approaches are continuing to evolve and more sophisticated and reliable versions are being developed.

MEMS Application: Lab-on-a-Chip

Another major area of advances that is made possible by the miniaturization capabilities of MEMS is Lab-on-a-Chip. As the name implies, this field is characterized by drastic miniaturization of forms and functions of a conventional chemical, biological, or clinical laboratory onto a chip-scale platform. It is a device enabled by MEMS that integrates one or several laboratory functions onto a chip from few millimeters to a few centimeters on a side (Volpatti and Yetisen 2014). The advantages of reducing a roomful of instruments in a conventional laboratory onto a chip with integrated smart electronics include drastic reduction in cost of the platform and the functions they perform, reduction in sample size and consumables, speed in achieving test results, automation, and massively deployable and widely accessible. The demand and rapid growth of the Lab-on-a-Chip field is captured by the establishment and success of one of the premier journals in the field, *Lab on a Chip*, published by the Royal Society of Chemistry (RSC.org). The journal was first introduced in 2001, and now carries an Impact Factor of 5.995. It comprehensively represents the field on novel micro- and nano-technologies and fundamental principles including (RSC.org)

- Micro- and nano-fabrication (including 3D printing, thin films).
- Micro- and nano-fluidics (in continuous and segmented multiphase flow, droplet microfluidics, new liquids).
- Micro- and nano-systems (sensor, actuator, reaction).
- Micro- and nano-separation technologies (molecular and cellular sorting).
- Micro- and nano-total analysis system (μTAS, nTAS).
- Digital microfluidics.

- Sample preparation.
- Imaging and detection.

It also publishes on significant biological, chemical, medical, environmental and energy applications such as (RSC.org)

- Nucleic acid biotechnology and analysis (DNA and RNA sequencing, genotyping, gene manipulation).
- Protein analysis (proteomics and metabolomics for targeted and global analysis).
- Medical diagnostics (for example point of care and molecular).
- Medical devices and treatments (including implantable and wireless).
- Drug development (screening and delivery).
- Cells, tissues, organs on chip and integrated tissue engineering.
- 3D cell culture.
- Single cell analysis.
- Cell and organism motility and interactions.
- Systems and synthetic biology and medicine.
- Energy, biofuels, fuel extraction.
- Environmental and food monitoring for health and security.

The above list of both the technology and applications are growing under the major driving force to provide medical cares cheaper, faster, and more readily available. Two of the most common used of Lab-on-a-chip are for gene sequencing and point-of-care diagnostics, and are described below.

Lab-on-a-chip for Gene Sequencing

One of the more prominent uses of lab-on-a-chip is for DNA analysis and gene sequencing. Since the completion of the Human Genome Project in 2001, with the publication of 90% complete sequence of all three billion base pairs in the human genome, the field of genomic research exploded. Dr. Francis Collins, the then Director of the National Center for Human Genome Research (NCHGR), wrote, "The information derived from the Human Genome Project, an international effort to decode the information embedded in the human genome, will revolutionize the practice of medicine in the 21st century by providing the tools to determine the hereditary component of virtually all diseases. This will lead to improved approaches to predict increased risk, provide early detection, and promote more effective treatment strategies" (Collins and Mansoura 2001). Since then, the possibility of performing fast and small-volume nucleic acid amplification and analysis on a single chip has attracted great interest (Zhang and Xing 2007). The ability to analyze a minute amount of genetic materials is greatly enhanced by duplicating the DNA segment in the materials up to thousands to millions times, a process called amplification. It is most commonly done by polymerase chain reaction (PCR), in which primers containing sequences complementary to the target DNA segment and a DNA polymerase are mixed with the sample followed by repeated thermal cycling. During each heating and cooling cycle, the DNA segments go through melting and enzyme-driven DNA replication, doubling in numbers. During the following cycle, the newly formed DNA segments themselves serve as template for replication, and so the number of copies grows exponentially with additional thermal cycles. Because of the need to repeatedly heat up and cool down the sample, a smaller sample chamber will be more efficient and faster for cycling than bigger one, and thus MEMS offers its key enabling advantage. Reports have shown that sample volumes as small as 0.45 nl were successfully amplified in a 72 parallel reverse transcription (RT)-PCRs fabricated with MEMS technology (Marcus et al. 2006). Also, since MEMS is inherently an integration technique, additional functions can be integrated with PCR chips. For example, capillary electrophoresis, a common approach for sample separation, has been one of the functions integrated with PCR on a single chip, greatly facilitating continuous operation from sample preparation to analysis for DNA contents (Easley et al. 2006). This "sample in, answer out" capability is only possible with the use of MEMS.

Lab-on-a-Chip for Point-of-Care (POC) Diagnostics

Another dramatic miniaturization of laboratory equipment onto a chip-scale platform is in blood analysis, which is perhaps the most common clinical laboratory work. By integrating sample preparation and fluidic processing on the microscale with highly-sensitive electrochemical detections, it has been demonstrated that blood analysis to quantify multiple analytes such as electrolytes, metabolites, and gases while performing immunoassays can be done with a hand-portable device (Davis et al. 2003). Such a device is commercially available (Abbott).

In addition to miniaturing testing equipment so that they become portable, MEMS also allows significant cost reduction. One of the great challenges in global health today is to develop technologies to improve the health of people in the poorest regions of the world (Chin et al. 2011). For example, enzyme-linked immunosorbent assay (ELISA), which is commonly available in developed countries and is a powerful tool in detecting the presence of antigen as an indicator for specific diseases, are out of reach in third-world countries. By using MEMS manufacturing process and the integration of multiple functions of fluidic processing and signal detection to realize an easy-to-use POC assay that performs ELISA at a much lower cost would be a significant step towards addressing this need. Chin et al. (2011) have demonstrated their "mChip" assay in Rwanda on hundreds of locally collected human samples, which showed very promising results of diagnosing HIV and syphilis co-infection with sensitivities comparable to reference standards.

Pushing the manufacturing cost even lower is the development of paper-based microfluidic POC diagnostic devices (Yetisen et al. 2013). The first example of paper-based diagnostics is the diabetes dipstick for quantifying glucose in urine samples. They offer rapid results at a low cost and is compact, easy to carry. However, the accuracy and specificity are never as high as a regimented laboratory procedure. The development of advanced use of paper as fluidic processing materials and quantitative biosensors offer the potentials to deliver diagnostic tools to third-world countries widely and cost effectively. Paper making is a very mature process and the quality is well-controlled. First, the capillary action in paper serves as a natural fluid transport mechanism without the use of pumps and valves. Second, patterning and printing certain region of the paper with wax or other similar materials transforms that region hydrophobic, and thus confine the flow of fluid in the unprinted regions. Third, clever designs and use of chemicals to functionalize paper can turn it into a sensing substrate. Fourth, the sample size needed to perform diagnosis is small. Fifth, the paper-based device is disposable by incineration, eliminating hazardous waste. Further, most of the readouts are based on color changes on the functionalized paper, development in using smart phones to perform color analysis may eliminate the subjectivity of manual comparison of the test results against a color chart.

Summary

The drive to miniaturize sensors and actuators has led to the use of photolithographic process to create ultra-miniaturized mechanical components. The field of MEMS leverages heavily the batch fabrication and micro-precision capabilities of the planar process in the mature IC industry. The broad categories of bulk micromachining, surface micromachining, soft lithography, wafer bonding, and integration techniques have been combined to create miniaturized biosensors and microfluidic devices. The two most prominent uses of MEMS for biomedical applications are in micro-implantable devices and lab-on-a-chip. Two important examples of micro-implants are cochlear prosthesis to restore hearing for profoundly deaf patients and retinal implants to allow limited vision for blind persons. Lab-on-a-chips are most used in gene sequencing and POC diagnostics because of the miniaturization and cost reduction capabilities of MEMS technologies.

References

[Abbott] Abbott Point of Care, Inc., www.pointofcare.abbott/us/en/offerings/istat/istat-handheld. Accessed 2018-08-24.

Bannon, III, F.D., J.R. Clark and C.T.C. Nguyen. 2000. High frequency micromechanical filters. IEEE J. Solid-State Circuits 35(4): 512–526.

Bartek, M., P.T.J. Gennissen, P.J. French and R.F. Wolffenbuttel. 1995. Confined selective epitaxial growth: Potential for smart silicon sensor fabrication. Proc. 8th Int. Conf. Solid-State Sensors and Actuators (Transducers'95/Eurosensors IX), Stockholm, Sweden, June 25–29, 1995, 91–94.

Böhringer, K.-F., K. Goldberg, M. Cohn, R.T. Howe and A.P. Pisano. 1998. Parallel microassembly with electrostatic force fields. Proc. IEEE 1998 Int. Conf. Robot. Automat., Leuven, Belgium, May, 1998, 1204–1211.

[BRAINYQUOTE] 2018a. Robert Moog Quotes. Brainy Quote. [https://www.brainyquote.com/authors/bobert_moog] accessed August 25, 2018.

[BRAINYQUOTE] 2018b. Amy Heckerling Quotes. Brainy Quote. [https://www.brainyquote.com/authors/amy_heckerling] accessed August 25, 2018.

[BRAINYQUOTE] 2018c. Brainy Quote. [https://www.brainyquote.com/search_results?q=technology] accessed August 25, 2018.

Burns, D.W. 1988. Micromechanics of Integrated Sensors and the Planar Processed Pressure Transducer, Ph.D. Thesis, Univ. Winsconsin, Madison.

Chen, C.S., M. MrKsich, S. Huang, G.M. Whitesides and D.E. Ingber. 1997. Geometric control of cell life and death. Science 278: 1425–1428.

Cheng, Y.T., L. Lin and K. Najafi. 2000. Fabrication and hermeticity testing of a glass-silicon package formed using localized aluminum/silicon-to-glass bonding. Proc. IEEE Micro Electro Mech. Syst. Workshop, Miyazaki, Japan, Jan. 23–27, 2000, 757–762.

den Besten, C., R.E.G. van Hal, J. Muñoz and P. Bergveld. 1992. Polymer bonding of micromachined silicon structures. Proc. IEEE Micro Electro Mech. Syst. Workshop, Travemünde, Germany, Feb. 4–7, 1992, 104–109.

Chin, C.D., T. Laksanasopin, Y.K. Cheung, D. Steinmiller, V. Linder, H. Parsa et al. 2011. Microfluidics-based diagnostics of infectious diseases in the developing world. Nature Med. 17: 1015–1019.

Collins, F.S. and M.K. Mansoura. 2001. The Human Genome Project: Revealing the shared inheritance of all humankind. Cancer. 91(1): 221–225.

Davis, G., I.R. Lauks, C. Lin and C.J. Miller. 2003. Apparatus and methods for analyte measurement and immunoassay. Patent WO2003076937A3.

Diem, B., M.T. Delaye, F. Michel, S. Renard and G. Delapierre. 1993. SOI (SIMOX) as a substrate for surface micromachining of single crystalline silicon sensors and actuators. Proc. 7th Int. Conf. Solid-State Sensors and Actuators (Transducers'93), Yokohama, Japan, June 7–10, 1993, 233–236.

Easley, C.J., J.M. Karlinsey, J.M. Bienvenue, L.A. Legendre, M.G. Roper, S.H. Feldman et al. 2006. A fully integrated microfluidic genetic analysis system with sample-in–answer-out capability. Proc. Natl. Acad. Sci. 103: 19272–19277.

[EMBS.org] www.embs.org. Accessed 2018-08-14.

[FDA.gov] www.fda.gov. Accessed 2018-08-21.

Franke, A.E., Y. Jiao, M.T. Wu, T.-J. King and R.T. Howe. 2000. Post-CMOS modular integration of poly-SiGe microstructures using poly-Ge sacrificial layers, Tech. Dig. Solid-State Sensor and Actuator Workshop, Hilton Head Island, SC, June 4–8, 2000, 18–21.

Gianchandani, Y.B. and K. Najafi. 1992. A bulk silicon dissolved wafer process for microelectromechanical devices. IEEE J. Microelectromech. Syst. 1(2): 77–85.

Gianchandani, Y.B., H. Kim, M. Shinn, B. Ha, B. Lee, K. Najafi et al. 1998. A MEMS-first fabrication process for integrating CMOS circuits with polysilicon microstructures. Proc. IEEE Micro Electro Mech. Syst. Workshop, Heidelberg, Germany, Jan. 25–29, 1998, 257–262.

[IEEE.org] www.ieee.org. Accessed 2018-08-14.

Jiang, G. and D.D. Zhou. 2010. Technology advances and challenges in hermetic packaging for implantable medical devices. pp. 28–61. In: Zhou, D.D., E.S. Greenbaum (eds.). Implantable Neural Prostheses 2: Techniques and Engineering Approaches. Berlin: Springer.

Joung, Y.-H. 2013. Development of implantable medical devices: From an engineering perspective. Int. Neurourol. J., DOI:10.5213/inj.2013.17.3.98.

Kendall, D.L. 1975. On etching very narrow grooves in silicon. Appl. Phys. Lett. 26: 195–198.

Kern, W. and C.A. Deckert. 1978. Chemical etching. pp. 401–496. In: Vossen, J.L. and W. Kern (eds.). Thin Film Processes, Orlando: Academic Press.

Kloeck, B., S.D. Collins, N.F. de Rooij and R.L. Smith. 1989. Study of electrochemical etch-stop for high-precision thickness control of silicon membranes. IEEE Trans. Electron Devices ED-36(4): 663–669.

Knecht, T.A. 1987. Bonding techniques for solid state pressure sensors. Proc. 4th Int. Conf. Solid-State Sensors and Actuators (Transducers '87), Tokyo, Japan, June 2–5, 1987, 95–98.

Ko, W.H., J.T. Suminto and G.J. Yeh. 1985. Bonding techniques for microsensors. pp. 41–60. In: Fung, C.D., P.W. Cheung, W.H. Ko and D.G. Fleming (eds.). Micromachining and Micropackaging of Transducers. The Netherlands: Elsevier.

Lärmer, F. and A. Schilp. 1996. Method of anisotropically etching silicon. U.S. Patent no. 5,501,893, Mar. 26, 1996.

Lärmer, F., A. Schilp, A. Funk and M. Offenberg. 1999. Bosch deep silicon etching: improving uniformity and etch rate for advanced MEMS applications. Proc. IEEE Micro Electro Mech. Syst. Workshop, Orlando, FL, USA, Jan. 17–21, 1999, 211–216.

Legtenberg, R., S. Bouwstra and M. Elwenspoek. 1991. Low-temperature glass bonding for sensor applications using boron oxide thin films. J. Micromech. Microeng. 1: 157–160.

Marcus, J.S., W.F. Anderson and S.R. Quake. 2006. Parallel picoliter RT-PCR assays using microfluidics. Anal. Chem. 78: 956–958.

Mastrangelo, C.H. and R.S. Muller. 1989. Vacuum-sealed silicon micromachined incandescent light source. Tech. Dig. Int. Electron Device Meeting, Dec. 1989, 503–506.

[MEDEL.com] MED-EL Corp. www.medel.com. Accessed 2018-08-20.

Nakakubo, T. and I. Shimoyama. 2000. Three-dimensional micro self-assembly using bridging flocculation. Sensors and Actuators A83(1–3): 161–166.

[NEI.NIH.gov] National Eye Institute. www.nei.nih.gov. Accessed 2018-08-20.

[NIDCD.gov] National Institute on Deafness and Other Communication Disorders. www.nidcd.nih.gov. Accessed 2018-08-20.

Nguyen, H., P. Patterson, H. Toshiyoshi and M.C. Wu. 2000. A substrate-independent wafer transfer technique for surface-micromachined devices", Proc. IEEE Micro Electro Mech. Syst. Workshop, Miyazaki, Japan, Jan. 23–27, 2000, 628–632.

Qin, D., Y. Xia and G.M. Whitesides. 2010. Soft lithography for micro- and nanoscale patterning. Nature Protocols 5(3): 491–502.

[RSC.org] Lab on a Chip by the Royal Society of Chemistry. www.rsc.org/journals-books-databases/about-journal/lab-on-a-chip/. Accessed 2018-08-24.

Schwartz, B. and H. Robbins. 1976. Chemical etching of silicon: IV. Etching technology. J. Electrochem. Soc. 123(12): 1903–1909.

[SECONDSIGHT.com] Second Sight Medical Products, Inc., www.secondsight.com. Accessed 2018-08-22.

Seidel, H., L. Csepregi, A. Heuberger and H. Baumgärtel. 1990. Anisotropic etching of crystalline silicon in alkaline solutions: I. Orientation dependence and behavior of passivation layers. J. Electrochem. Soc. 137(11): 3612–3626.

Shimbo, M., F. Kurukawa, F. Fukuda and K. Tanzawa. 1986. Silicon-to-silicon direct bonding method. J. Appl. Phys. 60(8): 2987–2989.

Singh, A., D.A. Horsley, M.B. Cohn, A.P. Pisano and R.T. Howe. 1997. Batch transfer of microstructures using flip-chip solder bump bonding. Proc. 9th Int. Conf. Solid-State Sensors and Actuators (Transducers '97), Chicago, IL, USA, June 16–19, 1997, 265–268.

Stingl, K., K.U. Bartz-Schmidt, D. Besch, C.K. Chee, C.L. Cottriall, F. Gekeler et al. 2015. Subretinal visual implant Alpha IMS—Clinical trial interim report. Vision Res. 111: 149–160.

Sugiyama, S., T. Suzuki, K. Kawahata, K. Shimaoka, M. Takigawa and I. Igarashi. 1986. Microdiaphragm pressure sensor. Tech. Dig. Int. Electron Device Meeting, Los Angeles, CA, Dec. 1986, 184–187.

Takao, H., H. Fukumoto and M. Ishida. 2000. Fabrication of a three-axis accelerometer integrated with commercial 0.8 μm-CMOS circuits, Proc. IEEE Micro Electro Mech. Syst. Workshop, Miyazaki, Japan, Jan. 23–27, 2000, 781–786.

Tang, W.C., T.-C.H. Nguyen and R.T. Howe. 1989. Laterally driven polysilicon resonant microstructures. Sensors and Actuators 20(2-Jan): 25–32.

Volpatti, L.R. and A.K. Yetisen. 2014. Commercialization of microfluidic devices. Trends Biotechnol. 32(7): 347–350.

von Philipsborn, A.C., Lang, S., Bernard, A., Loeschinger, J., David, C., Lehnert, D. et al. 2006. Microcontact printing of axon guidance molecules for generation of graded patterns. Nat. Protoc. 1: 1322–1328.

Webster, J.R., M.A. Burns, D.T. Burke and C.H. Mastrangelo. 2000. Electrophoresis system with integrated on-chip fluorescence detection. Proc. IEEE Micro Electro Mech. Syst. Workshop, Miyazaki, Japan, Jan. 23–27, 2000, 306–310.

Weiland, J.D. and M.S. Humayun. 2014. Retinal prosthesis. IEEE Trans. Biomed. Eng. 61(5): 1412–1424.

Whitesides, G.M., E. Ostuni, S. Takayama, X. Jiang and D.E. Ingber. 2001. Soft lithography in biology and biochemistry. Annu. Rev. Biomed. Eng. 3: 335–373.

[WHO.int] World Health Organization. www.who.int. Accessed 2018-08-22.

Xie, H., L. Erdmann, X. Zhu, K.J. Gabriel and G.K. Fedder. 2000. Post-CMOS processing for high-aspect-ratio integrated silicon microstructures. Tech. Dig. Solid-State Sensor and Actuator Workshop, Hilton Head Island, SC, June 4–8, 2000, 77–80.

Yeh, H.-J.J. and J.S. Smith. 1994. Fluidic assembly for the integration of GaAs light-emitting diodes on Si substrates. IEEE Photon. Technol. Lett. 6: 706–708.

Yetisen, A.K., M.S. Akram and C.R. Lowe. 2013. Paper-based microfluidic point-of-care diagnostic devices. Lab Chip 12: 2210–2251.

Zhang, C. and D. Xing. 2007. Miniaturized PCR chips for nucleic acid amplification and analysis: latest advances and future trends. Nucleic Acids Res. 35(13): 4223–4237.

QUESTIONS

1. What is the purpose of applying MEMS technology to biomedical engineering?

2. What are the key elements of MEMS processes?

3. What are microimplants and why are they needed? What roles do MEMS play in these devices?

4. What are POC diagnostics? What advantages are there to miniaturize and lower the cost of POC diagnostics?

5. Why is there a need to continue the exploration of novel MEMS?

PROBLEMS

The students will be grouped into 3–5 people each. Within each group, students will search the internet and brainstorm to find a real-life medical device that can benefit from using MEMS to miniaturize and/or save manufacturing cost. Collaboratively, the students within each group will perform a MEMS design and develop a fabrication process for the device and present their results to the class.

19

Biomedical Imaging
Magnetic Resonance Imaging

Xiaohong Joe Zhou

QUOTABLE QUOTES

"There is nothing that nuclear spins will not do for you, as long as you treat them as human beings."

Professor Erwin Hahn, 1921-2016; American Physicist (WILEY 2018)

"I wanted to be free to try any silly thing I decided to do."

Professor Paul C. Lauterbur, 1929-2007; 2003 Nobel Prize Laureate for Inventing MRI
(BRAINY QUOTE 2018a)

"To see the world for a moment as something rich and strange is the private reward of many a discovery."

Professor Edward M. Purcell, 1912-1997; 1952 Nobel Prize Laureate for Discovery of NMR
(BRAINY QUOTE 2018b)

Learning Objectives

This chapter is intended to introduce magnetic resonance imaging (MRI) to students who have had minimal exposure to this medical imaging technology.
After completing this chapter, the students should:

- Understand the origin of MRI signals.

- Be able to describe how magnetization interacts with radiofrequency (RF) pulses.

- Understand the three spatial encoding approaches: slice-selection, frequency-encoding, and phase-encoding.

- Understand the essential contrast mechanisms for anatomic and functional imaging.

- Become familiar with selected advanced applications.

- Recognize the present challenges in spatiotemporal resolution of MRI.

- Most importantly, appreciate MRI as an outstanding example of engineering and physics development that is highly impactful on medicine.

Center for MR Research and Departments of Radiology and Bioengineering, University of Illinois College of Medicine, MRI Center, MC-707, Suite 1A, 1801 West Taylor Street, Chicago, Illinois 60612, USA; xjzhou@uic.edu

Introduction

According to a survey among physicians, magnetic resonance imaging (MRI) and computed tomography (CT) are ranked the most important innovations over the past decades (Medscape 2001). Since their initial demonstrations in the early 1970's (Lauterbur 1973, Peeters et al. 1979), these two medical imaging modalities have not only revolutionized diagnostic radiology but also modernized other disciplines in biomedicine. Compared to CT, MRI features a number of additional advantages, including exquisite soft tissue contrast for visualizing anatomic structures, versatility in contrasts to reveal tissue physiologic, metabolic, and functional changes, and absence of harmful ionizing radiation. MRI also provides a fertile ground where engineers, scientists, mathematicians, and clinicians interact closely with each other to invent and innovate for improving human health. After more than four decades of development, MRI has become an indispensable tool in clinical diagnosis, treatment monitoring, as well as in biomedical research.

The goal of this chapter is to employ MRI as an example to illustrate how physics and engineering principles can be applied to medical imaging. A significant emphasis is placed on MRI signal generation and detection, spatial encoding and decoding, spatiotemporal resolution, and various contrast mechanisms. Throughout the chapter, we will present selected examples of MRI application and instrumentation. Given the depth and breadth of MRI technologies and the broad scope of their applications, it is not possible to thoroughly cover all facets of MRI in this chapter. In spite of this, it remains our intention to provide sufficient motivation for medical students to embrace physics and engineering principles to enhance their career as clinicians, and for engineering and science students to develop novel technologies with a focus on solving important biomedical problems.

MRI Signals

Origin of the Signal

In the human body, although hydrogen accounts for only ~9.5% of the body weight, its atomic percentage (63%) is much higher, far exceeding that of oxygen (26%) and carbon (9%) (Frausto Da Silva and Williams 2001). The vast majority of hydrogen atoms are in water, although many other molecules such as triglycerides (lipids) and proteins are also rich in hydrogen. The nucleus of a hydrogen atom contains a single proton surrounded by an electron. Similar to the electron, the proton spins with a specific angular frequency. This spinning property is often termed as proton spin, or more generally, nuclear spin or simply spin (Note that a neutron also spins). Within an imaging voxel of $1 \times 1 \times 1$ mm^3 of a representative biological tissue, approximately 10^{19} proton spins may coexist. Each spin produces a tiny magnetic moment $\vec{\mu}$, which is randomly distributed in the absence of any external magnetic field (Fig. 1a). Because of the random orientation, individual magnetic moments within the voxel cancel each other, producing no net magnetization. When an external magnetic field $\vec{B}_0$ is applied, the spins will align either parallel or anti-parallel to the $\vec{B}_0$ field (Fig. 1b), ending their random orientation. Quantum mechanically, the parallel and

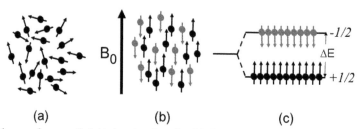

(a)	(b)	(c)

Fig. 1. (a) In the absence of a magnetic field, the spin orientation (black arrows) is random, resulting in a zero net magnetization. (b) When a $\vec{B}_0$ magnetic field is applied, the spins line up either parallel ($+\frac{1}{2}$; black arrows) or anti-parallel ($-\frac{1}{2}$; gray arrows) to the magnetic field with slightly more parallel spins than the anti-parallel spins. (c) Quantum mechanically, the $+\frac{1}{2}$ and $-\frac{1}{2}$ spins occupy different energy levels with the energy difference ΔE given by Eq. [2].

anti-parallel proton spins are denoted as ½ and - ½ spins, respectively. The ½ spins have a lower energy than -½ spins, as shown in Fig. 1c. The spin population in each energy state follows a Boltzmann distribution:

$$\frac{n_+}{n_-} = \exp\left(-\frac{\Delta E}{kT}\right), \tag{1}$$

where ΔE is the energy difference between the two energy states, τ is the absolute temperature in Kelvin (K), k is the Boltzmann constant (1.38×10^{-23} m²kg/s²/K), and n_+ and n_- are the number of spins in the ½ and -½ state, respectively. It is the population difference ($n_+ - n_-$) between the two energy states that gives rise to the MRI signal.

Equation [1] indicates that the population difference ($n_+ - n_-$) depends on the energy separation between the two energy levels (Fig. 1c). The energy difference, in turn, is proportional to the external magnetic field $\vec{B}_0$ amplitude (denoted by B_0):

$$\Delta E = \frac{1}{2\pi} \gamma h B_0, \tag{2}$$

where γ is the gyromagnetic ratio and h is the Planck constant (6.626×10^{-34} m²kg/s). Combining Eqs. [1] and [2], one can derive that the net magnetization $\vec{M}_0$, which arises from the vector summation of individual magnetic moments of ½ and -½ spins, can be expressed as:

$$M_0 \approx \frac{1}{4kT} \gamma^2 h^2 N_0 B_0, \tag{3}$$

where N_0 is the total number of the spins in a given system (e.g., water protons in a voxel, or total proton spins in an organ) and M_0 is the amplitude or modulus of $\vec{M}_0$. At the human body temperature (310 K) and in a magnetic field of 1.0 Tesla, the population difference between the ½ and -½ spin states is only ~ 7 ppm (parts per million), which explains the very low sensitivity of MRI. This sensitivity can be improved by increasing the magnetic field strength B_0, as indicated by Eq. [3].

The magnetic field strength used for clinical MRI studies is typically 1.5 Tesla or 3 Tesla. MRI scanners operating at a higher magnetic field strength, such as 7 Tesla, 9.4 Tesla, and 10.5 Tesla have been developed (Ugurbil 2014), with 7 Tesla MRI being adopted for clinical use. At these magnetic field strengths, low-temperature superconductors (such as niobium-titanium alloys) are used to produce a strong, stable, and homogeneous magnetic field needed for MRI. The costs of manufacturing superconducting magnets increase with the field strengths.

Although proton is the most frequently encountered nucleus in MRI, a number of other nuclei, such as ^{31}P, ^{23}Na, and ^{13}C, can also produce NMR signals. In general, any nuclei with an odd number of protons or neutrons are NMR active. Among those, some have ½ spins, others have an integer multiple of ½ spins, such as 1, 3/2, etc.

Manipulation of Magnetization

In the presence of a $\vec{B}_0$ magnetic field, the individual spins that give rise to the bulk magnetization are spinning about the direction of the magnetic field with an angular frequency, ω_0, described by the Larmor equation—one of the most important equations in MRI:

$$\omega_0 = \gamma B_0. \tag{4}$$

The frequency ω_0 is known as the Larmor frequency.

At equilibrium, the magnetization $\vec{M}_0$, arising from the vector summation of individual spins, is aligned along the direction of the magnetic field $\vec{B}_0$ (Fig. 2a), which is defined as the direction of the z-axis by convention. In order to detect the signal, the magnetization, known as the longitudinal magnetization M_z, must be tipped away from the z-axis to produce a component in the transverse plane (x, y), known as the transverse magnetization M_x, M_y, or M_{xy}. This is accomplished by applying a time-varying magnetic field $\vec{B}_1$ at a specific radiofrequency (RF), ω_{rf}, in the transverse plane, as shown in Fig. 2b. When the frequency of the $\vec{B}_1$ field equals the Larmor frequency (i.e., $\omega_{rf} = \omega_0$), nuclear magnetic resonance (NMR), or simply magnetic resonance (MR), occurs. The energy delivered by the $\vec{B}_1$ field causes a rotation of the longitudinal

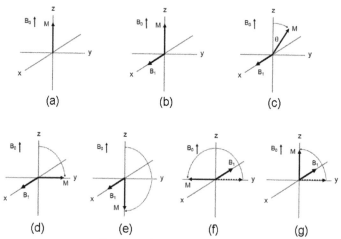

Fig. 2. Interaction between magnetization $\vec{M}$ and the $\vec{B}_1$ magnetic field. (a): In the absence of a $\vec{B}_1$ magnetic field, the equilibrium magnetization $\vec{M}$ is aligned along the $\vec{B}_0$ magnetic field (i.e., the direction of the z-axis). When a $\vec{B}_1$ magnetic field is applied along the x-axis (b), the magnetization is tipped towards the y-axis, producing a transverse component of the magnetization (c). (d)-(g): The effects of $\vec{B}_1$ magnetic field on magnetization for full excitation (d), inversion (e), refocusing (f), and restoration (g) are shown.

magnetization about the direction of the $\vec{B}_1$ magnetic field (Fig. 2c), and the rotation angle, θ, known as tip angle or flip angle, is given by:

$$\theta = \int_0^{T_0} \gamma B_1(t)dt, \qquad [5]$$

where T_0 is the duration, or the pulse width, of the applied $\vec{B}_1$ field.

When $0 < \theta \leq 90°$, the magnetization is partially or fully tipped from the longitudinal axis to the transverse plane, a process known as excitation (Fig. 2c for partial excitation and Fig. 2d for full excitation). The corresponding pulsed $\vec{B}_1$ magnetic field is referred to as an excitation RF pulse. When $\theta \approx 180°$, the magnetization is inverted from the $+z$-axis to the $-z$-axis (Fig. 2e), and the corresponding pulsed $\vec{B}_1$ magnetic field is referred to as an inversion RF pulse. If the magnetization is already in the transverse plane prior to the application of a $\vec{B}_1$ magnetic field (this situation can be created after an excitation RF pulse is already applied), a $\vec{B}_1$ magnetic field with $\theta \approx 180°$ results in the final magnetization remaining in the transverse plane at the end of a 180° rotation. Such pulsed $\vec{B}_1$ magnetic field is referred to as a refocusing RF pulse (Fig. 2f). In contrast, if the tip angle $\theta = -90°$, then the transverse magnetization is restored to the $+z$-axis as longitudinal magnetization. This pulsed $\vec{B}_1$ magnetic field is referred to as a restoring, fast-recovery, driven equilibrium, or flip-back RF pulse (Fig. 2g).

In practice, the $\vec{B}_1$ magnetic field is delivered by a coil operating at an RF frequency, known as an RF coil. The $\vec{B}_1$ field can be frequency or amplitude modulated to accomplish a variety of tasks to manipulate the magnetization for different applications (Bernstein et al. 2004). The power delivered by the RF coil can be absorbed by biological tissues being imaged, causing safety concerns such as local heating or even burning. The energy absorbed by tissue is quantified with a specific absorption rate (SAR; in units of watts per kilogram or W/kg), which must be carefully controlled to follow the regulatory guidelines (Bottomley 2008).

Essential MRI Signals and Relaxation Times

Free Induction Decay (FID) and T_2 and T_2^* Relaxations

After an excitation RF pulse is turned off, the resulting transverse magnetization will rotate about the $\vec{B}_0$ magnetic field in the (x,y)-plane at the Larmor frequency ω_0 (Fig. 3a). If an RF coil (which can be the same coil that produces the $\vec{B}_1$ magnetic field) is placed in the vicinity of the imaged object, then the

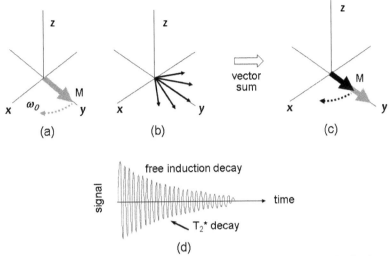

Fig. 3. Formation of free induction decay (FID). (a): Initial magnetization (gray arrow) immediately after an excitation RF pulse. (b): Dephasing among spin isochromats (thin arrows) in the magnetization due to $\vec{B}_0$ magnetic field inhomogeneity. (c): Reduced magnetization amplitude (black arrow) caused by dephasing. (d): FID and T_2^* decay.

rotating magnetization can induce an electric signal in the coil via electromagnetic induction (Fig. 3d). The amplitude of the signal, however, is not constant over time because of two transverse relaxation processes.

First, the transverse magnetization undergoes T_2 relaxation by which the individual spin isochromats in the magnetization dephase through spin-spin interaction. This causes the transverse magnetization (M_{xy}) to decay exponentially with a time constant of T_2 (Eq. [6] where M_0 is the initial magnetization). Intuitively, T_2 is the time required for the transverse magnetization to decay to ~ 37% (i.e., e^{-1}) of its initial value. T_2 relaxation is an inherent physical process that is independent of external factors such as $\vec{B}_0$ magnetic field inhomogeneity. The term of T_2 relaxation is used interchangeably with other terms such as transverse relaxation and spin-spin relaxation.

$$M = M_0 \exp\left(-\frac{t}{T_2}\right). \tag{6}$$

Second, the decay of the transverse magnetization can be accelerated by $\vec{B}_0$ magnetic field inhomogeneities (ΔB_0). In the presence of magnetic field inhomogeneities, the Larmor frequencies will be slightly different from each other among the spin isochromats in the magnetization at different locations, causing additional dephasing over time (Fig. 3b). The additional dephasing leads to a smaller vector sum as shown by the black arrow in Fig. 3c. Thus, the effective decay time constant, denoted by T_2^*, becomes shorter. T_2^* can be considered an "observed" or "apparent" T_2, whereas T_2 is the "ideal" or "inherent" transverse relaxation time of a tissue. T_2^* and T_2 are related by the following equation:

$$\frac{1}{T_2^*} = \frac{1}{T_2} + |\gamma \Delta B_0|, \tag{7}$$

which indicates that T_2^* is always shorter than or equal to T_2. The signal shown in Fig. 3d is subject to T_2^* relaxation and decays exponentially. This signal, which is produced in the absence of $\vec{B}_1$ magnetic field, is known as free induction decay (FID) (Brown et al. 2014). It is one of the major signals for image formation in MRI.

Spin Echo and T_2 Relaxation

The FID signal in Fig. 3 consists of contributions from many spin isochromats that experience difference magnetic fields due to the presence of local magnetic field inhomogeneity, as well as the intrinsic T_2 relaxation. To remove the influence of magnetic field inhomogeneity and reveal the true T_2-induced

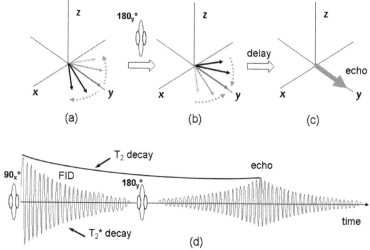

Fig. 4. Formation of spin echo. (a): Dephasing of spin isochromats (represented by the arrows with different shades of gray) following a 90° excitation RF pulse. (b): Rephasing or refocusing of the spin isochromats after a subsequent 180° refocusing RF pulse is applied. Note that the positions of the spin isochromats have been flipped by 180°, as represented by the reversed order of the gray shades in the arrows. (c): After a delay, a spin echo is formed (thick gray arrow). (d): 90° and 180° RF pulse pair showing the formation of FID and spin echo as well as the difference between T_2^* and T_2 decays. The subscript for the 90° and 180° pulses denotes the direction along which the pulse is applied.

signal decay that reflects the tissue intrinsic properties, a refocusing RF pulse can be applied following an initial excitation pulse (Fig. 4). After the excitation pulse (typically a 90° pulse along the x-axis), the spin isochomats dephase as shown in Fig. 4a. The subsequent 180° refocusing pulse rotates the dispersing spin isochromats about an axis in the transverse plane (e.g., the y-axis as shown in Fig. 4) such that the spin isochromats will converge or refocus (Fig. 4b). After a delay, the refocused magnetization forms a spin echo (Hahn 1950) (Figs. 4c and 4d). At the peak of the spin echo, phase dispersion (or dephasing) caused by magnetic field inhomogeneities among the spin isochromats is completely reversed. However, the intrinsic T_2 relaxation cannot be undone, resulting in a smaller echo amplitude than the initial amplitude of the FID. This signal decay reflects pure T_2 relaxation (Fig. 4d).

Refocusing RF pulses commonly have a flip angle of 180° as shown in Fig. 4. At this flip angle, the transverse magnetization is optimally refocused, and the largest spin echo signal is produced. Pulses with flip angles other than 180° are also capable of refocusing the transverse magnetization, although the refocusing is only partial.

Gradient Echo

Linear magnetic field gradient (or simply gradient) is crucial to MRI. In the presence of a gradient, magnetic field inhomogeneity is intentionally introduced in a controlled fashion. As shown in Fig. 5, the presence of a linear gradient results in one side of the imaged object experiencing a stronger magnetic field than the other side. Thus, the Larmor frequency varies linearly across the object along the direction of the applied gradient, $\vec{r}$:

$$\omega(r) = \gamma(B_0 + G \cdot r).$$ [8]

Following an excitation RF pulse that produces an FID signal, if a gradient is turned on for a fixed amount of time, then a phase dispersion will be produced across the object due to the spatially dependent frequency variation given by Eq. [8] (Fig. 6a). Such phase dispersion causes the signal to decay rapidly, essentially crushing out the signal when the gradient is sufficiently large and/or the duration sufficiently long. After the initial phase dispersion, if the gradient reverses its polarity, then an opposite phase dispersion is introduced (Fig. 6b). When the reversed gradient causes the same amount of phase "dispersion" as the previous gradient, the net phase dispersion becomes zero, resulting in a strong signal which is known as a

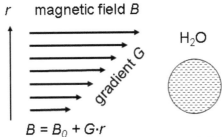

$$B = B_0 + G \cdot r$$

Fig. 5. Relationship between the overall magnetic field B and the linear magnetic field gradient G along the direction of $\vec{r}$. The presence of the gradient causes the spins at different locations along the $\vec{r}$ direction to resonate at different Larmor frequencies, leading to signal dephasing.

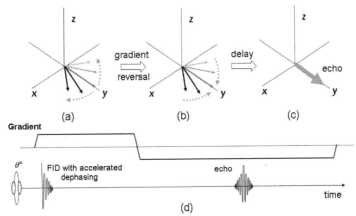

Fig. 6. Formation of gradient echo. (a): In the presence of a gradient, dephasing of spins is accelerated, resulting in a fast decay of the FID following a $\theta°$ excitation RF pulse applied along the x-axis. Note that dephasing of the spins is represented by the arrows with different shades of gray. (b): When the gradient polarity is reversed, the direction of spin "dephasing" is reversed accordingly, leading to rephrasing (shown by the curved arrows). Unlike the case of spin echo, the positions of the spin isochromats (arrows with different shades of gray) are not flipped. (c): After a delay, a gradient echo is formed. (d): Relationships among FID, gradient echo, and the gradient waveform.

gradient echo (Figs. 6c and 6d). Although the gradient echo illustrated in Fig. 6 arises from an FID signal, they can also be formed by utilizing a spin echo. Unlike spin echo which removes dephasing caused by $\vec{B}_0$ magnetic field inhomogeneities, gradient echo contains the effects of $\vec{B}_0$ magnetic field inhomogeneities, and thus is sensitive to T_2^* relaxation.

Inversion Recovery and T_1 Relaxation

So far, we have focused our discussion on the transverse magnetization. Now let us turn our attention to the longitudinal magnetization. As shown in Fig. 2a, the longitudinal magnetization is aligned along the $+z$-axis in the absence of the $\vec{B}_1$ magnetic field. Either an excitation or an inversion RF pulse perturbs this equilibrium by tipping the magnetization away from the $+z$-axis. In the case of a 180° inversion, the longitudinal magnetization is maximally perturbed (inset of Fig. 7). The inverted magnetization will return to its equilibrium via an exponential recovery process, also known as T_1 relaxation, longitudinal relaxation, or spin-lattice relaxation. T_1 is the exponential time constant of the recovery process, as shown in Eq. [9].

$$M(t) = M_0 \left[1 - 2exp\left(-\frac{t}{T_1} \right) \right]. \qquad [9]$$

Figure 7 displays the time evolution of the longitudinal magnetization in Eq. [9] following an inversion RF pulse. The curve shown in Fig. 7 is referred to as an inversion recovery (IR) curve. To measure the longitudinal magnetization, an excitation RF pulse must be applied after the inversion pulse to tip the

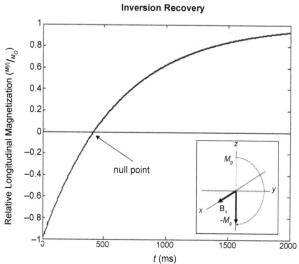

Fig. 7. An inversion recovery curve. A 180° inversion pulse maximally disturbs the longitudinal magnetization (inset), lead-ing to the longest recovery process described by Eq. [9]. The null point represents zero magnetization on the recovery curve.

longitudinal magnetization to the transverse plane. The time delay between the inversion and the excitation pulses is known as the inversion time denoted by TI. When $TI = T_1 ln2$, the magnetization becomes zero, which is often referred to as null point as shown in Fig. 7.

Spatial Encoding and Decoding

Gradient

Spatial information is encoded into MRI signals by linear magnetic field gradients. Generally, a "linear magnetic field gradient" refers to any linear spatial variation of the magnetic field. Specific to MRI, however, linear magnetic field gradient is defined as the spatial variation of the z-component of the $\vec{B}_0$ magnetic field, where the z-direction corresponds to the direction of $\vec{B}_0$.

The z-component of the magnetic field, B_z, can vary in any spatial directions. Mathematically, a gradient vector $\vec{G}$ is used to characterize the spatial variation:

$$\vec{G} = \frac{\partial B_z}{\partial x}\hat{x} + \frac{\partial B_z}{\partial y}\hat{y} + \frac{\partial B_z}{\partial z}\hat{z} = G_x\hat{x} + G_y\hat{y} + G_z\hat{z}, \tag{10}$$

where $\hat{x}$, $\hat{y}$ and $\hat{z}$ are the unit vectors in a Cartesian coordinate system, and G_x, G_y, and G_z are the three orthogonal components of $\vec{G}$. With the linear gradients, the overall magnetic field in the z-direction becomes

$$\vec{B}_z = (B_0 + G_x x + G_y y + G_z z)\hat{z}. \tag{11}$$

This is a generalization to 3D of what is depicted in Fig. 5.

In MRI, the linear magnetic field gradients are generated by a set of gradient coils. These coils are powered by gradient amplifiers to supply the needed current and voltage to control two important gradient parameters: gradient amplitude (or strength) and slew rate (or speed), respectively. The gradient amplitude describes how strong the magnetic field gradient is; and is measured in units of milliTesla per meter (mT/m). On commercial human scanners, the gradient amplitude typically ranges from 20–80 mT/m. The gradient slew rate, on the other hand, characterizes how fast the gradient amplitude can change; and has units of Tesla per meter per second (T/m/s). On state-of-the-art clinical MRI scanners, the gradient slew rate ranges from 50–200 T/m/s.

Although a high slew rate is desirable in fast imaging and other applications, rapid gradient switching can produce bioeffects, such as peripheral nerve stimulation, pain, or even ventricular fibrillation (Kanal et al. 2013). The slew rate must be controlled so that the maximally allowable *dB/dt* conforms to the safety guidelines. Rapid gradient switching with large changes in gradient amplitude can also exacerbate the acoustic noise during an MRI scan. In addition, gradient switching can generate eddy currents, leading to compromised image quality (Bernstein et al. 2004). These factors must be carefully considered before a fast slew rate and/or a strong gradient amplitude are used in an MRI examination.

As shown in Fig. 5, the primary function of a gradient is to introduce a spatially varying magnetic field so that spins at different locations can experience a slightly different magnetic field, and hence, a different Larmor frequency and/or phase. The spatially dependent frequency and phase are employed in MRI to perform 1D, 2D, or 3D spatial encoding as detailed below.

Slice Selection

A 2D section, or a slice, can be selected from a 3D object by using an RF pulse together with a magnetic field gradient applied along the slice-selection direction (i.e., normal to the slice; Fig. 8a). The RF pulse is amplitude-modulated (shaded in Fig. 8b) so that its frequency response closely mimics a rectangle function. In the presence of a concurrent magnetic field gradient, the rectangular frequency response Δf_{rf} is converted to a spatial response (Fig. 8c) where only spins within a selected spatial region, Δz, are affected by the RF pulse, achieving the goal of slice selection. Excitation, refocusing, inversion, and restoring RF pulses can all be made spatially selective, provided that they are properly amplitude-modulated. The amplitude modulation function can take many forms, such as a SINC function or a SINC function multiplied with a Hamming or Hanning window (Bernstein et al. 2004). In practice, analytical modulation functions do not produce an ideal boxcar-like slice profile. To improve the slice profile, numerically optimized or tailored RF pulses are often employed with various algorithms, such as the well-known Shinnar-Le Roux (SLR) algorithm (Pauly et al. 1991).

Unlike the amplitude-modulated RF pulse, the concurrent slice-selection gradient typically has a constant amplitude while the RF pulse is played out (Fig. 8b). Due to the limited slew rate, it takes time for the gradient to reach the desired amplitude at the beginning of the RF pulse and then return to zero at the end of the RF pulse. As such, the slice-selection gradient is straddled by ascending and descending ramps as shown in Fig. 8b. If the RF pulse performs excitation, then a subsequent slice rephasing gradient is used to ensure that the spins in the excited slice maintain their phase coherence. If the RF pulse performs refocusing, then a slice rephasing gradient is not needed. Instead, a pair of gradient lobes are often used to straddle the slice-selection gradient for the purpose of removing the FID signal inadvertently produced by an imperfect 180° RF pulse (e.g., the flip angle is not exactly 180° across the entire object) (Bernstein et al. 2004). For a given RF pulse waveform, the slice thickness is controlled by the gradient amplitude (i.e., the slope of the diagonal line in Fig. 8c). Increasing the amplitude of the slice-selection gradient G_z decreases the thickness of the slice Δz for a fixed RF bandwidth Δf_{rf} (Fig. 8c). The slice orientation is governed by the gradient direction (Fig. 8a). In order to produce a slice with any arbitrary orientation, two or three orthogonal gradients are used simultaneously to synthesize a gradient vector that is normal to the

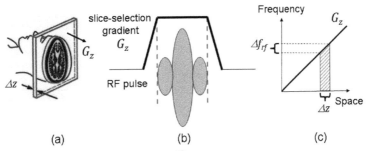

(a)   (b)   (c)

Fig. 8. Slice selection is achieved by applying a gradient $\vec{G}_z$ normal to the slice of interest (a). Concurrent to the slice-selection gradient G_z, an amplitude-modulated, band-limited RF pulse (shaded) is played out (b). The slice-selection gradient converts the frequency band Δf_{rf} of the RF pulse to the thickness of a slice Δz (c). By changing the gradient amplitude (i.e., the slope of the solid diagonal line in (c)), different slice thicknesses can be achieved.

desired slice. The slice location in a 3D object can be freely adjusted by changing the carrier frequency of the RF pulse.

Frequency Encoding

Frequency encoding is a common spatial encoding method in MRI. It is accomplished by applying a frequency-encoding gradient to the imaged object along a specific direction so that the Larmor frequency of the spins along that direction becomes proportional to their locations (Eq. [8]; Fig. 9). Under the influence of this gradient, time-domain signals (FIDs or spin echoes) contain a range of frequencies, each corresponding to a different spatial location. An inverse Fourier transform of the time domain signal reveals the signal amplitudes at different frequencies, and hence the spatial locations as indicated by Eq. [8].

To illustrate the effect of a frequency-encoding gradient on a spin system, let us consider a plate of water placed in a uniform magnetic field $\vec{B}_0$ as shown in Fig. 9a. Prior to the application of a frequency-encoding gradient, proton spins across the entire plate resonate with the same Larmor frequency, producing a single signal in the frequency spectrum (Fig. 9a) that does not contain the spatial information of the spins in the plate. After a frequency-encoding gradient, G, is applied along the vertical direction (Fig. 9b), the Larmor frequency of the spins becomes linearly related to their location along the direction of the gradient (Eq. [8]). Thus, the frequency spectrum reveals the number of spins at each frequency, thereby producing a projection of the objects (Fig. 9b). In doing so, spatial information is encoded into the signal.

A frequency-encoding gradient is often proceeded with a "pre-phasing" gradient that has the opposite polarity to that of the frequency-encoding gradient so that the phase dispersion introduced by the first half of the frequency-encoding gradient can be pre-compensated, leading to the formation of a gradient echo as described in Fig. 6. In a spin-echo acquisition, however, the pre-phasing gradient can have the same polarity as the frequency-encoding gradient when it is played out before the 180° refocusing pulse (Bernstein et al. 2004).

A frequency-encoding gradient can be applied along any direction. In a method known as projection reconstruction or radial sampling (Lauterbur and Lai 1980, Bernstein et al. 2004), the direction of the gradient is varied during the course of image acquisition, each producing a projection of the imaged object at a specific angle. Image reconstruction can be accomplished from all projections using algorithms similar to those used in CT. In another method known as Fourier imaging (Edelstein et al. 1980, Kumar et al. 2011), the frequency-encoding gradient is fixed to a specific direction (often known as the readout direction, frequency-encoding direction, or x-direction), and repeated while the phase-encoding value is incremented as discussed in the following subsection.

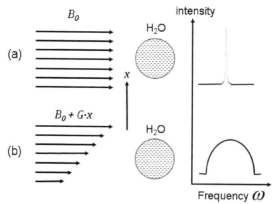

Fig. 9. Frequency encoding. In the absence of a linear magnetic field gradient, all spins in the water plate experience the same magnetic field B_0, leading to a single Larmor frequency in the spectrum (a). When a frequency-encoding gradient G_x is applied along the x-direction, spins in the water plate experience different magnetic field, and thus have different Larmor frequencies depending on their spatial locations. In doing so, the spatial information is frequency-encoded in the signal, leading to a frequency spectrum that corresponds to a projection of the water plate (b).

Phase Encoding

The fundamental effect of a magnetic field gradient on the signal is to change its precession frequency and broaden its frequency distribution. If the gradient is applied for only a short period of time, t_0, prior to the acquisition of the signal, then the phase of the signal can be changed because of the frequency variation. The amount of the phase change, ϕ, is proportional to not only the gradient amplitude and duration (i.e., the gradient area over time) but also the spatial location y along the direction of the applied gradient G_y:

$$\phi = \gamma y \int_0^{t_0} G_y(t)dt. \qquad [12]$$

Gradient G_y, which encodes the spatial information as spatially dependent phase into the signal, is called a phase-encoding gradient.

Phase encoding is applied after signal excitation, but before signal acquisition (Fig. 10). By varying the area under the phase-encoding gradient pulse, different amounts of phase variation can be introduced. Unlike frequency encoding in which many spatially dependent frequencies can be contained in a signal, each phase-encoding gradient introduces only one overall phase to the signal, although the overall phase contains contributions from multiple spin locations (Eq. [12]). As such, the phase-encoding gradient must be applied multiple times (Fig. 10), each with a different gradient amplitude, in order to obtain sufficient number of measurements to encode and decode the spatial locations. For example, for an image with 256 pixels along the phase-encoding direction, 256 phase-encoding gradient steps are needed in conventional MRI, which explains the lengthy scan time. The number of phase-encoding steps can be reduced in various fast imaging techniques, as discussed in the subsection on k-space.

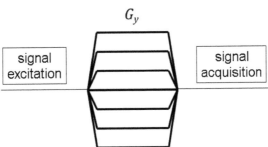

Fig. 10. Phase-encoding gradient G_y is applied after signal excitation, but before signal acquisition. Its amplitude varies in successive acquisitions in order to satisfy the sampling requirement.

2D Versus 3D Imaging

To image a 3D object, MRI can be performed in either a 2D multi-slice mode or a 3D volumetric mode. For 2D multi-slice imaging, a slice is first selected, followed by in-plane 2D spatial encoding using frequency encoding and/or phase encoding. A popular strategy is to employ frequency encoding along one direction and phase encoding along the other orthogonal direction within the 2D slice. This process is repeated for other slices so that multiple slices can be imaged. With the recent development of simultaneous multi-slice (SMS) imaging (Nunes et al. 2006, Bilgic et al. 2014), multiple (e.g., 2–8) slices can be simultaneously excited and spatially encoded, which can substantially reduce the scan time or improve the spatial coverage.

3D imaging can be accomplished in multiple ways. In a technique known as 3D radial sampling or 3D projection acquisition (Lauterbur and Lai 1980, Gu et al. 2005), only frequency encoding is employed by rotating a frequency-encoding gradient in 3D space to acquire a set of projections. A more prevalent 3D imaging technique is based on Fourier imaging. Typically, it uses frequency encoding in one dimension and phase encoding in the other two dimensions. There are also scenarios where phase encoding is used for all three dimensions so that the signal frequency can be reserved to measure other spin properties, such as the chemical shift (Brown et al. 1982). Although 3D imaging offers many advantages including higher signal-to-noise ratio (SNR) and contiguous spatial coverage, its acquisition time is typically much longer than 2D multi-slice imaging. Various fast imaging techniques have been developed to address this issue (see the next subsection). As a result, an increasing number of clinical MRI studies are now performed in 3D mode (Kijowski and Gold 2011). It is worth noting that 2D and 3D modes are not mutually exclusive. In some applications, such as magnetic resonance angiography (MRA), a slab (i.e., a thick slice) is first selected, followed by 3D imaging within the slab (Parker et al. 1991, Bernstein et al. 2004). Multiple slabs can be concatenated to form a 3D image in a manner similar to 2D multi-slice imaging.

k-Space

Irrespective of the specific image acquisition strategies, MRI signals are acquired in the spatial frequency domain which is related to the actual image (or image domain) by a 2D or 3D Fourier transform. The spatial frequency domain is known as k-space (Fig. 11). Under this notation, a frequency-encoded signal corresponds to a line in k-space. A set of k-space lines can be arranged rectilinearly in a Cartesian coordinate system (Fig. 11a) or as a set of radiating spokes in a polar coordinate system (Fig. 11b). The former corresponds to Fourier imaging where the separation between the parallel k-space lines is determined by the phase-encoding gradient, while the latter corresponds to radial sampling or projection acquisition. There are many other ways to sample k-space, each corresponding to a specific image acquisition technique. Spiral (Ahn et al. 1986) and PROPELLER (Pipe 1999) samplings are two examples that are used in some clinical MRI examinations.

The spatial resolution of an MR image is determined by the extent of k-space coverage (i.e., k_{max}), while the field-of-view (FOV) is determined by the separation of neighboring k-space points (i.e., Δk). A higher spatial resolution can be achieved by extending k-space coverage farther. To adequately sample k-space without excessively long scan times, various fast imaging techniques have been developed. These techniques are primarily based on two strategies. The first strategy focuses on improving k-space sampling efficiency by traversing k-space rapidly. Examples of this strategy include fast spin echo (FSE; also known as RARE or turbo spin echo) (Hennig et al. 1986), echo planar imaging (EPI) (Mansfield 2007), spiral (Ahn et al. 1986), twisted projection imaging (TPI) (Boada et al. 1997), etc. The second strategy is to acquire less data than what is conventionally required, followed by novel image reconstruction algorithms utilizing other information (such as RF coil sensitivity and data sparsity) and/or advanced mathematical tools. A large number of techniques in this category have been developed, as exemplified by SMASH (Sodickson and Manning 1997), SENSE (Pruessmann et al. 1999), GRAPPA (Griswold et al. 2002), compressive sensing (Lustig et al. 2007), etc.

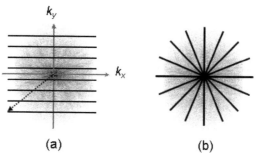

(a)   (b)

Fig. 11. *k*-Space sampling in a Cartesian coordinate system (a) and a polar coordinate system (b). Each line represents a frequency-encoded signal. The actual *k*-space data are shown in the background.

Image Reconstruction

After recognizing that the MRI data are acquired in the Fourier-conjugate space of the image space, image reconstruction (or spatial decoding) is reduced to a problem of a multi-dimensional Fourier transform. Although discrete Fourier transform (DFT) can be performed, the more time-efficient fast Fourier transform (FFT) algorithm is almost always employed. To use an FFT algorithm, MRI data are typically acquired with (or interpolated to) a matrix size of an integer power of 2 (i.e., 128×256, 256×256, or 512×512). For k-space data acquired rectilinearly in a Cartesian coordinate system, FFT can be applied directly. For k-space data acquired with non-Cartesian sampling, the raw data can be re-gridded to a rectilinear grid, followed by an FFT (Bernstein et al. 2004). A special case of non-Cartesian sampling is projection acquisition (Fig. 11b) where established projection reconstruction techniques can be used (Lauterbur and Lai 1980, Bernstein et al. 2004). In addition to these basic image reconstruction techniques, advanced reconstruction algorithms have also been developed for parallel imaging and other fast imaging methods (Sodickson and Manning 1997, Pruessmann et al. 1999, Griswold et al. 2002, Lustig et al. 2007).

MRI Pulse Sequences

MRI data acquisition is accomplished by using a pulse sequence which contains a series of events comprising RF pulses, gradient waveforms, and data acquisition in a coordinated fashion. Virtually all MRI applications, basic or advanced, are associated with one or more pulse sequences. Although a comprehensive description of MRI pulse sequences is beyond the scope of this chapter, we herein highlight two essential pulse sequences, based on which many other pulse sequences can be developed.

Gradient Echo Pulse Sequence

Figure 12 illustrates a simple 3D gradient echo (GRE) pulse sequence. The pulse sequence starts with an excitation RF pulse whose flip angle θ is less than 90° to partially tip the magnetization into the transverse plane. Following the excitation RF pulse, a frequency-encoding (or readout) gradient G_x is played out along the x-axis to produce a gradient echo. The time between the center of the excitation RF pulse and the peak of the echo is known as the echo time denoted by TE. Concurrent to the pre-phasing gradient (negative polarity) along the x-axis, two phase-encoding gradients are played out along the y- and z-axis, respectively, to phase-encode the spatial information in these two dimensions. Each phase encoding must step through multiple increments ($N_y \times N_z$ steps) to complete the phase-encoding process. The duration between two consecutive steps is known as repetition time denoted by TR (not shown in Fig. 12). At each phase-encoding step, a gradient echo signal is digitized to sample a line in k-space. After the sequence completes the $N_y \times N_z$ steps, each with a different phase-encoding value, a complete 3D k-space dataset is obtained.

The GRE pulse sequence shown in Fig. 12 can be modified to generate other 2D or 3D GRE pulse sequences for many applications. In GRE pulse sequences, the longitudinal magnetization can be repeatedly

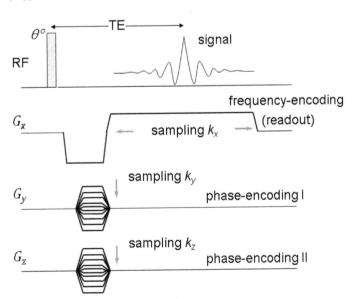

Fig. 12. A 3D gradient-echo (GRE) pulse sequence with frequency encoding along the x-axis, and phase encodings along the y- and z-axes.

used by subsequence excitation RF pulses because each RF flip angle is less than 90°. This feature, in conjunction with the fact that the longitudinal magnetization is never inverted due to the absence of RF refocusing pulses, allows a very short TR to be used for fast imaging.

Spin Echo Pulse Sequence

Figure 13 shows a simple 2D spin-echo pulse sequence that employs a 90° excitation and a 180° refocusing pulse to form a spin echo at each TR interval. To select a 2D slice, both 90° and 180° RF pulses are amplitude-modulated to excite and refocus spins only within the slice of interest. The slice-selection gradient G_z, concurrent with the 90° excitation pulse, requires a subsequent rephrasing gradient (negative polarity) to rephrase the spins within the slices, whereas the rephrasing gradient is not needed for the 180° pulse, as described in a previous subsection on slice-selection. The frequency-encoding gradient, G_x, and phase-encoding gradient, G_y, are applied along the *x*- and *y*-directions, respectively. On the *x*-axis, a pre-phasing gradient is played out before the 180° pulse to ensure that the dephasing effect of the first half of the frequency-encoding gradient is pre-compensated so that the maximal signal will occur when the spin echo is formed. Similar to the case of GRE, the time between the center of the 90° pulse and the peak of the echo is defined as TE.

Based on the simple spin-echo pulse sequence in Fig. 13, many other spin-echo pulse sequences can be derived. One particularly important sequence is FSE or turbo-spin echo (TSE) which employs multiple 180° refocusing pulses (e.g., 3–128) to produce a train of phase-encoded spin echoes so that the scan speed can be accelerated (Hennig et al. 1986).

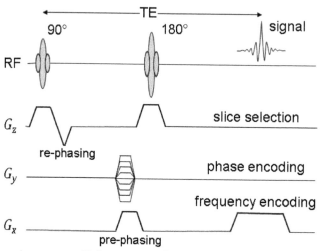

Fig. 13. A 2D spin-echo pulse sequence with frequency-encoding, phase-encoding, and slice-selection gradients along the *x*-, *y*-, and *z*-axes, respectively.

Contrast Mechanisms

Essential Contrasts for Anatomic Imaging

The essential contrasts for anatomic imaging include T_1-weighted, T_2-weighted, T_2*-weighted, and proton-density-weighted contrasts. Each contrast reflects different aspects of the signal characteristics among the tissues of interest. Multiple contrasts are often needed conjointly to take advantage of the complementary nature of the contrast mechanisms to achieve an optimal diagnostic accuracy. The contrast in an image depends on not only the type of the pulse sequence, but also specific pulse sequence parameters such as TR, TE, and the flip angle. Tissue contrasts can also be considerably enhanced by suppressing the signals from other tissues. Examples of this strategy include lipid suppression, cerebrospinal fluid (CSF) suppression, dark blood imaging, etc.

T_1-weighted Imaging

T_1-weighted contrast is typically obtained using a GRE pulse sequence because of its time efficiency. In a spoiled GRE pulse sequence (Bernstein et al. 2004), the signal intensity depends on three sequence parameters: TR, TE, and flip angle θ:

$$S = S_0 \frac{sin\theta \left(1 - e^{-\frac{TR}{T_1}} \right)}{\left[1 - (cos\theta)e^{-\frac{TR}{T_1}} \right]} e^{-\frac{TE}{T_2^*}}, \qquad [13]$$

where S_0 is the signal weighted by proton density. To achieve T_1-weighting, T_2^* effect must be minimized by using the shortest possible TE (e.g., 2–5 ms) and the T_1 effect emphasized by using a short TR (e.g., 5–10 ms) as well as a moderate flip angle (e.g., 30–50°).

T_1-weighted image contrast can also be obtained from a spin echo or a fast spin echo sequence. For spin echo, the signal intensity can be expressed as:

$$S = S_0 \left[1 - e^{-\frac{TR}{T_1}} \right] e^{-\frac{TE}{T_2}} . \qquad [14]$$

Similar to the case of GRE, a short TE and short TR are needed to produce a T_1-weighted contrast. For human brain tissues, for example, a TE of 10 ms and a TR of ~ 500 ms can be used in T_1-weighted imaging at 1.5 Tesla (Fig. 14a).

To enhance the T_1 contrast, two additional strategies are often employed. First, an inversion pulse can be added to the beginning of a pulse sequence. After an inversion time TI, improved T_1-contrast can be achieved between two tissues. Second, a contrast agent, which is typically based on gadolinium (Gd)

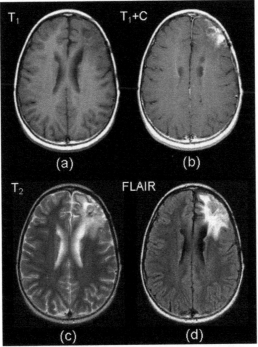

Fig. 14. Brain images of a glioma patient illustrating different contrasts. (a): T_1-weighted contrast without administration of Gd contrast agent; (b): T_1-weighted contrast with Gd contrast agent; (c): T_2-weighted contrast; and (d): FLAIR contrast.

chelators, is administered intravenously to shorten the T_1 values in tissues with a high contrast agent uptake. This method is very useful in detecting or characterizing lesions such as tumors (Fig. 14b).

T_2-weighted Imaging

T_2-weighted contrast can be obtained by using pulse sequences based on spin echoes. According to Eq. [14], a long TE and a long TR are needed to increase the sensitivity to T_2 while minimizing the influence of T_1. With a long TR, the scan time can increase considerably. In practice, this issue is addressed by using more time-efficient pulse sequences such as FSE. This sequence, in combination with parallel imaging, has made it possible to perform 3D volumetric T_2-weighted imaging in only a few minutes (Kijowski and Gold 2011).

Free water has a very long T_2 value (~ 2 s at 1.5 T), making CSF appear hyperintense in a T_2-weighted image (Fig. 14c). The intense CSF signals can sometimes mask T_2-weighted contrast between other tissues. An effective way to solve this problem is to apply an inversion pulse and wait until the inverted CSF signal to be at the null point (Fig. 7) before executing an FSE pulse sequence. Because CSF has a substantially longer T_1 (~ 4 s at 1.5 T) than the other brain tissues, when the CSF signal reaches the null point, the magnetization for other brain tissues is almost fully recovered, resulting in a T_2-weighted image with the CSF signal suppressed (Fig. 14d). This technique is known as fluid attenuated inversion recovery (FLAIR) (Hajnal et al. 1992).

T_2^*-weighted Imaging

Unlike T_2-weighted imaging, T_2^*-weighted imaging is almost always performed with a GRE pulse sequence because of its T_2^* sensitivity. When a spoiled GRE pulse sequence is employed (Eq. [13]), a long TE (20–50 ms), together with a long TR (e.g., 200–800 ms) and a small flip angle (e.g., 5–20°), is typically used to obtain T_2^* contrast. It is worth noting that the scan parameters often need to be adjusted for different B_0 field strengths, because tissue T_1 increases and T_2^* decreases with B_0. As such, a shorter TE and a longer TR are needed at 3.0 T than at 1.5 T.

Proton Density-weighted Imaging

Proton density-weighted imaging, or PD-weighted imaging, is intended to reflect the true proton contents in the tissue without being affected by the relaxation times and/or other factors. Many pulse sequences can be used for PD-weighted imaging. When a spin echo or FSE pulse sequence is employed, a long TR and a short TE must be selected to minimize T_1- and T_2-weighting, respectively. Similar to the case of T_2-weighted imaging, FSE is much more popularly used for PD-weighted imaging than spin-echo pulse sequences. When a GRE sequence is used (e.g., Eq. [13]), a long TR (200–500 ms), short TE (2–5 ms), and small flip angle (5–20°) should be chosen. Recently, pulse sequences capable of ultra-short TE (UTE), or even zero TE (ZTE), were developed to allow imaging of very short T_2 tissues such as the bone (Chang et al. 2015). These techniques rely on proton density contrast, and are an active area of research.

Lipid Suppression

Similar to CSF, lipids can also produce hyperintensity in an image, compromising the conspicuity of lesions. Protons in lipids have a different chemical environment than those in water, leading to different chemical and physical properties. Water is a small, agile molecule with its protons less "shielded" by the surrounding electrons due to the electronegativity of the oxygen atom. Lipids, on the other hand, are bulky molecules (primarily triglycerides) with most of their protons nested in aliphatic side chains among relatively electroneutral carbon atoms. As such, protons in lipids are more "shielded" by the surrounding electrons than those in water. The difference in electron shielding results in a slightly different magnetic field experienced by the protons, which is known as chemical shift (Rummens 1970, Brown et al. 1982). At 1.5 T, the chemical shift leads to a lower Larmor frequency of lipid protons than that of water protons by ~ 220 Hz. This very small difference is sufficient to enable lipid suppression techniques such as CHESS (chemical shift selective imaging) (Haase et al. 1985) and the Dixon method (Dixon 1984).

In CHESS, an excitation RF pulse is applied at the lipid proton frequency while leaving water protons intact. The excited lipid signal is immediately destroyed by a large "spoiling" gradient. The intact water signal is subsequently used for imaging. In the Dixon methods, the phase difference between water and lipid protons at a specific TE is exploited. After two or three measurements at different TEs to account for other factors affecting the phase, the water and lipid signals can be separated into different images.

The difference in molecular structures also leads to a much shorter T_1 value for lipid protons (~ 250 ms at 1.5 T) than for water protons (~ 900 ms for gray matter at 1.5 T). In a technique known as short TI inversion recovery (STIR) (Bydder and Young 1985), the difference in T_1 relaxation times is taken advantage of for lipid suppression. STIR applies an inversion pulse to invert the magnetization. After waiting for ~ 170 ms when the lipid signal is at the null point (Fig. 7), an imaging pulse sequence is applied to form an image in the absence of lipid signals. A major drawback of STIR is that the inverted water signal is not fully recovered at the time of imaging, resulting in a reduced SNR.

Essential Contrasts for Functional Imaging

Blood Oxygen Level Dependent (BOLD) Contrast

MRI is useful for imaging not only anatomy, but also functions. In neurofunctional imaging, BOLD contrast is a primary contrast mechanism based on which the vast majority of functional MRI (fMRI) studies have been performed. When neurons are activated, the increased need for oxygen is overcompensated by a larger increase in delivery of oxygenated blood supply. As a result, the venous oxyhemoglobin concentration increases and the deoxyhemoglobin concentration decreases. Because oxyhemoglobin is diamagnetic while deoxyhemoglobin paramagnetic, the T_2^* value in the activated areas increases, resulting in an elevated MRI signal intensity in a T_2^*-weighted image (Ogawa et al. 1990). Conversely, when there is no neuronal activation, a lower signal intensity is expected. This alternating signal pattern can be correlated to the task-on and task-off states according to a pre-designed paradigm for investigating a specific neurocognitive function (Kwong et al. 1992).

Diffusion Imaging

Diffusion is a "random walk" of molecules in a medium. In biological tissues, microstructures, cellular composition, and heterogeneity can substantially affect water molecular diffusion processes spatially and temporally. Measuring water diffusion is a viable way to probe tissue microstructures at a spatial scale much smaller than the voxel size (Le Bihan 2007). MRI signals are sensitive to diffusion, which provides an excellent window to study various aspects of tissue structures and the micro-environment. In the presence of a diffusion-weighting gradient, the MRI signal can be attenuated differently among tissues, introducing diffusion contrast to the image (Le Bihan 2007). The degree of attenuation depends on diffusion coefficient and an experimentally controllable parameter, known as the b-value (units: s/mm^2). The resultant diffusion-weighted images can be used to infer cellular density, fiber orientations, brain structural connectivity, tissue heterogeneity, and even the axonal diameter using various diffusion models (Tang and Zhou 2019).

Perfusion Imaging

Perfusion is a process that brings nutritive blood supply to the tissue through the arterial system and drains the metabolic byproducts into the veins. Perfusion measurement using MRI can be divided into two categories, those employing exogenous agents as a tracer, and those using water protons in the arterial blood as an endogenous label. Among exogenous agents for perfusion MRI, gadolinium chelates are most frequently used. To perform perfusion measurements, a bolus of gadolinium contrast agent is intravenously administered, followed by the rapid acquisition of a series of snap-shot images with either T_2^*-weighting or T_1-weighting. The former is known as dynamic susceptibility contrast (DSC) imaging, while the latter dynamic contrast-enhanced (DCE) imaging. In DSC, the time-series images are processed to extract perfusion-related parameters, such as cerebral blood volume, mean transient time, and time to peak. In

DCE, the images are analyzed with a pharmacokinetic model to yield a number of parameters relating to permeability, surface area, transfer constants, etc. (Jahng et al. 2014).

In a technique that relies on an endogenous tracer, the magnetization of arterial blood spins is labeled by an inversion RF pulse, causing a signal reduction when the labeled blood enters the imaging location. This method is known as arterial spin labeling (ASL) which can produce quantitative or semi-quantitative maps of cerebral blood flow and other parameters (Alsop et al. 2015). ASL can be implemented in a number of ways depending on the type of the labeling pulse, the imaging sequence employed, the strategies to compensate for magnetization transfer effects, and other factors. Despite the variations, a consensus has been developed for consistent use of ASL techniques (Alsop et al. 2015).

Selected Additional Imaging Contrasts and Applications

Blood Flow and MR Angiography

MRI signals are sensitive to blood flow. This flow sensitivity has led to a number of MR angiographic techniques. Some techniques (e.g., those based on FSE and/or inversion recovery) can produce dark blood signals, while others (e.g., those based on GRE) producing bright blood signals (Bernstein et al. 2004). Bright blood contrast can be obtained either with or without administration of exogenous contrast agents. Among the many angiographic techniques, time-of-flight (ToF) and phase-contrast (PC) are well established. ToF makes use of the inflow effect of fresh blood to the imaging plane or volume, while PC relies on flow-induced phase to discriminate flowing from static spins. The latter can also be used to quantitatively measure the flow rate.

Magnetic Resonance Elastography

Magnetic resonance elastography (MRE) is a technique that allows characterization of tissue mechanical properties, such as shear stiffness, in response to compression or vibration (Glaser et al. 2012). To perform MRE, a driver is used to deliver mechanical waves to the tissue. These mechanical waves can be imaged as they travel through the tissue using a phase-sensitive MR pulse sequence. Image acquisition can be accomplished within a breath hold. The ability to measure tissue mechanical properties can provide valuable information for characterizing a number of diseases, such as hepatic fibrosis (Glaser et al. 2012).

Susceptibility-weighted Imaging and Quantitative Susceptibility Mapping

When an object is placed in a $\vec{B}_0$ magnetic field, a tiny internal polarization $\vec{P}$ is produced that either weakens or strengthens the external $\vec{B}_0$ field. The ratio of P over B_0 is defined as magnetic susceptibility, χ, which is dimensionless. Diamagnetic substances have negative susceptibilities, while paramagnetic, superparamagnetic, and ferromagnetic substances all have positive susceptibilities. Although biological tissues are diamagnetic in general, they can become paramagnetic or superparamagnetic under some physiologic or pathologic conditions, such as focal accumulations of metals and iron-based protein conglomerates (e.g., ferritin and hemosiderin).

Susceptibility-weighted imaging (SWI) is a GRE-based technique with several enhancements to increase its sensitivity to tissue susceptibilities (Haacke et al. 2004). It can distinguish paramagnetic (e.g., hemorrhage/iron) from diamagnetic (e.g., calcification) lesions. It can also produce venograms based on the susceptibility differences. In quantitative susceptibility mapping (QSM), the phase information in GRE images is further analyzed to quantify susceptibility values using various mathematical methods (Wang and Liu 2015).

Selected Challenges

A primary challenge for MRI lies in its low sensitivity, which limits its spatiotemporal resolution. Conventional strategies to improve the sensitivity include increasing $\vec{B}_0$ magnetic field and/or the sensitivity of receiving RF coils. Recent development has also indicated that the use of hyperpolarization can be a viable approach to substantially increase the sensitivity of MRI (Kurhanewicz et al. 2011). With improved sensitivity, the achievable voxel size can be reduced, leading to a higher spatial resolution. To realize the high spatial resolution over a specific field-of-view, a large matrix size must be used, which requires time-efficient image acquisition strategies. Although considerable progress has been made in reducing the scan times using various advanced techniques such as parallel imaging, compressive sensing, and MR fingerprinting (Sodickson and Manning 1997, Pruessmann et al. 1999, Griswold et al. 2002, Lustig et al. 2007, Ma et al. 2013), MRI remains inadequate in capturing many biological processes due to the limitations in spatiotemporal resolution. Another challenge is that the majority of MRI contrast mechanisms are based on physical or chemical parameters. Significant gaps exist in relating these parameters to disease-sensitive and specific biomarkers. Advances in MR molecular imaging have demonstrated great potential in narrowing these gaps (Hengerer and Grimm 2006). However, continued interdisciplinary effort remains warranted to fully realize the potential of MRI. These challenges provide excellent opportunities for scientists, engineers and clinicians to work together in the fertile area of MRI to continue expanding its role in biomedicine.

Summary

MRI is one of the most versatile medical imaging modalities that does not involve harmful ionizing radiation. It requires a strong, homogeneous, and static magnetic field to polarize nuclear spins, a pulsed alternating magnetic field to transmit RF energy to the spin system and subsequently receive signals from it, three magnetic field gradients to encode spatial information into the signals, pulse sequences to acquire the signals with various contrasts, and image reconstruction algorithms to obtain the final image. MRI offers a long and growing list of contrast mechanisms, reflecting different aspects of tissue physical, chemical, physiological, functional, or metabolic features. A major challenge in MRI is the low sensitivity, which limits its spatiotemporal resolution. However, innovative solutions are continuously being developed to meet the challenge. MRI is an outstanding example of fruitful collaborations among scientists, engineers, and clinicians, and will continue providing a fertile ground for interdisciplinary development of this evergreen medical imaging technology.

Acknowledgements

The author is grateful to Professors Paul C. Lauterbur and G. Allan Johnson for introducing him to the field of MRI; to the faculty, staff, and students of the Center for MR Research, University of Illinois at Chicago for their assistance; and to Dr. Muge Karaman and Zheng Zhong for their suggestions and proof-reading this manuscript.

References

Ahn, C.B., J.H. Kim and Z.H. Cho. 1986. High-speed spiral-scan echo planar NMR imaging - I. IEEE Trans Med Imaging 5: 2–7.

Alsop, D.C., J.A. Detre, X. Golay, M. Günther, J. Hendrikse, L. Hernandez-Garcia et al. 2015. Recommended implementation of arterial spin-labeled perfusion MRI for clinical applications: A consensus of the ISMRM perfusion study group and the European consortium for ASL in dementia. Magn Reson Med 73: 102–116.

Bernstein, M.A., K.F. King and X.J. Zhou. 2004. Handbook of MRI Pulse Sequences. 1st ed. San Diego, California: Elsevier Academic Press.

Bilgic, B., B.A. Gagoski, S.F. Cauley, A.P. Fan, J.R. Polimeni, P.E. Grant et al. 2015. Wave-CAIPI for highly accelerated 3D imaging. Magn Reson Med 73: 2152–2162.

Boada, F.E., G.X. Shen, S.Y. Chang and K.R. Thulborn. 1997. Spectrally weighted twisted projection imaging: ReducingT2 signal attenuation effects in fast three-dimensional sodium imaging. Magn Reson Med 38: 1022–1028.

Bottomley, P.A. 2008. Turning up the heat on MRI. J Am Coll Radiol 5: 853–855.

[BRAINYQUOTE] 2018a. Paul Lauterbur Quotes. Brainy Quote. [https://www.brainyquote.com/authors/paul_lauterbur] accessed August 31, 2018.

[BRAINYQUOTE] 2018b. Brainy Quote. [https://www.brainyquote.com/search_results?q=Edward+M.+Purcell] accessed August 31, 2018.

Brown, R.W., Y.C.N. Cheng, E.M. Haacke, M.R. Thompson and R. Venkatesan. 2014. Magnetic resonance imaging: physical properties and sequence design. New York: Wiley Publishing.

Brown, T.R., B.M. Kincaid and K. Ugurbil. 1982. NMR chemical shift imaging in three dimensions. Proc Natl Acad Sci USA 79: 3523–3526.

Bydder, G.M. and I.R. Young. 1985. MR imaging: clinical use of the inversion recovery sequence. J Comput Assist Tomogr 9: 659–675.

Chang, E.Y., J. Du and C.B. Chung. 2015. UTE imaging in the musculoskeletal system. J Magn Reson Imaging 41: 870–883.

Dixon, W.T. 1984. Simple proton spectroscopic imaging. Radiology 153: 189–194.

Edelstein, W.A., J.M.S. Hutchison, G. Johnson and T. Redpath. 1980. Spin warp NMR imaging and applications to human whole-body imaging. Phys Med Biol 25: 751–756.

Frausto Da Silva, J.J.R. and R.J.P. Williams. 2001. The Biological Chemistry of the Elements: The Inorganic Chemistry of Life. New York: Oxford University Press.

Glaser, K.J., A. Manduca and R.L. Ehman. 2012. Review of MR elastography applications and recent developments. J Magn Reson Imaging 36: 757–774.

Griswold, M.A., P.M. Jakob, R.M. Heidemann, M. Nittka, V. Jellus, J. Wang et al. 2002. Generalized autocalibrating partially parallel acquisitions (GRAPPA). Magn Reson Med 47: 1202–1210.

Gu, T., F.R. Korosec, W.F. Block, S.B. Fain, Q. Turk, D. Lum et al. 2005. PC VIPR: a high-speed 3D phase-contrast method for flow quantification and high-resolution angiography. Am J Neuroradiol 26: 743–769.

Haacke, E.M., Y. Xu, Y.C.N. Cheng and J.R. Reichenbach. 2004. Susceptibility weighted imaging (SWI). Magn Reson Med 52: 612–618.

Haase, A., J. Frahm, W. Hänicke and D. Matthaei. 1985. 1H NMR chemical shift selective (CHESS) imaging. Phys Med Biol 30: 341–344.

Hahn, E.L. 1950. Spin Echoes. Phys Rev 80: 580–594.

Hajnal, J.V., B. De Coene, P.D. Lewis, C.J. Baudouin, F.M. Cowan, J.M. Pennock et al. 1992. High signal regions in normal white matter shown by heavily T2-weighted CSF nulled IR sequences. J Comput Assist Tomogr 16: 506–513.

Hennig, J., A. Nauerth and H. Friedburg. 1986. RARE imaging: A fast imaging method for clinical MR. Magn Reson Med 3: 823–833.

Jahng, G.H., K.L. Li, L. Ostergaard and F. Calamante. 2014. Perfusion magnetic resonance imaging: a comprehensive update on principles and techniques. Korean J Radiol 15: 554–577.

Kanal, E., A.J. Barkovich, C. Bell, J.P. Borgstede, W.G. Bradley, J.W. Froelich et al. 2013. ACR Guidance Document on MR Safe Practices. J Magn Reson Imaging 37: 501–530.

Kijowski, R. and G.E. Gold. 2011. Routine 3D magnetic resonance imaging of joints. J Magn Reson Imaging 33: 758–771.

Kumar, A., D. Welti and R.R. Ernst. 2011. NMR Fourier zeugmatography. J Magn Reson 213: 495–509.

Kwong, K.K., J.W. Belliveau, D.A. Chesler, I.E. Goldberg, R.M. Weisskoff, B.P. Poncelet et al. 1992. Dynamic magnetic resonance imaging of human brain activity during primary sensory stimulation. Proc Natl Acad Sci USA 89: 5675–5679.

Lauterbur, P.C. 1973. Image formation by induced local interactions: Examples employing nuclear magnetic resonance. Nature 242: 190–191.

Lauterbur, P.G. and C.M. Lai. 1980. Zeugmatography by reconstruction from projections. IEEE Trans Nucl Sci 27: 1227–1231.

Lustig, M., D. Donoho and J.M. Pauly. 2007. Sparse MRI: The application of compressed sensing for rapid MR imaging. Magn Reson Med 58: 1182–1195.

Mansfield, P. 2007. Echo-planar imaging. *In*: Encyclopedia of Magnetic Resonance. Chichester, UK: John Wiley & Sons, Ltd.

Medscape. MRI and CT Ranked the Top Medical Innovations by Physicians [Internet]. Medscape Med News 2001 Available from: https://www.medscape.com/viewarticle/411372.

Nunes, R.G., J.V. Hajnal, X. Golay and D.J. Larkman. 2006. Simultaneous slice excitation and reconstruction for single shot EPI . In: Proc Intl Soc Mag Reson Med, p. 293.

Ogawa, S., T.M. Lee, A.R. Kay and D.W. Tank. 1990. Brain magnetic resonance imaging with contrast dependent on blood oxygenation. Proc Natl Acad Sci USA 87: 9868–9872.

Parker, D.L., C. Yuan and D.D. Blatter. 1991. MR angiography by multiple thin slab 3D acquisition. Magn Reson Med 17: 434–451.

Pauly, J., P. Le Roux, D. Nishimura and A. Macovski. 1991. Parameter relations for the Shinnar-Le Roux selective excitation pulse design algorithm. IEEE Trans Med Imaging 10: 53–65.

Peeters, F., B. Verbeeten and H.W. Venema. 1979. Nobel Prize for medicine and physiology 1979 for A.M. Cormack and G.N. Hounsfield. Ned Tijdschr Geneeskd 123: 2192–2193.

Pipe, J.G. 1999. Motion correction with PROPELLER MRI: Application to head motion and free-breathing cardiac imaging. Magn Reson Med 42: 963–969.

Pruessmann, K.P., M. Weiger, M.B. Scheidegger and P. Boesiger. 1999. SENSE: sensitivity encoding for fast MRI. Magn Reson Med 42: 952–962.

Rummens, F.H.A. 1970. The chemical shift: a review. Org Magn Reson 2: 209.

Sodickson, D.K. and W.J. Manning. 1997. Simultaneous acquisition of spatial harmonics (SMASH): fast imaging with radiofrequency coil arrays. Magn Reson Med 38: 591–603.

Tang, L. and X.J. Zhou. 2019. Diffusion MR of cancer: from low to high b-values. J Magn Reson Imaging 49: 23–40.

Ugurbil, K. 2014. Magnetic resonance imaging at ultrahigh fields. IEEE Trans Biomed Eng 61: 1364–1379.

Wang, Y. and T. Liu. 2015. Quantitative susceptibility mapping (QSM): Decoding MRI data for a tissue magnetic biomarker. Magn Reson Med 73: 82–101.

[WILEY] 2018. Wiley Online Library. [https://onlinelibrary.wiley.com/doi/abs/10.1002/(SICI)1099-1476(19990725) 22:11%3C867::AID-MMA965%3E3.0.CO;2-D] accessed July 26, 2018.

QUESTIONS

1. Which of the following nuclei can produce an MRI signal? Please explain why.

    (a) ^{31}P

    (b) ^{23}Na

    (c) ^{12}C

    (d) ^{16}O

    (e) ^{14}N

2. Which of the following is/are the reason(s) for developing or performing MRI at a higher static magnetic field? Multiple reasons can be selected.

    (a) To achieve a higher SNR in the image.

    (b) To potentially increase the spatial resolution by reducing the voxel size.

    (c) To reduce the costs of an MRI scanner.

    (d) To obtain a longer T_2* relaxation time in biological tissues.

    (e) To obtain a shorter T_1 relaxation time in biological tissues.

PROBLEMS

After an inversion pulse is applied to tissues, the longitudinal magnetization recovers according to Eq. [9].

1. Lipid has a T_1 value of 250 ms at 1.5 T. What should be the inversion time (i.e., the time between the inversion pulse and the subsequent excitation pulse for imaging) in a STIR sequence to nullify the lipid signal?

2. CSF has a T_1 value of 2.2 s at 3.0 T. What should be the inversion time in a FLAIR sequence to nullify the CSF signal?

3. At 1.5 T, the T_1 values of brain white and gray matter are found to be 646 ms and 1,197 ms, respectively. What is the optimal inversion time to maximize the contrast between white and gray matter?

20

Biomedical Imaging
Molecular Imaging

Christian J. Konopka, [1,2,3] *Emily L. Konopka* [4] *and*
Lawrence W. Dobrucki [1,2,5,*]

QUOTABLE QUOTES

"Nothing in life is to be feared, it is only to be understood. Now is the time to understand more, so that we may fear less."

Marie Curie, Physicist and Chemist, Pioneer of Radioactivity Research (BRAINYQUOTE 2018a)

"Medicine is a science of uncertainty and an art of probability."

Sir. William Osler, Physician and Co-founder of Johns Hopkins Hospital (BRAINYQUOTE 2018b)

"Nature uses only the longest threads to weave her patterns, so that each small piece of her fabric reveals the organization of the entire tapestry."

Richard P Faynman, Theoretical Physicist (BRAINYQUOTE 2018c)

"Don't worry too much if you don't pass exams, so long as you feel you have understood the subject. It's amazing what you can get by the ability to reason things out by conventional methods, getting down to the basics of what is happening."

Godfrey Hounsfield, Electrical Engineer and Nobel Laureate in Physiology/Medicine 1979 (BRAINYQUOTE 2018d)

"Structure without function is a corpse. . . function without structure is a ghost."

Stephen Wainwright, Structural Biologist (Westhof 2014)

---

[1] Department of Bioengineering, University of Illinois at Urbana-Champaign 405 N Mathews Ave, Urbana, IL 61801;ckonop3@illinois.edu

[2] Beckman Institute for Advanced Science and Technology, Urbana, IL.

[3] University of Illinois College of Medicine.

[4] Division of Dermatology, Southern Illinois University School of Medicine, 800 E. Carpenter St. Springfield , IL 62769; ekonopka72@siumed.edu

[5] Biobanking and Biomolecular Resources Research Infrastructure Poland (BBMRI.PL).

[*] Corresponding author: dobrucki@illinois.edu

Learning Objectives

The learning objectives for this chapter are to teach students the fundamentals of molecular imaging, and how physics and engineering principles dictate the use of molecular imaging in the clinical and research realms. After completing this chapter students should:

- Understand the reason why medical imaging is an important tool for clinicians today.
- Understand the physical principles governing the source of signal and signal detection in PET and SPECT imaging, such as the type of radioactive decay necessary for nuclear imaging, and how gamma radiation interacts with matter.
- Understand the differences between PET and SPECT detection strategies and their respective advantages and disadvantages.
- Understand the rationale for what makes a radiotracer effective as a molecular imaging agent.
- Have an appreciation for the breadth of uses molecular imaging has in the clinic today.
- Be able to recognize the limitations of molecular imaging, particularly PET, SPECT and CT.
- Understand that molecular imaging is a rapidly evolving field with new technologies and tracers being constantly developed.

The Rationale of Medical Imaging

The field of medicine and biology is constantly progressing, and with it, also our understanding of the molecular, cellular, and anatomical changes that underlie disease processes. Genomics and proteomics have provided insight as to what genetic and molecular phenomena cause myriads of diseases. This knowledge is hoped to be used for the development of personalized, precision medicine. But what good can this knowledge of the mechanisms of human disease do for medicine, without the ability to peer into the human body and recognize the problem in the first place? Perhaps the most critical step in successfully treating disease is diagnosing, *in a timely manner,* the development of aberrant physiology or anatomy. Medical imaging is one of the top medical developments of the past 1,000 years, which has completely changed the practice of medicine. Medical imaging provides medical practitioners and researchers the ability to look inside the human body (without physically opening it) and see not only anatomy, but also physiological, metabolic, and molecular events. This incredible ability has now blossomed into a burgeoning field of research and medicine, at the forefront of precision and personalized medicine.

The Development of Molecular Imaging

It was a physicist, not a physician to utilize the first form of medical imaging. In the late 19th century, Wilhem Röntgen through his experimentation with cathode rays created the first 2-dimensional x-ray radiograph. Only weeks after his data was presented, its clinical potential began to be utilized in medicine, and mass production began only a few years later (Sullivan 2011). Not even a full century later, by the 1980s, 3-dimensional anatomical medical images were being constructed using Computed Tomography (CT), and Magnetic Resonance Imaging (MRI). These machines, are widely used in the clinic today to produce high resolution, 3-D datasets of human anatomy and physiology, in areas such as neurology, cardiology, gastroenterology and more (Hendee 1989). While these modalities offer great insight into anatomical features of the human body and even its kinetics, they generally lack the ability to measure concurrent molecular and cellular events.

To visualize molecular and cellular events happening *in vivo* is the current challenge faced by molecular imaging (MI) scientists. To detect molecular phenomena in a living body requires a system capable of detecting low concentrations of disease and patient specific molecular and cellular targets, and to locate them anatomically. MI is not an entirely new field, in fact it had its beginning as early as the 1920s when radioiodine was used to identify thyroid dysfunction and even cure thyroid cancers. However, what is new is the development of highly engineered tracers, used in conjunction with anatomical imaging strategies, to evaluate molecular events for making diagnoses, evaluating prognoses, and guiding therapeutics.

Molecular Imaging Fundamentals

The demands of MI to evaluate low concentration, disease specific molecular targets in an extremely vast and dynamic system, necessitate the use of extremely sensitive imaging systems, in conjunction with high affinity tracers. These demands gave rise to imaging devices such as Single Photon Emission Computed Tomography (SPECT), and Positron Emission Tomography (PET) which both make use of radiolabeled exogenous contrast agents (radiotracers) which home to specific disease states. Most MI techniques utilize these two imaging systems because their sensitivity of detection is in the picomolar range while at the same time providing virtually unlimited penetration depth into tissues. However, new research also is discovering uses for MI in other imaging modalities, such as ultrasound, CT, MRI, and optical imaging. To see an overview of common 3D imaging modalities, see Fig. 1 and Table 1.

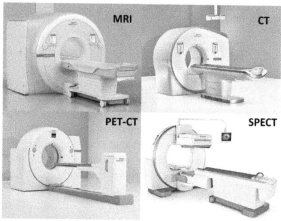

Fig. 1. Clinical imaging systems (MRI, CT, PET-CT, and SPECT) designed by Siemens. Source: https://usa.healthcare.siemens.com/

Table 1. Summary of clinical and preclinical imaging systems' spatial resolution, sensitivity for contrast agents, and common applications.

Imaging Modality	Spatial Resolution (Clinical)	Sensitivity for contrast agents (Clinical)	Spatial Resolution (Preclinical)	Sensitivity for contrast agents (Preclinical)	Common Applications
CT	0.5–1 mm	mM	10 µm	mM	anatomical imaging, bone imaging, angiography, GI imaging, cancer
MRI	1–2 mm	mM	15 µm	mM	soft tissue imaging, angiography, brain imaging, cancer
PET	4–6 mm	pM	1–2 mm	pM	cancer imaging, brain imaging, perfusion imaging, molecular imaging
SPECT	10–15 mm	nM-pM	0.5 mm	nM-pM	perfusion imaging, cardiac Imaging, bone imaging, molecular imaging

Types of Radiation in Medical Imaging

Radiation, defined simply as the transit of energy, is the fundamental principle underlying medical imaging. The utilization of imaging requires the use of energy transmitted, whether through electromagnetic waves (MRI, CT, PET, SPECT, optical), mechanical waves (ultrasound), or a combination of the two (photoacoustic), through tissues, and the properties of those specific waves determines the properties of the imaging modality. While MRI uses electromagnetic waves in the radio-wave energy levels, CT, PET, and SPECT all use radiation of higher energy levels (ionizing radiation), which, while useful, can be potentially harmful. CT uses an externally generated x-ray beam and the inherent properties of a tissue's interaction with the generated x-ray to produce images. CT takes these x-ray images at many angles around a body and computationally combines the final image to produce a 3-D dataset. In contrast, PET and SPECT both use radioactive substances, delivered within the body as the signal for image generation. Understanding the properties of the different types of radiation and how it affects the generation of images for each modality is critical to comprehending the information provided in the images and its inherent limitations (Karatas and Toy 2014).

X-Ray Formation—Bremsstrahlung and Characteristic X-Rays

When Wilhelm Röntgen was experimenting on the effects of cathode rays, he inadvertently discovered x-ray radiation, in a process that can be understood as essentially the inverse of the photoelectric effect. In the photoelectric effect, energy from photons is transferred into electrons, jumping them to higher energy states. The principle is similar for x-rays, except instead of photons transferring energy to electrons, fast moving cathode rays (electrons) interact with positively charged nuclei of glass (or Tungsten metal in the case for most CT systems) atoms in an x-ray tube, which results in the electrons losing speed and energy (acceleration), and the conversion of the electron's kinetic energy results in the emission of an electromagnetic wave in the form of an x-ray photon, termed Bremstrahlung radiation. It is this process which is the primary mechanism (80%) for producing an x-ray beam for CT or radiographic imaging (Vallabhajosula 2009). Other x-rays can be produced from the interaction of a cathode ray and an atom, whereby the bombarding electron collides with an orbiting electron of the positively charged anode atom. This collision can knock out the orbiting electron from its place in the atom, and subsequently a higher energy electron falls to fill that place. The energy change from the high energy electron moving to a lower energy results in the emission of a characteristic x-ray that is unique to the element which was bombarded (Agarwal 1979).

Radioactivity

Unlike x-ray production for CT and x-ray radiography, where the emission of radiation is induced by an external force (a bombarding electron), the radiation used for SPECT and PET imaging comes from an internal source. Both SPECT and PET rely on radioactivity, which is based on a fundamental property of the substances being used, called nuclear instability. Nuclear instability leads to a process known as radioactive decay, and the type of radioactive decay depends on the type of instability of the isotope in question. There are five types of radioactive decay known as alpha decay, beta-minus (β^-) decay, beta-plus (β^+) decay or positron emission, electron capture, and isomeric transition. While all are important physical events, discussion in this chapter will be limited to β^+ decay, electron capture, and isomeric transition for their uses in medical imaging, and we encourage readers to seek other resources to learn more regarding alpha and β^- decay. β^+ decay, electron capture, and isomeric transition are most important for medical imaging because for a radioisotope to be useful for medical imaging it must have a suitable half-life and with a decay pattern that results in photon emission with little to no associated particulate (subatomic emitting) radiation. These radioisotopes must also have photon energies which are high enough to penetrate the body and low enough energies for optimal detection by imaging systems (Hendee and Ritenour 2003).

Annihilation (B+ decay) (PET)

One specific method of radioactive decay useful for PET imaging is known as positron (β^+) decay. In β^+ decay, nuclides are deficient in neutrons, and consequently unstable. Therefore, the number of protons (positive charges) will decrease to reach a stable state by decaying through either β^+ emission or by electron capture (EC). During β^+ decay, a proton in the nucleus is transformed into a neutron, a positron and a neutrino. The neutron remains in the nucleus, but the positron and neutrino are ejected. This leads to a daughter nucleus of an atomic number decreased by 1, with the same mass number. The nuclear transition can be represented by:

$$p \rightarrow n + \beta^+ + v$$

where p represents a proton in the nucleus, n is a neutron in the nucleus, β^+ is a positron ejected from the nucleus during decay, v is a neutrino that accompanies the positron in being ejected from the nucleus. While positrons emitted from the nucleus can have a range of energies, one of the consequences of positron emission is a process known as annihilation radiation. Annihilation radiation is unique to β^+ emitting nuclides and the essential principle behind PET imaging. Essentially, annihilation is a result of the emitted β^+, from a β^+ decaying isotope, colliding with an orbiting electron of surrounding atoms. The collision between the β^+ and electron results in a pair of annihilation photons traveling 180° apart at the specific energy of 511-keV. It is these photons that PET detectors are measuring, in a process known as annihilation coincidence detection (ACD), which will be discussed later in the chapter. This process is illustrated in Fig. 2A.

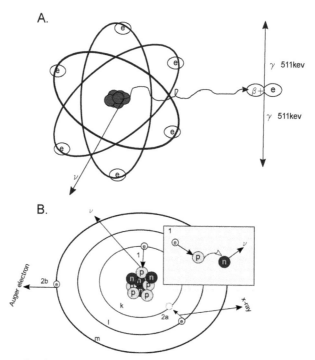

Fig. 2. Schematic diagram of positron decay and annihilation with an electron (A) and diagram of electron capture with subsequent x-ray emission (2a) or auger electron (2b) (B).

Electron capture

EC is the second method by which unstable neutron deficient nuclides decay. In general, neutron deficient isotopes which are of high atomic number decay more predominantly by electron capture, whereas those with lower atomic numbers will decay primarily by β^+ decay. During EC the positive charge in the nucleus is decreased when a K-shell electron is captured by the parent nucleus. This capture results in the conversion of a proton to a neutron and the ejection of a neutrino and can be represented by the equation:

$$p + e^- \rightarrow n + v$$

where p is a proton in the nucleus, e^- is an orbiting electron, n is a neutron in the nucleus, and v is an ejected neutrino. The result is a daughter nucleus with an atomic number lower by 1 and an unchanged mass number. The captured electron then leaves a vacancy in its electron shell, which is subsequently filled by a higher energy electron. The change in energy levels of the electron results in the emission of an x-ray (Fig. 2B, 2a) or Auger electron (Fig. 2B, 2b). Because of the change in the proton to neutron ratio that happens during electron capture, the nucleus of the daughter nuclide often exists in an excited state. When that nucleus relaxes to its ground state from the excited state, it causes the emission of a gamma photon. It is this production of gamma photons which provides the signal necessary for SPECT imaging.

Isomeric Transition

Of the three radioactive decay patterns discussed in this chapter, isomeric transition is characteristic of the most common radionuclide used in nuclear imaging, ^{99m}Tc. Isomeric transition happens when the daughter nuclide of one of the above-mentioned decay processes results in an unstable but relatively long-lived state known as a metastable state. During isomeric transition, the daughter nuclide is only different in energy levels from its parent, while atomic number and mass remain unchanged; these two nuclides are isomers of each other. It is the process of the metastable isomer transitioning to the lower energy stable isomer, known as isomeric transition, that results in the emission of a gamma photon useful for SPECT detection.

Detection and Image Formation

The fundamentals of PET and SPECT imaging rely heavily on the principles of high energy electromagnetic waves (gamma radiation) discussed above, and that radiation's interaction with matter. The way in which the gamma radiation interacts with both the matter in the human body, as well as the imaging system is critical for how SPECT and PET systems detect signals and ultimately construct useful images. There are two main interactions critical to understand how gamma radiation interacts with matter in SPECT and PET imaging: Compton scattering and the photoelectric effect. The combination of these two effects results in attenuation, a physical property of a substance that dictates how likely it is to disrupt the path of a traveling photon and thereby prevent that photon from reaching the detector (Nguyen et al. 2011).

Compton Scatter

Compton scattering is the predominant way in which high energy electromagnetic waves (such as gamma radiation) interact with tissue *in vivo*. Compton scattering consists of a collision between a photon (generated from radionuclide) and an outer orbital electron. The interaction between the electron and photon causes the photon to accelerate (changing its direction) and the electron is consequently ejected from the atom. The angle at which the photon is scattered is proportional to the energy lost by the photon (with maximum loss of energy occurring at 180°). This process is illustrated in Fig. 3A. Additionally, the probability of a high energy photon experiencing Compton scattering increases with decreasing atomic number, and increasing photon energy. This process is critical to understand when designing PET and SPECT systems as it affects the signal being detected by imaging systems by changing both the direction and energy of the photons being detected, thereby adding to the challenge of appropriately locating the source of the signal.

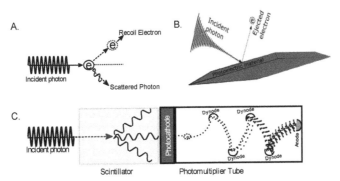

Fig. 3. Schematic representations of Compton scattering (A) and the Photoelectric Effect (B) and a Diagram illustrating the interaction between an incident gamma photon and scintillating material coupled to a PMT for signal detection (C).

Photoelectric effect

The other important type of radiation:matter interaction in nuclear imaging is the Photoelectric effect. Whereas Compton Scatter significantly affects the pathway of a traveling photon **in the body** by altering the path of traveling gamma radiation, the photoelectric effect plays a key role **in the imaging system** itself, by enabling the detection and recording of signal from a radionuclide. When electromagnetic radiation interacts with certain materials it has a high probability of colliding with inner shell electrons, and this collision stops the travel of the gamma photon and converts that energy into the emission or excitation of electrons (Fig. 3B). Furthermore, photons with higher energies cause electrons to be ejected with higher kinetic energies. PET and SPECT detectors utilize this property signal detection of radionuclides, by employing the use of collimators, scintillating crystals, and photomultiplier tubes (PMTs).

Collimators, Scintillating Crystals, and PMTs

Collimators are a fundamental element in SPECT detectors. Essentially, collimators are made of highly attenuating material which can stop incident gamma photons from reaching the detection element itself. Collimators exhibit high attenuation of gamma photons because they are made of a material that interacts with gamma photons predominantly in the form of the photoelectric effect. A collimator's purpose therefore is to filter out gamma photons incident to the detector at certain angles, thereby allowing the detector to only record data from photons traveling normal to the detector surface, thereby aiding in the locating of the signal.

Critical to the detection of these photons are scintillating crystals. Scintillating crystals are a unique substance that emits visible light in the form of luminescence, in response to interaction with gamma radiation photons. During this interaction, a scintillator will raise its electron to a high energy group, and when that electron returns to its ground state, this change in energy produces luminescence. The amount of light emitted from the crystal is proportional to the energy of the incident photon. In this way scintillating crystals can report signal from radionuclides to the PMT.

PMTs are vacuum tubes that contain a photocathode at their entrance followed by a series of dynodes (each of which is held at a greater voltage). The photon emitted from the scintillation crystal strikes the photocathode and causes the release of photoelectrons (via the photoelectric effect). The photoelectrons are then accelerated to the first dynode, which causes the release of more electrons. The electrons continue down the chain onto the subsequent dynode, and again the electron number is increased. This cascade of events continues along a chain of dynodes and thereby amplifies the current produced from the scintillation crystal photon. It is this current that can be recorded by PET and SPECT systems for the formation of images (Fig. 3C).

Engineering Principles of PET and SPECT Detectors

Most SPECT detectors are based on an Anger system. An Anger camera requires a few key components, including: Scintillating material, PMTs, and a method for collimation. PET systems are very similar in design to this, but generally lack physical collimators. Anger cameras utilize a large area scintillating material to detect incident high energy photons, and that scintillating material distributes photons to a comparatively small group of coupled PMTs for signal amplification. The incident photons on the Anger camera are collimated to filter out photons that are incoming at oblique angles. Collimation allows for positional determination, by limiting the angle at which photons strike the detector to ones traveling normal, or nearly normal to the scintillator. When the scintillator is activated by one of these high energy photons, it releases its own photons in proportion to the energy of its incident photon. The resulting scintillation photons are then distributed to a large group of PMTs. The intensity of the signal can then be analyzed by summing the response of the PMTs and its position by taking the ratio of the analogue signals being produced. While SPECT systems use physical collimators of high attenuating material, PET generally use electronic collimation for ACD. The type of collimation on SPECT systems also determines the sensitivity and resolution of the unit. For instance, pinhole collimators, which are collimators with a single narrow aperture, have decreased sensitivity compared to parallel hole collimators, which contain multiple parallelly drilled holes, because there are fewer incident photons allowed to interact with the detector. Alternatively, pinhole collimators allow for increased spatial resolution because the incident photons have very limited incident angles. There are a wide range of collimators for different applications, and their use is dictated by the resolution and sensitivity needs of the project (Peterson and Furenlid 2011). PET, however, because of its use of ACD, which is discussed in a subsequent section, uses a form of electronic collimation, so no physical collimators are necessary, and therefore generally has greater sensitivity than SPECT systems.

SPECT Image Acquisition

SPECT imaging as its name implies, relies on the detection of single photon events, and its subsequent tomographic reconstruction. Most SPECT systems consist of two rotating detectors positioned 180° apart, with parallel hole collimators. Planar images are then taken at fixed angle increments around a full 360° of rotation. At each angle 2D projections are taken of a 3D distribution of radioactivity in a body. Each projection consists of many pixels, and each pixel stores acquired counts of radioactivity. After the acquisition of all frames in a revolution, images are reconstructed using mathematical methods, such as iterative methods or filtered backprojection. Figure 4 demonstrates a simplified but typical workflow for SPECT or PET imaging.

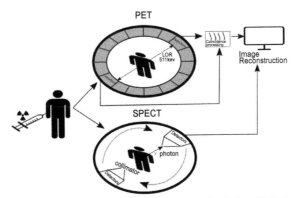

Fig. 4. A simplified diagram of PET or SPECT workflow. MI imaging studies begin with the injection of a radiotracer, followed by PET or SPECT imaging. The diagrams demonstrate coincidence detection in PET, with a solid ring of detectors, compared to rotating detectors in SPECT with physical collimators installed. Both imaging modalities require image reconstruction following the imaging protocol.

PET Image Acquisition

PET acquisition involves the use of β^+ emitting radionuclides and a full ring of pixelated scintillation crystals coupled to PMTs as detectors encircling the patient. PET imaging detects 2 coincident photons by two opposing detectors, at specific energies of 511-keV, generated from the annihilation event between a of β^+ and an electron (as opposed to a single photon detected by a single detector in SPECT). When coincident photons are detected, two opposing detectors are simultaneously activated (within 12 ns) and the originating positron annihilation event is then placed with the volume along a line of response (LOR) defined by the two detectors. The detection of photons in PET from this point is analogous to that of SPECT where scintillation crystals transmit their photons to shared PMTs, and that signal is interpreted to give positional and intensity information, followed by reconstruction protocols (Iniewski 2009). A simplified PET imaging system is illustrated in Fig. 4.

Benefits and Limitations of PET vs. SPECT

The design of SPECT and PET dictate the limitations and benefits of each imaging modality. Factors affected by the fundamental design of each scanner respectively include: spatial resolution, sensitivity, and temporal resolution.

Spatial Resolution

Spatial resolution refers to the ability of a system to distinguish between distinct objects of a certain size. The higher the spatial resolution, the greater the ability of a system to identify small objects, which can be significantly important in molecular imaging. The key difference between spatial resolution in SPECT and PET imaging is that for SPECT imaging spatial resolution is limited by technology, most importantly being collimator design, whereas for PET the spatial resolution is largely limited by the physical phenomena surrounding β^+ decay.

In general, because SPECT spatial resolution depends on knowing the incident angle of photons, as the distance from the camera to the source increases, spatial resolution deteriorates. This is partly why whole-body SPECT suffers from poorer resolution than brain SPECT, because detectors can come closer to the radiotracer signal in brain imaging than in whole body imaging. To combat this shortcoming, pinhole collimators can be used. These collimators greatly limit the number of photons reaching the detector to only photons traveling normal to the detector, thereby increasing the spatial resolution. The typical clinical system with opposing Anger cameras and parallel hole collimators has a spatial resolution of approximately 10 mm, however specialized systems have been developed that are able to achieve resolutions as low as 0.5 mm using multipinhole geometries.

PET spatial resolution is limited by two physical phenomena surrounding β^+ decay: non-colinearity and positron range. Non-collinearity refers to the property that a positron annihilation with an electron does not always result in a net momentum of 0, and as a result the positron can emit an annihilation photon at angles less than 180°, thereby altering the LOR. The effect of this angular uncertainty is proportional to the radius of the ring of detectors. Additionally, positrons can be emitted from the nucleus at different energies. The higher the energies, the greater the distance a positron can travel from the nucleus before it is annihilated by an electron. This distance that a positron can travel also leads to error in positioning of the source. PET systems cannot change the laws of physics that govern these sources for uncertainty in positioning, however, certain steps can be taken to mitigate their effects. Recently, researchers have found that applying high magnetic fields can limit positron range, and therefore decrease the uncertainty related to this effect. Additionally, reconstruction algorithms have been developed that are able to compensate for these effects. While the system cannot predict the effects of positron range or noncolinearity on individual events, these algorithms can calculate and incorporate their probability distributions into the system matrix and enhance spatial resolution. Typical clinical PET systems have resolutions of approximately 4–6 mm, and specialized systems as low as 1–2 mm for preclinical systems (Moses 2011).

While both PET and SPECT offer unique advantages as imaging modalities, their respective spatial resolution is considerably poor in comparison to CT or MRI used for anatomical imaging. The relative large spatial resolution, therefore, of PET and SPECT results in an artifact known as the partial volume effect. This effect is significant when regions smaller than the resolution of the system accumulate a large dose. The result is a significant bias in radiotracer concentration estimation and image quality. Methods have been devised to help correct this problem, but they often necessitate multimodality studies that coregister PET/SPECT images with MRI or CT (Rousset et al. 1998).

Sensitivity

Sensitivity in nuclear imaging relates to the proportion of a signal that a system can detect and record. Sensitivity is key in developing an imaging system because an increase in detected events results in an improved signal to noise ratio due to the Poisson nature of radioactive decay. This is one area in which PET systems offer clear advantages over SPECT. PET systems typically exhibit a sensitivity 1–2 orders of magnitude greater than SPECT systems. This is because PET systems do not require physical collimators, and therefore, the number of events reaching PET detectors is not restricted physically.

SPECT systems require physical collimation for accurate positioning of a source and therefore eliminate most photons travelling towards the detector, with only ~.01% reaching the detector to be recorded as events. SPECT systems can compensate for this decrease in sensitivity by adjusting the collimation method used, however increasing sensitivity generally requires compromising spatial resolution. Many researchers are developing novel collimator geometries to enhance sensitivity without compromising spatial resolution, however, each collimation method requires its own reconstruction protocol. Particularly promising are rotating multi-segment slant-hole (RMSSH) collimators which enhance sensitivity 2-3 fold and require very few camera positions to generate a tomographic image (Jingyan Xu et al. 2007). However, the SPECT imaging system has a distinct advantage compared to PET, because it is not detecting photons of only a specific energy (511 keV); instead, it can detect photons of a large range of energies. For this reason, SPECT is currently being used for dual energy studies, where two gamma emitting tracers that emit photons at different energies can be used simultaneously, enabling dual tracer imaging. This potential multiplexing of radiotracers is a property that PET imaging cannot achieve (Chapman et al. 2012).

PET detection relies on coincident events and therefore requires only electrical collimation. Therefore, the number of incidents reaching PET detectors is not physically limited as is the case for SPECT imaging and ~ 1% of events are accepted. The improved sensitivity of PET detection offers several benefits to PET imaging, including: improved signal to noise ratio, shorter scan times, possibility for multi-frame images, and improved temporal resolution. However, the method of coincident detection also has two main pitfalls which reduce image quality. Firstly, noncolinearity (discussed above) decreases spatial resolution. And secondly, random coincidences also decrease spatial resolution. Random coincidences occur when by chance two annihilation photons that are detected simultaneously have not originated from the same annihilation event, resulting in an incorrect LOR. This property causes an increased signal to noise ratio and detracts from image quality. While random coincidence is difficult to correct, fast scintillators are being developed to decrease the timing window for coincident events, and therefore decrease the number of random coincidences recorded (Vaquero and Kinahan 2015).

While there are differences in resolution between PET and SPECT sensitivity levels, both systems can detect picomolar concentrations of contrast agent. This detection efficiency is unrivaled when compared to other imaging systems, and it allows for the detection of extremely dilute molecular targets, which otherwise would be undetectable. Researchers have been exploring ways to enhance the sensitivity of other imaging modalities to compete with PET and SPECT systems. One promising strategy is the use of superparamagnetic tracers with MRI. These tracers have the benefit of not being radioactive, however they still do not achieve comparable detection levels of PET/SPECT and can be toxic (Yeh et al. 2017, Sinharay and Pagel 2016).

Temporal Resolution

Closely tied to the sensitivity of a system is its temporal resolution. Temporal resolution refers to the ability of a system to quickly acquire an acceptable image. This is an important quality when considering applications for MI, because researchers are finding that tracer dynamics, or how the tracer distribution changes over time, can be more important than looking at the distribution of a tracer at a certain time. To perform dynamic imaging, quick image acquisition is necessary to produce multiple frames in as little amount of time as possible. To do this, an adequate number of counts to reconstruct images of sufficient signal-to-noise ratio need to be acquired. PET imaging has the advantage in this arena due to its increased sensitivity and full ring of detectors, and dynamic PET imaging is capable in most systems. SPECT has the limitations of decreased sensitivity compared to PET, and the necessity to complete a rotation around a subject to produce an image. Because of these limitations, SPECT is not traditionally used for dynamic studies. Researchers have developed specialized SPECT systems however which can perform dynamic imaging, largely by modifying the detector gantry so that it is a fixed gantry with no need to rotate around the patient as in conventional SPECT imaging, or a ring of SPECT detectors as in PET. There are also dedicated cardiac SPECT imagers which rotate just enough to gain sufficient angular sampling, allowing for kinetic modeling of cardiac function (Ben-Haim et al. 2013, Farncombe et al. 1999, Furenlid et al. 2004).

PET-SPECT Imaging

Understanding the fundamentals, the benefits, and the limitations of MI techniques allows physicians and researchers greater appreciation for their use in medicine. Medical imaging is a rapidly evolving field of medicine, and PET and SPECT imaging applications are particularly growing. CT and MRI generally provide anatomical imaging; however, they are capable of providing more functional measures as well.

For instance, fMRI can give measures of cerebral blood flow, and CT can be used for the determination of cardiac function. Both these tools can also be enhanced by using contrast agents which typically illuminate vasculature. Despite the strengths of MRI and CT for measuring certain functional levels and detailed anatomy, in today's clinical practice they generally lack the ability to give molecular and functional information except for in a few specialized circumstances. While contrast agents can be administered, MRI and CT rely on endogenous contrast, or contrast that is a property of the tissue itself rather than an introduced contrast agent, as opposed to MI studies where an introduced tracer (radiotracer for PET and SPECT) provides the entire signal. This is a key difference between nuclear imaging and anatomical imaging studies. Where MRI and CT are mostly limited to observing gross-anatomy-related functions from endogenous tissue contrast, PET and SPECT, when paired with high performing contrast agents, have the sensitivity necessary to detect small molecular events relevant to a broad range of disease processes.

Contrast Agent Principles

The advancement of PET and SPECT imaging is a result not only from discoveries in physics and radiation, and progress in engineering complex systems, but also the development of novel radiotracers with a wide breadth of clinical utility. It is these tracers that provide PET and SPECT such a diverse portfolio of utility in medicine, from cancer to cardiology. Virtually all MI studies utilizing PET or SPECT require the use of contrast agents to provide the signal for imaging, but the choice of contrast agent dictates the imaging protocol and user's interpretation (McKinney and Raddatz 2006).

Types of Contrast Agents

There are two general categories of contrast agents, targeted or non-targeted. Most CT and MRI contrast agents are non-targeted as either blood pool, gastrointestinal, hepatobiliary, or extracellular agents. These agents typically use passive qualities by either being injected directly to the needed site of contrast (blood pool agents) or by relying on their natural secretory routes to bring them to the organs of interest (hepatobiliary and gastrointestinal agents). These are considered non-targeted or passively targeted agents because they do not require affinity for a specific receptor or molecule to direct their accumulation in the tissue of interest.

Active vs. passively targeted. PET and SPECT contrast agents are generally termed radiotracers. These radiotracers can also be targeted or non-targeted contrast agents. The key difference between MI and nuclear medicine is that MI uses radiotracers and contrast agents that are **specific** for the measurement of a certain biological process, whereas nuclear medicine also uses radiotracers that are non-specific, but still clinically useful. For instance, SPECT tracer ^{67}Ga-citrate is a SPECT radiotracer which can be useful for detecting malignant tumor tissue, but it is not specific for any molecular process; rather, it accumulates in tumor tissue because the ^{67}Ga is treated as a ferrous ion in the body and accumulates in sites of inflammation such as cancer or infection sites. ^{67}Ga-citrate uptake on SPECT or 2-D scintigraphy studies can therefore locate primary tumors, metastases, and infections. In contrast to ^{67}Ga-citrate, ^{18}F-fluoroestradiol (FES) is a MI radiotracer used for imaging breast cancers with PET. FES **specifically** binds to estrogen receptors and its uptake can predict the treatment effect of salvage hormonal therapy and guide therapeutic regimens for breast cancer patients (Linden et al. 2006).

Considerations for Constructing an MI Radiotracer

Developing new radiotracers is an ever-expanding field of research to bring new MI methods to the clinic. Designing tracers requires several considerations for it to be an effective MI imaging agent. Firstly, the target of the tracer whether it be a receptor, protein, or other molecule, must be relevant to a clinical biological process. The choice of target must be expressed differently in diseased tissue than normal tissue, usually at high levels in diseased tissue and very low to none in healthy. This target must be highly expressed, in an amount that allows it to be detectable using the imaging system. Secondly, the developed tracer must bind specifically to the target molecule. This is a critical component of any radiotracer, because if a tracer binds non-specifically to other molecules, its imaging data can be difficult to interpret because of areas of uptake not corresponding to the molecular events intended to be evaluated. Many researches use radiolabeled antibodies for this purpose as antibodies have highly specific targets. The drawback to antibody-based imaging is that antibodies have a long blood pool residence time, and therefore there is significant signal from the blood which causes a high background signal and generally poor image quality (low signal to noise ratio). Closely related to specificity is the radiotracer's affinity for the target. A high affinity is necessary for imaging, as the probe must be able to localize to the area of the molecular target in small enough concentrations not to perturb the disease state. The kinetics and biodistribution of radiotracers is also a significant consideration in the design and application of radiotracers. Antibodies also prove to be a good example for this, because while they exhibit high affinity and specificity for their targets, they also demonstrate a long blood pool residence time, as discussed above. This means that a significant amount of signal from antibody-based radiotracers is nonspecific, thereby reducing the quality of the image and increasing the background noise. Despite its limitations, antibody-based MI is a rapidly developing field, and a whole genre of contrast agents have been developed and termed immunoPET tracers. These tracers are positron-emitting labelled antibodies, generally targeted towards a cancer marker. These tracers could be used as supplements to theranostic approaches, aiding in patient selection for monoclonal antibody therapies (Bailly et al. 2016). Finally, a crucial quality of any radiotracer is that the tracer must also exhibit low toxicity, because it is to be injected into living systems.

CT and MRI struggle to be used in MI applications because typically the contrast agents used for these modalities must be present in such high concentrations to be visible under the imaging conditions that the tracers cannot accumulate enough in the sites of the molecular expression; additionally, if the injected dose was increased to improve image contrast, the tracer would be too toxic to the system. This is why, taking all these factors of targeted contrast agents into consideration, MI is predominantly based in PET and SPECT imaging.

Quantification

Tracers in PET and SPECT imaging provide a unique feature in medicine in that they allow for quantification of uptake values. This means that imaging scientists and clinicians are able to contour regions of interest (ROIs) on the image itself and receive a quantitative value for how much of the tracer accumulated in that site. To do this, one must know the initial injected dose, correct for the decay of the isotope (from the time of injection to that of imaging), and convert the counts/pixel to a known radiation dosage (determined by the camera efficiency), using the following formula:

$$\frac{\%I.D.}{8} = Ct * \frac{Vt}{Wt} * \frac{1}{Dinj} * 100\%$$

where $\%I.D./g$ is the percent injected dose per gram tissue, C_t is the activity per volume, V_t is tissue weight, and D_{inj} is the does injected. Typically, $\%I.D./g$ is used for preclinical studies on animals, but in humans where body mass can vary greatly, standard uptake value (SUV) is typically used. SUV normalizes the $\%I.D./g$ value by the patient's body weight using the following formula:

$$SUV = \left(\frac{\%I.D.}{g}\right) * \frac{Wp}{100\%}$$

Where the W_p is the patient's weight. SUV is critical to use in clinical imaging studies because a patient's weight and body composition can have a significant effect on the %I.D./g that is observed. Therefore, without this correction, significant bias can be introduced into quantitative analyses, unless body weights are consistent among all subjects in a study. The ability to quantify imaging results is highly valuable in longitudinal studies and comparative studies between patients. However, quantifying PET and SPECT imaging can be difficult, as PET and SPECT lack anatomical features, meaning it can be difficult to accurately contour an ROI for a specific organ or tissue of interest. To combat this shortfall researchers have begun to do multimodal imaging, by combining PET and SPECT systems with CT, so that CT can provide anatomical landmarks for the PET and SPECT functional imaging. Recent advances in electronics have also enabled PET-MRI systems to be developed, which provide the coregistration of

PET imaging with superior soft tissue resolution from MRI. Unfortunately, these systems are expensive and not yet popular in clinics or research. Furthermore, quantification of MI studies must be done cautiously as many factors could affect the resulting values, including: ROI effects (partial volume effect, over or under estimation due to ROI size or positioning), body composition, quality of radiotracer delivery, and standardized measurement times. If these parameters are not carefully controlled for, quantitative values may be skewed and give erroneous or nonsensical results (Keyes 1995).

Current and Future Clinical Applications of Molecular Imaging

While new radiotracers and applications of MI are being explored every day in laboratories, the process in the clinic moves more slowly as FDA regulation on new tracers is held under extremely tight regulations. Yet, advances in MI are still taking place in the clinic today with new tracers being introduced to aid in the diagnosis and treatment planning of many diseases. The most common tracer in nuclear medicine is ^{99m}Tc, which is a SPECT tracer. Various radiotracers have been designed using ^{99m}Tc, which range from being used in tagging blood cells to find sites of bleeding, performing perfusion imaging to identify areas of ischemia or hypoperfusion, as well as in molecular imaging tracers such as ^{99m}Tc-Bombosin, which can play a role in staging prostate cancer (De Vincentis et al. 2004). PET tracers also find routine clinical use, the most common of which is [^{18}F]Fluorodeoxyglucose (FDG). This tracer is a glucose analogue and serves as an MI radiotracer for the enzyme hexokinase, a key enzyme in the metabolism of glucose. FDG uptake therefore correlates with high levels of glucose metabolism and can be used to detect primary cancer tumors, metastases, or infections. A recent addition to MI radiotracers in the clinic for PET also include agents targeted towards amyloid-β as a marker for Alzheimer's disease (AD). ^{18}F-labeled pharmaceuticals such as florbetaben, florbetapir and flutemetamol are now clinically being used for the detection of beta-amyloid in the brain. These radiotracers are able to determine with negative test results that AD is very unlikely to be the cause of cognitive dysfunction. However, positive test results do not predict AD as the diagnosis (Morris et al. 2016). These promising new tracers exemplify the utility MI can have in the clinical setting with a proper target and a well-designed tracer.

Traditionally, radiotracers have been single molecule imaging agents. New advancements in nanotechnology have now enabled more complex design to be considered in probe developments. Researchers can utilize nanotechnology to enhance delivery of contrast to tissues, carry therapeutic agents, and contain multimodal signaling (e.g., radiolabel and fluorescent label). Even the shapes and sizes of the nanoparticles can be engineered and have significant impact on the tracer's properties. For instance, many nanoparticle radiotracers in cancer imaging exploit a property of cancer endothelium called enhanced permeability and retention (EPR). EPR is a property of cancer tissue in which tumor vessels become hyperpermeabilized. This essentially enables the passive accumulation of nanoparticles which do not pass through the ~ 10 nm pores of healthy vascular endothelium to accumulate through the leaky tumor vasculature. Thus, well-designed nanoparticles could passively target tumor tissue through their size, while additionally carrying targeting moieties and therapeutics. It is evident that the applications of nanotechnology for MI are truly vast, and will be of significant impact in future research and medicine (Stylianopoulos 2013, Jokerst and Gambhir 2011).

Further utility in MI is beginning to be explored by mining imaging data of a variety of modalities. Texture analysis, is a growing field in radiology, under the premise that there is far more data in a medical image that can be interpreted by simple visual analysis. Texture analysis takes medical imaging data and analyzes it for many different parameters, many of which cannot be easily visualized or described by a trained eye, but can be quantified by a computer, and the results can correspond to biologically important phenomena. Because of the vast amount of data obtained in this method and the personalized quality of the data, this research has been compared to genomics and termed *radiomics* (Brooks and Grigsby 2013). Typically, texture analysis has been used for MRI and CT, and researchers have been able to use this data to associate texture patterns with cancer outcomes (Alic et al. 2014). More recently, researchers have been applying this method to MI methods such as FDG tumor imaging, and have been able to correlate texture findings with specific genomic profiles of breast cancer, which are relevant to treatment outcomes and prognosis of the patients (Fig. 5) (Ha et al. n.d.).

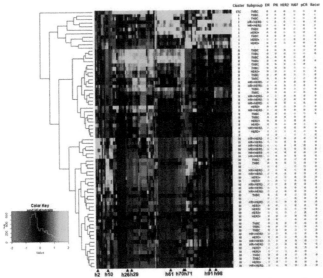

Fig. 5. Radiomics analysis of FDG imaging of breast cancer patients reveals specific clusters of patients with clinically relevant gene findings. "Unsupervised radiomics heat map with 109 texture features. Three individual tumor clusters (TCs) with distinctive metabolic radiomics patterns were identified after unsupervised clustering. Notes: Row, cases; column, texture features or clinical information; green circle, positive or high expression; red circle, negative or low expression; h2, ZP^{GL-SZM}; h10, NL_Homogeneity^{GLCM}; h26, NL_Dissimilarity^{GLCM}; h29, SUV$_{max}$; h61, TLG; h70, CV; h71, NL_Entropy^{GLCM}; h91, HILZE^{GLSZM}; h98, MTV. Abbreviations: TNBC, triple negative breast cancer; HR, hormone receptor; ER, estrogen receptor; PgR, progesterone receptor; human epidermal growth receptor 2, HER2; ZP, zone percentage; GLSZM, gray level size zone matrix; NL, normalized; GLCM, gray level co-occurrence matrix; SUV$_{max}$, maximum of standardized uptake value; HILZE, high-intensity large-zone emphasis; MTV, metabolic tumor volume."

Challenges in Molecular Imaging

Despite the value MI can play in providing optimal care for patients, it is not without drawbacks. Cost is a significant hurdle hampering the routine use of MI. The significant cost to performing a PET or SPECT scans deters physicians, hospitals, and patients from using a MI protocol when other methods could be an alternative. Therefore, the use of traditional blood tests, biopsies, or other conventional imaging methods as screening and diagnostic tests are utilized preferentially to MI, unless MI can *drastically* improve the quality of care given. Consequently, "lab-on-a-chip" technologies and other cheap and quick tests for diseases often outcompete new imaging probes as screening methods. However, imaging can provide an invaluable quality of information with spatial and quantitative data that will hold its value in medicine, despite its cost.

Another disadvantage to MI with radiotracers is the exposure to ionizing radiation. Other modalities such as ultrasound and MRI use sound or electromagnetic waves at energy levels that do not damage living tissue, but PET, SPECT, and CT all use ionizing radiation which can cause cellular and DNA damage and are often prohibited in conditions such as pregnancy. Therefore, all doses must be monitored carefully and minimized wherever possible (Sinusas et al. 2011).

Summary

In this chapter, the principles, applications, limitations, and challenges of MI are detailed. Current technologies in MI exploit fundamental principles of physics, within cleverly engineered systems, while utilizing precisely designed tracers to yield clinically relevant data, which is invaluable to patient care. Despite the difficulty and cost associated with designing new MI tracers and systems, MI remains a booming field in research and medicine. MI has the unprecedented ability to locate and quantify molecular phenomena happening within a living system, providing clinicians the ability to develop personalized and effective treatment plans. MI's many future technologies and applications of these technologies have the potential to shape the future practice of medicine.

Abbreviations: PET, SPECT, MI, CT, MRI, EC, LOR, ROI, ACD, PMT, RMSSH, EPR, FES, FDG, AD

References

Agarwal, B.K. 1979. Characteristic X-rays. pp. 53–119. *In*: X-Ray Spectroscopy. Berlin, Heidelberg: Springer Berlin Heidelberg.

Alic, L., W.J. Niessen and J.F. Veenland. 2014. Quantification of heterogeneity as a biomarker in tumor imaging: A systematic review. PLoS ONE 9(10): 1–15.

Bailly, C. et al. 2016. Immuno-PET for clinical theranostic approaches. International Journal of Molecular Sciences 18(1).

Ben-Haim, S., V.L. Murthy, C. Breault, R. Allie, A. Sitek, N. Roth et al. 2013. Quantification of myocardial perfusion reserve using dynamic SPECT imaging in humans: A feasibility study. Journal of nuclear medicine : official publication, Society of Nuclear Medicine 54(6): 873–9.

[BRAINYQUOTE] Brainy Quote. 2018a. Marie Curie Quotes. [https://www.brainyquote.com/authors/marie_curie] accessed June 5, 2018.

[BRAINYQUOTE] Brainy Quote. 2018b. William Osler Quotes. [https://www.brainyquote.com/authors/william_osler] accessed June 5, 2018.

[BRAINQUOTE] Brainy Quote. 2018c. Richard P. Feynman Quotes. [https://www.brainyquote.com/authors/richard_p_feynman] accessed June 5, 2018.

[BRAINYQUOTE] BrainyQuote. 2018d. Godfrey Hounsfield Quotes. [https://www.brainyquote.com/authors/godfrey_hounsfield] accessed June 5, 2018.

Brooks, F.J. and P.W. Grigsby. 2013. Quantification of heterogeneity observed in medical images. Brooks and Grigsby BMC Medical Imaging, 13.

Chapman, S.E., J.M. Diener, T.A. Sasser, C. Correcher, A.J. González, Avermaete T. Van et al. 2012. Dual tracer imaging of SPECT and PET probes in living mice using a sequential protocol. American Journal of Nuclear Medicine and Molecular Imaging 2(4): 405–14.

Farncombe, T., A. Celler, D. Noll, J. Maeght and R. Harrop. 1999. Dynamic SPECT imaging using a single camera rotation (dSPECT). IEEE Transactions on Nuclear Science 46(4): 1055–1061.

Furenlid, L.R., D.W. Wilson, Yi-chun Chen, Hyunki Kim, P.J. Pietraski, M.J. Crawford et al. 2004. FastSPECT II: a second-generation high-resolution dynamic SPECT imager. IEEE Transactions on Nuclear Science 51(3): 631–635.

Ha, S., S. Park, J.I. Bang, E.K. Kim and H.Y. Lee. 2017. Metabolic Radiomics for Pretreatment 18F-FDG PET/CT to Characterize Locally Advanced Breast Cancer: Histopathologic Characteristics, Response to Neoadjuvant Chemotherapy, and Prognosis. Sci Rep 7(1). doi:10.1038/s41598-017-01524-7.

Hendee, W.P. 1989. Cross Sectional Medical Imaging: A History. Radiographics 9(6): 1155–1180.

Hendee, W.R. and E.R. Ritenour. 2003. Radioactive Decay. Medical Imaging Physics Fourth Edition.

Iniewski, K. 2009. Medical Imaging: Principles, Detectors, and Electronics, Wiley.

Jingyan, Xu, Chi Liu, Yuchuan Wang, Frey EC and Tsui BMW. 2007. Quantitative Rotating Multisegment Slant-Hole SPECT Mammography With Attenuation and Collimator-Detector Response Compensation. IEEE Transactions on Medical Imaging 26(7): 906–916.

Jokerst, J.V. and S.S. Gambhir. 2011. Molecular imaging with theranostic nanoparticles. Accounts of Chemical Research 44(10): 1050–60.

Karatas, O.H. and E. Toy. 2014. Three-dimensional imaging techniques: A literature review. European Journal of Dentistry 8(1): 132–40.

Keyes, J.W. 1995. SUV: Standard uptake or silly useless value? Journal of Nuclear Medicine 36(10): 1836–1839. doi:10.1016/j.physletb.2007.04.051.

Linden, H.M., S.A. Stekhova, J.M. Link, J.R. Gralow, R.B. Livingston, G.K. Ellis et al. 2006. Quantitative fluoroestradiol positron emission tomography imaging predicts response to endocrine treatment in breast cancer. Journal of clinical oncology : official journal of the American Society of Clinical Oncology 24(18): 2793–9.

McKinney, M. and R. Raddatz. 2006. Practical Aspects of Radioligand Binding. In Current Protocols in Pharmacology. Hoboken, NJ, USA: John Wiley & Sons, Inc., pp. 1.3.1–1.3.42.

Morris, E., A. Chalkidou, A. Hammers, J. Peacock, J. Summers, S. Keevil et al. 2016. Diagnostic accuracy of (18)F amyloid PET tracers for the diagnosis of Alzheimer's disease: a systematic review and meta-analysis. European Journal of Nuclear Medicine and Molecular Imaging 43(2): 374–85.

Moses, W.W. 2011. Fundamental limits of spatial resolution in PET. Nuclear Instruments and Methods in Physics Research Section A: Accelerators, Spectrometers, Detectors and Associated Equipment 648: S236–S240.

Nguyen, M.K., T.T. Truong, M. Morvidone and H. Zaidi. 2011. Scattered radiation emission imaging: principles and applications. International Journal of Biomedical Imaging, pp. 913893.

Peterson, T.E. and L.R. Furenlid. 2011. SPECT detectors: the Anger Camera and beyond. Physics in Medicine and Biology 56(17): R145–82.

Rousset, O.G., Y. Ma and A.C. Evans. 1998. Correction for partial volume effects in PET: principle and validation. J Nucl Med 39(5): 904–911.

Sinharay, S. and M.D. Pagel. 2016. Advances in magnetic resonance imaging contrast agents for biomarker detection. Annual review of Analytical chemistry (Palo Alto, Calif.) 9(1): 95–115.

Sinusas, A.J., J.D. Thomas and G. Mills. 2011. The Future of Molecular Imaging. JACC: Cardiovascular Imaging 4(7): 799–806.

Stylianopoulos, T. 2013. EPR-effect: utilizing size-dependent nanoparticle delivery to solid tumors. Therapeutic Delivery 4(4): 421–423.

Sullivan, P.J. 2011. Seeing the bones of things: a scan of x-rays' early history. Canadian Family Physician Medecin De Famille Canadien 57(10): 1174–5.

Vallabhajosula, S. 2009. Science of Atomism: A Brief History. In Molecular Imaging. Berlin, Heidelberg: Springer Berlin Heidelberg, pp. 11–23.

Vaquero, J.J. and P. Kinahan. 2015. Positron emission tomography: current challenges and opportunities for technological advances in clinical and preclinical imaging systems. Annual Review of Biomedical Engineering 17: 385–414.

De Vincentis, G. et al. 2004. Role of ^{99m}Tc-Bombesin scan in diagnosis and staging of prostate cancer. Cancer Biotherapy & Radiopharmaceuticals 19(1): 81–84.

Westhof, E. 2014. RNA structure and folding. RNA. 20: 1843. Doi: 10.1261/rna.048322.114.

Yeh, B.M. et al. 2017. Opportunities for new CT contrast agents to maximize the diagnostic potential of emerging spectral CT technologies. Advanced Drug Delivery Reviews 113: 201–222.

QUESTIONS

1. What benefits does medical imaging, especially molecular imaging, provide as opposed to other diagnostic methods?

2. What types of radioactive decay are there? How is radiation used for SPECT imaging? How is it used for PET imaging? How is the radiation used for PET and SPECT different from that of CT?

3. Describe the advantages and disadvantages of PET vs SPECT imaging in terms of sensitivity, spatial resolution, temporal resolution, and choice of radiotracers.

4. Describe the difference between targeted and non-targeted tracers.

5. What considerations must be made in the design of a targeted radiotracer?

6. Typical SPECT imaging systems are not optimized for imaging PET radioisotopes. Describe what elements of a SPECT system would need to be optimized for detection of PET radioisotopes.

PROBLEMS

1. Imagine you are a researcher developing a novel radiotracer for a new biomarker in cancer. What imaging modality would you choose to use and why? Imagine your probe has made it into clinical trials, and you have a series of imaging studies on patients with recorded weight, tumor volume and mean %I.D./g values depicted in the table below. You notice that patient 3 and patient 7 exhibit significant differences from the mean values of the rest of the group. How do you explain those outliers? Show any calculations if necessary. Assume tissue volume has a density of 1 g/mm^3.

Patient (#)	Weight (kg)	Tumor Volume (mm³)	Mean uptake %I.D./g
1	80	8.9	14.375
2	82	20	12.07317073
3	65	1.1	53.84615385
4	60	12.9	14.16666667
5	77	15.6	13.11688312
6	85	11.3	13.05882353
7	200	10.4	4.5
8	81	14.6	10.98765432
9	78	9.8	13.07692308
10	80	30.7	12.25

21

Emerging Biomedical Imaging
Optical Coherence Tomography

Lawrence S. Chan

QUOTABLE QUOTES

"Music is the arithmetic of sounds as optics is the geometry of light."

Claude Debussy, French Composer (BRAINYQUOTE 2018a).

"Optics, developing in us through study, teach us to see."

Paul Cezanne, French Artist and Post-impressionist Painter (BRAINYQUOTE 2018a).

"There is nothing worse than a sharp image of a fuzzy concept."

Ansel Adams, American Landscape Photographer and Environmentalist (BRAINYQUOTE 2018b).

"Modern technology has taken the angst out of achieving the perfect shot. For me, the only thing that counts is the idea behind the image: what you want to see and what you're trying to say. The idea is crucial."

Martin Parr, British Documentary Photographer and Photojournalist (BRAINYQUOTE 2018b).

"Sometimes you have to delete characters from a scene just to keep from overcrowding the image."

Scott Westerfeld, American Fiction Writer (BRAINYQUOTE 2018b).

"The best cartoons have no words at all—just the image pops out."

Jeff MacNelly, American Editorial Cartoonist (BRAINYQUOTE 2018b).

Learning Objectives

The learning objectives of this chapter are to introduce the students to an emerging biomedical imaging technology optical coherence tomography, its operation principles, its current and future applications, and its limitations.

After completing this chapter, the students should:

- Understand the mathematical and physical principles governing the technology of optical coherence tomography.

University of Illinois College of Medicine, 808 S. Wood Street, R380, Chicago, IL 60612; larrycha@uic.edu

- Understand the current applications of optical coherence tomography in medicine, including its usage in ophthalmological, gastrointestinal, cardiovascular, neurological, and dental care.
- Understand the limitation of optical coherence tomography.
- Able to apply optical coherence tomography in real-life clinical situation.

Optical Coherence Tomography: An Introduction

Optical coherence tomography (OCT) is a relative new biomedical imaging technique that utilizes coherence light source to generate biological and medical images. OCT technique is similar to ultrasonography in principle, but it utilizes light instead of sound waves as illuminating energy source and recording source for image generation. Whereas ultrasonography captures the sound wave scattered back from a sound wave source, OCT captures the light scattered back from a light wave source. In the process, OCT can generate biological image with resolution near the microscopic level when ultra-resolution OCT technique is utilized (Tsai et al. 2014). While most OCT technology utilize near-infrared light, some use visible light (Shu et al. 2017). The primary focus of this chapter will be on OCT technology that utilizes near-infrared light. Although OCT technology has been applied to many medical practices, including ophthalmology, cardiology, gastroenterology, dentistry, and many other medical fields, its applications are being investigated into new areas and its future applications are expected to expand.

Fundamental Principles of OCT

The fundamental principles of OCT were described in great details by several recent publications (Fercher et al. 2003, Popescu et al. 2011, Tsai et al. 2014). These principles are briefly described here. The principle of OCT is perhaps better to be understood by first visualizing a basic physical set up of OCT.

There are many OCT configurations and we will first discuss a fiber-based OCT system in a Michelson configuration as an illustrative example of OCT basic concept. The schematic diagram of such configuration is depicted in Fig. 1 (Popescu et al. 2011). In this system, the light from a low-coherence source delivers to a coupler, which split it into two arms: one arm is designated as reference arm, and the other arm is termed sample arm. When the split light arrives at the reference arm, it will be backscattered or back-reflected by a reference mirror and returns into the system through the same path but in the opposite (reverse) direction. Similarly, when the light reaches at the sample arm, it will be backscattered by the sample and returns through the same path in the reverse direction. The returning light from both reference arm and sample arm will meet and combine at the coupler, generate an interference pattern that is delivered to and recorded by the detector of a single point type (Fig. 1). Analogous to B-mode ultrasound imaging methodology, OCT measures and records the echo time delay and intensity or magnitude of backscattered light (Popescu et al. 2011, Tsai et al. 2014).

Mathematically, the axial resolution, which is determined by the coherence length of the light source and is independent of the beam focusing conditions, is stated as below, if the light source has a Gaussian spectral distribution (Fujimoto et al. 2000):

{Axial Resolution} = $\Delta z = (2 \ln 2/\pi) (\lambda^2/\Delta\lambda)$

Where λ is the source center wavelength

The transverse resolution, which is determined by the focused spot size is stated as below (Fujimoto et al. 2000):

{Transverse Resolution} = $\Delta x = (4\lambda/\pi) (f/d)$

Where f is objective lens focal length and d is its spot size

Therefore, a higher transverse resolution can be obtained by a longer objective lens focal length and/or smaller spot size. Furthermore, the relationship between the depth of focus b and the transverse resolution is expressed as (Fujimoto et al. 2000):

{Depth of Focus} = $b = \pi\Delta x^2/2\lambda$

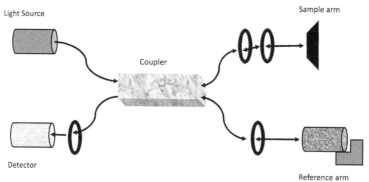

Fig. 1. Schematic diagram of an OCT set up (Derived from Popescu et al. 2011).

So as the transverse resolution becomes finer (smaller Δx), the depth of focus decreases. The Signal to noise detection can also be mathematically expressed (Fujimoto et al. 2000):

(Signal to Noise Ratio} = 10 Log ($\eta P/2E p Neb$)

Where η is the detector's quantum efficiency, P is the optical power, Ep is photon energy, and Neb is the noise equivalent bandwidth of the detected light. Thus, a higher signal to noise ratio requires a higher optical power P since the other factors cannot be easily modified for a given set up.

OCT Systems

Two main OCT systems that are utilizing the point-scanning/point-detection technology, as described above, are Time-Domain type and Fourier-Domain type.

Time-Domain OCT (TD-OCT)

TD-OCT system is detection technique based on the utilization of low-coherent light source and a scanning reference delay. The typical light source is a semiconductor super-luminescent diode (SLD) or pulse laser. As the first developed OCT technology, TD-OCT system measures the light echoed back sequentially by the step-movement of the reference mirror (Popescu et al. 2011).

Fourier-Domain OCT (FD-OCT)

Compared to TD-OCT system, FD-OCT system can achieve a higher data acquisition speed and produce real-time data (Tsai et al. 2014). The FD-OCT system utilizes a fixed reference mirror instead of moving mirror used in the TD-OCT system and this feature is in part responsible for the higher speed (Popescu et al. 2011). Unlike the image collected sequentially by the TD-OCT system, the FD-OCT system captures all the spectral components simultaneously, since the echoed light arrive from all axial depths at the same time in the FD-OCT system (Popescu et al. 2011). FD-OCT technique also carry out a "Fourier transform" process in its image processing, a mathematical method that converts a function as the sum of sinusoidal (sine waves) functions (Dror 2015). Typically, FD-OCT devices can capture 18,000 to 40,000 A-scans per second, 45–100 times faster than that of TD-OCT system. In addition, the faster acquisition speed also translates to reduction of movement artefacts and better signal-to-noise ratio (Popescu et al. 2011). FD-OCT imaging can be obtained in one of the two ways (spectral-domain and swept-source):

Spectral-Domain OCT (SD-OCT): SD-OCT is one of the FD-OCT system methods. The core makeup of SD-OCT is a Michelson-type interferometer and a stationary reference mirror. However, the spectrum of the combined return beams in interference pattern is dispersed and then immediately detected by a high-speed CCD line camera, rather than a point detector. A Fourier transformation is conducted on the spectrally resolved interference pattern and the reflectivity is detected as a function of depth (A-scan) with the upper acquisition speed bound by the read-out rate of the line senor of the camera, usually in the kHz range (Popescu et al. 2011). Typically operated at 800 nm or 1 mm wavelengths with imaging speed at 29,000 to 300,000 lines per second, SD-OCT system has substantially impacted the field of ophthalmology for its ability to generate ultrahigh resolution and 3-dimensional images of *in vivo* retinal pathology. In addition, SD-OCT provides an important resolution advantage of less than 3 mm, very critical for retinal image even though the speed is acceptably slower, when compared to SS-OCT (Tsai et al. 2014).

Swept-Source OCT (SS-OCT): SS-OCT, also known as optical Fourier domain imaging (OFDI) and under the FD-OCT system, on the other hand, utilizes a broadband wavelength-swept laser light scanning method and a fixed reference mirror. As such, SS-OCT could operate at longer wavelength of 1 to 1.3 mm, which reduces optical scattering and enhances depth of penetration (Tsai et al. 2014). A Fourier transformation is performed on detected signal over one sweep of the laser across the available broadband. SS-OCT utilizes the point detector, which has a higher signal-to-noise ratio than line camera used in SD-OCT (Popescu et al. 2011). SS-OCT technology is especially well suited for biological tissues that tend to scatter light, such as gastrointestinal track and breast ducts and is also a great device for clinical situation where speed is essential, since it is faster than SD-OCT in image acquisition (Tsai et al. 2014).

Full-Field OCT (FF-OCT)

FF-OCT is yet another imaging system where 2-dimensional OCT image can be directly acquired. FF-OCT produces tomographic images in an *en face* orientation, i.e., images made are perpendicular to the optical axis of the sample arm and are acquired in one shot rather than scanning across the sample surface. Usually, FF-OCT can achieve images with an axial resolution of 1 micro meter and a transverse spatial resolution of about 1 micro meter, close to that of a conventional microscope level (Popescu et al. 2011).

Comparing OCT to some other commonly used medical diagnostic techniques, OCT has a relative finer resolution, with its resolution down to 1 micro meter range, near the microscopic level for the ultrahigh resolution OCT. On the other hand, OCT has a low penetration depth of just about 1 mm, which is much shallower than ultrasonography (just short of 10 cm), high-resolution CT (entire body), and magnetic resonance imaging (entire body). Figure 2 depicts these comparisons and illustrates a nearly inverse relationship between penetration depth and resolution among these 6 commonly used diagnostic techniques. It is like the biological body put a limit to these tests, so we can get either the resolution or the penetration, but not both.

increasing in resolution power

	Confocal Microscopy	Ultra High Resolution OCT	Conventional OCT	Ultrasonography	High Resolution CT Scan	Magnetic Resonance Imaging
Resolution	*1 micro meter*	*1 micro meter*	*10 micro meter*	*150 micro meter*	*300 micro meter*	*1 mm*
Penetration Depth	*< 1 mm*	*1 mm*	*1 mm*	*< 10 cm*	*Entire Body*	*Entire Body*

Increasing in penetration depth

Fig. 2. Comparison of imaging methods' penetration depth and resolutions (Derived from Popescu et al. 2011).

Applications of OCT Technology

Ophthalmological Applications

Ophthalmology field is an early adaptor of OCT and its applications are continuing to expand (Ang et al. 2018, Borrelli et al. 2018, Coscas et al. 2018, Dastiridou and Chopra 2018, Iorga et al. 2018, Kim and Park 2018, Lauermann et al. 2018, Lee et al. 2018, Liu et al. 2018, Mateo et al. 2017, Mwanza and Budenz 2018, Mwanza et al. 2018, Pan et al. 2018, Spaide et al. 2018, Tan et al. 2018, Tatham and Medeiros 2017,

Venkateswaran et al. 2018, Vincent et al. 2018, Wang et al. 2018). At the followings the major applications in the ophthalmology field are discussed.

Anterior Segment Assessment

The improvement of histology-like OCT imaging technique has provided an excellent tool for examining the normal structure and pathology of the eye's anterior segment. Some of the active research subjects and clinical practices utilized OCT include imaging the ocular surface, cornea, anterior chamber structures, aqueous flow, vascular network, tear film evaluation, angle assessment, and scleral contact lens assessment (Ang et al. 2018, Venkateswaran et al. 2018, Vincent et al. 2018).

Diabetic Retinopathy Assessment

Diabetic retinopathy is a common complication of a very common disease diabetes mellitus and is a major cause of blindness. In one European study, it is ranked as the most common cause of blindness (Bourne et al. 2014). One of the excellent and cost-effective ways of prevent blindness from diabetic retinopathy is to assess early disease and control the progression of retinal neovascularization (Romero-Aroca et al. 2016). Fluorescein angiography has been the "gold standard" for such assessment but without the ability to examine vessels of different layers. Here, OCT angiography (OCTA) can help provide that 2- and 3-dimensional visualization of the retinal and choroidal vascular network, including micro-aneurysm, non-perfusion region, intraretinal vascular abnormality, neovascularization, vascular density quantification, and foveal avascular zone, without the need for intravenous dye administration (Coscas et al. 2018, Liu et al. 2018, Spaide et al. 2018). Moreover, OCTA could be useful for assessment of other retinal and choroidal vascular diseases such as age-related macular degeneration, central serous chorioretinopathy, uveitis, and inherited retinal disorders (Borrelli et al. 2018, Lauermann et al. 2018, Tan et al. 2018). OCTA photos taken from normal and diabetic retinas are shown in Figs. 3, 4 and Fig. 5, respectively.

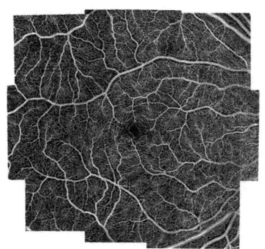

Fig. 3. A OCTA wide-field montage of a normal retina (Source: de Carlo et al. 2015, Open Access).

Glaucoma Assessment

As a leading cause of irreversible blindness in the global scale, glaucoma is a highly prevalent disease, accounting for 3.5% for population aged between 40 to 80 years. The number of people affected by glaucoma was estimated to be 64 million worldwide in 2013 and is projected to increase to 76 million in 2020 and 111 million in 2040 (Tham et al. 2014). Glaucoma is caused by increase of intra-ocular pressure which in turn cause the damages in the retinal nerve, especially the vision signal conductors "retinal ganglion cells" (RGC), leading to vision loss (Xiang et al. 1996). In some patients, the loss of vision is due to diagnosis

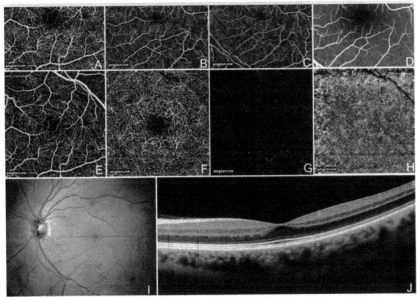

Fig. 4. OCTA images of a normal left eye retina: (A). Full thickness 3 × 3 mm (internal limiting membrane to Bruch's membrane; (B). Full-thickness 6 × 6 mm; (C). Full-thickness 8 × 8 mm; (D). Fluorescein angiography cropped 8 × 8 mm; (E). Superficial inner retina 3 × 3 mm; (F). Deep inner retina 3 × 3 mm; (G). Outer retina 3 × 3 mm; (H). Choriocapillaris 3 × 3 mm; (I). En-face intensity image, with red line corresponding to b-scan image in J; (J). b-scan (Source: de Carlo et al. 2015, Open Access).

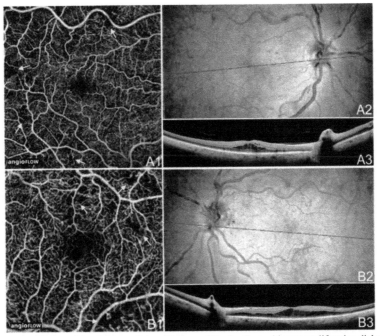

Fig. 5. OCTA images of retina of the right eye (A) and left eye (B) of patient with non-proliferative diabetic retinopathy: (A1). Full-thickness 6 × 6 mm; (A2). En-face image with red line corresponding to b-scan image in A3; (A3). 12 mm b-scan; (B1). Full-thickness 3 × 3 mm; (B2). En-face image with red line corresponding to b-scan in B3; and (B3). 12 mm b-scan. Yellow arrows point to non-perfusion capillary areas (Source: de Carlo et al. 2015, Open Access).

delay attributing to its relative asymptomatic nature at the early stage of glaucoma (Weinreb et al. 2014). Therefore, the ability of assessing early development of glaucoma will help physicians to start timely treatment to stop the progression of vision loss. Moreover, periodic assessment of glaucoma progression will also help physicians to adjust treatment in optimizing therapeutic outcome. OCT has been studied to examine its potential role in assessing glaucoma (Tatham and Medeiros 2017, Kim and Park 2018). OCT has been utilized increasingly to objectively measure retinal nerve fiber layer (RNFL), optic nerve head and macula to evaluate glaucoma progression. The theoretical basis behind these measurements for glaucoma assessment are that more than 50% of RGCs are located at the macula and that macular thickness complementary to peripheral RNFL thickness is a good reflection of glaucoma damage to the retinal nerves (Kim and Park 2018). Recently, a new measurement of the minimum distance band, a 3-dimensional optic nerve head parameter was introduced as an assessment for glaucoma (Mwanza and Budenz 2018). In addition, OCTA provides more structural and functional evaluation tools for glaucoma, as evidences have shown the relationship between retinal vascular density reduction at the optical nerve head, peripapillary and macula and early glaucoma (Dastiridou and Chopra 2018, Mwanza and Budenz 2018).

Optic Neuropathy Assessment

As a more recent development, OCT has been examined for a potential role of evaluating neurological diseases that affect optical nerve and other central nervous system (CNS). For example, OCT has been investigated for potential role in assessing neuromyelitis optica spectrum disorders (NMOSD), which comprise a group of CNS disorders that are of inflammatory or autoimmune in nature. This group of diseases can result in severe visual and general disability with clinical signs resembling multiple sclerosis (MS). OCT technique can be useful in the diagnosis of NMOSD, as well as for the differentiation of NMOSD from MS (Mateo et al. 2017, Iorga et al. 2018).

Gastrointestinal (GI) Applications

Since one of OCT technical abilities is the generation of cross-sectional image, GI field also utilizes OCT as an endoscopic examination tool (Tsai et al. 2014, 2017). Some of the diseases that physicians have utilized are depicted below.

Colorectal Cancer

As a common GI cancer with high morbidity and mortality, colorectal cancers account for nearly 10% of all cancer death and being the third leading cancer mortality in the US, despite its early detection by routine endoscopy would be easily treatable. The failure of early detection in many cases are due to nearly 50% of these early cancers are missed during routine endoscopy in major part because they are flat and subtle rather than protruding and easily visible like polyps. This is especially true for patients with inflammatory bowel diseases, where the early cancers often present as flat lesions rather than pudunculated lesions (Kiesslich et al. 2001, Tsai et al. 2014). Moreover, random sampling of colon with biopsies do not provide significant outcome in early cancer detection for two major reasons. First, the random sampling of a broad area could easily miss the cancerous area. Second, the histologic features of inflammatory bowel diseases could interfere the proper pathological interpretation even if biopsies were obtained during conventional endoscopy (Tsai et al. 2014). Thus, OCT may be a good choice in the detection of early cancers, as it could provide histology-like optical image.

Esophageal Cancer

With a five-year survival rate of only 16%, esophageal cancers demand attention for early detection (Jemal et al. 2007). Similar to its use in detecting GI cancer, OCT could be very useful in detecting esophageal cancer (Tsai et al. 2014).

Cardiovascular Applications

Cardiovascular applications of OCT have been under investigation by many groups (Boi et al. 2018, Fujii et al. 2018, Ha et al. 2017, Kim et al. 2018, Shlofmitz et al. 2018). Some of the studies are depicted below.

Coronary Arterial Assessment

Since coronary arterial plaque formation is the key element of arterial stenosis and cardiac ischemia and heart attack, and the potential cardiac damage is related to the degree of arterial occlusion, the ability to assess and quantify the coronary arterial atherosclerosis is highly relevant to patient care. Comparing the only two methods that could detected the specific atherosclerotic plaque components such as cap fibroatheroma, fibrous cap, macrophage infiltration, large necrotic core, and thrombus, OCT provides better visualization of plaque tissue layers than the only other method "intravascular ultrasound" (IVUS) (Boi et al. 2018).

OCT-guided Coronary Arterial Stent Placement

Not only that OCT can assist physicians in assessing the pathology of coronary arterial blockage, it can also help guiding the implantation of coronary arterial stent (Ha et al. 2017, Shlofmitz et al. 2018). Potentially, OCT can assist coronary arterial stent placement in the setting of acute myocardial infarction by its abilities to assess baseline lesion characteristics and post-placement outcome, including visualization of procedural dissections, malapposition, tissue prolapse, and thrombus (Kim et al. 2018).

Neurological Disease Applications

In addition to ophthalmological, gastrointestinal, and cardiovascular diseases, OCT has been applied to examine neurological diseases, especially for multiple sclerosis (Alonso et al. 2018, Britze and Frederiksen 2018, Lambe et al. 2018) and Alzheimer's disease (Doustar et al. 2017). Multiple sclerosis (MS) is characterized by neural inflammation and degeneration. Significantly, study have documented that near 90% of patients affected by MS had visual lesions with or without optic neuritis, which is the onset symptom in about 20% patients affected by MS (Britze and Frederiksen 2018). OCT could play a role in MS assessment as many studies have shown that both the RNFL and RGC layers are reduced significantly in MS patients in compare to health individuals and these changes were well correlated with other parameters such as visual function, disability, and MRI assessment. Thus, potentially, OCT could be utilized to predict disability progression and visual function in MS (Britze and Frederiksen 2018).

Dental Applications

The applications of OCT in dental field have been reviewed by some field experts (Katkar et al. 2018, Machoy et al. 2017). Among them, OCT applications in dental field include microleakage detection at restoration site, assessment of tooth cracks and fractures, periodontal tissue evaluation, early oral cancer detection, and locating pulp canal for endodontic operation (Katkar et al. 2018).

Other Medical Applications

In addition to the applications in ophthalmological, GI, cardiovascular, Neurological and dental specialties, OCT has also been applied to extramammary Paget's disease (Bayan et al. 2018), oral pathology (Reddy and Sai Praveen 2017, Yang et al. 2018), precancerous and cancerous skin diseases (Casari et al. 2018, Levine et al. 2017), and some urological (Huang et al. 2018) and gynecological (Kinillin et al. 2017) diseases.

Limitation and Challenge of Clinical OCT Applications

Just like any other medical imaging techniques, OCT also has its limitation and faces some challenges. Some of these identified issues are discussed below.

Penetration Depth Limitation

One of limitation of OCT is its shallow penetration depth (Fig. 2, Popescu et al. 2011). Although it works well with internal organs that have a lumen, such as coronary artery and gastrointestinal track, it will not work with other deep body structures due to its penetration depth limitation.

Correlation of Structural Changes with Real Disease Progression in Glaucoma

While OCT technique can assess structural changes in patients with glaucoma, it is still unclear if some of those changes are directly related to the disease or just some natural phenomena of aging. Medical investigators are currently raising questions like these: What is the best structure to measure? What quantity of change should be considered significant? What is the relevance of these changes to the patients? How are the longitudinal changes affected by aging? Will it be beneficial to combine OCT results with visual field data for better assessment? What should be the frequency of OCT examination? Some of these questions are being addressed by the academic communities and others will hopefully be answered in the near future (Tatham and Medeiros 2017).

Conflicting and Subtle Data in Early Glaucoma Assessment

Challenges when using single OCT parameter to evaluate glaucoma are that the changes in early glaucoma could be so subtle that escape the OCT detection and that conflicting data could arise. So to counter these drawbacks, investigators have suggested the utility of combining multiple OCT parameters from the same test into a composite parameter, for the purpose of improving diagnostic accuracy for early glaucoma (Mwanza et al. 2018).

Summary

In summary, OCT is a very versatile biomedical imaging technique. Thus far, its applications have been expanded to the specialty fields of ophthalmology, cardiovascular diseases, neurological diseases, gastrointestinal disorders, dental practices, as well as multiple sclerosis, Alzheimer's disease, and mucocutaneous pathology. With its ability to penetrate 1 mm and resolution as fine as 1 micro meter, OCT has established itself as an excellent research and clinical diagnostic tool. In addition, OCT, in conjunction with other technology, has been applied in certain interventional medical procedures. As the volumes of new medical discoveries on OCT continue to flow into the literatures, it strongly suggests that OCT will become a highly useful biomedical instrument for sometimes into the future.

References

Alonso, R., D. Gonzalez-Moron and O. Garcea. 2018. Optical coherence tomography as a biomarker of neurodegeneration in multiple sclerosis: a review. Mult Scler Relat Disord 22: 77–82.
Ang, M., M. Baskaran, R.M. Werkmeister, J. Chua, D. Schmidl, V. Aranha Dos Santos et al. 2018. Anterior segment optical coherence tomography. Prog Retin Eye Res Apr 7. Pii: S1350–9462(17)30085-X. Doi: 10.1016/j.preteyeres.2018.04.002.
Bayan, C.Y., T. Khanna, V. Rotemberg, F.H. Samie and N.C. Zeitouni. 2018. A review of non-invasive imaging in extramammary Paget's disease. J Eur Acad Dermatol Venereol May 15. Doi: 10.1111/dv.15072.
Boi, A., A.D. Jamthikar, L. Saba, D. Gupta, A. Sharma, B. Loi et al. 2018. A survey on coronary atherosclerotic plaque tissue characterization in intravascular optical coherence tomography. Curr Atheroscler Rep May 21; 20(7): 33. Doi: 10.1007/s11883-018-0736-8.
Borrelli, E., D. Sarraf, K.B. Freund and S.R. Sadda. 2018. OCT angiography and evaluation of the choroid and choroidal vascular disorders. Prog Retin Eye Res July 27. Pii: S1350-9462(18)30022–3. Doi: 10.1016/j.preteyeres.2018.07.002.
Bourne, R.R., J.B. Jonas, S.R. Flaxman, J. Keeffe, J. Leasher, K. Naidoo et al. 2014. Vision loss expert group of the global vurden of disease stuy. Prevalence and causes of vision loss in high-income countries and in Eastern and Central Europe: 1990–2010. Br J Ophthlmol 98: 629–638.
[BRAINYQUOTE] 2018a. Brainy Quote. Optics Quotes. [https://www.brainyquote.com/topics/optics] accessed August 8, 2018.

[BRAINYQUOTE] 2018b. Brainy Quote. Image Quotes. [https://www.brainyquote.com/topics/image] accessed August 8, 2018.

Britze, J. and J.L. Frederiksen. 2018. Optical coherence tomography in multiple sclerosis. Eye 32: 884–888.

Casari, A., J. Chester and G. Pellacani. 2018. Actinic keratosis and non-invasive diagnostic techniques: an upate. Biomedicine Jan 8;6(1). Pii: E8. Doi: 10.3390/biomedicines6010008.

Coscas, G., M. Lupidi, F. Coscas, J. Chhablani and C. Cagini. 2018. Optical coherence tomography angiography in health subjects and diabetic patients. Ophthalmologica 239: 61–73.

Dastiridou, A. and V. Chopra. 2018. Potential applications of optical coherence tomography angiography in glaucoma. Curr Opin Ophthalmol 29: 226–233.

De Carlo, T.E., A. Romano, N.K. Waheed and J.S. Duker. 2015. A review of optical coherence tomography angiography (OCTA). Int J Retina and Vitreous 1: 5 doi: 10.1186/s40942-015-0005-8.

Doustar, J., T. Torbati, K.L. Black, Y. Koronyo and M. Koronyo-Hamaoui. 2017. Optical coherence tomography in Alzhemer's disease and other neurodegenerative diseases. Front Neurol Dec 19; 8: 701. Doi: 10.3389/neur.2017.00701.

Dror, R. 2015. The Fourier transform. Fall 2015. CS/CME/BIOPHYSICS/BMI 279 [https://web.stanford.edu/class/cs279/notes/FT-notes.pdf] accessed August 14, 2018.

Fercher, A.F., W. Drexler, C.K. Hitzenberger and T. Lasser. 2003. Optical coherence tomography – principles and applications. Rep Prog Phys 66: 239–303.

Fujii, K., R. Kawakami and R. Hirota. 2018. Histopathological validation of optical coherence tomography findings of the coronary arteries. J Cardiol 72: 179–185.

Fujimoto, J.G., C. Pitris, S.A. Boppart and M.E. Brezinski. 2000. Optical coherence tomography: an emerging technology for biomedical imaging and optical biopsy. Neoplasia 2: 9–25.

Ha, F.J., J.P. Giblett, N. Nerlekar, J.D. Cameron, I.T. Meredith, N.E.J. West et al. 2017. Optical coherence tomography guided percutaneous coronary intervention. Heart Lung Circ 26: 1267–1276.

Huang, J., X. Ma, L. Zhang, H. Jia and F. Wang. 2018. Diagnostic accuracy of optical coherence tomography in bladder cancer patients: a systematic review and meta-analysis. Mol Clin Oncol 8: 609–612.

Iorga, R.E., A. Moraru, M.R. Ozturk and D. Costin. 2018. The role of optical coherence tomography in optic neuropathies. Rom J Ophthalmol 62: 3–14.

Jemal, A., R. Siegel, E. Ward, T. Murray, J. Xu and M.J. Thun. 2007. Cancer statistics, 2007. CA Cancer J Clin 57: 43–66.

Katkar, R.A., S.A. Tadinada, B.T. Amaechi and D. Fried. 2018. Optical coherence tomography. Dent Clin North Am 62: 421–434.

Kiesslich, R., M. von Bergh, M. Hahn, G. Hermann and M. Jung. 2001. Chromoendoscopy with indigocarmine improves the detection of adenomatous and nonadenomatous lesions in the colon. Endoscopy 33: 1001–1006.

Kim, K.E. and K.H. Park. 2018. Macular imaging by optical coherence tomography in the diagnosis and management of glaucoma. Br J Ophthalmol 102: 718–724.

Kim, Y., T.W. Johnson, T. Akasaka and M.H. Jeong. 2018. The role of optical coherence tomography in the setting of acute myocardial infarction. J Cardiol 72: 186–192.

Kinillin, M., T. Motovilova and N. Shakhova. 2017. Optical coherence tomography in gynecology: a narrative review. J Biomed Opt 22: 1–9.

Lambe, J., O.C. Murphy and S. Saidha. 2018. Can optical coherence tomography be used to guide treatment decisions in adult or pediatric multiple sclerosis? Curr Treat Options Neurol Mar 21; 20(4): 9. Doi: 10.1007/s11940-018-0493-6.

Lauermann, J.L., N. Eter and F. Alten. 2018. Optical coherence tomography angiography offers new insights into choriocapillaris perfusion. Ophthalmologica 239: 74–84.

Lee, M.J., A.G. Abrham, B.K. Swenor, A.R. Sharrett and P.Y. Ramulu. 2018. Application of optical coherence tomography in the detection and classification of cognitive decline. J Curr Glaucoma Pract 12: 10–18.

Levine, A., K. Wang and O. Markowitz. 2017. Optical coherence tomography in the diagnosis of skin cancer. Dermatol Clin 35: 465–488.

Liu, G., D. Xu and F. Wang. 2018. New insights into diabetic retinopathy by OCT angiography. Diabetes Res Clin Pract 142: 243–253.

Machoy, M., J. Seeliger, L. Szyszka-Sommerfeld, R. Koprowski, T. Gedrange and K. Wozniak. 2017. The use of optical coherence tomography in dental diagnostics: a state-of-the-art review. J Healthc Eng 2017:7560645. Doi: 10.1155/2017/7060645.

Mateo, J., O. Esteban, M. Martinez, A. Grzybowski and F.J. Ascaso. 2017. The contribution of optical coherence tomography in neuromyelitis optica spectrum disorders. Front Neurol Sep 29; 8: 493. Doi: 10.3389/fneur.2017.00498.

Mwanza, J.C. and D.L. Budenz. 2018. New development in optical coherence tomography imaging for glaucoma. Curr Opin Ophthalmol 29: 121–129.

Mwanza, J.C., J.L. Warren and D.L. Budenz. 2018. Utility of combining spectral domain optical coherence tomography structural parameters for the diagnosis of early glaucoma: a mini-review. Eye Vis (Lond) Apr 15; 5: 9. Doi: 10.1186/s40662-018-0101-6.

Pan, T., Y. Su, S.T. Yuan, H.C. Lu, Z.Z. Hu and Q.H. Liu. 2018. Optic disc and peripapillary changes by optic coherence tomography in high myopia. Int J Ophthalmol 11: 874–880.

Popescu, D.P., L.P. Choo-Smith, C. Flueraru, Y. Mao, S. Chang, J. Disano et al. 2011. Optical coherence tomography: fundamental principles, instrumental designs and biomedical applications. Biophys Rev 3: 155–169.

Reddy, R.S. and K.N. Sai Praveen. 2017. Optical coherence tomography in oral cancer: a transpiring domain. J Cancer Res Ther 13: 883–888.

Romero-Aroca, P., S. de la Riva-Fernandez, A. Valls-Mateu, R. Sagarra-Alamo, A. Moreno-Ribas, N. Soler et al. 2016. Cost of diabetic retinopathy and macular oedema in a population, an eight year follow up. BMC Ophthalmol 16: 136. Doi: 10.1186/s12886-016-0318-x.

Shlofmitz, E., R.A. Shlofmitz, K.K. Galougahi, H.M. Rahim, R. Virmani, J.M. Hill et al. 2018. Algorithmic approach for optical coherence tomography-guided stent implantation during percutaneous coronary intervention. Interv Cardiol Clin 7: 329–344.

Shu, X., L. Beckmann and H. Zhang. 2017. Visible-light optical coherence tomography: a review. J Biomed Opt 22: 1–14.

Spaide, R.F., J.G. Fujimoto, N.K. Waheed, S.R. Sadda and G. Staurenghi. 2018. Optical coherence tomography angiography. Prog Retin Eye Res 64: 1–55.

Tan, A.C.S., G.S. Tan, A.K. Denniston, P.A. Keane, M. Ang, D. Milea et al. 2018. An overview of the clinical applications of optical coherence tomography angiography. Eye (Lond) 32: 262–286.

Tatham, A.J. and F.A. Medeiros. 2017. Detecting structural progression in glaucoma with optical coherence tomography. Ophthalmology 124: S57–s65.

Tham, Y.C., X. Li, T.Y. Wong, H.A. Quigley, T. Aung and C.Y. Cheng. 2014. Global prevalence of glaucoma and projections of glaucoma burden through 20140: a systematic review and meta-analysis. Ophthalmology 121: 2081–2090.

Tsai, T.H., J.G. Fujimoto and H. Mashimo. 2014. Endoscopic optical coherence tomography for clinical gastroenterology. Diagnostics 4: 57–93.

Tsai, T.H., C.L. Leggett, A.J. Trindade, A. Sethi, A.F. Swager, V. Joshi et al. 2017. Optical coherence tomography in gastroenterology: a review and future outlook. J Biomed Opt 22: 1–17.

Venkateswaran, N., A. Galor, J. Wang and C.L. Karo. 2018. Optical coherence tomography for ocular surface and corneal diseases: a review. Eye Vis (Lond) Jun 12; 5: 13. Doi: 10.1186/s40662-018-0107-0.

Vincent, S.J., D. Alonson-Caneiro and M.J. Collins. 2018. Optical coherence tomography and scleral contact lenses: clinical and research applications. Clin Exp Optom July 30. Doi: 10. 1111/cxo.12814.

Wang, L., Q. Murphy, N.G. Caldito, P.A. Calabresi and S. Saidha. 2018. Emerging applications of optical coherence tomography angiography (OCTA) in neurological research. Eye Vis (Lond) May 12;5:11. Doi: 10.1186/s40662-018-0104-3.

Weinreb, R.N., T. Aung and F.A. Medeiros. 2014. The pathophysiology and treatment of glaucoma: a review. JAMA 311: 1901–1911.

Xiang, M., H. Zhou and J. Nathans. 1996. Molecular biology of retinal ganglion cells. Proc Natl Acad Sci USA 93: 596–601.

Yang, E.C., M.T. Tan, R.A. Schwarz, R.R. Richards-Kortum, A.M. Gillenwater and N. Vigneswaran. 2018. Noninvasive diagnostic adjuncts for the evaluation of potentially premalignant epithelial lesions: current limitations and future directions. Oral Surg Oral Med Oral Pathol Oral Radiol 125: 670–681.

QUESTIONS

1. What is source of image generation in OCT?
2. What are the advantages of OCT over ultrasonography?
3. What two principle features the OCT measurement are based on?
4. What are the penetration depth and resolution of conventional OCT and ultrahigh resolution OCT?
5. Can you explain the essential difference between TD-OCT and FD-OCT?
6. Can you explain the essential difference between SD-OCT and SS-OCT?
7. The use of OCT in coronary arterial study is considered to be invasive. What is the maximum time that can be used for the OCT imaging inside the coronary artery and why?
8. What are some of the limitations and challenges of OCT for medical applications?

PROBLEMS

Students will be grouped into 3–5 people in each. Students will brainstorm to select a real-time medical case for which there is no good non-invasive method to reach a diagnosis. Together the students within the group will deliberate and determine if OCT could help solving the diagnostic challenge and state the logical conclusion. The groups will present their discussion results in the class discussion session.

22

Emerging Biomedical Imaging
Photoacoustic Imaging

Lawrence S. Chan

QUOTABLE QUOTES

"It's an amazing feeling to know that life is actually growing inside your body. The first time you see the ultrasound and you see the little bones and you realize that it's part of you and it's in your care is life changing and this sort of protective instinct has taken over."

Halle Berry, American Actress (ULTRASOUND QUOTES 2018)

"The fingers and toes and beating hearts that we can see on an unborn child's ultrasound come with something that we cannot see: a soul."

Geroge W. Bush, 43rd President of the USA (ULTRASOUND QUOTES 2018)

"Indeed, we often mark our progress in science by improvements in imaging."

Martin Chalfie, American Scientist and 2008 Nobel Laureate in Chemistry (IMAGING QUOTES 2018)

"The fundamental principle behind medical imaging is to noninvasively depict an underlying tissue property based on the interaction of tissue with some form of radiation."

Valluru, Wilson, and Willmann, Stanford University Researchers (Valluru et al. 2016)

Learning Objectives

The learning objectives are to help students to understand mathematically the physics principles governing the imaging techniques derived from photoacoustic effect and to introduce students to clinical situations that photoacoustic imaging may have potential usage.
After completing this chapter, the students should:

- Understand the physics principles of photoacoustic wave generation.
- Understand the two major platforms of photoacoustic imaging: microscopic and tomographic.
- Understand the ability of photoacoustic imaging technique in enhancing biological detectability when combining with other imaging methods or tracers.

University of Illinois College of Medicine, 808 S Wood Street, R380, Chicago, IL 60612; larrycha@uic.edu

- Understand the advantage of photoacoustic imaging.
- Understand the limitation of photoacoustic imaging.
- Able to consider real-life clinical situation where photoacoustic imaging may be beneficial.

History of Photoacoustic Imaging

The photoacoustic (also termed optoacoustic) effect was first discovered by American inventor Alexander Graham Bell in 1880, when he observed the generation of sound during a process of absorption of modulated sunlight (Meiburger 2016). Possibly due to the unavailable high intense light source, little investigation or development progress have been made to further the benefit of this discovery until 1960s, when laser light source, which could provide the high peak power, spectral purity, and directionality needed for photoacoustic signal generation, became available. Photoacoustic imaging as a technique has surfaced in Pubmed-based medical literature in the late 1970s. Some of the earliest publications enlightening the potential medical applications of photoacoustic appeared in the late 1970s and the early 1980s (Campbell et al. 1977, Rosencwaig and Pines 1977, Rosencwaig 1982). For one reason or another, the academic publications on photoacoustic imaging at the Pubmed data base during the 1990s were low in number and only 7 papers were published. Starting in 2000s, research in both fundamental and clinical aspects of photoacoustic images skyrocketed, resulting in many outstanding publications, including some papers published in the top-ranked journals like Nature, Science, and Physical Review Letters (Zhang et al. 2006, Fang et al. 2007, Wang 2009, Wang 2012). Several textbooks in the subject of photoacoustic imaging that cover the general principles and biomedical applications have been recently published (Jiang 2015, Meiburger 2016, Wang 2017, Yang and Xing 2018). At the moment, we are at a very exciting time for discovery and application of photoacoustic principles to the real-life clinical situations.

Principles of Photoacoustic Imaging

Having described the history of photoacoustic imaging, we will now discuss the principles of photoacoustic imaging. First, a general framework of how photoacoustic imaging is generated will be discussed, followed by two key steps of the imaging processes: (1). The Signal generation and (2). The image formation.

General Concept of Photoacoustic Imaging

To make the concept very simple and easy to understand, let us just consider a schematic diagram depicted in Fig. 1.

Essentially, a pulse laser delivers a pulsed light onto the image target through a living tissue. The pulsed light results in generation of a heat (temperature change), which is then transformed to a pressure change. Lastly this changing pressure is converted to an acoustic (sound) wave that will be picked up by an ultrasound transducer or probe. In short, it is a "light goes in sound comes out" process (Fig. 1).

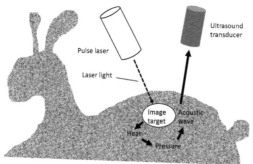

Fig. 1. Schematic diagram of a simple concept of photoacoustic imaging.

Principles of Initial Signal Generation

In this session, the physics of the signal generation will be explained through several condensed mathematical equations based on recent publications (Xia et al. 2014, Kim et al. 2015, Zhou et al. 2016). More detailed explanation of this principle is available through a recently published textbook (Wang 2017). According to experts of the field of photoacoustic imaging, the signal generation involves three sequential steps. First, the object receives and absorbs an intense light (and the accompanied optical energy). Next, the absorbed optical energy is then converted into heat, leading to a temperature increase. Lastly, the thermoelastic expansion as a result of temperature rise leads to a pressure rise. And this change of pressure then propagates to an emission of acoustic waves (Xia et al. 2014, Zhou et al. 2016). Since the generation of acoustic wave requires that the thermoelastic expansion to be in a "time variant" form, i.e., it changes with time, the most common light energy source nowadays is the pulsed laser, which fulfills the time variant requirement. Other light sources include continuous-wave laser with intensity modulation at a constant or variable frequency. Pulsed laser is preferred since it has a higher signal to noise ratio if maximum allowable fluence or power set by the American National Standards Institute is used (Xia et al. 2014).

Following a short pulse of laser excitation,

$$\{\text{Local Fractional Volume Expansion}\} = dV/V = -\kappa P + \beta T \qquad [1]$$

where κ is isothermal compressibility (Pa^{-1})

β is the thermal coefficient of volume expansion (K^{-1})
P is change in pressure (Pa, Pascal)
T is change in temperature (K, Kelvin)

Consider two essential time scales in the generation of photoacoustic signal, the thermal relaxation time (τ_{th}) and the stress relaxation time (τ_s), with the given equations:

$$\tau_{th} = d_c/\alpha_{th} \qquad\qquad \tau_s = d_c/\upsilon_s$$

Where d_c is the desired spatial resolution; α_{th} is the thermal diffusivity (m^2/s); and υ_s is the speed of sound (m/s).

Now if the laser pulse duration is set to be shorter than τ_{th} and τ_s, the excitation satisfies both thermal and stress confinements and the dV/V becomes negligible (or zero). Therefore, the left side of Eq. 1 will be set at zero and the equation can be rearranged to be:

$$\{\text{Initial pressure}\} = P_0 = \beta T/\kappa \qquad [2]$$

The other expression of T can be written as:

$$T = (n_{th})\,(A_e)/\rho\,C_v \qquad [3]$$

Where n_{th} is percent light energy converted to heat ($0 < n_{th} < 1$)

A_e is specific optical energy deposition (J/m^3) and $= \mu_a F$
ρ is mass density
C_v is the specific heat capacity at constant volume
μ_a Is absorption coefficient (1/cm)
F Is the local optical fluence (J/cm^2)

Rearrange equations 2 and 3, by substituting equation 3 into Eq. 2, we will have:

$$P_0 = \beta/(\rho C_v \kappa)\,n_{th}\,\mu_a F \qquad [4]$$

Lastly, by defining the temperature-dependent, dimensionless, Grueneisen Parameter as $\Gamma = \beta/(\rho C_v \kappa)$, Eq. 4 can then be rewritten as:

$$P_0 = \Gamma\,n_{th}\,\mu_a F \qquad [5]$$

Therefore, according to equation No. 5, the initial pressure is proportion to photo/heat converting efficiency (percent light converted to heat (n_{th})), the absorption coefficient (μ_a), and the local optical fluence (F).

Though Γ, n_{th}, and F are generally considered to be constants, they may be varied depending on tissue types (Xia et al. 2014). In theory, higher laser fluence (expressed as J/cm^2) will yield higher photoacoustic transient production in tissue and better imaging depths. To ensure patient safety in photoacoustic imaging, the maximum skin exposure energy is currently set at 20–100 mJ/cm^2 at 700–1050 nm laser wave length (Valluru et al. 2016, LASER 2007). At the present time, the most common type of laser used in photoacoustic image is the solid-state Q-switched lasers (Valluru et al. 2016).

Principles of Photoacoustic Wave Propagation

Following the discussion of initial signal generation, the next step of the process is formation of the photoacoustic wave, the process of converting the initial pressure change into a photoacoustic wave. Like the above session on signal generation, a condensed physics equation will also be utilized to illustrate the process, based on recent publications (Xia et al. 2014, Kim et al. 2015, Zhou et al. 2016). Again, the readers are encouraged to seek more detailed explanation on this subject matter in a recent textbook (Wang 2017).

The photoacoustic wave equation can be expressed as equation 6a or equation 6b below (Kim et al. 2015):

$$(1/(v_s^2))\,(\partial^2\,P(r_0,t))/\,\partial\,t^2 - \nabla^2\,P(r_0,t) = \beta/C_p\,(\partial\,H(r,t))/\partial\,t \tag{6a}$$

$$[\partial^2/(v_s^2\,\partial\,t^2) - \nabla^2]\,P(r_0,t) = \beta/C_p\,(\partial\,H(r,t))/\partial\,t \tag{6b}$$

Where ∂ is partial differentiation, $P(r_0,t)$ is the acoustic pressure at position r and time t, ∇ is gradient (or grad) which is the sum of the partial derivatives of the function with respect to the 3 orthogonal spatial coordinates, C_p is the specific heat capacity at constant pressure, $H(r,t)$ is the heating function of thermal energy conversion, i.e., it increases with the rate of variation of the heat source.

As mentioned above, effective photoacoustic wave generation requires that the width of the laser pulses to be sufficiently smaller than both the thermal and stress relaxation times. As a common example for illustration, a spherical object of 15 mm in diameter would have a thermal and stress relaxation times of 1.7 milliseconds and 10 nanoseconds, in soft tissue respectively, thus requiring an effective laser pulse width to be shorter than 10 nanoseconds (Kim et al. 2015). The generated photoacoustic wave, represented by the left side of equation No. 6, is therefore a function of heat source ($H(r,t)$), represented by the right side of equation No. 6.

Principles of Photoacoustic Image Formation

Following the generation of photoacoustic wave, we then need to format the captured wave and convert it into a humanly visible image. Currently two major platforms useful for medical analysis include the microscopic and tomographic applications of the photoacoustic principles (Kim et al. 2015). For microscopic application, photoacoustic images are usually generated by one-dimensional manner on a single location by using a single-element focused ultrasound transducer. If three dimensional (volumetric) image is the objective, then mechanical or optical raster scanning method is needed. Regarding tomographic application, photoacoustic image capturing process usually involves array-type transducers as well as a mathematical image reconstruction algorithm to form image (Kim et al. 2015). The typical photoacoustic images are usually captured by a polyvinylidene fluoride broadband ultrasound transducer, as the corresponding signals from biological absorbers usually contain frequency in the ranges of 1–10 MHz (Valluru et al. 2016). Since mathematical reconstruction algorithm is needed to form tomographic imaging, image artifact could be introduced if the algorithm is imperfect. Thus, one of the major challenges of photoacoustic tomography is the improvement of reconstruction algorithm to reduce formation of artifacts (Kim et al. 2015).

Advantages of Photoacoustic Imaging

Having explaining the signal generation and image formation process, we may ask what are the advantages of photoacoustic imaging compared to other currently existed imaging techniques. To understand the advantage of photoacoustic technique, we need to first know what hinder our abilities to look into deep-tissue with high "spatial resolution", which is defined as the ability to distinguish two points as separate in space. Fundamentally the optical physics govern the spatial resolution in deep tissue imaging boils down to the key issue of light diffusion. Since light diffusion will lead to blurriness at the edge of image and loss of signal concentration/intensity, higher tissue diffusion results in lower spatial resolution and signal intensity, and lower tissue diffusion leads to higher spatial resolution and signal intensity. Because photoacoustic effects result in converting the absorbed optical energy into acoustic energy and acoustic waves scatter much less in tissue in comparison to optical wave in tissue, photoacoustic method therefore can generate signal resulting in images of higher spatial resolution and signal strength (Xia et al. 2014). In addition to being able to acquire higher spatial resolution, photoacoustic imaging technique can also provide functional and molecular information enabling non-invasive soft-tissue characterization, since it delineates a spatial map of optical absorption of both endogenous components (hemoglobin, fat, melanin, water, etc.) and exogenous compounds (dyes, fluorescence, nanoparticles, etc.) (Valluru et al. 2016). In comparison to imaging techniques using ionizing radiation, photoacoustic imaging method, which devoids ionizing radiation, clearly has the advantage of being a safer modality. Comparing to magnetic resonance imaging, photoacoustic imaging is less expensive. A detailed comparison of technical parameters important for imaging with non-ionizing methodologies is depicted in Table 1 below (Kim et al. 2015). Furthermore, photoacoustic imaging techniques can be combined with other existing imaging techniques to deliver more novel methods of *in vivo* tissue imagine. Researchers have reported many combined usages including photoacoustic/ultrasound, photoacoustic/fluorescence, photoacoustic/optical coherence tomography, and photoacoustic/multiphoton microscopy (Kim et al. 2015). As an emerging biomedical technology,

Table 1. Comparison of non-ionizing radiation-based image quality in biological tissues.[a]

Technical Issue	Photoacoustic	Ultrasound	Flourescein	OCT[b]	MPM[c]
Speckle Artifact	Absent	Present			
Spatial Resolution	Excellent		Poor		
Image Depth	8 cm	< 10 cm	~ 1 mm	~ 10 mm	~ 1 mm
Hgb[d] concentration	Detected	No	No	No	No
Hgb oxygen Saturation	Detected	No	No	No	No
Angiogenesis	Detected	No	No	No	No

[a] Information primarily derived from Kim et al. 2015

[b] OCT = optical coherence tomography

[c] MPM = multiphoton microscopy

[d] Hgb = hemoglobin

photoacoustic imaging is going through a process of continuous research and improvement, its resolution is being enhanced constantly. With tracer, the resolution of photoacoustic imaging may reach a nm level.

Research Applications of Photoacoustic Imaging

Having delineated the principles governing photoacoustic imaging, let us now discuss some of potential applications. In many areas of biomedical research, photoacoustic imaging techniques have been utilized with greater benefits over other existing methodology (Xia et al. 2014). Due to its advantage of higher spatial resolution, greater frame rate, and improving detection sensitivity, the recent photoacoustic advancements have been progressed rapidly. These scientific and technological achievements are the results of contributions of multiple fields including biology, chemistry, physics, computer science, and nanotechnology. As this chapter's focus is on potential clinical applications, readers are encouraged to search for current literatures updated textbook for details of these research-related advancements (Xia et al. 2014, Hai et al. 2015, Rich and Seshadri 2015, Zhou et al. 2016, Wang 2017, Zhao et al. 2018).

Potential Clinical Applications of Photoacoustic Imaging

With regard to clinical applications, photoacoustic imaging is still in its infancy, so to speak. However, there are many potential usages and some of them will now be discussed based on the updated reports on pre-clinical investigations, involving both *ex vivo* and *in vivo* samples.

Applications in Oncology

Photoacoustic imaging technique has gradually entered the clinical applications of cancer detection, characterization, therapeutic monitoring, and prognosis (Valluru et al. 2016, Ho et al. 2014). Several cancers have been intensively studied with promising results and will be discussed in the sections below. However, this list is by no ways all inclusive, as new applications are being reported continuously. Readers are therefore encouraged to seek recent updates from medical literatures. Figure 2 illustrates an *in vivo* experiment where photoacoustic imaging in combination with nanoparticles produces very promising results. Taking advantage of increased blood half-life and high vascular tumor accumulation of the conjugate between an iron oxide particle and a hydrophilic polymer polyoxazoline derivative (A), the tumor implanted in the mouse is clearly detected 24 hours after the conjugate administration with the photoacoustic imaging technique (B) (Saji 2017).

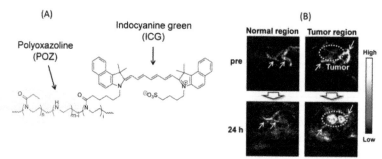

Fig. 2. Photoacoustic imaging detection of *in vivo* tumor implanted in mouse (B) enhanced by administration of fluorescence molecule indocyanine green (ICG) and water-soluble polymer polyoxazoline (POZ) conjugate (A). (Saji 2017, Free Acess).

Breast Cancer

Being the most common cancer occurred in American women and the second leading cause of cancer mortality in women, breast cancer is logically in the front of new diagnostic technique development (Valluru et al. 2016). Additional conditions that place photoacoustic imaging as a potentially ideal detection method for breast cancer include the superficial location of breast tissue and the low optical absorption and ultrasound scattering of lipo (fat)-dominant breast tissue (Valluru et al. 2016). Researchers recently generated a biodegradable marker suitable for photoacoustic imaging purpose for breast tumor (Xia et al. 2017). Multispectral optoacoustic tomography, which utilizes several spectra of laser wavelengths for illumination, has been investigated for potential improvement of detection ability on breast cancer (Razansky et al. 2011, Diot et al. 2017). Multispectral optoacoustic tomography is also known as functional photoacoustic tomography. Recently, a combination of multispectral optoacoustic tomography and ultrasound method has been studied for the potential clinical use in breast cancer diagnostic, with some promising data (Becker et al. 2018). Photoacoustic imaging has also been investigated as a potential tool for breast cancer staging. Since it is well suited for detecting blue dye chromophore, photoacoustic tool has the potential to help guiding sentinel lymph node biopsy and therefore avoid the use of radioactive colloid that is currently required for conventional procedure (Valluru et al. 2016, Erpelding et al. 2010). Taking advantage of the important knowledge on the role of epidermal growth factor receptor in breast cancer, researchers have investigated the potential of molecular diagnosis of lymph node metastases without the need of sentinel lymph node biopsy and could achieve this goal by the use of molecularly activated plasmonic nanosensors with 40 nm gold nanoparticles targeting epidermal growth factor receptor, in conjunction with photoacoustic

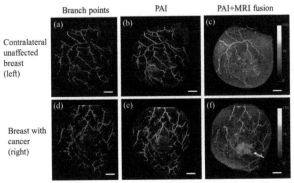

Fig. 3. Number of vessel branching points (in red) are identified in unaffected (a) and affected (cancer-containing) breast (d), are viewed in parallel with unmarked photoacoustic images of unaffected (b) and affected breast (e), and compared with fusion MRI/photoacoustic images of unaffected (c) and affected breast (f). Arrow points to cancer area (f) (Yamaga et al. 2018, Open access).

imaging (Valluru et al. 2016, Masuda et al. 2012, Luke et al. 2014). Additional advantage of photoacoustic imaging for breast cancer detection is the ability of photoacoustic imaging in vascular visualization and in identifying tumor vasculature abnormality. In a study of 22 patients with unilateral breast cancers, 3-d photoacoustic imaging at 7 mm depth detected significant increase of vessel branching points in cancer-containing breast compared to contralateral unaffected breast of the same patient (p < 0.01) (Yamaga et al. 2018). Figure 3. Depicts such a patient case as reported by Yamaga and colleagues.

Ovarian Cancer

The advantage of high sensitivity provided by photoacoustic imaging technique logically encourages the investigation of its potential application on detection of ovarian cancer, for which there is no adequate non-invasive technology of early detection. When most ovarian cancers are detected, they already metastasized and the treatments at that point would be very difficult if not impossible. Thus, there is a great urgency to develop a reliable method to detect early cancers before they become metastasized. Using *ex vivo* excised human samples, researchers have developed combined photoacoustic and ultrasound methods to quantify significantly higher level of light absorption by malignant ovaries than normal ones, with a sensitivity and specificity of 83 percent and 83 percent, respectively (Aguirre et al. 2011). Subsequently, the same group of investigators further examined *ex vivo* and *in vivo* real patients' ovaries by the same combined photoacoustic and ultrasound methods. These researchers found from 25 excised ovary and 15 patient samples that went through the diagnostic testing that they could achieve sensitivity for excised samples (*ex vivo*) and patient samples (*in vivo*) to be 70 percent and 88 percent, respectively and could achieve specificity for excised samples and patient samples to be 96 percent and 98 percent, respectively (Salehi et al. 2016).

Prostate Cancer

Presented very superficially, prostate and the cancer occurs in it, would logically be excellent targets of investigation for photoacoustic imaging applications (Levi et al. 2014, Bell et al. 2015, Valluru et al. 2016, Zhang et al. 2018). In a report of 2014, researchers have developed a photoacoustic agent, called AA3G-740, which targeted a surface protein named gastrin releasing peptide receptor (GRPR) that is highly expressed in prostate cancer. Using a mouse model of prostate cancer, the researchers demonstrated an excellent detection of AA3G-740-bond cancer with photoacoustic imaging instrumentation (Levi et al. 2014). Subsequently, other researchers have developed a transurethral laser light delivery method and found an improvement of image acquisition (Bell et al. 2015). More recently, researchers investigated a molecular photoacoustic imaging method by using a known prostate specific membrane antigen-targeted fluorescence agent, YC-27, on an experimental prostate model, since this antigen is overexpressed in majority of prostate cancers. These investigators found a 60-fold increase of detectability by using this YC-27 agent in the mouse model (Zhang et al. 2018).

Skin Cancer

Situated at the outpost of body surface, skin cancers as a group are excellent candidates for examining the potential usefulness for photoacoustic imaging. Although experienced dermatologists usually can make very accurate diagnoses on most skin cancers just by visual inspection or using a hand-held dermatoscope, a small percent of skin cancers could still escape the physicians' detection and could potentially be life-threatening to the patients, particularly for melanoma which is one of most aggressive cancers. Using a combined photoacoustic and optical coherence tomography methods, researchers examined the dual parameters of vascular patterns and scattering structures of basal cell carcinomas and melanomas and were able to characterize the *in vivo* vasculatures and tissue structures in high resolution (Zhou et al. 2017). Since hematogenous spread is the most common path for melanoma metastasis, other investigators looked for way to utilize photoacoustic imaging technique to eliminate circulating melanoma cells before they can invade distant tissues. Taking advantage of the intrinsic optical absorption contrast of circulating melanoma cells under photoacoustic imaging, researchers tested a novel nanosecond-pulsed melanoma-specific laser treatment of melanoma cells in artery and vein of mice, coupled with a dual-wavelength photoacoustic flow cytography. These investigators were able to detect rare single melanoma cells and were able to eliminate those cells on the spot, without causing collateral damage (He et al. 2016). Utilizing a molecular photoacoustic technique, other researchers examined its diagnostic capability for squamous cell carcinoma, the second most common non-melanoma skin cancer. Taking advantage of the known fact of overexpression of $\alpha_v\beta_6$ integrin in this cancer, investigators construct an anti-$\alpha_v\beta_6$ antibody tagged with indocyanine green probe. This fluorescence-labeled antibody probe in conjunction with photoacoustic imaging was able to detect squamous cell carcinoma *in vivo* with high specificity and achieved a penetration depth of 1 cm (Zhang et al. 2017).

Thyroid Cancer

Thyroid, due to its superficial location, also became one of the target organs for investigators who were interested in testing the potential application of photoacoustic imaging for diagnosis and disease management purposes (Levi et al. 2013, Dogra et al. 2014, Sinha et al. 2017, Yang et al. 2017b). As early as 2013, investigators have developed molecular photoacoustic method to examine a specific kind of follicular thyroid cancer. Utilizing a matrix metalloproteinase-activatable photoacoustic probe injected into nude mice implanted subcutaneously with FTC133 type of thyroid cancers, photoacoustic imaging revealed significantly higher signals in tumors injected with this probe and those tumors injected with nonactivatable agent (Levi et al. 2013). Another article, published in 2014, reported the use of *ex vivo* multispectral photoacoustic imaging on thyroid cancer, benign thyroid nodule, and normal thyroid tissue. Their investigations resulted in statistically significant differences in mean intensity of deoxyhemoglobin between thyroid cancer and benign thyroid nodule, between thyroid cancer and normal thyroid tissue, and between benign thyroid nodule and normal thyroid tissue. Overall, the sensitivity, specificity, and positive and negative predictive values of tests in differentiating thyroid cancer from non-cancerous tissue were determined to be 69 percent, 97 percent, 82 percent, and 94 percent, respectively (Dogra et al. 2014). Additional study from the same group of investigators later confirmed their findings reported in 2014 (Dogra et al. 2014, Sinha et al. 2017). In a clinical study of *in vivo* human thyroid by a dual modality system, investigators determined that photoacoustic imaging is superior than color Doppler ultrasound as it reveals many vasculatures information that are not detected by Doppler ultrasound method (Yang et al. 2017b).

Application in Brain Injury Monitor

Photoacoustic imaging has the potential to reach far beyond the field of oncology. Recently researchers have investigated the potential utilization for determining and monitoring the progression of brain injury (Yang et al. 2017a, Li et al. 2018). Utilizing dual wavelength method to monitor the development of the hemorrhagic area, photoacoustic imaging has the potential to track the dynamic progression of post-intracerbral hemorrhage (Yang et al. 2017a). In another aspect of brain injury, combining high-efficient near-infrared dye labeled mesenchymal stem cells and *in vivo* photoacoustic imaging, researchers developed a non-invasive and high-resolution imaging approach to traumatic brain injury that can potentially be used to monitor post-injury recovery process, particularly important for neural regenerative medicine (Li et al. 2018).

Application in Tendon Injury Monitor

Tendon injury accounts about 50% of all sport injuries and none of the current non-invasive imaging techniques provide a clear three-dimensional evaluation important for accurate clinical assessment. Taking advantage of the unique protein content of tendon, which is composing of high density of type I collagen in a cross-linked triple-helical configuration that can provide a sharp endogenous absorption contract in near infrared spectrum, researchers were able to utilize label-free photoacoustic microscopy with a 780 nm wavelength pulse laser light to visualize *in vivo* tendon images in three-dimension (Lee et al. 2018).

Application in Synovial Tissue Imaging

Recently, a pilot study has been conducted to investigate the ability of photoacoustic method to examine synovial tissues. Researchers have detected increasing angiogenesis and synovial hypoxia in synovial tissues of osteoarthritis compared to normal tissues and found correlation between synovial hypoxia and severity of cartilage loss in arthritic tissues (Liu et al. 2018).

Application in Vascular Imaging

The ability of photoacoustic method in delineating clear absorption contract between intravascular hemoglobin and the surrounding tissues has encouraged, not surprisingly, a huge array of investigations into its potential applications of vascular imaging and many promising findings have been reported in the medical literatures (Hai et al. 2015, Rich and Seshadri 2015, Horiguchi et al. 2017, Ogunlade et al. 2017, Matsumoto et al. 2018, Zhao et al. 2018). Below, just a few of these reports will be discussed.

In an attempt to determine human vascular elastic properties, researchers constructed a vascular elastic photoacoustic tomography and used it to assay human blood vessel compliance, which is the ability of blood vessel wall expand in response to increase blood pressure. Mathematically, the vascular compliance (C) is written as $C = \Delta V / \Delta P$. ΔV denotes the change of blood vessel volume and DP indicates the change of blood pressure. This study is based on the knowledge that in general mechanical properties of biological tissue are directly related to its underlying structure which can be altered in pathological conditions such as arteriosclerosis or tumor. In particular, the elastic properties of blood vessels are linked to hemodynamic changes in the system and can indicate presence of vascular blockage or narrowing from thrombosis. As can be deduced logically, when blood vessel is hardened from arteriosclerosis, thrombosis, or tumor growth, the increase blood pressure will not result in normal level of expansion but will instead lead to abnormally low level of expansion. In this study, investigations found a decrease of compliance in simulated thrombosis and when blockage occurred downstream from the measurement point of blood vessels, suggest a potential usage of photoacoustic imaging for clinical thrombosis detection (Hai et al. 2015).

Application in Carotid Artery Thrombosis Detection

Recognizing the main cause of cardiovascular event is thrombosis and the challenge in detecting it noninvasively, researchers sought to evaluate the potential role of photoacoustic imaging. Using an experimental mouse model of carotid artery thrombosis, researchers were able to detect the formation of vascular thrombi of different sizes (Li et al. 2017). Thus, photoacoustic technology can potentially become a fast, inexpensive, and noninvasive procedure to monitor carotid artery thrombosis and facilitate medical decision making to prevent cerebral events such as strokes.

Application in Chorioretinal Oxygen Gradient Monitor

Chorioretinal vascular diseases, such as age-related macular degeneration involves neovascularization. Thus, chorioretinal imaging, especially on oxygen gradient, is essential for monitoring disease progression in those patients. Researchers recently examined the utility of photoacoustic imaging in measuring the oxygen saturation in chorioretinal vasculatures. Their preliminary data suggest that photoacoustic imaging technique can detect changes in chorioretinal vascular oxygenation in certain disease state and has the potential for real-life clinical use (Hariri et al. 2018).

Application in Placental and Fetal Oxygen Monitor

It is well established that accurate analysis of placental and fetal oxygenation is a critical medical process during pregnancy, especially for high-risk pregnancies. Using a mouse model, investigators have successfully monitored the placental and fetal oxygenation levels at physiologic and pathologic pregnant conditions (Yamaleyeva et al. 2017).

Limitations of Photoacoustic Imaging

Just as in other newly developed medical technology, photoacoustic imaging has some imitations. We will discuss some of these limitations and how the scientists are finding ways to solve these limitation challenges.

The limitation of Depth

Currently, the maximum photoacoustic image depth has reached to about 8.4 cm. Researchers have created new ways to increase the image depth. One of examples was to illuminate the target from both sides of the target, with a potential image depth reach of 16.4 cm. Other method has been tried including internal illumination for study target far beneath the skin (Xia et al. 2012, 2014, Yao et al. 2012, Bell et al. 2015).

The limitation of Image Speed

Another limitation of photoacoustic image is the image speed as the current speed of both microscopic and tomographic applications of this technique is limited by the pulse repetition rate of lasers. With the improvement of laser technology, the future of the photoacoustic rate will improve accordingly (Xia et al. 2014).

The limitation of Quantitation

The issue of quantitation using photoacoustic imaging method is due to the difficulty of measuring local fluence distribution. Researchers are looking at ways to deal with this challenge, including new contract agents and advanced spectral separation algorithms (Xia et al. 2014).

Summary

Although originally discovered in 1880, the photoacoustic effect was not further developed for the benefit of medicine until 1960s, due to the lack of high intensity light source necessary for useful photoacoustic imaging applications. The development of laser technology in recent decades provides the driving force for further advance the research and construction of photoacoustic instrumentation for medical applications and rapid progresses have been made in this aspect. The significant advantage of photoacoustic imaging over other optical imaging rests on the unique characteristics of acoustic wave in minimum light scattering, thus result in higher spatial resolution of image production. The basic principle centers on the physics of sequence of events: A pulse laser light energy illuminates on a living tissue results in a temperature change (heat), which leads to a pressure change that subsequently converted to an acoustic wave to be captured by an ultrasound probe. In this chapter, the mathematical equations governing the photoacoustic wave formation are delineated and explained. The potential clinical applications in assisting disease diagnosis and treatment are discussed. Although some limitations, such as depth, speed, and quantitation, existed for photoacoustic imaging technique, its ability to provide a mapping for both endogenous substances (hemoglobin, melanin, fat, water, etc.) and exogenous agents (such as dye, fluorescence, nanoparticles, etc.) gives photoacoustic imaging a substantial advantage over other optical imaging techniques as it can provide functional and molecular information that other methods cannot. Furthermore, photoacoustic imaging can also be combined with other imaging techniques to give even greater power of medical imagery. We are indeed at a very exciting time for the development of this novel imaging technique.

Reference

Aguirre, A., Y. Ardeshirpour, M.M. Sanders, M. Brewer and C. Zhu. 2011. Potential role of coregistered photoacoustic and ultrasound imaging in ovarian cancer detection and characterization. Translational Oncology 4: 29–37.

Becker, A., M. Masthoff, J. Claussen, S.J. Ford, W. Roll, M. Burg et al. 2018. Multispectral optoacoustic tomography of the human breast: Characterisation of health tissue and malignant lesions using a hybrid ultrasound-optoacoustic approach. Eur Radiol 28: 602–609.

Bell, M.A.L., X. Guo, D.Y. Song and E.M. Boctor. 2015. Transurethral light delivery for prostate photoacoustic imaging. J Biomed Opt Doi: 10.1117/1.JBO.20.3.036002.

Campbell, S.D., S.S. Yee and M.A. Afromowitz. 1977. Two applications of photoacoustic spectroscopy to measurements in dermatology. J Bioeng 1: 185–88.

Diot, G., S. Metz, A. Noske, E. Liapis, B. Schroeder, S.V. Ovsepian et al. 2017. Multispectral optoacoustic tomography (MSOT) of human breast cancer. Clin Cancer Res November 2017 Doi: 10. 1158/1078-0432.CCR-16-3200.

Dogra, V.S., B.K. Chinni, K.S. Valluru, J. Moalem, E.J. Giampoli, K. Evans et al. 2014. Preliminary results of *ex vivo* multispectral photoacoustic imaging in the management of thyroid cancer. Am J Roentgenol 202: W552–8.

Erpelding, T.N., C. Kim, M. Pramanik, L. Jankovic, K. Maslov, Z Guo et al. 2010. Sentinel lymph nodes in the rat: Non-invasive photoacoustic and US imaging with a clinical US system. Radiology 250: 102–110.

Fang, H., K. Maslov and L.V. Wang. 2007. Photoacoustic Doppler effect from flowing small light-absorbing particles. Physical Review Letters. 99: 184501.

Hai, P., Y. Zhou, J. Liang, C. Li and L.V. Wang. 2015. Photoacoustic tomography of vascular compliance in humans. J Biomedical Opt December 2015. Doi: 10.1117/1. JBO.20/12.126008.

Hariri, A., J. Wang, Y. Kim, A. Jhunjhunwala, D.L. Chao and J.V. Jokerst. 2018. *In vivo* photoacoustic imaging of chorioretinal oxygen gradients. J Biomed Opt 23: 1–8.

He, Y., L. Wang, J. Shi, J. Yao, L. Li, R. Zhang et al. 2016. *In vivo* label-free photoacoustic flow cytography and on-the-spot laser killing of single circulating melanoma cells. Scientific Report Doi: 10. 1038/srep38616.

Ho, C.J.H., G. Balasundaram, W. Driessen, R. McLaren, C.L. Wong, U.S. Dinish et al. 2014. Multifunctional photosensitizer-based contrast agents for photoacoustic imaging. Scientific Reports 4: 5342. Doi: 10.1038/srep05342.

Horiguchi, A., M. Shinchi, A. Nakamura, T. Wada, K. Ito, T. Asano et al. 2017. Pilot study of prostate cancer angiogenesis imaging using a photoacoustic, imaging system. Urology 108: 212–219.

[IMAGING QUOTES] Imaging quotes. 2018. Brainy Quotes. [https://www.brainyquote.com/topics/imaging] accessed April 11, 2018.

Jiang, H. 2015. Photoacoustic Tomography. CRC Press. Boca Raton, FL.

Kim, J., D. Lee, U. Jung and C. Kim. 2015. Photoacoustic imaging platforms for multimodal imaging. Ultrasonography 34: 88–97.

[LASER] Laser Institute of America. 2007. American National Standard for safe use of lasers: ANSI Z136.1–2000. New York, NY: Laser Institute of America.

Lee, H.D. J.G. Shin, H. Hyun, B.-A. Yu and T.J. Eom. 2018. Label-free photoacoustic microscopy for *in-vivo* tendon imaging. Scientific Reports 8: 4805. Doi:10.1038/s41598-018-23113-y.

Levi, J., S.R. Kothapalli, S. Bohndiek, J.K. Yoon, A. Dragulescu-Andrasi, C. Nielsen et al. 2013. Molecular photoacoustic imaging of follicular thyroid carcinoma. Clin Cancer Res 15: 1494–1502.

Levi, J., A. Sathirachinda and S.S. Gambhir. 2014. A high-affinity, high-stability photoacoustic agent for imaging gastrin-releasing peptide receptor in prostate cancer. Clin Cancer Res 20: 3721–3729.

Li, B., C. Fu, G. Ma, Q. Fan, Y. Yao. 2017. Photoacoustic imaging: A novel tool for detecting carotid artery thrombosis in mice. J Vasc Res 54: 217–225.

Li, W., R. Chen, J. Lv, H. Wang, Y. Liu, Y. Peng et al. 2018. *In vivo* photoacoustic imaging of brain injury and rehabilitation by high-efficient near-infrared dye labeled mesenchymal stem cells with enhanced brain barrier permeability. Adv Sci doi: 10.1002/advs.201700277.

Liu, Z, M. Au, X. Wang, P.B. Chan, P. Lai, L. Sun et al. 2018. Photoacoustic imaging of synovial tissue hypoxia in experimental post-traumatic osteoarthritis. Prog Biophys Mol Biol March 28. Doi. 10. 1016/j.pbiomolbio.2018.03.009.

Luke, G.P., J.N. Myers, S.Y. Emelinnov and K.V. Sokolov. 2014. Sentinel lymph node biopsy revisited: ultrasound-guided photoacoustic detection of micrometastases using molecularly targeted plasmonic nanosensors. Caner Res 74: 5397–5408.

Masuda, H., D. Zhang, C. Bartholomeusz, H. Doihara, B.N. Hortobagyi and N.T. Ueno. 2012. Role of epidermal growth factor receptor in breast cancer. Breast Cancer Res Treat 136: 331–345.

Matsumoto, Y., Y. Asao, A. Yoshikawa, H. Sekiguchi, M. Takada, M. Furu et al. 2018. Label-free photoacoustic imaging of human palmar vessels: a structural morphological analysis. Sci Rep Jan 15. 8: 786. Doi: 10.1038/s4 1598-018-19161-z.

Meiburger, K.M. 2016. Quantitative ultrasound and photoacoustic imaging for the assessment of vascular parameters. Springer, Berlin, Germany.

Ogunlade, O., J.J. Connell, J.L. Huang, E. Zhang, M.F. Lythgoe, D.A. Long et al. 2017. *In vivo* 3-dimensional photoacoustic imaging of the renal vasculature in preclinical rodent models. Am J Physiol Renal Physiol Dec.20. Doi: 10.1152/ajprenal.00337.

Razansky, D., A. Buehler and V. Ntziachristos. 2011. Volumetric real-time multispectral optoacoustic tomography of biomarkers. Nat Protocol 6: 1121–1129.

Rich, L.J. and M. Seshadri. 2015. Photoacoustic imaging of vascular hemodynamics: Validation with blood oxygenation level-dependent MR imaging. Radiology 275: 110–118.

Rosencwaig, A. and E. Pines. 1977. Photoacoustic study of newborn rat stratum corneum. Biochim Biophys Acta 493: 10–23.

Rosencwaig, A. 1982. Potential clinical applications of photoacoustics. Clin Chem 28: 1878–81.

Salehi, H.S., H. Li, A. Merkulov, P.D. Kumavor, M. Vavadi, M. Sanders et al. 2016. Coregistered photoacoustic and ultrasound imaging and classification of ovarian cancer: ex vivo and in vivo studies. J Biome Opt April 2016. Doi: 10.1117/1.JBO.21.4.046006.

Saji, H. 2017. *In vivo* molecular imaging. Biol Pharm Bull 40:1605–1615.

Sinha, S., V.S. Dorga, B.K. Chinni and N.A. Rao. 2017. Frequency domain analysis of multiwavelength photoacoustic signals for differentiating among malignant, benign, and normal thyroids in an *ex vivo* study with human thyroids. J Ultrasound Med 36: 2047–2059.

[ULTRASOUND QUOTES] Ultrasounds Quotes. 2018. AZ Quotes. [www.azquotes.com/quotes/topics/ultrasounds.html] accessed April 11, 2018.

Valluru, K.S., K.E. Wilson and J.K. Willmann. 2016. Photoacoustic imaging in oncology: Translational preclinical and early clinical experience. Radiology 280: 332–349.

Wang, L.V. 2009. Multiscale photoacoustic microscopy and computed tomography. Nat Photon 3: 503–509.

Wang, L.V. 2012. Photoacoustic tomography: *In vivo* imaging from organelles to organs. Science 355: 1458–1462.

Wang, L.V. 2017. Photoacoustic imaging and spectroscopy. CRC Press. Boca Raton, FL.

Xia, J., M.R. Chatni, K.I. Maslov, Z. Guo, K. Wang, M.A. Anastasio et al. 2012. Whole-body ring-shaped confocal photoacoustic computed tomography of small animals *in vivo*. Journal of Biomedical Optics 17: 050506.

Xia, J., J. Yao and L.V. Wang. 2014. Photoacoustic tomography: Principles and advances. Electromagn Waves (Camb). 147: 1–22.

Xia, J., G. Feng, X. Xia, L. Hao and Z. Wang. 2017. NH_4HCO_3 gas-generating liposomal nanoparticle for photoacoustic imaging in breast cancer. Int J Nanomedicine 12: 1803–1813.

Yamaga, I., N. Kawaguchi-Sakita, Y. Asao, Y. Matsumoto, A. Yoshikawa, T. Fukui et al. 2018. Vascular branching point counts using photoacoustic imaging in the superficial layer of the breast: A potential biomarker for breast cancer. Photoacoustics 11: 6–13.

Yamaleyeva, L.M., Y. Sun, T. Bledsoe, A. Hoke, S.B. Gurley and K.B. Brosnihan. 2017. Photoacoustic imaging for *in vivo* quantification of placental oxygenation in mice. FASEB J 31: 5520–5529.

Yang, J., D. Wu, G. Zhang, Y. Zhao, M. Jiang, X. Yang et al. 2017a. Intracerebral hamemorrhage-induced injury progression assessed by cross-sectional photoacoustic tomography. Biomedical Optics Express December 1, 2017. Doi.org/10.1364/BOE.8.005814.

Yang, M., L. Zhao, X. He, N. Su, C. Zhao, H. Tang et al. 2017b. Photoacoustic/ultrasound dual imaging of human thyroid cancers: an initial clinical study. Biomed Opt Express 26: 3449–3457.

Yang, S. and D. Xing. 2018. Biomedical photoacoustics. Pan Stanford Publishing. Singapore.

Yao, J., K.J. Maslov, K.J. Rowland, B.W. Warner and L.V. Wang. 2012. Double-illumination photoacoustic microscopy. Optics Letters 37: 659–661.

Zhang, C., Y. Zhang, K. Hong, S. Zhu and J. Wan. 2017. Photoacoustic and fluorescence imaging of cutaneous squamous cell carcinoma in living subjects using a probe targeting integrin $\alpha_v\beta_6$. Scientific Reports Doi: 10. 1038/srep42442.

Zhang, H.F., K. Maslov, G. Stoica and L.V. Wang. 2006. Functional photoacoustic Microscopy for high-resolution and noninvasive *in vivo* imaging. Nat Biotech 24: 848–851.

Zhang, H.K., Y. Chen, J. Kang, A. Lisok, I. Minn, M.G. Pomper et al. 2018. Prostate specific membrane antigen (PSMA)-targeted photoacoustic imaging of prostate cancer *in vivo*. J Biophotonics April 13. Doi: 10.1002/jbio/201800021.

Zhao, H., G. Wang, R. Lin, X. Gong, L. Song, T. Li et al. 2018. Three-dimensional Hessian matrix-based quantitative vascular imaging of rat iris with optical-resolution photoacoustic microscopy *in vivo*. J Biomed Opt 23: 1–11.

Zhou, Y., J. Yao and L.V. Wang. 2016. Tutorial on photoacoustic tomography. J Biomed Opt April 18. Doi: 10.1117/1.JBO.21.6.061007.

Zhou, W., Z. Chen, S. Yang and D. Xing. 2017. Optical biopsy approach to basal cell carcinoma and melanoma based on all-optically integrated photoacoustic and optical coherence tomography. Opt Lett Doi: 10. 1364/OL.42.002145.

QUESTIONS

1. Who was the first person discovering the photoacoustic effect?

2. Why was there a long gap between the discover of the photoacoustic effect and the application of this effect in medicine?

3. What is another term for photoacoustic imaging?

4. What are the two major platforms for photoacoustic imaging?

5. What is the key factor that makes photoacoustic effect unique and beneficial for medical imaging?

6. What are the endogenous tissue substances that photoacoustic method can provide a high contrast image?

7. What are the exogenous substances that can be used to enhance the usefulness of photoacoustic method?

8. Why did pulse laser become the most used light source for photoacoustic imaging?

9. What is maximum pulse laser duration beyond which no photoacoustic effect will occur?

10. What other imaging method could photoacoustic technique combine to form another unique medical imaging?

11. What are the known limitations of photoacoustic imaging?

12. What physical properties make breast tissue suitable for photoacoustic imaging?

PROBLEMS

Working in small groups of 3–5 students, discuss and propose a real-life medical situation for which photoacoustic imaging would be the ideal diagnostic technique or therapeutic aid. Provide evidence to support the proposal.

23

Emerging Biomedical Analysis
Mass Spectrometry

Rui Zhu[1],* and *Lawrence S. Chan*[2]

QUOTABLE QUOTES

"Believe you can and you're halfway there."

Theodore Roosevelt, the 26th President of the United States (BRAINY 2018a)

"Keep your face always toward the sunshine – and shadows will fall behind you."

Walt Whitman, American Poet (BRAINY 2018b)

"The first wealth is health."

Ralph Waldo Emerson, American Poet (BRAINY 2018c)

"Nothing great was ever achieved without enthusiasm."

Ralph Waldo Emerson, American Poet (BRAINY 2018c)

Learning Objectives

The learning objectives of this chapter are to familiarize the students to the principles, applications, and limitations of mass spectrometry in medicine.
After completing this chapter, the students should:

- Understand the engineering, physics and chemistry principles of mass spectrometry.
- Understand the three major steps of mass spectrometry.
- Understand the potential applications of mass spectrometry in medicine.
- Understand the application of mass spectrometry in cancer diagnostics. Able to understand the mechanism of intraoperative tumor margin identification by mass spectrometry.
- Understand the limitations and challenges of popularizing the utilization of mass spectrometry in medicine.
- Be able to apply mass spectrometry in a real-life medical situation.

[1] AbbVie Bioresearch Center, 100 Research Dr., Worcester, MA, 01605.
[2] University of Illinois College of Medicine, 808 S. Wood Street, R380, Chicago, IL60612; larrycha@uic.edu
* Corresponding author: rui.zhu@abbvie.com

Historical Development of Mass Spectrometry (MS)

In 1913, J.J. Thomson, the Nobel Prize laureate in Physics in 1906 for his work on the conduction of electricity in gases, built an instrument to measure the mass of charged atoms. With the help of his assistant, Francis W. Aston, who received his own Nobel Prize in Chemistry in 1922 for his discovery of isotopes, J.J. Thomson observed two light patches when he measured the deflection of charged neon ions and made the conclusion that neon was a mixture of two gases, one of which had an atomic weight of 20 and the other 22 (Thomson 1913). This experiment was later considered as the first evidence of isotopes for a stable element and the instrument he constructed was recognized as the first mass spectrometer in the world (Griffiths 2008).

In the following three decades, the mass spectrometer was redesigned by Francis W. Aston, Arthur J. Dempster, Kenneth T. Bainbridge, Ernest O. Lawrence and other scientists to improve the resolving power (Audi 2006). However, at that stage, mass spectrometers were predominantly used by physicists to prove the existence of elemental isotopes or to separate an isotope from its natural mixture. The most famous application during this period was the use of the mass spectrometer in the Manhattan Project in World War II. In the Manhattan Project, Ernest O. Lawrence, who received the Nobel Prize in Physics in 1939 for his invention of the cyclotron, developed the calutron to prepare high-purity uranium-235 (Parkins 2005).

In the 1940s, mass spectrometers were commercialized and used by industrial chemists as a quality control tool. The substance of interest was shifted from atoms to organic compounds. Klaus Biemann and other scientists showed that the fragmentation of molecules that were identified in a mass spectrometer could be used for the structural elucidation of unknown complex organic compounds. These pioneering works dramatically extended the use of mass spectrometry (MS) as a research tool for chemists (Biemann 2007). The growing needs for resolving power then led to the birth of high-resolution MS. In 1974, Melvin B. Comisarow and Alan G. Marshall developed the Fourier transform ion cyclotron resonance FT-ICR MS (Marshall and Comisarow 1974). This enabled ultrahigh mass resolving power (> 100,000) and mass accuracy (sub-part-per-million) to become routinely available in commercial instruments.

In the 1980s, MS was routinely applied in the field of small organic molecule analysis. However, analysis of biomolecules such as proteins and DNA by MS was rarely reported. Lack of ionization techniques that could prevent fragmentation of large molecules became the major hurdle. This obstacle was subsequently removed by John B. Fenn and Koichi Tanaka in the late 1980s when Fenn first developed electrospray ionization (ESI) in MS and Tanaka first reported matrix-assisted laser desorption/ionization (MALDI) in MS (Koichi et al. 1988, Yamashita and Fenn 1984). They later shared the 2002 Nobel Prize in Chemistry for their work on the development of soft desorption ionization methods for mass spectrometric analyses of biological macromolecules.

Currently, MS is widely applied in many different fields from the petroleum industry to a biological research laboratory and on many substances from a single atom to complex biofluids. A modern mass spectrometer is usually a tandem mass spectrometer that is frequently joined to separation techniques such as high performance liquid chromatography (HPLC), gas chromatography (GC) and capillary electrophoresis (CE) for enhanced resolution. The improvement in MS instrumentation allows scientists to more thoroughly understand complex mixtures. For example, according to a recent report, over 90% of proteins from yeast lysate were identified within 1.3 hours by using ultra high performance liquid chromatography (UHPLC) interfaced with an advanced Orbitrap™ hybrid mass spectrometer (Hebert et al. 2014).

Principles of Mass Spectrometry

Having described the history of MS development, we now turn to delineate the physical and chemical principles in MS instrumentation. The basic concept of MS consists of three major steps:

- The first step is the generation of analyte ions (ionized sample/molecule).
- The second step involves sorting of analyte ions by their respective mass-to-charge ratio (*m/z*).
- The last step requires detecting the intensity of sorted ions and eventually identifying or quantifying the analyte.

A mass spectrometer, in turn, consists of three major corresponding components:

- An ionization source that converts the sample into ions.
- A mass analyzer that sorts the ions by their m/z.
- A detector that measures the abundance of each type of ion.

Ionization Techniques in Mass Spectrometry

The ion separation in a mass analyzer is affected by electric or magnetic fields. Therefore, the formation of gaseous analyte ions is necessary for MS analysis. Many ionization techniques have been invented over the past century for different applications. Based on the integrality of ionized analyte molecules, ionization techniques are commonly classified into two classes: hard ionization and soft ionization.

Hard ionization is a process that imparts sufficient internal energy to induce fragmentation of analyte molecules. Hard ionization sources generate in-source ion fragments, which provide the chemical structure information for the molecules of interest. This feature makes hard ionization approaches suitable for elemental analysis and small molecule analysis. Common examples of hard ionization are electron ionization (EI) and inductively coupled plasma (ICP).

On the other hand, the in-source fragments usually interfere with the analysis of large biological molecules. The excessive in-source fragments generated from large molecules often result in complicated and insolvable mass spectra. Therefore, soft ionization techniques that minimize in-source fragmentation are required for large-molecule analysis. The two soft ionization techniques most commonly used in current mass spectrometers are MALDI and ESI, as described in the above section.

The MALDI technique employs a matrix that transfers protons to large molecules by a laser source. Ionization by MALDI consists of three steps (Karas and Kruger 2003):

- In the first step, the analyte is mixed with the selected matrix material and transferred to a sample plate.
- In the second step, short laser pulses of UV or IR radiation is applied to the sample-matrix mixture for a few nanoseconds to trigger the desorption process, where the mixture plume leaves the surface at a velocity of several hundred m/s.
- In the final step, the analyte molecules are ionized by a proton transferred from the matrix during the desorption process.

Organic compounds that absorb at the laser wavelength, sublime readily and co-crystallize with the analyte are chosen as the matrix. The most commonly used materials are 2,5-dihydroxybenzoic acid (DHB); 4-hydroxy-3-methoxycinnamic acid (Ferulic acid); 3,5-dimethoxy-4-hydroxycinnamic acid (sinapinic acid) and α-cyano-4-hydroxycinnamic acid (CHC). The MALDI method has been successfully applied to protein, peptide and DNA analyses in biological research for many years. More recently, there has been a rapid increase in the interest in mass spectral imaging directly from biological tissue. Currently, MALDI is the method of choice in MS imaging applications due to its superior physical compatibility with imaging experimentation (Gessel et al. 2014). However, since the sample preparation for MALDI is time consuming and not compatible with commonly used chromatographic separation techniques, it is now generally replaced by ESI for analysis of biomaterials.

ESI is a technique that applies a high voltage (usually several kV) on a heated capillary needle containing the analyte solution (Fenn et al. 1989). The heated solvent inside the capillary is dispersed by the high voltage to form finely-charged droplets. As the evaporation process continues, the droplet becomes smaller and the electrostatic repulsion of like charges becomes more powerful than the surface tension holding the droplet together. Unstable large droplets explode into smaller more stable droplets, and eventually generate a mixture of analyte molecules with various numbers of cations attached (Fig. 1). Common cations are protons (H^+), sodium ions (Na^+) or ammonium ions (NH_4^+). As a result, the ions observed by amass spectrometer from an ESI source are normally denoted as $[M + nH]^{n+}$, $[M + nNa]^{n+}$, $[M + nNH_4]^{n+}$ or sometimes as a mixed adduct $[M + xH + yNa + zNH_4]^{(x+y+z)+}$. In order to improve the ionization efficiency, a heated inert gas, often nitrogen, is generally used as a drying gas to accelerate the

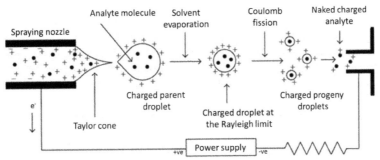

Fig. 1. Schematic representation of the electrospray ionization process. Reprinted from Int J Anal Chem, 2012: Article ID 282574. (Open Access Journal).

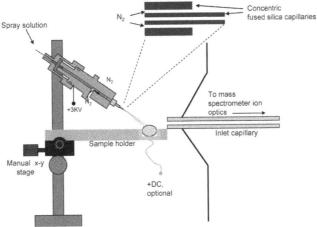

Fig. 2. An example of a DESI ion source. Reprinted with permission from Anal Chem, 79: 2150-57. Copyright 2007 American Chemical Society.

evaporation process. Using a lower liquid flowrate also increases the sensitivity of instruments because it reduces the solvent volume. Compared to pulsed laser activation in MALDI, ESI is a continuous soft ionization technique. More importantly, ESI is compatible with liquid-phase separation techniques such as HPLC and CE that are commonly used for biomaterial analysis. Therefore, a mass spectrometer equipped with an HPLC or CE and an ESI source is currently the most common configuration in a modern bioanalytical laboratory for analysis of biological molecules (Banerjee and Mazumdar 2012).

Many variants of ESI have been developed in recent years. Ambient ionization is a family of techniques that was derived from ESI to enable rapid MS analysis at lower cost. Ambient ionization techniques generate ions under ambient conditions for subsequent MS analysis directly on the sample with minimum sample preparation (Cooks et al. 2006, Nyadong et al. 2007). Representative ambient ionization techniques are desorption electrospray ionization (DESI) and direct analysis in real time (DART). In the DESI technique, an electrospray of charged solvent droplets hits the sample surface and extracts analyte molecules to form secondary droplets (Fig. 2). The secondary droplets undergo a similar ionization process with ESI and eventually are delivered to the inlet of a mass spectrometer (Takats et al. 2004). In the DART methodology, an electrical potential is applied to a helium gas stream to generate metastable species. These excited-state gas molecules subsequently react with the analyte surface to form ions (Cody et al. 2005). Due to the simple and cost-effective experimental setup, DESI, DART and other ambient ionization techniques, such as paper spray ionization, liquid microjunction surface sampling probe (LMJ-SSP) and rapid evaporative ionization mass spectrometry (REIMS), were used in several clinical studies (Li et al. 2017). Examples of clinical applications of ambient ionization MS will be discussed later in this chapter.

Mass Analyzer of Mass Spectrometry

The performance of a mass analyzer determines the ability to distinguish different ions. The capability of a high-performance mass spectrometer is in turn dependent on four key characteristics of the analyzer: scan speed, mass-resolving power, mass range, and mass accuracy, and these four characteristics form the basis of specification for a given mass analyzer. Mass analyzers separate ions based on their *m/z* in a vacuumed electric or magnetic field. The resolving forces are the Lorentz force in a magnetic field or the Coulomb force in an electric field. The force **F** exerted on a particle of charge *q* with the velocity **v**, in an external electric field **E** and magnetic field **B**, is given by:

$$\mathbf{F} = q\mathbf{E} + q\mathbf{v} \times \mathbf{B}$$

Key Analyzer Characteristics

Mass-resolving power. Mass-resolving power is defined as $m/\Delta m_{50\%}$, where $\Delta m_{50\%}$ is the mass spectral peak full-width at half-maximum peak height. As the mass-resolving power increases, the number of separated ions in a mass spectrum increases; therefore, new chemical information becomes accessible for identification (Marshall et al. 2002). Mass-resolving power is a key parameter for evaluating mass analyzer performance because it represents the precision of a mass analyzer. It is important to note that the mass-resolving power would change as the *m* (or *m/z*) value changes over the mass range. Therefore, the mass-resolving power is always reported at a certain *m* (or *m/z*) value for a given mass analyzer.

Mass accuracy: Mass accuracy is the difference between measured *m/z* and theoretical *m/z*. It is usually presented as part-per-million (ppm) with a smaller number indicating higher accuracy.

Mass range: Mass range is the range of m/z that can be measured by a mass analyzer with sufficient resolution to differentiate adjacent peaks (McNaught and Wilkinson 1997).

Scan speed: Scan speed is the rate of mass spectrum acquisition, generally given in mass unit per unit time.

In order to collect sufficient information, some of these parameters are usually required to reach a certain level. For example, analysis of biological molecules such as DNA and protein often requires a high-resolution mass spectrometer with a fast scan speed, a mass-resolving power greater than 10,000, and a mass accuracy less than 100 ppm.

Types of Mass Analyzers

In modern mass spectrometers, several different types of mass analyzers are routinely used, including a quadrupole mass filter, ion trap, time of flight, Fourier transform ion cyclotron resonance, and an Orbitrap.

Quadrupole mass filter analyzer: A quadrupole mass filter consists of four parallel metal electrodes that have both a fixed direct current (DC) and an alternating radio frequency (RF) potential applied to them (Fig. 3). Opposite electrodes are applied the same potentials. One pair of electrodes is applied a positive DC and RF potential while the other pair is applied to a negative potential. Ions generated in the ion source of the instrument are focused and passed between the electrodes. The trajectories of ions are then affected by the voltages applied to the quadrupole. For a given voltage setup, only ions of a particular *m/z* will have a stable trajectory and eventually reach the detector, while other ions will collide with the electrode.

Ion trap analyzers: Ion traps (IT) are built based on the same mechanism as the quadrupole mass filter. Instead of continuously passing ions through the quadrupole, ion traps are designed with ring electrodes (in a 3D ion trap) or a two-dimensional quadrupole field (in a linear ion trap) to trap and eject ions sequentially.

The quadrupole mass filter and IT are inexpensive and compact. Nonetheless, they have the major disadvantages of low mass-resolving power (generally < 4000) and low scan speed (< 4000 u/sec), thus are less used as individual mass analyzers in advanced mass spectrometers. However, due to their ion filtration/selection ability, they are commonly configured in a tandem mass spectrometer as an ion isolation unit or collision cell.

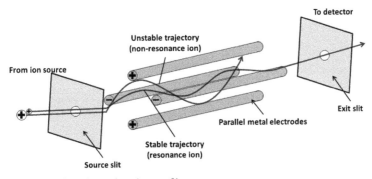

Fig. 3. A schematic representation of a quadrupole mass filter.

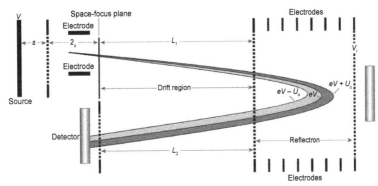

Fig. 4. Schematic diagram of a reflectron time-of-flight mass analyzer. Adapted with permission from Anal Chem, 71: 445a–51a. Copyright 1999 American Chemical Society.

Time of flight analyzer: In the time of flight (TOF) mass analyzer, m/z of ions is measured by the flight time of accelerated ions in a fixed electric field. In the vacuumed electric field, the velocity of the ion is determined by its m/z. Subsequently, the m/z of ions can be calculated from the known accelerating energy, the electric field strength and the length of the flight tube. Compared with a quadrupole mass filter and IT, a TOF is faster in scan speed and more sensitive because all ions are measured simultaneously (Cotter 1999).

The way to improve the mass-resolving power in a TOF is to increase the length of the ion path. Due to the physical limitation of an instrument, an extended ion path is generally achieved by using a reflectron. A reflectron is a reversed electric field that is placed on the opposite side of the accelerating electric field in the flight tube. Ions are "reflected" (slowed, stopped and accelerated to the reverse direction) in the reflectron. Therefore, the ion path is doubled in a given flight tube by using one reflectron. Theoretically, there is no upper limit of the mass-resolving power in a TOF mass analyzer as long as a sufficient number of reflectrons are configured within the mass spectrometer. However, the excessive length of the ion path results in a significant reduction of sensitivity and mass range (Toyoda et al. 2003). Current commercialized reflectron TOF instruments are usually equipped with one or two reflectrons and can routinely attain a mass-resolving power of > 20,000 and a mass accuracy of 5 ~ 10 ppm (Fig. 4).

Fourier transform ion cyclotron resonance analyzer: In Fourier transform ion cyclotron resonance analyzer (FT-ICR), ions are trapped in a cell by an electric field and a high magnetic field is applied to the electric field. Ions are excited by an oscillating electric field orthogonal to the magnetic field. Affected by the Lorentz force, excited ions take larger circular trajectories. The frequencies of ion rotational motions, which are inversely proportional to their m/z, induce an electronic signal on a pair of receiver electrodes as the ion passes near the plates. By using Fourier transform, the electronic signal containing all frequencies is converted to individual m/z values.

FT-ICR MS has become the highest-resolution broadband mass analyzer since its introduction in 1974 by Melvin B. Comisarow and Alan G. Marshall (Marshall et al. 1998, Marshall and Comisarow 1974). The

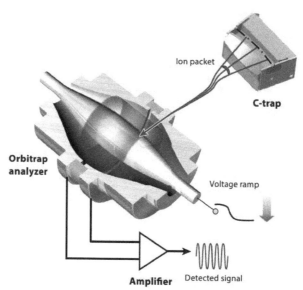

Fig. 5. Cross section of the C-trap ion accumulation device and the Orbitrap mass analyzer with an example of an ion trajectory. Reprinted with permission from Anal Chem, 85: 5288–96.Copyright 2013 American Chemical Society.

high resolving power is obtained by the ultra-long ion path traveled during the ion cyclotron resonance, since the mass-resolving power of FT-ICR is proportional to the ion path traveled during the acquisition (Scigelova et al. 2011). For example, in an FT-ICR mass spectrometer equipped with a 9.4 Tesla magnetic field, ions with an m/z of 1000 have a cyclotron frequency of 144,346 Hz and travel 9070 meters in one second. The observation of the same 1000 m/z ions for 3 s yields a potential mass-resolving power of up to 400,000 (Marshall and Hendrickson 2008). In recent reports, a 21 Tesla FT-ICR instrument was able to resolve the isotopic distribution of the 48[+] charge state of bovine serum albumin (molecular weight 66 kDa) by use of a 0.38 s detection period (approximate mass-resolving power of 150,000) (Hendrickson et al. 2015). The highest mass-resolving power reported on the same instrument was 2,700,000 at m/z = 400 (Smith et al. 2018). The ultrahigh mass-resolving power allows FT-ICR MS to act as a useful tool in many fields, especially in large intact biomolecule analysis.

Orbitrap analyzer. The Orbitrap™ mass analyzer consists of a spindle-like central electrode that has a high voltage applied to it and is surrounded by an outer barrel-like electrode kept at ground (Scigelova et al. 2011). The electrodes are precisely shaped to maintain a radial and an axial electric field inside the Orbitrap. During the data acquisition, ions are accumulated in an RF-only bent quadrupole, which is called the C-trap, and ejected into the Orbitrap (Zubarev and Makarov 2013) (Fig. 5). Affected by the strong electric field inside of the Orbitrap, ions rotate around the central electrode and oscillate along the axis of the central electrode. The trajectory of ions is similar to a complicated spiral. Similar to FT-ICR MS, ion motions are recorded by the outer electrode as image currents, and converted to m/z by using Fourier transform. Currently, the Orbitrap mass analyzer is exclusively produced by Thermo Fisher Scientific.

The Orbitrap MS is one of the most popular mainstream mass spectrometry techniques, the other being TOF-MS. Commercial Orbitrap mass spectrometers routinely provide a mass-resolving power of > 60,000, and reliably deliver mass accuracy below 1 ppm (Makarov et al. 2006), which meet the needs for biomolecular analysis. Although the reported maximum mass resolving power of the Orbitrap MS (approximately 1,000,000) is slight lower than the FT-ICR MS (Denisov et al. 2012), in practice, the Orbitrap MS avoids the short comings of FT-ICR MS. Compared to FT-ICR MS, an Orbitrap requires less laboratory space and site preparation, and is relatively inexpensive in price and cost of daily maintenance.

Since the first commercial Orbitrap mass spectrometer was released by Thermo Electron (now Thermo Fisher Scientific) in 2005, several series of Orbitrap mass spectrometers have been introduced. Specialized designs allow Orbitrap MS to meet different analytical needs. The Orbitrap MS has been widely applied in almost all analytical research areas. Representative applications include proteomics, metabolomics, and environmental analyses (Perry et al. 2008).

Detector in Mass Spectrometry

The detection of an ion in a modern mass spectrometer is achieved by creating an electric signal. The basic ion detector is the Faraday cup. The principle of the Faraday cup is that a current is induced when a packet of ions hits the dynode surface. The number and charge of ions are determined by measuring the current.

In quadrupole, IT and TOF instruments, electron multiplier and microchannel plates are used as detectors. The electron multiplier was designed by extending the principles of the Faraday cup. It utilizes either multiple dynodes or a horn-shaped continuous dynode to amplify the electric signal and improve the sensitivity of detection. Microchannel plates are compact electron multipliers. Typically, the diameter of each channel is around 10 micrometers, and a microchannel plate usually consists of 10,000 to 10,000,000 closely packed channels (Wiza 1979). This design allows microchannel plates to provide additional spatial resolution.

In Fourier transform MS (FT-ICR and Orbitrap), the detector is built in the mass analyzer. The movement of ions induces currents on the receiver electrodes (FT-ICR) or the outer electrode (Orbitrap). Image currents are subsequently generated, recorded and converted into readable m/z and intensity values.

Tandem Mass Spectrometry

Soft ionization techniques generate minimum in-source fragments of the analyte in order to reduce the complexity of the mass spectrum. However, the identification of intact molecule ions (especially large biological molecules) is difficult due to their large molecular weight. Fragmentation of ions is essential for the structural elucidation of analyte molecules. In practice, tandem mass spectrometry techniques are used to create gas-phase fragments of analyte molecules.

Tandem mass spectrometry (MS/MS or MS^2) refers to the combination of two or more mass analyzers in a single mass spectrometer. In a tandem mass spectrometer, ions in a wide mass range are first measured in the first stage of mass spectrometry (MS1); based on the MS1 data, ions within a narrow mass range (precursor ions) are isolated and dissociated; the fragment or product ions are then identified in the second stage of mass spectrometry (MS2). This concept can be extended to multi-stage mass spectrometry (MS^n) if the product ions are further isolated, fragmented and measured in the nth stage of mass spectrometry (MSn).

Fragmentation Techniques

Several techniques based on different mechanisms have been developed to generate different types of fragment ions, with collision-induced dissociation, electron transfer dissociation, and electron capture dissociation being the most commonly used. Many other techniques, such as photodissociation and surface-induced dissociation (SID), are also employed for MS/MS applications.

Collision-Induced Dissociation (CID)

CID is probably the most commonly used fragmentation technique in tandem mass spectrometry. During the CID process, ions produced in the ion source are accelerated by an electric field to gain kinetic energy and then collided with collision gas (usually helium or argon). The conversion from kinetic energy to vibrational energy occurs in the collision and subsequently results in the dissociation of chemical bonds, thus breaking the ion into fragments (Fig. 6). The acceleration mechanism of CID can vary depending on the instrument. For example, in a quadrupole mass filter, precursor ions are accelerated by the electric field between two parallel lenses before entering the quadrupole. On the other hand, in a linear ion trap, precursor ions are fragmented through resonant-excitation when they are trapped in the ion trap. Therefore, CID is also known as collisionally activated dissociation (CAD).

CID is a favorable fragmentation technique for peptide identification. Typically, CID results in ion dissociation occurring at amide bonds along the peptide backbone, generating certain types of fragment ions, namely, b- and y-type ions. CID also leads to loss of small neutral molecules, such as water and/ or

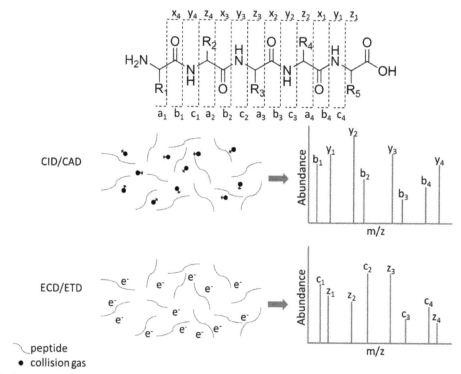

Fig. 6. Common peptide fragmentation methods and different type of fragment ions. Reprinted with permission from Chem Rev, 113: 2343–94. Copyright 2013 American Chemical Society.

ammonia or other fragments derived from side chains (Huang and McLuckey 2010). Thus, the fragmentation pattern is relatively simple and easy to solve. In MS-based proteomics or large-scale protein mixture analyses, CID is the most preferred fragmentation technique. However, CID is less effective for large, highly-charged peptides and is not suitable for fragmentation of peptides with labile post-translational modifications (PTMs), such as phosphorylation and S-nitrosylation (Frese et al. 2011). Therefore, other fragmentation techniques are often used in tandem with CID to provide complementary information.

Electron Capture Dissociation (ECD) and Electron Transfer Dissociation (ETD)

In the ECD process, the attachment of a free electron to a multiple positively-charged molecule forms an odd-electron ion. The odd-electron species then undergoes rearrangement with subsequent cleavages of chemical bonds. If the electron is transferred from a radical anion, the process is called ETD.

In peptide analysis, ECD and ETD cleave the amide bonds to generate c- and z-type ions (Fig. 6). These types of ions can provide complementary information for the structural elucidation of certain kinds of peptides. Additionally, the labile PTMs tend to remain intact in ECD and ETD processes. Since both sequence information and localization of the modification sites are available in one spectrum, ECD and ETD techniques are useful in protein PTMs and proteomics research (Zhurov et al. 2013).

Configurations of Tandem Mass Spectrometers

Many combinations of mass analyzers have been designed and produced for different applications over the past century. In general, quadrupole mass filters and IT are used as mass filter/isolation units, while the TOF, FT-ICR and Orbitrap are used as high-resolution mass measurement units. Currently, commercial tandem mass spectrometers are primarily in the following configurations: triple quadrupole mass spectrometer, quadrupole time-of-flight mass spectrometer, quadrupole Orbitrap mass spectrometer, ion trap Orbitrap mass spectrometer, and quadrupole Orbitrap iontrap mass spectrometer.

Triple Quadrupole Mass Spectrometer (QqQMS)

QqQMS is a tandem mass spectrometer that has two quadrupoles serving as mass filters with an RF-only quadrupole between them to act as a collision cell for CID. Selected reaction monitoring (SRM), also known as multiple reaction monitoring (MRM), can be performed in this type of instrument. In SRM mode, the first quadrupole (Q1) scans for predefined specific precursor ions which are subjected to CID with helium or argon in the second quadrupole (Q2) to produce fragments. Specific fragments of the precursor ion are filtered in the third quadrupole and detected by electron multipliers. This dual-step ion selection allows for highly specific detection with high sensitivity and low background interference. Thus, QqQMS is often the instrument of choice for quantitative analysis of both small and large molecules.

Quadrupole Time-of-flight Mass Spectrometer (Q-TOFMS) & Quadrupole Orbitrap Mass Spectrometer (Q-OTMS)

Q-TOFMS and Q-OTMS are operated with similar principles. The quadrupole can be used as a path to transfer all of the ions or as a mass filter to select specific ions. The TOF or Orbitrap serves as a high-resolution mass analyzer for accurate mass measurement. In these types of instruments, data-dependent acquisition (DDA) and data-independent acquisition (DIA) are the two commonly used acquisition modes. In DDA mode, a high-resolution survey scan of all precursor ions is first acquired in the TOF or Orbitrap. Several MS/MS scans of the selected ions are then acquired based on the survey scan according to a certain rule, typically on ions exhibiting the highest intensities in the survey scan. DDA allows the data acquisition of certain ions without any information prior to the analysis. It is very useful for large-scale unknown sample analysis when combined with front-end separation techniques, such as HPLC or CE. DIA involves data acquisition within a predefined mass range. SRM is one of the DIA strategies used for quantitatively measuring samples with known information. Other DIA strategies, such as MS^E and sequential window acquisition of all theoretical fragment ion spectra (SWATH™), can be applied to analyze large-scale protein mixtures.

Ion Trap Orbitrap Mass Spectrometer (IT-OTMS)

In the tandem IT-OTMS configuration, the ion isolation and fragmentation processes are performed in the IT to allow DDA. Since both ion isolation and fragmentation are conducted in the IT, the MS/MS scan speed is generally not as fast as in Q-TOFMS or Q-OTMS. Therefore, high throughput DIA strategies for scanning a large list (SRM) or a wide range (MS^E and SWATH) of precursor ions are generally not available in these types of instruments. However, the involvement of IT allows the usage of multiple fragmentation techniques including CID, ETD and higher energy collisional dissociation (HCD). The combination of various fragmentation approaches is especially useful in proteomics and PTM analysis.

Quadrupole Orbitrap Iontrap Mass Spectrometer (Q-OT-ITMS)

The Q-OT-ITMS, in which the Quadrupole, IT and Orbitrap are all configured together to exploit the advantage of each mass analyzer, is a recent release from Thermo Fisher Scientific (Fig. 7). In addition to the aforementioned features of Q-OTMS and IT-OTMS, another fragmentation technique, ultraviolet photodissociation (UVPD), is also an option in this type of instrument (Senko et al. 2013).

Utilization of Tandem Mass Spectrometry in Proteomics

Proteins perform numerous functions in living organisms, including but not limited to: catalyzing metabolic reactions, replicating DNA, responding to stimuli, and transporting molecules (Zhang et al. 2013). The alteration of a protein may be directly correlated with one or multiple biological processes and associated with a disease state. Therefore, the comprehensive characterization of proteins in a proteome, so-called proteomics studies, becomes readily apparent with aberrant cells, tissues or bodily fluids. On the other hand, accurate proteome characterization remains tremendously challenging due to the high complexity

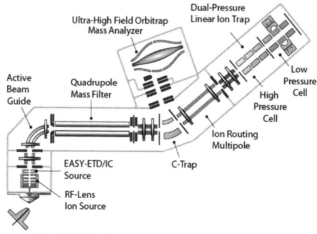

Fig. 7. Schematic diagram of the Q-OT-IT hybrid mass spectrometer. Reprinted with permission from Anal Chem 85: 11710–14. Copyright 2013 American Chemical Society.

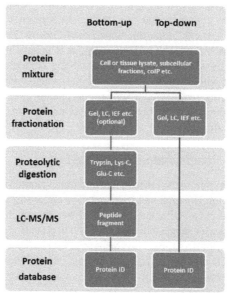

Fig. 8. Proteomic Strategies: bottom-up vs. top down. Reprinted and modified with permission from Chem Rev 113: 2343–94. Copyright 2013 American Chemical Society.

of proteoforms and PTMs, the wide dynamic range of protein abundances and the absence of protein amplification methods.

The development of tandem mass spectrometry technologies dramatically improved the depth of proteomics studies by providing highly sensitive and specific analytical approaches. Coupled with a separation front end, usually LC or CE, tandem mass spectrometry is exclusively used as the analytical platform for different proteomic strategies. There are two major classes of proteomic strategies: bottom-up proteomics and top-down proteomics (Fig. 8). In bottom-up strategies, the characterization of proteins is achieved by the analysis of peptides released from proteins through proteolysis. In contrast, top-down strategies mainly focus on the analysis of intact proteins. Compared with intact proteins, peptides are more easily fractionated, ionized and fragmented. Therefore, bottom-up proteomics is more universally used in practice.

DDA and DIA acquisitions are widely used in bottom-up proteomics for both qualitative and quantitative purposes. Peptides are identified by comparing the experimental tandem mass spectra with theoretical tandem mass spectra generated from *in silico* fragmentation or a spectral library collected in previous experiments. Without any derivatizations, proteins can be quantitatively compared using precursor ion intensities, peak areas, spectral counts or the transition ion intensities (SRM/MRM only) of peptides. Several labeling techniques including stable isotopic labeling of amino acids in cell culture (SILAC), ^{18}O, demethylation, isobaric tags for relative and absolute quantitation (iTRAQ) and tandem mass tags (TMT) are also available for relative and absolute quantitative analysis of peptides and proteins (Zhang et al. 2013).

MS-based proteomics approaches are powerful tools for protein biomarker discovery (Wang et al. 2016). In typical MS-based proteomics studies, identification and quantification of tens of thousands of proteins and PTMs are routinely achieved. This allows for the identification of useful biomarkers of disease in high throughput using small amounts of material. Similar approaches can also be applied to metabolomics, lipidomics and glycomics studies for biomarker discovery of metabolites, lipids and glycans, respectively.

Clinical Applications of Mass Spectrometry

As illustrated in the applications of MS below, the successful development of clinically-applicable MS, as in other areas of clinical instrumentation, required close collaboration between the engineers and healthcare providers. Whereas the engineers have the expertise of building technically competent devices, the participation of healthcare providers helped make the devices practical, user-friendly, and truly beneficial to the patients in real-life situations.

Improvement of Early Disease Detection

One of the potential uses of MS in clinical application is to detect early disease markers. Many diseases, especially some of the more highly aggressive cancers such as pancreatic cancer and ovarian cancer, are usually diagnosed when the disease has reached an advanced stage. By then, therapeutic intervention is often not successful. Thus, the best medical management strategy for these diseases is to detect the cancer at its earliest stage, at which time the treatment success rate is significantly higher. Therefore, identification of biomarkers that are easily detected (very sensitive) and highly specific for early-stage cancer is key for this management approach. MS instrumentation has been a leader in this aspect owing to its highly sensitive and specific detection capabilities (Wang et al. 2016, Flores-Morales and Iglesias-Gato 2017). To illustrate this point, several clinical examples are provided in the following sections.

Discovery of Novel Protein Markers in Ovarian Cancer

As early as 2003, a group of researchers utilized surface-enhanced laser-desorption and ionization-MS profiling to investigate serum samples from 80 ovarian cancer patients and 91 normal individuals in order to search for novel biomarkers. They identified a candidate protein of 11,700 Da in size, which was subsequently purified by affinity chromatography and sequenced by liquid chromatography-tandem MS to be haptoglobin-alpha subunit. In combination with another ovarian cancer biomarker CA 125, this biomarker achieves 91% sensitivity and 95% specificity for early diagnosis of ovarian cancer (Ye et al. 2003).

To look into the possibility of identifying protein biomarkers in ovarian cancer patients, another group of investigators targeted urine, the biological sample collected by the most non-invasive method. In this pilot study, urine samples from 20 ovarian cancer patients and 20 patients with benign ovarian diseases were analyzed by MS using label-free quantification. This study found that the levels of 23 different proteins were significantly elevated between the patient sets. Using another method, the levels of several proteins (LYPD1, LYVE1, PTMA, and SCGB1A1) were confirmed to be increased (Sandow et al. 2018). This discovery will pave the way for further investigation into large populations of patients so as to identify the definitive biomarkers for ovarian cancer.

Development of Biomarkers for Pancreatic Cancer

Due to the lack of reliable biomarkers suitable for early pancreatic cancer detection, investigators looked into non-invasive or nearly non-invasive methods for such biomarker identification. The rationale for non-invasive method development is the logic that this kind of test can be safely performed repeatedly, which is needed for a highly aggressive cancer requiring continuous monitoring. These investigators utilized capillary electrophoresis-MS to analyze salivary samples of patients with pancreatic cancers (N = 39), chronic pancreatitis (N = 14), and no disease (N = 26), with a focused target on polyamines. Their examination discovered that several polyamines including spermine, N1-acetylspermidine, and N1-acetylsperine, showed levels in saliva that were significantly different between cancer patients and normal individuals. In addition, the combined results of these four metabolites would provide a highly accurate distinction between pancreatic cancers and the other two groups (pancreatitis and normal). Pending additional study with a large patient population including early pancreatic cancer patients, this polyamine test could be very helpful in detecting early pancreatic cancer (Asai et al. 2018).

Another group of scientists used a combined 2-D image-converted analysis of the liquid chromatogram and MS ion chromatogram to identify the pancreatic cancer biomarkers Prolyl-hydroxylated α-fibrinogen (result from 43 cancer and 43 normal samples) and CXC chemokine ligand 7 (CXCL-7) (result from 24 cancer and 21 normal samples) in blood plasma samples collected from pancreatic cancer patients and depleted of plasma-abundant proteins. The biomarkers' identities were verified with an immunoassay (Ono et al. 2012). Very recently, another group of researchers also reported the identification of a new panel of pancreatic cancer biomarkers using MS technology (Lu et al. 2018).

Intraoperative Cancer Margin Identification

Another potential use of MS in a clinical situation is to provide therapeutic aid during a real-life surgical operation. When a patient with a solid cancer tumor is determined to be a candidate for surgery, i.e., no metastasis has been detected by diagnostic imaging or other tests, the surgeon will perform an excisional procedure to remove the cancer. Intra-operatively, surgeons can generally access if metastasis has occurred at a gross level, however they cannot reliably determine if the cancer has spread to the surrounding areas microscopically. It would be highly beneficial if the operating surgeons can, during the procedure, quickly access if the cancer has spread to adjacent tissues, nearby blood vessels or draining lymphatic tissues, so that they can remove those issues immediately at the operating table to reduce the possibility of cancer metastasis in the future. A clinical situation that would especially benefit from surgical intervention would be an operation for brain cancer, since re-operation of brain cancer will require technically challenging procedures with higher possibility of morbidity and even mortality. Since the surgeons cannot wait for a long period of time for such determination, speed and accuracy are essential for such an approach. The current approach is to analyze surgically excised specimens that are frozen, but microscopic examination of these sections is labor-intense, time-consuming, scope-limiting and often has less than optimal accuracy in margin identification. MS could greatly assist with this need (Pope 2014, Santagata et al. 2014).

DESI-MS Method

The development of intraoperative tumor margin detection owes a great deal to the scientist R. Graham Cooks and his Purdue University collaborators, who constructed a device called desorption electrospray ionization-mass spectrometry (DESI-MS) (Cooks et al. 2006, Pope 2014). The DESI-MS is capable of ionizing untreated surgical samples in an open air ambient environment, allowing for MS analysis of the ionized sample with nearly instant diagnostic result (Pope 2014). The key elements for the detection of tumor margin include the following:

- An exposed clinical sample to allow access to the testing instrument.
- A tumor-specific metabolite to allow distinction between tumor and normal surrounding tissue.

- DESI, which is able to ionize small tissue samples ready for MS analysis.
- An MS inlet to facilitate travel of the ionized sample to the MS.
- A MS to analyze the ionized sample and report to the healthcare providers regarding the presence or absence of tumor-specific metabolites.

As one can see, one of the key elements allowing for the development of this diagnostic method was the discovery of a tumor-specific metabolite. A recent finding revealed that there was a mutation of the isocitrate dehydrogenase-1 (IDH1) gene present in a large majority of the common brain tumor glioma, with elevated levels up to 100-fold compared to wild type IDH1. This mutation and resulting gain-of-function leads to the production of the tumor-specific metabolite, 2-hydroxyglutarate (2-HG), which is not present in normal tissue. Besides glioma, the IDH1 mutation is also found in acute myeloid leukemia, cholangiocarcinoma, chondrosarcoma, and T cell lymphoma. These findings gave a substantial advantage for the development of tumor margin detection using MS (Pope 2014). Alternatively, other chemicals, if substantially different in quantity between tumor and normal tissues, can also be used for this identification purpose.

A schematic illustration of how intraoperative margin identification works is depicted in Fig. 9 below.

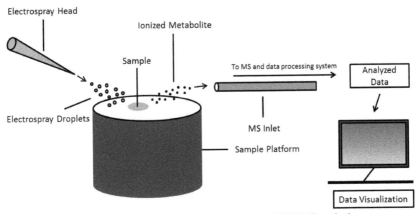

Fig. 9. Schematic representation of intraoperative tumor margin detection by DESI-MS method.

The following clinical examples illustrate the significance of this new technological development.

Using this DESI-MS method, researchers examined intraoperative brain tissues for tumor-type determination. In the validation process, they built a classifier to distinguish between gliomas and meningiomas based on histopathology-certified gliomas (N = 36) and meningiomas (N = 19). A subsequent examination of 32 surgical brain tumor specimens including oligodendroglioma, astrocytoma, and meningioma with various histological grades and tumor concentrations found that the molecular diagnoses acquired through DESI-MS were in very good agreement with the histological diagnoses. Thus, this paved the way for further examination of the margin detection purpose (Eberlin et al. 2013).

Subsequently, medical investigators at Brigham and Women's Hospital in Boston, MA, were able to successfully incorporate this DESI-MS method in real-life brain tumor surgeries by defining the tumor margin based on the tumor-specific metabolite 2HG detection by MS. In addition, they were able to overlay the molecular information from MS to histological specimens and thus further validate the DESI-MS method. Furthermore, they were able to map the tumor-specific metabolite 2HG concentration with the tumor concentration in a 3-dimensional MRI (magnetic resonance imaging) tomography, further enhancing their ability to identify the tumor margin (Santagata et al. 2014).

REIMS Method

Because the DESI-MS method was initially developed requiring the preparation of the clinical sample in a way that was technically difficult and time-consuming in addition to being expensive and low-throughput, it led to a new method that could perform a direct analysis of clinical tissues without the need for sample preparation (St. John et al. 2016, 2017, Phelps et al. 2018). This method, termed REIMS for rapid evaporative ionization mass spectrometry, utilizes a monopolar hand piece on both *ex vivo* and *in vivo* tissues. Also termed as "iKnife" (or intelligent knife), the step-by-step process of the REIMS method is explained below (St. John et al. 2017):

- The operating surgeon applies electrosurgical (Bovie) tools for the cutting or coagulating of the clinical tumor margin.
- The electrosurgical process of the cutting mode (continuous radiofrequency wave) or coagulating mode (pulsed radiofrequency wave) on tissues causes heat to dissipate inside the tissue, leading to cellular explosion and release of cellular content in gaseous form as an aerosol or vapor.
- The extracted vapor is ionized for rapid MS chemical identification.
- The application of artificial intelligence with a machine learning-trained computational algorithm allows identified data to be distinguished between cancerous and benign, and reported to the surgeon within 2 seconds.
- The surgeon makes an intraoperative decision based on the MS determination.

Utilizing this REIMS method, St. John and colleagues examined breast cancer and benign breast tissues with this iKnife. The impetus for investigation of breast cancer derived from the knowledge that breast cancer is the most common cancer affecting women and that the re-operation rate for breast cancers reached 20–25% due to margin involvement detected on tissues from the initial surgery. Thus, delineating a cancer-free margin on the initial surgery is a significant winning healthcare strategy, both from the economic perspective and from the personal/emotional perspective. In the validation process, this iKnife achieved 93% sensitivity and 95% specificity in histopathology-determined breast cancers (N = 226) and normal breast tissues (N = 932). In a subsequent examination of fresh and frozen breast tissues (99 cancers and 161 normal), this iKnife achieved 91% sensitivity and 99% specificity, and the average time requirement for the interpretation of the MS result, including data acquisition and analysis, was only 1.8 seconds (St. John et al. 2017).

Another group of investigators looked into the utilization of this iKnife in ovarian cancer determination on 335 tissue samples. The tumor-specific metabolites/chemicals, five phophatidic acid and phosphatidyl-ethanolamine lipid species, were significantly more abundant in ovarian cancers than in normal or borderline ovarian tumors (p < 0.001). Their validation process found that ovarian cancers can be easily distinguished from benign tissues with 97.4% sensitivity and 100% specificity, that ovarian cancers can be distinguished from borderline ovarian tumors with 90.5% sensitivity and 89.7% specificity, and that there is a very good determination agreement between iKnife reports and histopathology results (kappa = 0.84, p < 0.001, z = 3.3) (Phelps et al. 2018).

Hand-held "MasSpec Pen" Method

As recently reported, scientists at the University of Texas at Austin, led by Livia Eberlin, Associate Professor of Chemistry, have developed a handheld mass spectrometry device that can identify cancer in near real-time. Using this "MasSpec Pen," which connects to a mass spectrometry instrument, surgeons can easily search residual cancer at the surgical margin and adjacent tissues during their cancer surgeries within seconds. More importantly, the process does not cause tissue damage that may become a physical healing problem and an ethical concern (Waltz 2017, Zhang et al. 2017). The step-by-step intraoperative procedure of this "MasSpec Pen" is described below (Waltz 2017):

- The operating surgeon holds the "MasSpec Pen" against the questionable tissue *in vivo*.
- Water-soluble molecules, including tumor-specific metabolites, are dissolved by the water droplets suspended at the tip of the "MasSpec Pen".
- Dissolved and ionized molecules in the droplet travel to the MS for analysis.
- The molecular information, the mass-to-charge ratio representing the quantitative relationship between mass and electrical charge that is unique for each molecule, is identified.

- The identified data is analyzed by a statistical classifier that is trained and validated with hundreds of histopathology-certified cancer and normal tissue samples and will determine and report if the questionable tissue is cancerous or normal, within seconds, with greater than 95% accuracy.
- The surgeon makes an intraoperative decision depending on the finding by the "MasSpec pen".

This group of investigators published their first findings utilizing this "MasSpec Pen" in their 2017 paper with the following findings (Zhang et al. 2017). On the *ex vivo* examinations of 20 thin slides and 253 human tissue samples of both normal and cancerous breast, lung, ovary, and thyroid gland, they were able to obtain a substantial amount of data on tumor-specific and tumor-enhanced molecular information (metabolite, lipid, protein) useful for future tumor margin identification. In conjunction with the statistical classifier built with the histopathology-certified database, this "MasSpec Pen" method predicted cancers in tissue with 96.4% sensitivity and 96.2% specificity, and an overall accuracy of 96.3%. In addition, applying this "MasSpec Pen" on tumor-barring animals showed no observable physical injury or distress to the animals, further supporting the non-destructive intraoperative function of this device (Zhang et al. 2017). Pending further clinical studies to confirm its usefulness and accuracy in cancer margin determination, this diagnostic pen could be utilized in many intra-operative margin assessment, including a skin cancer

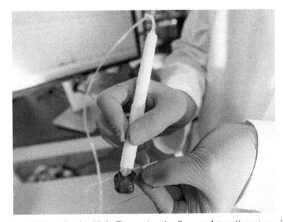

Fig. 10. MasSpec Pen: Photo by Vivian Abagiu, Univ Texas, Austin. Source: https://spectrum.ieee.org

removal procedure called Mohs Micrographic Surgery, eliminating the need for tissue sectioning and microscopic examination. A photograph of MasSpec Pen is showed in Fig. 10 (Photo by Vivian Abagiu, Univ Texas at Austin, Source: https://spectrum.ieee.org).

Limitations and Challenges

At the present time, the challenges for wide-spread use of this state-of-the-art intraoperative cancer margin identifying method are two-fold: cost and size. Currently, a commercially available mass spectrometer costs at least a half of a million dollars and its size is large and not suitable for mobile use in a typical operating room. Going forward, production of smaller and less expensive mass spectrometers will be important for popularizing the use of this high-tech engineering instrument (Waltz 2017). In fact, the engineers at George Tech are currently working on this. By replacing the currently used high-voltage power source with a nanogenerator, the future size of the MS could be much reduced with the possibility of lowering the cost of MS as well (Johnson 2017).

Summary

In this chapter, we delineated the engineering, physical, and chemical principles of mass spectrometry. We discussed the key steps of mass spectrometry and some of its potential applications in medicine. One of the most exciting implementation of mass spectrometry might be the hand-held MasSpec Pen that could deliver intra-operative cancer margin detection, a highly efficient and useful technique in clinical medicine.

References Cited

Asai, Y., T. Itoi, M. Sugimoto, A. Sofuni, T. Tsuchiya, R. Tanaka et al. 2018. Elevated polyamines in saliva of pancreatic cancer. Cancers (Basel) Feb 5; 10: pii: E43. Doi: 10.3390/cancers10020043.

Audi, G. 2006. The history of nuclidic masses and of their evaluation, Int J Mass Spectrom 251: 85–94.

Banerjee, S. and S. Mazumdar. 2012. Electrospray Ionization Mass Spectrometry: A Technique to Access the Information beyond the Molecular Weight of the Analyte, Int J Anal Chem 2012: Article ID 282574.

Biemann, K. 2007. Laying the groundwork for proteomics—Mass spectrometry from 1958 to 1988, Int J Mass Spectrom 259: 1–7.

[BRAINY] 2018a. Brainy Quote. [https://www.brainyquote.com/authors/Theodore_roosevelt] accessed February 17, 2018.

[BRAINY] 2018b. Brainy Quote. [https://www.brainyquote.com/authors/walt_whitman] accessed February 17, 2018.

[BRAINY] 2018c. Brainy Quote. [https://www.brainyquote.com/authors/ralph_waldo_emerson] accessed February 17, 2018.

Cody, R.B., J.A. Laramee and H.D. Durst. 2005. Versatile new ion source for the analysis of materials in open air under ambient conditions. Anal Chem 77: 2297–2302.

Cooks, R.G., Z. Ouyang, Z. Takats and J.M. Wiseman. 2006. Detection technologies. Ambient mass spectrometry. Science 311: 1566–1570.

Cotter, R.J. 1999. The new time-of-flight mass spectrometry. Anal Chem 71: 445a–451a.

Denisov, E., E. Damoc, O. Lange and A. Makarov. 2012. Orbitrap mass spectrometry with resolving powers above 1,000,000. Int J Mass Spectrom 325: 80–85.

Eberlin, L.S., I. Norton, D. Orringer, I.F. Dunn, X. Liu, J.L. Ide et al. 2013. Ambient mass spectrometry for the intraoperative molecular diagnosis of human brain tumors. Proc Natl Acad Sci 110: 1611–1616.

Fenn, J.B., M. Mann, C.K. Meng, S.F. Wong and C.M. Whitehouse. 1989. Electrospray ionization for mass spectrometry of large biomolecules. Science 246: 64–71.

Flores-Morales, A. and D. Iglesias-Gato. 2017. Quantitative mass spectrometry-based proteomic profiling for precsion medicine in prostate cancer. Frontier Oncol November 2017. Doi: 10.3389/fonc.2017.00267.

Frese, C.K., A.F.M. Altelaar, M.L. Hennrich, D. Nolting, M. Zeller, J. Griep-Raming et al. 2011. Improved Peptide Identification by Targeted Fragmentation Using CID, HCD and ETD on an LTQ-Orbitrap Velos. J Proteome Res 10: 2377–2388.

Gessel, M.M., J.L. Norris and R.M. Caprioli. 2014. MALDI imaging mass spectrometry: Spatial molecular analysis to enable a new age of discovery. J Proteomics 107: 71–82.

Griffiths, J. 2008. A brief history of mass spectrometry. Anal Chem 80: 5678–5683.

Hebert, A.S., A.L. Richards, D.J. Bailey, A. Ulbrich, E.E. Coughlin, M.S. Westphall et al. 2014. The One Hour Yeast Proteome. Mol Cell Proteomics 13: 339–347.

Hendrickson, C.L., J.P. Quinn, N.K. Kaiser, D.F. Smith, G.T. Blakney, T. Chen et al. 2015. 21 Tesla Fourier Transform Ion Cyclotron Resonance Mass Spectrometer: A National Resource for Ultrahigh Resolution Mass Analysis. J Am Soc Mass Spectrom 26: 1626–1632.

Huang, T.Y. and S.A. McLuckey. 2010. Gas-phase chemistry of multiply charged bioions in analytical mass spectrometry. Annu Rev Anal Chem (Palo Alto Calif) 3: 365–385.

Johnson, D. 2017. Mass Spectrometry Gets a New Power Source and a New Life: Replacing high-voltage power source with nanogenerators increases sensitivity to new records. IEEE. Spectrum. February 27, 2017. [https://spectrum.ieee.org/nanoclast/semiconductors/materials/mass-spectrometry-gets-a-new-power-souce-and-a-new-life] accessed June 29, 2018.

Karas, M. and R. Kruger. 2003. Ion formation in MALDI: the cluster ionization mechanism. Chem Rev 103: 427–440.

Koichi, T., W. Hiroaki, I. Yutaka, A. Satoshi, Y. Yoshikazu, Y. Tamio et al. 1988. Protein and polymer analyses up to m/z 100 000 by laser ionization time-of-flight mass spectrometry. Rapid Commun Mass Spectrom 2: 151–153.

Li, L.H., H.Y. Hsieh and C.C. Hsu. 2017. Clinical Application of Ambient Ionization Mass Spectrometry. Mass Spectrom (Tokyo), 6: S0060.

Lu, X., W. Zheng, W. Wang, H. Shen, L. Liu, W. Lou et al. 2018. A new panel of pancreatic cancer biomarkers discovered using a mass spectrometry-based pipeline. Br J Cancer Feb 13. Doi: 10.1038/bjc.2017.365.

Makarov, A., E. Denisov, O. Lange and S. Horning. 2006. Dynamic range of mass accuracy in LTQ Orbitrap hybrid mass spectrometer J Am Soc Mass Spectrom 17: 977–982.

Marshall, A.G. and M.B. Comisarow. 1974. Fourier transform ion cyclotron resonance spectroscopy. Chem Phys Lett 25: 282–283.

Marshall, A.G. and C.L. Hendrickson. 2008. High-resolution mass spectrometers. Annu Rev Anal Chem (Palo Alto Calif) 1: 579–599.

Marshall, A.G., C.L. Hendrickson and G.S. Jackson. 1998. Fourier transform ion cyclotron resonance mass spectrometry: a primer. Mass Spectrom Rev 17: 1–35.

Marshall, A.G., C.L. Hendrickson and S.D. Shi. 2002. Scaling MS plateaus with high-resolution FT-ICRMS. Anal Chem 74: 252A–259A.

McNaught, A.D. and A. Wilkinson. 1997. Compendium of Chemical Terminology, 2nd ed. Wiley, Indianapolis, IN.

Nyadong, L., M.D. Green, V.R. De Jesus, P.N. Newton and F.M. Fernandez. 2007. Reactive desorption electrospray ionization linear ion trap mass spectrometry of latest-generation counterfeit antimalarials via noncovalent complex formation, Anal Chem 79: 2150–57.

Ono, M., M. Kamita, Y. Murakoshi, J. Matsubara, K. Honda, B. Miho et al. 2012. Biomarker Discovery of Pancreatic and Gastrointestinal Cancer by 2DICAL: 2-Dimensional Image-Converted Analysis of Liquid Chromatography and Mass Spectrometry. Int J Proteomics 2012: 897412. doi: 10.1155/2012/897412.

Parkins, W.E. 2005. The Uranium bomb, the calutron, and the space-charge problem Phys Today 58: 45–51.

Perry, R.H., R.G. Cooks and R.J. Noll. 2008. Orbitrap Mass Spectrometry: Instrumentation, Ion Motion and Applications. Mass Spectrom Rev 27: 661–699.

Phelps, D.L., J. Balog, L.F. Gildea, Z. Bodai, A. Savage, M.A. El-Bahrawy et al. 2018. The surgical intelligent knife distinguishes normal, borderline and malignant gynecological tissues using rapid evaporative ionization mass spectrometry (REIMS). Br J Cancer April 19. Doi: 10.1038/s41416-018-0048-3.

Pope, W.B. 2014. Intraoperative mass spectrometry of tumor metabolites. Proc Natl Acad Sci USA 111: 10906–10907.

Sandow, J.J., A. Rainczuk, G. Infusini, M. Makanji, M. Bilandzic, A.L. Wilson et al. 2018. Discovery and validation of novel protein biomarkers in ovarian cancer patient urine. Proteomic Clin ApplFeb 9. Doi: 10.1002/prca.201700135.

Santagata, S., L.S. Eberlin, I. Norton, D. Calligaris, D.R. Feldman, J.L. Ide et al. 2014. Intraoperative mass spectrometry mapping of an onco-metabolite to guide brain tumor surgery. Proc natl Acad Sci USA 111: 11121–11126.

Scigelova, M., M. Hornshaw, A. Giannakopulos and A. Makarov. 2011. Fourier transform mass spectrometry. Mol Cell Proteomics 10: M111 009431.

Senko, M.W., P.M. Remes, J.D. Canterbury, R. Mathur, Q.Y. Song, S.M. Eliuk et al. 2013. Novel Parallelized Quadrupole/Linear Ion Trap/Orbitrap Tribrid Mass Spectrometer Improving Proteome Coverage and Peptide Identification Rates. Anal Chem 85: 11710–11714.

Smith, D.F., D.C. Podgorski, R.P. Rodgers, G.T. Blakney and C.L. Hendrickson. 2018. 21 Tesla FT-ICR Mass Spectrometer for Ultrahigh-Resolution Analysis of Complex Organic Mixtures. Anal Chem 90: 2041–2047.

St. John., E.R., M. Rossi, P. Pruski, A. Darzi and Z. Takats. 2016. Intraoperative tissue identification by mass spectrometric technologies. Trends Analyt Chem 85: 2–9.

St. John, E.R., J. Balog, J.S. McKenzie, M. Rossi, A. Covington, L. Muirhead et al. 2017. Rapid evaporative ionisation mass spectrometry of electrosurgical vapours for the identification of breast pathology: towards an intelligent knife for breast cancer surgery. Breast Cancer Res 19: 59. Doi: 10.1186/s13508-017-0845-2.

Takats, Z., J.M. Wiseman, B. Gologan and R.G. Cooks. 2004. Mass spectrometry sampling under ambient conditions with desorption electrospray ionization. Science 306: 471–473.

Thomson, J.J. 1913. Rays of positive electricity. Proc R Soc Lond A 89: 1–20.

Toyoda, M., D. Okumura, M. Ishihara and I. Katakuse. 2003. Multi-turn time-of-flight mass spectrometers with electrostatic sectors. J Mass Spectrom 38: 1125–1142.

Waltz, E. 2017. Handheld mass-spectrometry pen identifies cancer in seconds during surgery. IEEE Spectrum. September 7. [https://spectrum.ieee.org/the-human-os/biomedical/devices/handheld-mass-spectrometry-pen-identifies-cancer-in-seconds-during-surgery] accessed February 17, 2018.

Wang, H., T.J. Shi, W.J. Qian, T. Liu, J. Kagan, S. Srivastava et al. 2016. The clinical impact of recent advances in LC-MS for cancer biomarker discovery and verification. Expert Rev Proteomics 13: 99–114.

Wiza, J.L. 1979. Microchannel plate detectors. Nucl Instr Meth 162: 587–601.

Yamashita, M. and John B. Fenn. 1984. Electrospray ion source. Another variation on the free-jet theme. J Phys Chem 88: 4451–4459.

Ye, B., D.W. Cramer, S.J. Skates, S.P. Gygi, V. Pratomo, L. Fu et al. 2003. Haptoglobin-alpha subunit as potential serum biomarker in ovarian cancer: identification and characterization using proteomic profiling and mass spectrometry. Clin Cancer Res 9: 2904–2911.

Zhang, Y., B.R. Fonslow, B. Shan, M.C. Baek and J.R. Yates, 3rd. 2013. Protein analysis by shotgun/bottom-up proteomics. Chem Rev 113: 2343–2394.

Zhang, J., J. Rector, J.Q. Lin, J.H. Young, M. Sans, N. Katta et al. 2017. Nondestructive tissue analysis for ex vivo and in vivo cancer diagnosis using a handheld mass spectrometry system. Sci Transl Med 9:(406), eaan3958. Doi: 10.1125/scitranslmed.aan3968.

Zhurov, K.O., L. Fornelli, M.D. Wodrich, U.A. Laskay and Y.O. Tsybin. 2013. Principles of electron capture and transfer dissociation mass spectrometry applied to peptide and protein structure analysis. Chem Soc Rev 42: 5014–5030.

Zubarev, R.A. and A. Makarov. 2013. Orbitrap Mass Spectrometry. Anal Chem 85: 5288–5296.

QUESTIONS

1. What is the basic principle of mass spectrometry?
2. Describe the 3 major steps that the basic mass spectrometry is consisted of.
3. What are the advantages mass spectrometry could provide?
4. Why is intraoperative cancer margin detection so important?

5. Describe the step-by-step intraoperative procedures utilizing the DESI-MS method. What are the key elements for DESI-MS to be useful in cancer margin detection?

6. Describe the step-by-step intraoperative procedures utilizing the REIMS method in cancer margin detection. What does it differ from the DESI-MS method?

7. Describe the step-by-step intraoperative procedures utilizing the "MasSpec Pen" method in cancer margin detection. What does it differ from the DESI-MS method and the REIMS method?

8. What is the biggest advantage of the "MasSpec Pen" method and why?

9. What are the limitations/challenges for the popularization of intraoperative usage of MS?

PROBLEMS

- Students will form groups of 3–5 student each.
- Group gathering will brainstorm a real-life medical or surgical problem that mass spectrometry may help resolving.
- Delineate each and every individual step your group will solve this medical or surgical problem.
- Present your group's delineation to the class.

24

Robotic Technology and Artificial Intelligence in Rehabilitation Medicine

Kun Yan

QUOTABLE QUOTES

"In the midst of difficulty lies opportunity."

Albert Einstein, Scientist, Nobel Laureate of Physics, 1921 (EINSTEIN 2018)

"However difficult life may seem, there is always something you can do and succeed at."

Stephen Hawking, Theoretical Physicist and Cosmologist, Recipient of the Presidential Medal of Freedom 2009 (BRAINYQUOTE 2018a)

"Intelligence is the ability to adapt to change."

Stephen Hawking, Theoretical physicist and cosmologist, a recipient of the Presidential Medal of Freedom 2009 (BRAINYQUOTE 2018a)

"There are many ways of going forward, but only one way of standing still."

Franklin D. Roosevelt, 32nd President of the United States (BRAINYQUOTE 2018b)

Learning Objectives

The learning objectives are to help students to understand the concepts of disability, assistive technology, rehabilitation and assistive technology services, and rehabilitation engineering; and the applications, limitations, and future opportunities of robotic technology and artificial intelligence in the field of rehabilitation medicine.

Northern California Veterans Administration Healthcare System, 150 Muir Road, Martinez, CA 94553; Kun.yan@va.gov

After completing this chapter, the students should:

- To have a general understanding of the definition of disabilities and its evolution.
- To have a general understanding of the definitions of assistive technologies, assistive technology devices, rehabilitation and assistive technology services and rehabilitation engineering.
- To develop a general understanding how do assistive technologies help individuals with disabilities, and support participation and independence throughout the lifespan.
- To be able to describe the principles of assistive technology device provision.
- To obtain a general understanding of the current state of development of assistive technology devices and robotic assistive technology in different disease conditions.
- To acquire a general understanding of the benefits and limitations of the existing assistive and robotic technologies.
- To be able to describe the importance of assistive technologies and environment modifications in the lives of individuals with disabilities in terms of accessibility and affordability.

Introduction

Throughout history, many inventions and innovations in the medical field have been inspired by the need to improve or replace an existing body part that has a deficient or a lost function due to different conditions. Alexander Graham Bell's mother lost her hearing when he was twelve years old. This tragic event encouraged Bell to become interested in acoustics, and later led to the development of telegraph and telephone (BELL 2018). During the two World Wars, many soldiers lost their limbs. As a consequence, prosthetic technologies have made significant strides, therefore benefited people with traumatic and non-traumatic amputations. Louis Braille became completely blind by age five due to a childhood accident. He worked diligently to develop a code, dots and dashes impressed into thick paper, for the French alphabet for night writing and later musical notations. Eventually in 1837, he invented the first small binary writing and reading system, called Braille, which blind people still use today (LOUIS BRAILLE 2018).

Presently in the United States, 57 million people live with either a physical or mental disability according to the Center for Disease Control (2018). More than 20 million families have at least one family member who is disabled (Service and Inclusion 2018). Moreover, 76 million baby boomers are aging (US Census Bureau Report 2014). These facts have a significant impact on healthcare needs and cost, employment, education, recreation, and caregiver burden. Thanks to decades of courageous and tenacious advocacy for people with disabilities, our awareness about them, and particularly their needs for rehabilitation services has improved significantly.

The medical rehabilitation service covers a wide range of conditions, including, but not limited to, mobility, cognition, communication, and sensory impairments such as visual and hearing. The primary goals of rehabilitation are to restore, improve, or maintain function. By doing so, rehabilitation services aim to maximize individuals' independence, improve quality of life, preserve self-esteem and dignity, and ensure that people live productive lives regardless of their physical limitations.

The concept of disability has evolved over the years. It was originally defined mainly as a result of a medical or physical condition(s) or impairment(s), leading to limitation of function or activities, hence the term disability. But disability, which when taken in context, could result in disadvantages and handicap in real-life situations (WHO 1980). In 1990, Mike Oliver, a British academic and disability advocate, developed a social model that widened the viewpoint of disability to include participation barriers and their elimination. In 2001, the International Classification of Functioning, Disability and Health (ICF) integrated the medical and social models of disability into a biopsychosocial model which was later adopted by the WHO (WHO 2001). The ICF now regards disability as multifactorial and dynamic interactions between body functions and structures, activities and participation influenced by persons' conditions, environment, and personal factors. This conceptual change has caused a paradigm shift for people with disabilities and society at large, from *impossible* to *I'm-possible*. The improvement and utilization of rehabilitation services and assistive technologies have become the cornerstone of these changes. They have made the intuitive assumption, such as, "he can't read, because he is blind; or he can't run, because he lost his leg," no longer a definitive or even acceptable status quo. Therefore, we must not consider disability as a simple vertical (i.e., personal), unchangeable, and intrinsic process, but as a horizontal, modifiable, and environment-dependent course. It is our responsibility to enable people with *disabilities* to become *able* and to live a fulfilling life. To achieve this, we must involve families, schools, work places, and communities, in addition to medicine.

The focus of this chapter is the application of advanced assistive and robotic technologies, and artificial intelligence (AI) in the field of rehabilitation. It provides introductions to the concepts of disabilities, assistive technologies (AT), assistive technology devices (ATDs), rehabilitation technologies and engineering, rehabilitation services, AT services, robotic assistive devices, assistive robotic manipulators (ARMs), and several examples of their applications (such as limb prostheses, smart wheelchairs, exoskeletons, environmental control units, gadgets for blind people, and smart electronics, etc.).

Assistive Technologies

Advanced assistive and robotic technologies have emerged as an integral part of our lives, regardless of whether people have conventionally defined disabilities or not. For instance, the use of household robot cleaners, global positioning system (GPS), iPhone Apps, self-driving cars, or socially interactive pets, etc., has played increasingly important roles in our daily activities. While many of these provide convenience, or can be considered luxuries, they are necessities for certain individuals to perform basic as well as instrumental activities of daily living (IADLs). The line between convenience and mandatory requirement can be indistinguishable at times.

The Assistive Technology Act of 2004, (P.L. 108–364) defines an (ATD) as "any item, piece of equipment, or product system, whether acquired commercially, modified, or customized, that is used to increase, maintain, or improve functional capabilities of individuals with disabilities" (GPO 2004).

WHO identified six principles for AT provision, that were described at the Convention on the Rights of Persons with Disabilities (Encarnação and Cook 2017). These six principles are:

- Acceptability (i.e., efficiency, reliability, simplicity, safety, and aesthetics)
- Accessibility
- Adaptability (i.e., meeting the needs)
- Affordability
- Availability
- Attribute (i.e., quality)

There are many ways to classify ATDs. One way is to divide them into two major categories: low level ATDs (such as a cane, a walker, a portable ramp, or a transfer bench, etc.), versus high level ATDs. The robotic assistive technologies and the application of AI belong to the high level ATDs. ATDs can also be classified more comprehensively into ten domains: (1) architectural; (2) sensory aids; (3) computer software; (4) environmental control units; (5) personal care items; (6) prosthetics and orthotics; (7) personal mobility; (8) modified furniture; (9) adaptive sports; and (10) services.

Traditional *rehabilitation service* utilizes four conventional strategies: (1) training individuals with disability to achieve certain functional capacities; (2) providing environment modification to reduce activity requirements; (3) relying on assistive technology devices to perform required tasks; and (4) receiving assistance from a caregiver or helper.

As the need for advanced technology devices has increased in the field of rehabilitation, the *assistive technology service* has emerged as an important field in rehabilitation. Assistive technology services require in-depth understanding of both technology and of the human with disability. The assistive technology service gathers the expertise from a team of professionals, including rehabilitation engineers, physical therapists, occupational therapists, speech language pathologists, recreational therapists, physicians, and/ or prosthetists. The team works together to provide (1) assessment of the individual's needs of ATDs; (2) acquisition of the devices; (3) selecting, designing, fitting, customizing, adapting, applying, maintaining, repairing or replacing them; and (4) training of health care providers, individuals with disabilities, and their caregivers.

Rehabilitation Technology and Engineering

Technologies are the scientific facts and principles that engineers use to make tools and products. The National Institute of Biomedical Imaging and Bioengineering describes Rehabilitation Engineering as "the use of engineering principles to (1) develop technological solutions and devices to assist individuals with disabilities and (2) aid the recovery of physical and cognitive functions lost because of disease or injury. Rehabilitation engineers systematically apply engineering sciences to design, develop, adapt, test, evaluate, apply, and distribute technological solutions to problems confronted by individuals with disabilities in functional areas, such as mobility, communications, hearing, vision, and cognition; and in activities associated with employment, independent living, education, and integration into the community."

The Rehabilitation Act of 1973 defines "Rehabilitation Technology" as "the systematic application of technologies, engineering methodologies, or scientific principles to meet the needs of and address the barriers confronted by individuals with disabilities in areas which include education, rehabilitation, employment, transportation, independent living, and recreation. The term includes rehabilitation engineering, assistive technology devices, and assistive technology services (Rehabilitation Act of 1973, P.L. 93–112 (19))." (EEOC 1973).

Robotic technologies are part of ATDs. The International Organization for Standardization (ISO) defines a robot as an "actuated mechanism programmable in two or more axes with a degree of autonomy, moving within its environment to perform intended tasks (ISO 2012)." A robot comprises a control system and an interface with the environment and the user. Robotic assistive devices are used increasingly to improve the independence and quality of life of people with disabilities.

Human Robot Interface (HRI)

The HRI is the foremost factor to be considered when designing any ATDs. To design a person-centered HRI, a multidisciplinary approach is essential. It requires expertise from a variety of professional disciplines, including computer science, engineering rehabilitation, psychology, behavioral and social science, even anthropology, philosophy, and ethics. There are six main HRI principles in designing, implementing, and evaluating robotic systems. The descriptions of these principles are depicted below:

Level of Autonomy

There are ten levels of autonomy, ranging from completely controlled by human to totally independent of human input. These levels reflect a robot's intelligence, learning capabilities, reliability, and the ability to communicate with the user. The level of autonomy of choices is affected by users' acceptance, situational awareness, trust, and ease of use. Autonomy is very important in physical HRI in wheelchair use, assistive manipulators, and prosthetics.

HRI Interface

The HRI interface is a bidirectional interaction between a human user and a robot. The most commonly used four modes of interaction are: auditory, tactile, visual, and brain-computer interface (BCI). The selection of the mode depends on users' motor function, cognitive and mental status, and situational awareness, as well as user preferences and social norms.

Interactive Structure

The interactive structure can be either a single robot interacting with one or multiple users (e.g., classroom or senior center setting), or multiple robots interacting with one or multiple users (e.g., game playing). The role and number of robots involved are factors to consider when implementing a structural design.

Table 1. Definitions for Levels of Autonomous Decision Making.

Level	Decision for Actions	Implement Actions
1	Human makes all the decisions	Robot
2	Robot helps	Robot
3	Robot helps and suggests an action which the human need not to follow.	Robot
4	Robot selects an action, and the human may or may not do it.	Robot
5	Robot selects an action, but waits for the human to approve.	Robot
6	Robot selects an action and gives the human a fixed time to stop it.	Robot
7	Robot	Robot implements the task and informs the human its action.
8	Robot	Robot implements the task and informs the human only if being asked.
9	Robot	Robot implements, and informs the human if it feels the need to.
10	Robot	Robot

Adaptation, Training, and Learning

Adaptation, training and learning are bidirectional, which require both human and robot to adapt and engage. This is widely used in cognitive and communication training as well as social activities. Machine learning and AI application in robotic assistive technology play increasingly important roles to make ATDs more adaptable to individual's needs.

Aesthetics

The physical appearances of robots can depend on the preferences of distinct user groups, including but not limited to age, gender, ethnicity, and spiritual believes, etc.

Length of Exposure

This is particularly relevant for social assistance robots. The robot utilization rate can be initially high due to its novelty, but tend to decline as the users' interest lessens.

Assistive Robotic Manipulators (ARMs)

An ARM is any robotic arm that is not attached to user's body, with a minimum of six degree of freedom (DoF) and a gripper. It interacts with and assists humans with disabilities, such as people with amyotrophic lateral sclerosis (ALS, a disease affected the famed British theoretical physicist Stephen Hawking), muscular dystrophy, spinal muscular atrophy, high level spinal cord injuries (SCI), stroke, and individuals with spasticity or strong hand tremors.

The first principle of ARMs' design is safety. One should always keep in mind regarding the need of the end user, the place where the robot will be used, the desired speed and force of the robot movement by the user. These essential questions must be answered to ensure a safe robot and human interaction.

The second principle of ARMs' design is that the human must have the necessary control to move the robot in the intended orientation to accomplish various tasks. ARMs should be able to grasp and handle objects needed to perform most of the basic and instrumental activities of daily living, such as self-feeding and drinking, dressing, bathing, personal hygiene, certain housework, meal preparation, shopping, and phone use, etc. An ARM can also perform certain light office work, such as taking a book from a shelf, removing paper from the printer, opening and closing doors, or starting a computer. At the social and recreational level, an ARM can assist a person in playing cards, dining at a restaurant, or fishing.

Technical specifications of ARMs can be complex. The following aspects should be considered when determining if an ARM is indicated for individual use:

- Lifting capabilities (heavy versus light)
- Accuracy (fine versus gross motor control)
- Speed of motion
- Gripper (size and sturdiness of the objects)
- Complex tasks (such as toileting)
- Two handed tasks
- Environment

Depending on the individual's needs, many user interfaces and control systems are available, including joysticks, keypads, touch screens, personal computer control, microlight switches, eye-tracking, voice control, brain- or body-robot interface, and or a combination of them (Krishnaswamy and Oates 2015).

Brain Computer Interface

Brain computer interface enables severely impaired individuals, such as people with advanced ALS who are not able to use tactile and voice modes of interface, to communicate and access information. This type of interface involves implantation of small electrodes (sensors) into the brain. The sensors pick up the electric impulses from neurons. The signals are translated into commands and transferred to a robotic arm (Schiatti et al. 2017). This allows the individual to pick up a cup of water, feed oneself independently, or move a dot on the screen without moving the limbs. New advancements include improved wireless power, lower power integrated circuits (which generate less heat), better mathematics decoding systems, and improved sensitivity nanoscience electrodes (measures 10,000 neurons).

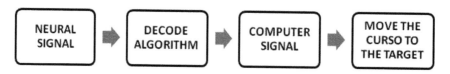

The above process, so called "read out", illustrates how does the BCI work. Future improvement of the BCI is aimed at not only to "read out" from the command (motor) area of the brain to move the arm, but also to sense the object and provide "feedback" back to the receiving (sensory) area of the brain. This is done by placing electrodes in the sensory territory of the brain. These electrodes are connected to a robotic arm with tactile sensors through wires. The user can tell which finger touches the object without looking. Based on the user perspectives, ongoing researches have been developed in effort to improve sensory feedback systems (visual and force), impedance control, gripper designs, and increased DoF (Brose et al. 2010, Encarnação and Cook 2017). Non-invasive EEG recording of visual evoked potentials has been studied as an alternative method of BCI (Spüler 2017, Chen et al. 2015). Ongoing researches have been aimed at to improve its lower signal-to-noise-ratio, time-consuming application, reliability and speed of output.

Robotic Prostheses for people with limb loss

In contrast with other assistive robots, robotic limb prostheses have a direct human body interface, and their motions are initiated and modulated by users, e.g., muscle movement or BCI. The intrinsic factors of an individual with limb loss, such as length of amputation, level of amputation (below or above elbow or knee), type of amputation (upper or lower limb), and desired activity levels, are essential to determine the types of prostheses needed.

The goal of a robotic prosthesis is to replace a lost limb in form and in function. In recent years, tremendous amount of progress has been made in prosthetic device configuration, materials used, and cosmetic appearance (skin color, texture, etc.). The invention of newer multi-grip powered hands with anthropomorphic form and several grasp patterns are examples of these advancements. Nonetheless, despite all of the progress, at present, there is still a trade-off between the exterior appearance and functional capacities of a prosthesis. The functionality of robotic limb prostheses is intricate and demanding, especially for an upper limb prosthesis because of the diverse range of activities performed by hands. A prosthetic hand's function varies widely from delicate movements of dressing (including buttoning, tying shoes, or putting on socks), to holding a glass or paper cup without crashing or slipping, to opening a jar of pickles, throwing a baseball, or carrying a gallon of milk. Our human hands are capable of performing not only tasks of various ranges, but also movements of multi-directions, e.g., bending and rotation occur simultaneously (flexion, extension, supination or pronation). It is not surprising that the acceptance and utilization rate of upper extremity prostheses remains low due to the inconvenience of donning and doffing, inadequate dexterity, and lack of reliability and sensory feedback (Resnik 2011). To improve prosthetic hand dexterity and functionality, some upper-limb myo-electric prosthetic feedback systems have been developed such as the add-on feedback system (Fallahian et al. 2017, Markovic et al. 2018) and the build-in vibrotactile system (Rosenbaum-Chou et al. 2016). However, generally speaking, the more complex the design, the heavier and less durable the prosthesis tends to become, thus challenging the future improvement of robotic upper-limb prostheses.

Compared to upper limb prostheses, lower limb prostheses are more readily acceptable because of the relatively simpler movement patterns required for ambulation and human's natural desire and need to stand and walk. Over the years, there have been several generations of microprocessor knees which provide better balance, prevent falls, and normalize gait patterns (Kaufman et al. 2007). Powered robotic lower limb prostheses add the ability to generate movements actively, thereby assist in activities, such as standing up and climbing stairs. Powered lower limb prostheses can reduce energy spent during activities and improve the biomechanics of gaits.

Unfortunately, similar to upper limb prostheses, lower limb prostheses also have limitations and are not yet able to completely satisfy the ideal requirements of negotiating different terrains (such as hills or beaches), curbs, incline, decline, and stairs; at various cadences; changing directions (forward, backward, or turns); and performing diverse types of activities (running, swimming, or squatting, etc.) as naturally as one desires. Highsmith and colleagues (Highsmith et al. 2016) recently studied the functional performance differences between two kinds of microprocessor knee users and non-amputees. They demonstrated that subjects with transfemoral amputations using microprocessor knees (Genium or C leg) did not equal or surpass non-amputees in any gait functional domains (including pot carry one meter, don/doff jacket, vertical reach, picking up scarves from floor, floor sweep, laundry, stairs ascend/descend, carry groceries 70 meters, and six-minute walk).

One of the major challenges of prosthetic functionality is the delay between the intention and the actual action time. Recent reports suggest the ideal delay between the intent of movement to the motion execution can be as short as 100–125 ms (Peerdeman et al. 2011). Researchers continue to make advances in design and function of prostheses to better simulate normal limb movements and meet user objectives.

Prosthetic control relies on the acquisition of signals by mechanical trigger, electrical recording, or a combination of the two (Fougner et al. 2012). The robotic prosthesis uses electromyographic (EMG) recording from the residual limb muscles to gather the user's intent. This switch process, called myoelectric control, has been used clinically since the late 1970's (Childress 1985, Micera et al. 2010, Oskoei and Hu 2007, Parker et al. 2006, Scheme and Englehart 2011). Signals generated by the user's contraction of certain groups of muscles can be recorded by electrodes located inside of the socket, which are amplified, and then trigger the intended movement of the prosthesis.

Compared to the traditional mechanically controlled prosthesis, a myoelectric controlled prosthesis is heavier, costly, and less durable. However, it provides wider range of motion, functional range, and is more cosmetic and easier to don and doff without the need for a harness. As mentioned above, this type of device has its limitations. First, it depends on the level and length of the amputations (i.e., the amount of available muscles to provide signals). Second, the stability of the socket fitting affects signal collection precision and can be unreliable in detecting a user's commands, especially fine motor control. Lastly, the delay in response time remains long.

A recent commercially available prosthesis control, called Complete Control, is based on machine learning principle and pattern recognition. This prosthetic control system records and learns myoelectric patterns of movements of a particular user, then automatically generates these movements once a pattern is identified. This system increases the intelligence, adaptability and situational awareness of the prosthesis (Castellini et al. 2014), however, it is still unable to control more than one, or at the most two, motions at the same time. More research is needed to improve robotic hand's orientation and intuitiveness, including using better orientation control algorithm in the robotic arm (Vu et al. 2017).

To improve the precision of myoelectric recording, the Alfred Mann Foundation developed and deployed implantable wireless EMG electrodes (Merrill et al. 2011, Pasquina et al. 2015). These Implantable Myoelectric Sensors (IMES[R]) are small cylindrical devices, 2.5 by 16 mm, that sample EMG signals via multi-channels. The IMES system is durable, accurate, and provides improved usability and functionality. Other implantable signal recording systems, such as Longitudinal Intrafasicular Electrodes (LIFEs) (Rossini et al. 2010), Transverse Intrafasicular Electrodes (TIMEs) (Raspopovic et al. 2014), the Utah Slant Array, and the Flat Interface Electrode (FINE), can be inserted into the peripheral nerves and nerve cuff. To bypass poor reliability and unsatisfying long term stability issues of implantable devices, Ortiz-Catalan and colleagues (Ortiz-Catalan et al. 2014) pioneered a percutaneous osseointegrated (bone-anchored) interface system that allows a permanent and bidirectional communication of the prosthesis with user's body. The osseointegrated electrodes were able to provided more precise and reliable control than surface electrodes. The stability of myoelectric recording serves as a foundation for pattern recognition and sensory feedback, therefore more closely mimicking natural and intuitive limb movements.

McFarland et al. attempted to implant electrodes into the areas of the spinal cord or the brain. By doing so, electrical signals can be detected and converted into digital commands, and drive the motors in a bionic hand prosthesis. This BCI technology has the potential for better neural control and feedback signals, and ultimately improved hand function. More recently, another technique was developed by Marasco and colleagues (Marasco et al. 2018), called Kinesthesia, which uses vibration sense from muscles and joints to provide feedback to the brain. This technique demonstrated improved proprioception sense, and movement accuracy of the elbow and hand.

Other investigations have been done to improve bidirectional control and feedback, such as surgical reconstruction of bones, muscles and nerves, and transfers of original hand and wrist nerves to new muscle sites, creating biological amplifiers of neural signals for myoelectric control. This procedure is called *targeted motor and sensory reinnervation* (Young et al. 2013). It has shown to provide better motor control, and even has the potential for sensation restoration (Hebert et al. 2014, Zuo et al. 2018). In addition, some novel non-invasive recording techniques such as ultrasound, high density EMG arrays, topographic force-mapping inside of the socket, accelerated measurement, and mechanomyography (measurement of muscle movement), have demonstrated promising results for better control systems (Radman et al. 2016).

Osseointegration refers to the direct structural connection between the residual limb bone and the prosthetic metal implant. This technique has been accepted in Europe (Hagberg and Branemark 2009) and Australia since 1990 (Al Muderis et al. 2016), and was recently approved in the United States as an investigational procedure. The technique eliminates the need for a socket and provides better functional leverage of the prosthetic limb. A titanium fixture is placed in the residual limb bone and a replaceable titanium abutment extends through a skin opening. The abutment end serves as the mounting point for the prosthetic limb. The most significant complication with this procedure is recurrent infection, which can be as high as 34–55%, and the surgical survival rate is 92%–96% (Al Muderis et al. 2016, Branemark 2014). This technique offers greater range of motion and control without socket fitting complications. Patients reports improved mobility, function, prosthetic utilization, quality of life and tactile proprioception feedback (Al Muderis et al. 2016, Branemark 2014, Jönsson et al. 2011, Lundberg et al. 2011, van de Meent et al. 2013).

Exoskeletons for Mobility and Manipulation

Mobility typically implies locomotive function, while manipulation includes performance of activities requiring the use of upper limbs and hands, such as activities of daily living.

Exoskeletons have established as an emerging sector of assistive technology which enables people with paralyzed limb(s) to restore function (Brewster 2016). These devices are powered robotic orthoses, which are wearable and portable, and which operate autonomously. Lower limb exoskeletons (LLEs) focus on standing and walking abilities, while upper extremity exoskeletons (ULEs) assist in manipulation. These devices should be distinguished from non-portable exoskeletons, which are robotic manipulators that substitute limb movements, rather than assist.

Lower Limb Exoskeletons

Below are some examples and characteristics of the lower limb exoskeletons.

Table 2. Lower limb exoskeleton characteristics.

Name & Company	Movement supported	Target users	Weight	Walking Speed	Control	Battery Life (Hours)	Cost*
ReWalk ^{R} by ReWalk Robotics	Sit to stand Stand to sit Walking Stairs	Personal use C7-L5 SCI Rehab setting SCI T4 to L5	29.9 kg	Max 0.6 m/s	COG	8	$70,000–$85,000
Ekso ^{TM} GT by Ekso Bionics	Sit to stand Stand to sit Walking	Complete SCI C7 or below Any level incomplete SCI Stroke	20.4 kg	Max 0.89 m/s	COG	4	$100,000
HAL^{R}by CYBERDYNE	Sit to stand Stand to sit Walking Stairs	SCI Stroke	15 kg	Min 0.11 m/s	COG/ EMG	2.5	$14,000–$19,000
Indego^{R}by Parker Hannifin	Sit to stand Stand to sit Walking	Personal use C7-L5 SCI Rehab setting SCI T4 to L5	12.3 kg	Mean 0.22 m/s	COG	4	$80,000
Rex^{R}by Rex Bionics	Sit to stand Stand to sit Walking Stairs	SCI up to C4/5 level Muscular dystrophy Multiple sclerosis Post-polio syndrome	38 kg	0.08 m/s	Joystick	2	$150,000
Phoenix by SuitX			12.3 kg	1.1 m/s	Buttons on the crutches	8	$40,000

*Costs are subject to changes.

All LLEs have hip and knee joints, and one of several types of ankles (spring assisted, variable stiffness, or fixed). When applying the six principles of ATDs, i.e., acceptability (especially safety, portability, and reliability), accessibility, adaptability, affordability, and quality, one can understand why independent applications of these devices have been limited. Most LLEs are FDA-approved for rehabilitation use as training tools; only Re-Walk and Indego are approved for home use for patients with SCI level at T7 or below with the supervision of a trained individual. Since most individuals with these levels of injuries are independent at light-weight wheelchair level, the use of the LLEs has not been efficient enough for functional use. Limited battery life span is also a barrier to make donning and doffing the device worthwhile. Because of these concerns, the technology readiness level of the LLEs remains low for most except for Rex[R] at this time (level 1 out of 9; level 9 as the most mature).

Phoenix, made by SuitX', consists of small motors attached to standard orthotics; wearers control the movement of each leg by pushing buttons integrated into a pair of crutches. This device is made of carbon-fiber material, one of the lightest and cheapest LLEs. It can also adjust to different heights and sizes of wearers, so it can potentially be used by children or patients with hemiplegia. An App is included in the device and can be used to track patient's walking data.

Upper Limb Exoskeletons

Similar to LLEs, several upper limb exoskeletons (ULEs) have been developed. These devices are multi-joint orthoses powered by batteries, to create two to three joint motions. Most of ULEs have reaching and grasping functions only and weight about 2–3 pounds. Their control systems vary from EMG sensors to eye tracking or BCI. These devices have been used in individuals with stroke, TBI, SCI, ALS, multiple sclerosis, or ataxia.

The MyoPro (Motion-G and MyoPro2) is one of the examples of ULE. It is commercially available, portable elbow-wrist-hand orthoses. It assists individuals with incomplete paralysis to perform elbow flexion, feeding, reaching, and lifting activities. The device requires a certain level of muscle contraction to create EMG detectable signals to initiate a movement (Stein et al. 2007). Mundus (Multimodal Neuroprosthesis for Daily Upper Limb Support) was a research project funded by the European Commission. It is a three-joint device, with shoulder, elbow, and wrist controls. It has two DoF at the shoulder and one DoF at the elbow (Pedrocchi et al. 2013). Three control system prototypes were implemented: EMG sensors, eye-tracking, or EEG-based BCI. WOTAS (wearable orthosis for tremor assessment and suppression) is another portable, powered orthosis aiming to suppress tremor by sensing joint angle and movement velocities (Rocon et al. 2007).

The utilization rate of ULEs is low since these devices do not meet the normal upper extremity functional requirements, meaning to have thirty DoF, with seven at the shoulder level, and complex hand joint movements. In addition, for patients who still have one sound upper limb, e.g., stroke patients, they tend to compensate more efficiently without the use of an orthosis.

Similar to other types of prosthesis or orthosis, current ULEs are not able to recognize the HMI at an acceptable speed to produce the most natural gait or upper limb movements. Different research approaches have been developed to solve this challenge. For instance, Long and colleagues has showed an online sparse Gaussian Process algorithm which is able to learn HMI in real time and can be applied in other exoskeleton systems (Long et al. 2017). Lu and colleagues demonstrated that advanced myoelectric pattern recognition technique has the potential to improve ULE control accuracy and efficiency (Lu et al. 2017).

Smart Wheelchairs

Wheelchairs are another commonly used ATDs for many people with physical disabilities. They are one of the first pieces of adaptive equipment prescribed for individuals with spinal cord injuries. They help to provide a sense of freedom and allow users to go to places that are accessible to wheelchairs. Wheelchairs enable people with disability to interact with their living and social surroundings including going to work, grocery shopping, and participating in social and community events (Bourret et al. 2002). There is a wide variety of commercially available wheelchairs that feature unique styles, weights, and performance capabilities. The choice of a wheelchair is largely dependent on the user's needs, and his or her physical and cognitive conditions. Moreover, one must address the following questions: where the chair will be used, on what type of terrain, where the wheelchair will be stored, how will the user transport it, is the user involved in sports and recreational activities, can the user propel the chair independently, and what other function does the user want/need the chair to have?

Primary care doctors should be knowledgeable about the conditions of people with disability, including their prognosis and whether they are progressive, and feel comfortable making referrals to physical medicine and rehabilitation specialists. Patients with disabilities often have complex physical, medical and psychosocial needs; therefore, a power wheelchair assessment is often treated by a team that includes a physical therapist or an occupational therapist, a physician, and often a rehabilitation engineer.

Wheelchair Design

Most long term wheelchair users are at risk for developing repetitive trauma injuries. Over the years, wheelchair design has improved in both weight (by using lighter materials, such as aluminum and titanium instead of steel, and carbon fiber) and ergonomics. Manual, power electric, and several hybrid-type wheelchairs are commercially available. The hybrid chairs are standard manual chairs equipped with a powerful motor designed to give the chair a boost when needed, such as when going up hills. The most important advance for electric-powered wheelchairs is the greater use of microprocessors and computer technology. Computer technology allows wheelchair users to adjust the acceleration, maximum speed, and rate of turning. This has made wheelchairs safer, functionally more reliable and precise, and more user friendly. It has improved independent mobility for people with severe impairments. A recent survey regarding clinical providers' opinions on the future needs of ATD research (Dicianno et al. 2018) showed more improvements in advanced wheelchair design are needed, particularly in navigation in a variety of environments and travel, better power systems, application of the assistive robotic technology and AI to assist caregivers, transfers, and aid ambulation; as well as development of new smart devices. In the next section, we will focus on the integration of robotics technologies into electric powered wheelchairs.

Components of a Smart Wheelchair

The components of a smart wheelchair include sensors for identifying features in the environment, input from users, feedback systems for the user, and a control algorithm that coordinates all information and decides the system's behavior.

Sensors

The sensor component functions to help the wheelchair user to avoid obstacles. They should be accurate, inexpensive, small, lightweight, and energy efficient (loPresti et al. 2002).

Inputs

The input components vary from traditional joystick, pneumatic switches, voice recognition (Pineau et al. 2011, Simpson and Levine 2002), to eye tracking (Rofer et al. 2009) and brain-computer interfaces (Carlson and Del R Millán 2013, Mandel et al. 2009).

Feedback

Feedback modalities are available depending on the users' ability to perceive them. They can be auditory, visual (using a flashing light to warn dangerous proximity to an object), vibratory sensation, or touch screen (which provides information about user's destination, GPS, and scheduled events, etc.) (Viswanathan et al. 2016, Wang et al. 2011).

Control algorithms

Control algorithms can be automatic, and therefore not affected by the user's input, or semiautomatic, where the user can override the intelligent system. As technology advances in self-driving cars, collision avoidance systems, and GPS, the affordability and accessibility of automatic or semiautomatic wheelchair has much improved (IHS 2014, Mosquet et al. 2015, Davies 2015).

Using robotic technology, several companies and academic organizations have created "powerchairs" that can climb stairs and balance on two wheels, to allow people to stand to reach higher objects or to have eye-level conversations. For recreational use, some specialty chairs, such as racing chairs, are designed for maximum speed, and others can be driven on sand and gravel. One of the unsolved limitations of robotic power wheelchairs is the electromagnetic interference between devices (such as hearing aids or cell phones). More research needs to be done to resolve this issue.

Driving

Today, driving is an integral part of our lives for personal, social, recreational, and vocational use. While driving in most cases requires the use of hands and feet, for people with limited use of their limbs, there are adaptive vehicles and driving equipment which enables them to manage their own vehicles. Modifications are varied and depend upon the needs of the individual. There are hand controls that allow braking and acceleration for those who have lost lower body control; easy-touch pads for ignition and shifting; and joysticks and spinner knobs for those who have limited use of their hands.

Self-driving cars are tools, which have the potential to revolutionize the way blind people, and people with limb loss or limb paralysis navigate in their communities and travel long distances (IHS 2014). The application of AI and machine learning in autonomous vehicle technology can provide better safety, communication, and eventually independent mobility for people with disabilities.

Blind Rehabilitation

Navigation through familiar and unfamiliar locations is a hugely challenging task for people with blindness or advanced low vision. Individuals with vision loss have traditionally used white canes and service guide dogs to walk outside in the community. In recent years, different versions of commercially available GPSs have improved accessibility for blind people to travel, participate in outdoor activities, and enjoy social and community events.

Depending on the needs of person with a visual impairment, the selection of a GPS should consider the following variables:

- User interface
- Multifunctionality
- Portability
- Price.

User interface includes information input and output. There are four types of inputs currently, including Braille keyboard, QWERTY keyboard, mobile phone keypad, and voice; and two output means: Braille and voice. Voice control provides convenience, but sacrifices privacy in certain public spaces and can be less effective in a noisy environment. Many Personal Digital Assistants (PDAs) have add-on GPS software, but they are less portable. One of the benefits of PDAs is the inclusion of the Braille display, which helps blind people obtain names of the street addresses and points of interest. GPS systems on iPhone or Android cost only $5.00, whereas on other platforms, such as BrailleNote, VoiceNote, BrailleSense, VoiceSense, cost range from $1388.00 to $1977.00. Of course, the ultimate questions are what does the user need and how often do they use the device.

Talking Signs[R] remote infrared audible signage (RIAS) is a wayfinding device with a handheld receiver and infrared transmitters that can scan and inform the user about his or her surroundings, including location of elevators, bus stops, information windows, stairs, intersections, etc. Its beam width is adjustable from six degrees to 360 degrees. The messages from the device are simple and text message-like.

Computer vision (or machine vision) is a form of AI, which uses a computer as the individual's eyes. It uses several existing algorithms, including optical character recognition (OCR); object recognition (e.g., facial recognition) and 3-D reconstruction of objects and shapes as well as video analysis and event recognition. Computer vision aims to provide access for blind people with descriptive visual information. This technique is limited by its inability to read less standard signs or prints, or to respond to specific needs as simple as, "Where can I sit down"? These challenges remind rehabilitation engineers to be user-centered when designing assistive devices.

There are several types of wearable live streaming devices available for people with blindness or low vision. Artificial vision glasses comprised of a lightweight smart camera technology that reads text aloud from any surfaces and recognizes faces and products, are commercially available. Some of them can read either digital or printed materials, such as newspapers, books, restaurant menus, and signs. They can provide real-time face recognition, and identify products, credit cards, etc. Apps such as, AIPoly, and SeeAI, etc., are handheld devices that use AI to identify objects, street signs, and scenery for people who are blind or visually impaired. There are also Apps and systems of way finding maps for blind travelers, which provide both tactile and narrative maps of colleges, hotels, transit systems and other places. They offer step-by-step guidance for indoor and outdoor walking directions (including public transportation, buses, trains, traffic lights, crosswalks, and gasoline stations) and wheelchair accessibility information. These devices serve as visual interpreters, and enable blind people or people with advanced low vision to navigate more independently in the community and perform a wide range of basic and instrumental activities of daily living. Screen-readers have also been developed by many companies and are important tools to provide web and internet accessibility for people with vision impairments.

Researchers have developed white canes for blind people that are enhanced with computer vision and vibration feedback. These canes use advanced image processing to map the structure of a room, identify important features (e.g., doors, stairs, and obstacles), and create a walking plan to guide the user towards the desired destination. It provides feedback in the form of either voice or vibrations through the handle. The use of vibration as a means of feedback has not been well accepted because adding vibration to a constantly moving cane could be confusing at times.

As a result of advances in our understanding of our brain and technology, we can now place sensors and small circuits into the brain, to "write in" visual information. The system consists of a retinal implant, a pair of glasses containing a small camera, and a cell phone-sized device to convert the camera's images into digital signals. The system provides electrical stimuli to the retina, which sends nerve impulses through the optic nerve to the occipital lobe, an area of the brain which normally receives visual information. This procedure, in its early stage of development, allows people with profound visual deficit to see white and black patches and general shapes, and to walk outside (da Cruz et al. 2016).

Environmental Control Units (ECUs)

Most survivors of SCI, or ALS, who have lost the use of their limbs, have the desire to be independent in everyday activities and live a dignified life. This means being able to control the things that they used to be able to do, such as adjust the thermostat, turn on the TV, computers, DVD player, room lights; open and close doors, and raise the head of the bed.

There are a wide variety of environmental control technologies designed to make home life easier for individuals living with SCI or ALS. The ECUs give the person more control over their living space, allowing them to have more privacy and independence.

Many of these technologies are "hands-free", allowing an individual to use one of the following control systems: puffs of air, voice, head movement, blinking, or eye tracking. The user's home can generally be set to be operated with one remote control unit (also see the Smart Home Assistants section below).

Smart Home Assistants

This domain is one of the fastest growing aspects of home services due to the emerging terminology of "internet of things (IoT)". Per Wikipedia definition, "the Internet of Things is the network of physical devices, vehicles, home appliances, and other items embedded with electronics, software, sensors, actuators, and connectivity which enables these objects to connect and exchange data."

At-home digital assistants like Google Home and Amazon's Alexa and Echo can make home significantly more wireless and convenient. For people who have limited use of their upper limbs or visual loss, this technology is extremely useful. People can talk with the virtual assistants and get responses appropriately. These hands-free assistive devices can make appointments, answer telephones, set timers and thermostats, tell the weather, and make grocery lists. They can also handle more sophisticated tasks like selecting and playing a song from YouTube, ordering a product online, or making movie recommendations. The person at home can see if the aide or caregiver has finished grocery shopping or even check on whether a desired item has been picked up. Family members at work can monitor what's going on at home, or who is at the front door. Other features include searching for recipes, turning on and off lights and TV, controlling home appliances, or cars, or taking pictures. It is noteworthy to mention that as technologies advance, concerns regarding privacy, security and confidentiality of these devices have heightened as well, and therefore need to be addressed accordingly.

Robotic Technologies for Seniors

The United States population is aging along with the rest of the world. Because of this, the need for and application of robotic and AI systems in assistive technology has grown rapidly, especially in the ICF domains of mobility, self-care, and social interaction (Bedaf et al. 2013, 2015). In recent years, non-robotic Information and Communication Technologies (ICT) (e.g., Skype, tele-home care system, and cell phone, etc.) have become widely accessible and affordable to seniors. Their applications have provided safe and convenient support systems to help them connected with others, thus preventing social isolation.

Robotic systems for seniors fall into two categories, physical assistive robot (PAR), and social assistive robot (SAR) (Feil-Seifer and Matarić 2005). PAR physically assists or supports people in their needs, whereas, SAR does so by socially interacting with the users. SAR can be used the same as a service dog guiding a blind person. SARs are similar to SIRs (socially interactive robots) in their interaction with users; by assisting users, they also aim to achieve measurable progress in learning and rehabilitation (Fong et al. 2003).

Social Assistive Robot

The seal-like robot *PARO* is an example of an SAR. It is covered with soft fur, can move limbs, eye lids, and head, and make sounds. It has sensors to detect sound, light, touch, temperature, and position (being held or not), and is able to learn user preferred behavior and its given name. Other research prototypes of SARs are also being developed.

A project termed "Accompany" (acronym for Acceptable RobotiCs COMPanions for AgeiNg Years) uses a sensor, a 3-D camera, and three laser scanners for navigation. The torso has four DoF and its hand has seven DoF. It aims to support cognitively unimpaired elderly to remain independent at home by serving as a caregiver in a socially acceptable manner (Duclos and Amirabdollahian 2014).

VictoryaHome and Giraff are two robotic systems that can monitor the elderly's health and safety. They include activity monitors, a fall detector, an automatic pill dispenser and a mobile telepresence robot. Using a smart phone App, the family member or health professional can monitor the user. They are examples of user- and caregiver-oriented robotic systems.

VGo is another commercially available SAR which has both handheld and remote internet controls. Its inclusion of a camera, a microphone, speakers, and a video display makes distant communication between the user, the caregiver, and healthcare team easier.

Physical Assistive Robot

An example of PAR is the mealtime assistant robot MySpoon, which provides the user with feeding assistance.

Apart from user-oriented robots, caregiver-oriented robots are also on the horizon, including RIBA and ROBEAR (Robot for Interactive Body Assistance). ROBEAR, covered with soft materials, has a friendly teddy bear appearance. Its main function is to transfer a person from bed to wheelchair or vice versa.

Cleaning Robots

Home use cleaning robot technologies are becoming more popular for people with busy lives, seniors, and people with physical disabilities. There are many types of cleaning technologies for both vacuuming and mopping. Most models work well on carpets and swerve their little dirt-munching bodies under every available surface. Newer models have sensors that detect which areas of the house the cleaning robot has tidied. Most modern models allow users set cleaning schedules, and have options to program the robot to return to its dock after cleaning. The floor-mopping robots are scrubbers which perform vacuuming and have a water reservoir to scrub clean hardwood, linoleum, or tile floors.

Robotic Systems for Children with Disabilities

Physical activities and object manipulation play a vital role in children's cognitive, perceptual, communication, and social skills (Flanagan et al. 2006, McCarty et al. 2001, Affolter 2004). Robotic systems have been developed to interact with children with disabilities due to physical limitations, and to assist them in accomplishing tasks that they would not otherwise be able to do.

Robotic systems promote the active engagement of children to learn through playing. Playing is intrinsically motivating, enjoyable, process-oriented with potential rewarding challenges, and gives a sense of being in control (Blanche 2008, Skard and Bundy 2008). The robotic augmentative systems improve children's creativity, problem-solving ability, social, cognitive, and functional performance at home and at school (Chang et al. 2010). They are also cognitive assessment tools for children with physical disabilities. The control mechanisms for these systems include keyboards, joysticks, gaze tracking, voice control, switches, etc. Robotic systems for children with disabilities are mostly flexible, easy to learn and use, portable, affordable, and can pick up and place objects (Cook et al. 2005, Poletz et al. 2010). The limitations of these are lack of sensory feedbacks, and potential decrease in motor involvement and/or function of the limbs. More research needs to be done regarding cognitive and functional outcomes with the use of these devices.

Impaired auditory function is a major barrier for children to learn, express themselves, and communicate. Some studies have demonstrated the benefits of early age, especially pre-linguistic age of cochlear implant surgery in language development and voice quality (Nicholas and Geers 2018, Wang et al. 2017).

Advanced Technologies in Rehabilitation Services and Other Medical Domains

Advanced technologies and AI have evolved not only as assistive technologies, but also in the rehabilitation services. Smart rehabilitation robots are used in mobility training for individuals with impaired motor function in upper or lower extremities, such as following a stroke or SCI.

Tangential-Normal Varying-Impedance Controller (TNVIC) is an intelligent assistive robotic system that uses robot learning from demonstration techniques (RLfD). It can follow both the therapist's instructed motion and motion impedance. Its learning ability makes task-specific robot less useful (Najafi et al. 2017).

Virtual rehabilitation uses virtual reality simulation exercises for physical and cognitive rehabilitation. These tools can be entertaining, therefore motivate patients to exercise or practice communication skills. They provide objective measures, such as range of motion or improve pronunciation accuracy, etc. The exercises can be performed at home by the patient and monitored by a therapist over the Internet (known as tele-rehabilitation), which offers convenience as well as reduced costs. They are also used in senior centers, schools, and other group settings.

The App Factory is another fast growing innovative model for development and application of new assistive and technology solutions. Using diverse disciplines of expertise, including clinical, technical, policy making, and advocacy, this approach explores ways to translate technology into flexible, adaptable, and cost-effective products for people with disabilities (Jones et al. 2017).

Although conventional assistive technology devices that we have illustrated in the previous sections emphasize predominantly on rehabilitation for people with physical disabilities. One cannot ignore the fast growing application of artificial intelligence in the innovation of pacemakers, leadless implementable cardioverter defibrillators (ICD) (Burke 2017), subcutaneous ICD, and implementable insulin pumps (Buch et al. 2018), etc. These ongoing efforts and progress will significantly advance disease management, prevent complications, and improve quality of life.

Summary

In this chapter, we briefly discussed traditional rehabilitation services and assistive technology services. With the application of robotic technologies and AI, the nature of these services has evolved at an accelerating pace. Technological advances, such as IoTs, 5G (with faster and multi-inputs and outputs capabilities), machine learning and AI will continue to transform the way we think and practice medicine. The traditional way of defining ADLs into basic (such as dressing, grooming, or eating, etc.) and instrumental ADLs (such as doing laundry and cooking, etc.) is no longer sufficient. A new category of ADLs should be considered as technological ADL (tADL).

In the past, people provided services, presently machines (ATDs) help people to do the work. In the near future, machines may replace people to provide the majority of care, and ultimately, people will in turn assist machines (ATDs).

This above trend is happening in front of our eyes. In this chapter, you can see only some of the highlights of the applications and limitations of current robotic technologies and AI in the field of rehabilitation. It is our responsibility to maintain the principle of people-centeredness in the application of AI and machine learning in ATD development, and to further improve or invent new ATDs to meet the needs of people with disabilities. A continued advocacy effort for people with disabilities will benefit the society at large.

In this new era of medicine, clinicians are at the cross-roads of human and technology interface, and we should make ourselves more knowledgeable not only about human conditions but also the state-of-the art of available assistive technologies to meet their needs.

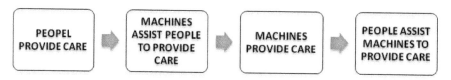

Acknowledgments

The author would like to acknowledge the academic supports of the Northern California VA Healthcare System, Martinez, CA, and the following individuals for their kind and professional prove reads: Dianne Damole-Fua (physical therapist), Amanda Dyrek (speech language therapist), Kathryn Firestone (blind rehabilitation specialist, Susan Raich (kinesiotherapist), and Randi Woodrow (physical therapist),

References

Affolter, F. 2004. From action to interaction as primary root for development. *In*: Stockman, I.J. (ed.). Movement and Action in Learning and Development. Elsevier Academic Press, San Diego, USA.

Al Muderis, M., A. Khemka, S.L. Lord, H. Van de Meent and J.P. Frölke. 2016. Safety of Osseointegrated Implants for Transfemoral Amputees: A Two-Center Prospective Cohort Study. J Bone Joint Surg Am. 98(11): 900–9. doi: 10.2106/JBJS.15.00808.

Al Muderis, M., K. Tetsworth, A. Khemka, S. Wilmot, B. Bosley, S.J. Lord et al. 2016. The Osseointegration Group of Australia Accelerated Protocol (OGAAP-1) for two-stage osseointegrated reconstruction of amputated limbs. Bone Joint J. 98-B(7): 952–60.

Al Muderis, M., W. Lu, K. Tetsworth, B. Bosley and J.J. Li. 2017. Single-stage osseointegrated reconstruction and rehabilitation of lower limb amputees: the Osseointegration Group of Australia Accelerated Protocol-2 (OGAAP-2) for a prospective cohort study. BMJ Open. 2017 Mar 22; 7(3): e013508. doi: 10.1136/bmjopen-2016-013508.

Bedaf, S., G.J. Gelderblom, D.S. Syrdal, H. Lehmann, H. Michel, D. Hewson et al. 2013. Which activities threaten independent living of elderly when becoming problematic: inspiration for meaningful service robot functionality. Disability & Rehabilitation: Assistive Technology 9: 445–452.

Bedaf S., G.J. Gelderblom and L. de Witte. 2015. Overview and categorization of robots supporting independent living of elderly people: What activities do they support and how far have they developed. Assistive Technology 27: 88–100.

[BELL] 2018. Alexander Graham Bell. Wikipedia. [https://en.wikipedia.org/wiki/Alexander_Graham_Bell] accessed May 21, 2018.

Blanche, E.I. 2008. Play in children with cerebral palsy: Doing with-not doing to. pp. 375–393. *In*: L.D. Parham and Fazio L.S. (eds.). Play in Occupational Therapy for Children. 2nd ed. Elsevier.

[BRAINYQUOTE] 2018a. Hawkins, Stephen. Brainy Quote 1942–2018. [https://www.brainyquote.com/authors] Accessed May 21, 2018.

[BRAINYQUOTE] 2018b. Franklin D. Roosevelt. Brainy Quote. 1882–1945. [https://www.brainyquote.com/authors] Accessed May 21, 2018.

Brewster, S. 2016. This $40,000 robotic exoskeleton lets the paralyzed walk. MIT Technology Review Feb. 1, 2016.

Brose, S.W., D.J. Weber, B.A. Salatin, G.G. Grindle, H. Wang, J.J. Vazquez et al. 2010. The role of assistive robotics in the lives of persons with disability. Am J Phys Med Rehabil 89: 509–521.

Buch, V., G. Varughese, M. Maruthappu. 2018. Commentary: Artificial intelligence in diabetes care. Diabet. Med. 35: 495–497.

Burke, M.C. 2017. Modular cardiovascular device communication: A step towards artificial intelligence to treat cardiovascular disease. Cardiac Rhythm News. March 854. https://cardiacrhythmnews.com/modular-cardiovascular-device-communication-a-step-towards-artificial-intelligence-to-treat-cardiovascular-disease/.

Carlson, T. and J.D.R. Millán. 2013. Brain-controlled wheelchairs: A robotic architecture. IEEE Robotics and Automation Magazine 20: 65–73.

Castellini, C., P. Artemiadis, M. Wininger, A. Ajoudani, M. Alimusaj, A. Bicchi et al. 2014. Proceedings of the first work shop on Peripheral Machine Interfaces: going beyond traditional surface electromyography. Frontiers in Neurorobotics 8: 1–17.

Center for Disease Control. 2018. https://www.cdc.gov/ncbddd/disabilityandhealth/disability-inclusion.html. Accessed Nov. 13.

Chang, C., J. Lee, P. Chao, C. Wang and G. Chen. 2010. Exploring the possibility of using humanoid robots as instructional tools for teaching a second language in primary school. Educational Technology & Society 13: 13–24.

Chen, X., Y. Wang, M. Nakanishi, X. Gao, T.P. Jung and S. Gao. 2015. High-speed spelling with a noninvasive brain–computer interface. Proceedings of the National Academy of Sciences. 112(44): E6058–E6067. pmid:26483479.

Childress, D.S. 1985. Historical aspects of powered limb prostheses. Clin. Prosthet. Orthot. 9: 2–13.

Cook, A.M., B. Bentz, N. Harbottle, C. Lynch and B. Miller. 2005. School-based use of a robotic arm system by children with disabilities. IEEE Transactions on Neural Systems and Rehabilitation Engineering 13: 452–460.

da Cruz, L., J.D. Dorn, M.S. Humayun, G. Dagnelie, J. Handa, P.-O. Barale et. al. 2016. Five-Year Safety and Performance Results from the Argus II Retinal Prosthesis System Clinical Trial, HYPERLINK "https://www.sciencedirect.com/science/journal/01616420" \o "Go to Ophthalmology on ScienceDirect" Ophthalmology 123(10): 2248–2254.

Davies, A. 2015. Self-driving cars really will revolutionize the economy. Here's how. Slate.com. March 9, 2015. [http://www.slate.com/blogs/future_tense/2015/03/09/self_driving_cars_will_revolutionize_economy_mckinsey_company_report_shows.html] accessed May 24, 2018.

Dicianno, B.E., J. Joseph, S. Eckstein, C.K. Zigler, E.J. Quinby, M.R. Schmeler et al. 2018. The future of the provision process for mobility assistive technology: a survey of providers. Disabil Rehabil Assist Technol. 20: 1–8.

Duclos, A. and F. Amirabdollahian. 2014. Deliverable 7.3. Economic Evaluation. ACOMPANY Project. [https://cordis.europa.eu/docs/projects/cnect/4/287624/080/deliverables/001-ACOMPANYD73EconomicModel.pdf] accessed May 24, 2018.

[EEOC] 1973. Equal Employment Opportunity Commission. Rehabilitation Act of 1973. P.L 93-112. [https://www.eeoc.gov/eeoc/history/35th/thelaw/rehab_act-1973.html] accessed May 19, 2018.

[EINSTEIN] 2018. Albert Einstein Quote. 1879–1955. Confluence Wellness. [http://confluencewellness.com/natural-health-care-philosophy] Accessed May 21, 2018.

Encarnação, P. and A. Cook. 2017. Robotic Assistive Technologies: Principles and Practice. CRC Press, Boca Raton, FL.

Fallahian, N., H. Saeedi, H. Mokhtarinia and F.T. Ghomshe. 2017. Sensory feedback add-on for upper-limb prostheses. Prosthet Orthot Int 41: 314–317.

Feil-Seifer, D. and M. Matarić. 2005. Defining socially assistive robotics. pp. 465–468. *In*: Proceedings of the International Conference on Rehabilitation Robotics (ICORR). Chicago, IL.

Flanagan, J.R., M.C. Bowman and R.S. Johansson. 2006. Control strategies in object manipulation tasks. Current Opinion in Neurobiology 16: 650–659.

Fong T., I. Nourbakhsh and K. Dautenhahn. 2003. A survey of socially interactive robots. Robotics and Autonomous Systems 42: 143–166.

Fougner, A., O. Stavdahl, P.J. Kyberd, Y.G. Losier and P.A. Parker. 2012. Control of upper limb prostheses: terminology and proportional myoelectric Control-A review. IEEE Trans. Neural Syst Rehabil Eng 20: 663–677.

[GPO] 2004. Government Publishing Office. The US Assistive Technology Act of 2004. Public Law 108-364, 118 STAT. 1709-1710. [https://www.gpo.gov/fdsys/pkg/STATUTE-118/pdf/STATUTE-118-Pg1707.pdf] Accessed May 21, 2018.

Hagberg, K. and R. Branemark. 2009. One hundred patients treated with osseointegrated transfemoral amputation prostheses-Rehabilitation perspective. J Rehabil Res Dev 46: 331–344.

Hebert, J.S., K. Elzinga, K.M. Chan, J. Olson and M. Morhart. 2014. Updates in targeted sensory reinnervation for upper limb amputation. Curr Surg Rep 2: 45–53.

Highsmith, M.J., J.T. Kahle, R.M. Miro, M.E. Cress, D.J. Lura, W.S. Quillen et al. 2016. Functional performance differences between the Genium and C-Leg prosthetic knees and intact knees. J Rehabil Res Dev 53: 753–766.

[IHS] 2014. Self-driving cars moving into industry's driver's seat. IHS News Release. Jan. 2, 2014. [http://news.ihsmarkit.com/press-release/automotive/self-driving-cars-moving-industrys-drivers-seat] Accessed May 21, 2018.

[ISO] 2012. International Organization for Standardization (ISO). ISO 8373: 2012. Robots and Robotic Devices—Vocabulary. [https://www.iso.org/standard/55890.html] accessed May 24, 2018.

Jones, M., J. Mueller and J. Morris. 2017. App factory: A flexible approach to rehabilitation engineering in an era of rapid technology advancement. Assist Technol 29: 85–90.

Jönsson, S., K. Caine-Winterberger and R. Brånemark. 2011. Osseointegration amputation prostheses on the upper limbs: methods, prosthetics and rehabilitation. Prosthet Orthot Int 35: 190–200.

Kaufman, K.R., J.A. Levine, R.H. Brey, B.K. Iverson, S.K. McCrady, D.J. Padgett et al. 2007. Gait and balance of transfemoral amputees using passive mechanical and microprocessor-controlled prosthetic knees. Gait and Posture 26: 489–493.

Krishnaswamy, K. and T. Oates. 2015. Robotic assistive devices for independent living. Robohub. June 8, 2015. [http://robohub.org/robotic-assistive-devices-for-independent-living/] accessed May 24, 2018.

Long, Y., Z.J. Zou, C.F. Chen, W. Dong and W.D. Wang. 2017. Online sparse Gaussian process based human motion intent learning for an electrically actuated lower extremity exoskeleton. IEEE Int Conf Rehabil Robot. Jul. 919–924. doi: 10.1109/ICORR.2017.8009366.

[LOUIS BRAILLE] 2018. Louis Braille. Wikipedia. [https://en.wikipedia.org/wiki/Louis_Braille] accessed May 21, 2018.

Lu, Z., X. Chen, X. Zhang, K.Y. Tong and P. Zhou. 2017. Real-Time Control of an Exoskeleton Hand Robot with Myoelectric Pattern Recognition. Int J Neural Syst. 27: 1750009. doi: 10.1142/S0129065717500095.

Lundberg, M., K. Hagberg and J. Bullington. 2011. My prosthesis as a part of me: a qualitative analysis of living with an osseointegrated prosthetic limb. Prosthetic and Orthotic InternationalShow all authors 35: 207–214.

Mandel, C., T. Luth, T. Laue, T. Rofer, A.I. Graser and B. Krieg-Bruckner. 2009. Navigating a smart wheelchair with a brain-computer interface interpreting steady-state visual evoked potentials. Proceedings of the 2009 IEEE/RSJ International Conference on Intelligent Robots and Systems 1118–1125.

Marasco, P.D., J.S. Hebert, J.W. Sensinger, C.E. Shell, J.S. Schofield and Z.C. Thumser. 2018. Illusory movement perception improves motor control for prosthetic hands. Science Translational Medicine 14 Mar 2018. 10(432): eaao6990 DOI: 10.1126/scitranslmed.aao6990.

Markovic, M., H. Karnal, B. Graimann, D. Farina and S. Dosen. 2017. GLIMPSE: Google Glass interface for sensory feedback in myoelectric hand prostheses. Journal of Neural Engineering 14: 036007. doi: 10.1088/1741-2552/aa620a.

Markovic, M., M.A. Schweisfurth, L.F. Engels, T. Bentz, D. Wüstefeld, D. Farina et al. 2018. The clinical relevance of advanced artificial feedback in the control of a multi-functional myoelectric prosthesis. J Neuroeng Rehabil 27. 15: 28 Doi: 10.1186/s12984-018-0371-1.

McCarty, M.E., R.K. Clifton and R.R. Collard. 2001. The beginnings of tool use by infants and toddlers. Infancy 2: 233–256.

Merrill, D.R., P.R. Troyk, R.F. Weir and D.L. Hankin. 2011. Development of an Implantable Myoelectric Sensor for Advanced Prosthesis Control. Artificial Organs 35: 249–252.

Micera, S., L. Citi, J. Rigosa, J. Carpaneto, S. Raspopovic, G. Di Pino et al. 2010. Decoding information from neural signals recorded using intraneural electrodes: Toward the development of a neurocontrolled hand prosthesis. Proc. IEEE. 98: 407–417.

Mosquet, X., T. Dauner, N. Lang, M. Rußmann, A. Mei-Pochtler, R. Agrawal et al. 2015. Revolution in the Driver's Seat: The road to autonomous vehicles. The Boston Consulting Group. April 21, 2015. [https://www.bcg.com/publications/2015/automotive-consumer-insight-revolution-drivers-seat-road-autonomous-vehicles.aspx] accessed May 24, 2018.

Najafi, M., K. Adams and M. Tavakoli. 2017. Robotic learning from demonstration of therapist's time-varying assistance to a patient in trajectory-following tasks. IEEE Int Conf Rehabil Robot. Jul. 888–894.

Nicholas, J.G. and A.E. Geers. 2018. Sensitivity of expressive linguistic domains to surgery age and audibility of speech in preschoolers with cochlear implants. Cochlear Implants Int. 19(1): 26–37. doi: 10.1080/14670100.2017.1380114.

Oliver, M. 1990. The Politics of Disablement. MacMillan Education LTD, Basingstoke, UK.

Ortiz-Catalan, M., B. Håkansson and R. Brånemark. 2014. An osseointegrated human-machine gateway for long-term sensory feedback and motor control of artificial limbs. Science Translational Medicine 6:257re6. doi: 10.1126/scitranslmed.3008933.

Oskoei, M.A. and H. Hu. 2007. Myoelectric control systems—A survey. Biomedical Signal Processing and Control J 2: 275–294.

Parker, P., K. Englehart and B. Hudgins. 2006. Myoelectric signal processing for control of powered limb prostheses. J Electromyogr Kinesiol 16: 541–548.

Pasquina, P.F., M. Evangelista, A.J. Carvalho, J. Lockhart, S. Griffin, G. Nanos et al. 2015. First-in-man demonstration of a fully implanted myoelectric sensors system to control an advanced electromechanical prosthetic hand. J Neurosci Methods 15: 85–93.

Pedrocchie, A., S. Ferrante, E. Ambrosini, M. Gandolla, C. Casllato, T. Schauer et al. 2013. MUNDUS Project: Multimodal Neuroprosthesis for Daily Upper Limb Support. Journal of Neuroengineering Rehabilitation 10: 66. [https://jneuroengrehab.biomedcentral.com/articles/10.1186/1743-0003-10-66#Declarations] accessed May 24, 2018.

Peerdeman, B., D. Boere, H. Witteveen, H. Hermens, S. Stramigioli, H. Rietman et al. 2011. Myoelectric forearm prostheses: State of the art from a user-centered perspective. J Rehabil Res Dev 48: 719–737.

Pineau, J., R. West, A. Atrash, J. Villemure and F. Routhier. 2011. On the feasibility of using a standardized test for evaluating a speech-controlled smart wheelchair. Int J Intell Control and Syst 16: 124–131.

Poletz, L., P. Encarnação, K. Adams and A. Cook. 2010. Robot skills and cognitive performance of preschool children. Technology and Disability 22: 117–126.

Radmand, A., E. Scheme and K. Englehart. 2016. High-density force myography: A possible alternative for upper-limb prosthetic control. J Rehabil Res Dev 53: 443–456.

Raspopovic, S., M. Capogrosso, F.M. Petrini, M. Bonizzato, J. Rigosa, G. Di Pino et al. 2014. Restoring natural sensory feedback in real-time bidirectional hand prostheses. Science Translational Medicine 6: 19–23.

Resnik, L. 2011. Development and testing of new upper-limb prosthetic devices: Research designs for usability testing. J Rehabil Res Dev 48: 697–706.

Robitaile, S. 2010. The Illustrated Guide to Assistive Technology and Devices. Demo Medical Publishing, New York, USA.

Rocon, E., M. Manto, M.J. Pons, S. Camut and J.M. Belda. 2007. Mechanical suppression of essential tremor. Cerebellum 6: 73–78.

Rofer, T., C. Mandel and T. Laue. 2009. Controlling an automated wheelchair via joystick/head-joystick supported by smart ariving assistance. Proc. of 2009 IEEE 11th Int. Conf. on Rehabilitation Robotics Kyoto, Japan. [https://www.tib.eu/en/search/id/ieee%3Adoi~10.1109%252FICORR.2009.5209506/Controlling-an-automated-wheelchair-via-joystick/] accessed May 25, 2018.

Rosenbaum-Chou, T., W. Daly, R. Austin, P. Chaubey and D.A. Boone. 2016. Development and real world use of a vibratory haptic feedback system for upper-limb prosthetic users. J. Prosthet Orthot 28: 136–144.

Rossini, P.M., S. Micera, A. Benvenuto, J. Carpaneto, G. Cavallo, L. Citi et al. 2010. Double nerve intraneural interface implant on a human amputee for robotic hand control. Clin Neurophysiol 121: 777–783.

Schiatti, L., J. Tessadori, G. Barresi, L.S. Mattos and A. Ajoudani. 2017. Soft brain-machine interfaces for assistive robotics: A novel control approach. IEEE Int Conf Rehabil Robot 863–869.

Service and Inclusion. 2018. http://www.serviceandinclusion.org/index.php?page=basic Accessed Nov. 13, 2018.

Sherrod, B.A., D.A. Dew, R. Rogers, J.H. Rimmer and A.W. Eberhardt. 2017. Design and validation of a low cost, high-capacity weighing device for wheelchair users and bariatrics. Assist Technol 29: 61–67.

Simpson, R.C. and S.P. Levine. 2002. Voice control of a powered wheelchair. IEEE Trans. Neural Syst Rehabil Eng 10: 122–125.

Singh, H., J. Unger, J. Zariffa, M. Pakosh, S. Jaglal, B.C. Craven et al. 2018. Robot-assisted upper extremity rehabilitation for cervical spinal cord injuries: A systematic scoping review. Journal of Disability and Rehabilitation: Assistive Technology 15: 1–12.

Skard, G. and A.C. Bundy. 2008. Play and playfulness. pp. 71–93. *In*: Parham, L.D., L.S. Fazio (eds.). Play in Occupational Therapy for Children. 2nd ed. Mosby, St. Louis, MO.

Spüler, M. 2017. A high-speed brain-computer interface (BCI) using dry EEG electrodes. PLoS ONE 12(2): e0172400. https://doi.org/10.1371/journal.pone.0172400.

Stein, J., K. Narendran, J. McBean, K. Krebs and R. Hughes. 2007. Electromyography-controlled exoskeletal upper-limb-powered orthosis for exercise training after stroke. Am J Phys Med Rehabil 86: 255–261.

Tsikandylakis, G., Ö. Berlin and R. Brånemark. 2014. Implant survival, adverse events, and bone remodeling of osseointegrated percutaneous implants for transhumeral amputees. Clin Orthop Relat Res 472: 2947–2956.

Tyler, D. 2016. Creating a Prosthetic Hand That Can Feel. IEEE. Spectrum April 28, 2016. [https://spectrum.ieee.org/biomedical/bionics/creating-a-prosthetic-hand-that-can-feel] accessed May 25, 2018.

Van de Meent, H., M.T. Hopman and J.P. Frolke. 2013. Walking ability and quality of life in subjects with transfemoral amputation: A comparison of osseointegration with socket prostheses. Archives of Physical Medicine and Rehabilitation 94: 2174–2178.

Viswanathan, P., E.P. Zambalde, G. Foley, J.L. Graham, R.H. Wang, B. Adhikari et al. 2016. Intelligent wheelchair control strategies for older adults with cognitive impairment: User attitudes, needs, and preferences. Autonomous Robots 41(2).

Vu, D.S., U.C. Allard, C. Gosselin, F. Routhier, B. Gosselin and A. Campeau-Lecours. 2017. Intuitive adaptive orientation control of assistive robots for people living with upper limb disabilities. IEEE Int Conf Rehabil Robot 795–800.

Wang, R.H., A. Mihailidis, T. Dutta and G.R. Fernie. 2011. Usability testing of multimodal feedback interface and simulated collision-avoidance power wheelchair for long-term-care home residents with cognitive impairments. J Rehabil Res Dev 48: 801–822.

Wang, Y., F. Liang, J. Yang, X. Zhang, J. Liu and Y. Zheng. 2017. The Acoustic Characteristics of the Voice in Cochlear-Implanted Children: A Longitudinal Study. J Voice 31(6):773.e21–773.e26. doi: 10.1016/j.jvoice.2017.02.007.

[WHO] 1980. World Health Organization (WHO). 1980. The International Classification of Impairments, Disabilities. and Handicaps (ICIDH). Geneva: WHO. [http://www.who.int/classifications/en/] accessed May 25, 2018.

[WHO] 2001. World Health Organization (WHO). 2001. The International Classification of Functioning, Disability and Health (ICF). Geneva: WHO. [http://www.who.int/classifications/icf/en/] accessed May 25, 2018.

Young, A.J., L.H. Smith, E.J. Rouse and L.J. Hargrove. 2013. Classification of simultaneous movements using surface EMG pattern recognition. IEEE Trans. Biomed Eng 60: 1250–1258.

Zuo, K.J., M.P. Willand, E.S. Ho, S. Ramdial and G.H. Borschel. 2018. Targeted muscle reinnervation: considerations for future implementation in adolescents and younger children. Plast. Reconstr. Surg. Mar 19. doi: 10.1097/PRS.0000000000004370.

QUESTIONS

1. How do you define disability? Why does the social model of disability matter?
2. What are the six principles of assistive technology devices? Why each one of them is important?
3. What are the six principles of human robot interface? Why each one of them is important?
4. What are the similarities and differences between Assistive Robotic Manipulators, upper or lower limb prostheses and upper or lower limb exoskeletons?
5. What are the similarities and differences between Social Assistive Robots and Social Interactive Robot?
6. What are the potential negative impacts of using robotic assistive and/or AI technologies?

PROBLEMS

To have a better appreciation of how assistive technology work, consider trying to use a power wheelchair or pretend to be visually impaired or lost the use of one arm:

Identify issues that need to be improved when using a wheelchair: both about wheelchair itself or the environment that you navigate in. If you don't have access to a wheelchair, you can also interview someone who currently uses one.

Or,

Identify issues that need to be improved when using an assistive technology device for visually impaired people, such as a white cane, Talking-GPS, or simply walk inside of a handicap accessible hospital or hotel.

Or

Identify issues that need to be improved for people with impaired use of one upper limb due to either a stroke, an amputation, etc. Try a few activities of daily living (such as cooking, dressing, or taking a shower) with the use of one hand only.

Then

Think of one or more solutions to solve above issue(s).

Does your solution meet the six principles of an ATD?

Do some of the solutions require environment modification(s)?

How the use of robotic technology or AI may help to solve your issue(s)?

25

Role of Academic Health Center Programs and Leadership in Enhancing the Impact of Engineering-Medicine

Jay Noren

QUOTABLE QUOTES

"The way to get good ideas is to get lots of ideas and throw the bad ones away."

Linus Pauling, American Chemist, and Nobel Laureate in Chemistry in 1954, Nobel Laureate in Peace in 1962 (BRAINY QUOTE 2018a)

"Never doubt that a small group of thoughtful, committed citizens can change the world. Indeed, it is the only thing that ever has."

Margaret Mead, American Cultural Anthropologist (BRAINY QUOTE 2018b)

"If I have a thousand ideas and only one turns out to be good, I am satisfied."

Alfred Noble, Swedish Chemist, Engineer, and Inventor (AZ QUOTES 2018a)

"I failed my way to success."

Thomas Edison, American Inventor and Business Man (AZ QUOTES 2018b)

"To win big, you sometimes have to take big risks."

Bill Gates, Entrepreneur and Principle Founder of Microsoft (AZ QUOTES 2018c)

"Success today requires the agility and drive to constantly rethink, reinvigorate, react, and reinvent."

Bill Gates, Entrepreneur and Principle Founder of Microsoft (AZ QUOTES 2018c)

"Leaders are those who empower others."

Bill Gates, Entrepreneur and Principle Founder of Microsoft (AZ QUOTES 2018c)

Associate Dean, University of Illinois College of Medicine, 131 CMW, 1853 W. Polk Street; jnoren@uic.edu

Learning Objectives

These learning objectives are intended to focus the students on understanding the role of academic health centers in enhancing the impact of engineering-medicine.
After completing this chapter, the students should:

- Understand the complexity of the role academic health centers serve in the broader community.
- Understand the impact of the academic health center on the future health care clinical delivery workforce.
- Understand the magnitude of the research enterprise of academic health centers.
- Appreciate the economic impact of academic health centers.
- Understand the breadth of the educational mission of academic health centers through the undergraduate, graduate, and professional degree programs.
- Be able to discuss potential future challenges and opportunities through collaboration in engineering, medicine, and all health professions.

Introduction

Academic health centers (AHCs) serve a critical leadership role in the current and future applications of engineering in medicine (engineering-medicine). Academic health centers are defined as accredited, degree-granting institutions of higher education that educate a wide variety of healthcare professionals; offer comprehensive basic and advanced patient care; and conduct a broad spectrum of biomedical and health services research. They must include: an allopathic or osteopathic school of medicine; at least one additional health professions school or program (allied health, dentistry, graduate studies, nursing, pharmacy, public health, or veterinary medicine); and one or more owned or affiliated teaching hospitals, health systems, or other organized healthcare services. There are 120 academic health centers and 151 medical schools in the United States. Thirty medical schools are free-standing, not components of academic health centers (AAHC 2018).

The most significant contributions to engineering-medicine that academic health centers make are system leadership, education, economic development, research, health professions workforce, healthcare delivery advancement, and advocacy for public policy changes. The leadership and collaborative roles AHCs play in the complex health care delivery system have far-reaching impact culturally and economically. Among the most important products of AHCs are multiple health professions disciplines essential to the healthcare system's future. The approaches that AHCs take to innovative production of the future health professions workforce, including education of medical students, biomedical sciences and engineering graduate students, medical residents, and all health professions, will have profound impacts on the degree to which collaboration occurs at the interface of engineering and medicine, particularly in terms of the efficiency and effectiveness of health services delivery. Similarly, of major importance to future engineering-medicine development is the nature of the collaborative research enterprise in academic health centers. Of course the AHCs enterprise is just one of the contributions to the development of engineering-medicine. Much of that development occurs outside of the academic health center, but collaboration between academic health centers and the engineering production and development community is essential. Ultimately implementation of innovative models of clinical delivery of services that are significantly derived from engineering innovations depends upon the hospitals and clinics that are operated by and affiliated with academic health centers. Critical to that experimentation and implementation, or the results thereof, are the AHCs impacts on accreditation and licensure of healthcare professionals. Finally, ultimately much of the accomplishments of these various contributions by the academic health center must result in substantial changes in national and state health policy across the broad spectrum of the healthcare delivery system. Those policy changes will be essential to facilitating the enhanced role of engineering-medicine.

System Leadership Impact

Leadership of academic health centers is a critical element in facilitating inter-disciplinary collaboration. Academic health centers often are characterized as operating somewhat independently of the broader university. Substantial collaboration between the health professions and the broader university depends upon very innovative leadership of academic health centers. This is particularly true with regard to research, education, and innovation partnerships between medicine and engineering. A recent publication of the Association of Academic Health Centers (Wartman 2015) underscored the challenge noting that:

> "…the potential that exists for advantageous synergy among the component parts—undergraduate and graduate professional schools, health science and non-health science colleges and departments, the clinical enterprise, research function, clinical and nonclinical faculty and administration—is, in many institutions not fully realized. Significant differences in culture, mission, and financial structure drive separation and siloing; these are trends that institutional leaders must consciously and deliberately counter with strategies for cross-institutional alignment to achieve the greatest possible success" (Wartman 2015).

There are several key elements of effective leadership essential to achieving medicine and engineering collaboration. Among the most important are inclusion of disruptive technologies at the engineering-medicine interface in medical student and biomedical sciences and engineering education, collaboration in business incubators and research parks between engineering and medicine, emphasizing the value of technological innovation in graduate medical education (residencies) in all disciplines from primary care to surgical subspecialties, collaborative research at the interface of medicine and engineering, early implementation of engineering-medicine technologies in the clinical setting, and health policy advocacy to enhance approval and implementation of innovative technologies in patient care.

Undergraduate Medical Education Impact

Education of the most effective future professional workforce in health care requires collaboration among medicine, engineering and all health professions. Career-long continual education and adaptation to future disruptive innovation at the interface of medicine, engineering and health professions will depend upon increasing substantive interactive experiences among students from all these disciplines. The commitment of Academic Health Centers to the importance of the medicine-engineering collaboration in medical education was emphasized in a recent paper by Dr. Steven A. Wartman, President and CEO of the Association of Academic Health Centers, in the journal, Academic Medicine. The paper focused particularly on emphasizing artificial intelligence in undergraduate medical education, noting that:

> "… the practice of medicine is rapidly transitioning from the information age to the age of artificial intelligence. The consequences of this transition are profound and demand the reformulation of undergraduate medical education programs…"

> "…Whether physicians use decision support software based on reliable artificial intelligence or manage robots deployed in hospitals, patients' homes, or within the human body, they will need to be educated in this new paradigm" (Wartman and Combs 2018).

Additionally educational delivery technology has far-reaching implications for the methods used to provide access to education in all disciplines. This is particularly true in undergraduate medical education (MD degree program). Although traditionally medical student education has been facilities-based, face-to-face on campus, and laboratory and classroom focused, emerging educational technologies in higher education will increasingly, of necessity, transform medical student education. While education of medical students must be linked to the realities of professional medical practice, emerging educational technology will enhance that linkage. A recent publication of the Association of Academic Health Centers commented on the impact of emerging technology for curricula:

"The characteristics of curricula that are successful in the future will... embrace... ubiquitous digital devices and social networks;... take advantage of big data and cloud computing to increase the customization and personalization of educational programs;... achieve balance between the constant connectivity afforded by digital devices and the need for off-line reflection on the abiding philosophical questions of what is real" (Wartman 2015).

Numerous emerging educational technologies have current and future major implications for medical student and health professions education. Among the increasingly utilized approaches are flipped classrooms. The traditional classroom lecture format in the basic sciences is increasingly replaced by the flipped classroom. Students, individually and on their own time schedules, access recorded lectures, videos, and other supplemental materials and then engage in active, team-based learning sessions to expand their understanding of the topics. The supplementary materials provide extensive access to presentations by international experts. Other emerging educational technologies include web based software providing self-initiated student testing and feedback, virtual dissection replacing traditional anatomy labs, physiologic simulations, use of ultrasound in teaching, exposure to artificial intelligence as applied to medical diagnostics and decision-making, and many other technologically based educational methodologies.

Graduate Medical Education (GME) Impact

Critically important to the application of engineering in medicine is current and future graduate medical education programs (residencies) and the degree to which they include experience for trainees in the application of engineering to their specialty disciplines. Engineering graduate students will also achieve much enhanced educational value through increased interaction with physicians in residency programs and graduate students in all health professions. Engineering applications are relevant to every discipline in medicine from primary care to surgical subspecialties to endocrinology to psychiatry and all other disciplines. A recent effort of the Association of Academic Health Centers sought to address current and future needs for changes in graduate medical education. The effort convened seven regional roundtables throughout the United States with representatives from 30 states and numerous professional organizations including the Association of American Medical Colleges, the American Medical Association, the American Hospital Association, and the National Institutes of Health. These regional roundtables addressed the charge (AAHC 2016a):

"...reforms to the GME system must be considered as part of a national health strategy that considers all aspects of patient care, including not only the education of physicians, but also the other health professions... and the accelerating development of health-related discoveries and technologies, will bring substantive changes to all aspects of the healthcare delivery system" (Wartman 2016).

Among the key conclusions of the roundtables was the need to address new technologies in day-to-day clinical practice. Among the examples of these new technologies emphasized by the roundtables were: wearable health technology, bio banks, telehealth, artificial intelligence, monitors of patient status based upon digital media, smart phones, and diagnostics through chips applied to patients in their homes. And this is just a short list of the multiple examples important to the education of future physicians and specialty training through graduate medical education programs (residencies).

An important challenge to graduate medical education is in fact funding. It will be essential that funding provides opportunities for residents and specialty training programs to experience and gain insights into the applications of disruptive technologies and innovative medicine through engineering. This will require new approaches to national funding for graduate medical education as well as innovative accreditation and licensure. The academic health center has a major role in influencing these factors. Traditionally the accreditation and licensure process is not typically flexible or innovative. In the future that process of accreditation and licensure must become more flexible and innovative, if engineering-medicine is to become an accepted part of the graduate medical education enterprise (AAHC 2016b).

Economic Impact

Academic health centers often have a major role economically in the communities in which they are located. The average annual payroll for an academic health center is $430 million. Additionally employment in academic health centers averages 6,000, and 20% of AHCs employ more than 10,000 people. The employment trend line and economic impact is steadily increasing, and labor costs for academic health centers increases faster than inflation. Academic health centers are expanding with 32% of them developing branch campuses away from the central campus. The impact is also important in terms of service to the community. On average academic health centers provide $44 million annually in uncompensated patient care, and 14% of academic health centers provide more than $100 million annually in uncompensated patient care. Many AHCs include numerous community health centers providing broad clinical services.

Furthermore a very important part of AHCs economic development impact is through closely affiliated and owned/operated business incubators and research parks. In fact 44% of academic health centers have at least one research park or business incubator as an essential part of the organization. Clinical trials that AHCs implement have broad impact because they are often geographically broadly distributed with 25% of AHCs engaged in more than 10 separate sites for clinical trials. AHCs international impact is far reaching with 75% engaging in research partnerships with academic institutions internationally. With this tremendous amount of resources, these business development efforts and partnerships are essential to future applications and enhanced development of engineering-medicine in the United States and internationally (AAHC 2009).

The 151 medical schools and teaching hospitals in the United States (of which 120 are components of academic health centers) also have a large economic impact.

These institutions contributed more than $562 billion in gross domestic product (GDP). This amount translates to about 3.1% of the U.S. GDP, making the economic impact of these medical schools and teaching hospitals comparable in size to other important sectors such as transportation, warehousing, accommodation and food services. On a per capita basis, these medical schools and teaching hospitals generate approximately $1,750 in economic impact per person.

These institutions' education, research, and patient care work supports more than 6.3 million jobs in the United States across multiple industries. These jobs are approximately 3.3% of jobs nationwide, paying more than $387 billion in aggregate annual wages, salaries, and benefits—an average of about $61,000 in wages, salaries, and benefits per job (AAMC 2018).

Research Impact

The academic health centers impact on business development is strongly enhanced by extensive research contributions. Each academic health center averages $137 million annually in research expenditures, of which about $100 million is funded by the National Institutes of Health. Additionally AHCs enhance research and development infrastructure dramatically in the community. Currently 90% have new research space under construction and about half of those enterprises are for space in excess of 250,000 square feet. Much of that support comes from the private sector with only about 40% coming from the federal sector. Important elements of these research enterprises currently and in the future are derived from engineering-medicine partnerships. Moreover, the funding of biomedical engineering programs by the National Institutes of Health substantially increased from 2005 to 2010, whereas NIH funding for traditional biomedical programs, with the exception of neuroscience, remained essentially flat-lined (Richardson 2013). This strong engineering funding by NIH will also impact the development of engineering-medicine.

Healthcare Workforce Production Impact

In terms of long-term impact, among the most important contributions of academic health centers is education of the future workforce. The future workforce will necessarily reflect the integration and application of engineering to medicine and the impact on future healthcare delivery will be profound. Consequently engineering has become increasingly integrated not only with medical education but also with education in many other healthcare professions that exist in AHCs. This is particularly important given that 72% of academic health centers have three or more health professions schools and 60% have four or more schools including dentistry, pharmacy, nursing and allied health disciplines. The enrollment of students in health professions nationally totals over 200,000 annually. The average student enrollment in each AHC is over 2,000 and 15% of AHCs enroll more than 4,000 students. Additionally new academic programs are emerging continually. During the last decade 60% of academic health centers added new clinical doctoral degrees in the health professions and 85% are engaged in cross-disciplinary programs beyond the health professions. Important among the cross-disciplinary programs in medical schools are joint or dual degree programs combining the MD degree with the Master's or PhD degree in Engineering. In some medical schools biomedical engineering is fully incorporated into the school organizational/departmental structure.

And of much significance to engineering-medicine is the extent of faculty resources. Academic health centers have over 100,000 full-time faculty averaging about 1,400 in each center and more than 2,000 in 12% of centers. Increasingly education programs in academic health centers are interdisciplinary and increasingly that includes engineering.

Of particular importance is the total healthcare workforce impact (beyond the AHCs) on the economy. Substantial future employment in the US economy will be derived in healthcare areas and at the interface of engineering and healthcare. The health sector now accounts for 19% of the US economy and that will become 25% projected by 2025. This is a dramatic change from 1960 when the health sector comprised only 5.2% of the US economy. Healthcare employment now comprises 11% of the total US workforce and in 2014 20% of new jobs were created in healthcare.

A critically important element of collaboration between medicine and engineering is the need for increased efficiency of the healthcare workforce in the future. Specifically physician supply is projected to face serious shortages in the next decade. Current projections indicate a shortage of between 61,000 and 95,000 physicians by 2025. Improved methods of delivering health services will be an important part of addressing the shortage. An element of improved methods will be the use of engineering and information systems technology to make physicians more readily accessible to patients in nontraditional settings outside of hospitals and clinics through delivery of care in patients' home settings. Additionally, artificial intelligence serving to augment the physician role will be critically important in the future. Furthermore, demographic trends will exacerbate the need for greater efficiency in the use of the healthcare workforce and the importance of engineering innovation in enhancing that workforce. In the United States the population over 65 years old will increase by 41% by 2025 and that compares to an increase of only 8.6% in the general population. This poses a serious challenge to access to healthcare services for populations in need because of increasing prevalence of chronic illness related to aging of the population.

Healthcare Delivery Impact

Engineering and engineering-medicine innovation is increasingly transforming clinical healthcare delivery provided by Academic Health Centers and providing access to specialized services far beyond the facilities and location of those services on the Academic Health Center campus. This accessibility enhancement is the result of ongoing collaboration of the engineering expertise with the clinical and medical research resources of the university. Among the transformative innovations are telemedicine, mobile health, tele-therapy, virtual physician visits, E-health, health information technology including electronic health records, and tele-ICUs (Shi and Singh 2018).

Telemedicine provides medical services at a distance including diagnostic approaches and specialty consultation. For example tele-radiology provides for review of x-rays by specialty radiologists at great distance from the patient. Pathology specimens can be viewed at a distance using tele-microscopy and thereby taking advantage of specialty pathologists for complicated diagnostic determinations not available in the patient's immediate location. Tele-surgery can provide surgical interventions at a distance through surgical specialists controlling robots to perform surgical procedures. Mobile health takes advantage of wireless communication such as smart phones to support clinical diagnosis, monitoring, and advice to the individual patient at a distance. E-health offers resources available on the Internet providing patients with extensive information and through secure internet portals providing readily accessible consultation from the physician. Tele-therapy provides readily accessible mental health consultation by psychiatrists and psychologists to patients in their homes. The approach provides substantially increased specialty mental health services access to children and adults previously not readily available for a whole range of problems including the full range of complex mental health challenges requiring the specialized expertise available only in the academic health center. Virtual physician visits provide enhanced access in particular for patients with chronic illness who receive major benefit from more readily available consultations in a manner much more efficient than through office visits. The approach has the potential to effectively enhance control of chronic diseases such as diabetes. Tele-ICUs provide more continuous monitoring of complex patients in intensive care units that augment the monitoring provided by clinical professionals on-site. It provides a particularly valuable resource when the staff at the intensive care unit are engaged in emergency procedures and thus distracted from the ongoing monitoring of other patients. Electronic health records (EHRs) provide much more rapid access to a patient's health status than the historical paper record. This is particularly important in an emergency situation in which a patient is being seen in a facility other than where they commonly receive their care. Additionally, EHRs provide access of the patient to their own personal health records immediately at their request. Finally, electronic health records are a major source of data for research on quality of healthcare and clinical outcomes assessment in which academic health centers regularly engage.

All of these innovations underscore the critical importance of continuing and increased collaboration at the interface of medicine and engineering in the research, teaching, and clinical services mission of academic health centers. Furthermore energetic pursuit of that collaboration has potential for significantly increased economic efficiency in healthcare delivery. The importance of increased efficiency is dramatically underscored by comparisons of US healthcare expenditures to other developed countries. A recent international healthcare expenditures analysis based upon 2016 data concluded:

> "The United States spent approximately twice as much as other high-income countries on medical care, yet utilization rates in the United States were largely similar to those in other nations. Prices of labor and goods, including pharmaceuticals, and administrative costs appeared to be the major drivers of the difference in overall cost between the United States and other high-income countries."

The analysis more specifically noted:

> "In 2016, the US spent 17.8% of its gross domestic product on health care, and spending in the other countries ranged from 9.6% (Australia) to 12.4% (Switzerland)… (and) … The US did not differ substantially from the other countries in physician workforce (2.6 physicians per 1000; 43%primary care physicians), or nursing workforce (11.1 nurses per 1000)."

Increased utilization of the multiple technologies at the medicine-engineering interface has the potential to diminish the high administrative and professional workforce costs through enhanced access and delivery of clinical services (Papanicolas et al. 2018).

Health Policy Impact

A recent report of the National Academy of Medicine strongly advocated policy changes that 1-facilitate health professions education emphasizing engineering related expertise, 2-increase efficiency in regulatory processes for review and approval of medical care and related engineering innovations, 3-support implementation of digital technology in the delivery of clinical medical care, and 4- Increases investment in research at the interface of engineering and medicine (Dzau et al. 2017a, 2017b, AAHC 2013).

Education Policy

Regarding health professions education the National Academy report emphasized the importance of integrating informatics requirements in residency (graduate medical education) programs to assure that the future physician workforce has the skills to fully utilize in clinical practice the rapid innovations in health information technology. Given that major funding for graduate medical education programs depends heavily on federal health policy, updating current federal legislation and regulations is essential.

Additionally, new educational pathways for engineering and related health sciences is essential to future U.S. global leadership in engineering-medicine research and innovation (Dzau et al. 2017a, 2017b, IOM 2014). The National Academy of Medicine report noted:

> "Training the science workforce for the future will require new models, new partners, and cross-disciplinary thinking. Our new workforce will need to be diverse, multidisciplinary, team-oriented, and possess strong skills in data analytics and informatics. Recruiting and retaining the most talented will necessitate innovative education pathways and programs to assemble and support a cutting-edge biomedical science workforce."

Regulatory Policy

Although the United States has historically been a world leader in biomedical sciences and innovation, regulatory review processes have become cumbersome and updating those processes to more streamlined approaches deserves serious analysis. The National Academy report underscores the fact that the cost of drugs and medical device development has risen substantially in recent years. The report notes that the cost of bringing a new drug to market is estimated at $2.6 billion. Addressing these challenges will require facilitation of more collaborative approaches involving medicine, engineering, and related fields across academia, private industry, government regulatory agencies, and legislative bodies.

Clinical Service Policy

Digital technology has current, rapidly increasing and far-reaching impact on clinical services from multiple perspectives including the health care system and its organization and management, the practicing clinician, and the individual patient. The broad implementation of the electronic health record (EHR), partly mandated by federal policy through the Affordable Care Act, has entailed major investment by health systems and provided substantive enhancements in information valuable to effective and efficient deployment of resources. Nonetheless planning and implementation of the EHR has posed significant challenges to health care systems. From the clinician's perspective, EHRs have affected the nature of the physician-patient interaction, a change which will require adaptation over time. Mobile devices increasingly provide new approaches to physician-patient interaction including telemedicine and e-prescribing as well as increased accessibility to patient health records through digital transmission. The patient impact includes increased access to the health care system and clinicians as well as more continual and effective monitoring of patient health status, while changing somewhat the nature of the clinician/patient personal interaction (Dzau et al. 2017a, 2017b). The National Academy of Medicine report summarized the impact succinctly:

> "Health and health care are being fundamentally transformed by the development of digital technology with the potential to deliver information, link care processes, generate new evidence, and monitor health progress."

Research Policy

Global biomedical leadership of the United States will depend predominantly on federal policy that sustains investment in research at the interface of engineering, medicine, and all biomedical sciences. An important element of the investment is facilitation of multi-disciplinary collaboration in basic and applied research. Particularly important is recognition of the value embodied in clinical translational sciences focused on efficient translation of basic and applied research to clinical applications, "bench to bedside". The National Academy report emphasizes the importance of taking advantage of rapidly emerging large datasets, collaboration among rapidly developing new science and technology, and cognitive computing to:

> "…ensure the most effective and appropriate interventions for the best possible clinical outcomes."

Conclusion

Academic health centers must continue to lead innovation and implementation of engineering and medicine collaboration in the rapidly evolving transformation that has far-reaching impacts on the health of the population. In order to achieve optimal impact, academic health centers must vigorously pursue much enhanced partnerships with private industry and government policymakers to stimulate an increasing emphasis on multi-disciplinary approaches to health and biomedical sciences. That pursuit will require continuing progress of universities to diminishing the historical organizational focus on specific academic disciplines in departments and schools/colleges and increasing emphasis on multi-disciplinary approaches to education, research, and leadership both within the academic institution and externally. The continuing global leadership of the United States in health, medical, engineering, and broad biomedical sciences will depend upon this transformed model of multi-disciplinary innovation.

References Cited

[AAHC] 2009. Academic Health Centers-Creating the Knowledge Economy. Association of Academic Health Centers. Washington, DC.

[AAHC] Association of Academic Health Centers. 2013. Health Workforce Out of Order Out of Time. 2008 Updated 2013. [http://www.aahcdc.org/Publications-Resources/Books-.Reports/View/ArticleId/15926/Out-of-Order-Out-of-Time] accessed June 28, 2018.

[AAHC] Association of Academic Health Centers. 2016a. Report: Regional Roundtable Series on Graduate Medical Education (GME) Reform. [http://www.aahcdc.org/Portals/41/Publications-Resources/BooksAndReports/AAHC-GME-Roundtable-Report-2016.pdf] accessed June 28, 2018.

[AAHC] Association of Academic Health Centers. 2016b. Health Workforce Issues Summary 2016. [http://www.aahcdc.org/Initiatives/Health-Workforce] accessed June 26, 2018.

[AAHC] Association of Academic Health Centers. 2018. Academic Health Centers. [http://www.aahcdc.org/About/Academic-Health-Centers] accessed June 26, 2018.

[AAMC] Associate of American Medical Colleges. 2018. The Economic Impact of Medical Schools and Teaching Hospitals. AAMC News. April 2, 2018. [https://news.aamc.org/medical-education/article/economic-impact-medical-schools-and-teaching-hospi/] accessed June 28, 2018.

[AZ QUOTES] 2018a. Alfred Nobel Quotes. AZ Quotes. [www.azquotes.com/author/10847-Alfred_Nobel] accessed June 26, 2018.

[AZ QUOTES] 2018b. Thomas A. Edison. AZ Quotes. [www.azquotes.com/quote/522812] Accessed June 26, 2018.

[AZ QUOTES] 2018c. Bill Gates. AZ Quotes. [www.azquotes.com/authors/5382-Bill-Gates] Accessed May 11, 2018.

[BRAINY QUOTE] 2018a. Linus Pauling Quotes. Brainy Quote. [https://www.brainyquote.com/authors/linus_pauling] accessed June 26, 2018.

[BRAINY QUOTE] 2018b. Brainy Quote. [https://www.brainyquote.com/search_results?q=Margaret+mead] accessed June 26, 2018.

Dzau, V.J., M. McClellan, J.M. McGinnis, S.P. Burke, M.J. Coye, A. Diaz et al. 2017a. Vital Directions for Health and Healthcare-Priorities from a National Academy of Medicine Initiative. JAMA. 317: 1461–1470.

Dzau, V.J., M. McClellan, S. Burke, M.J. Coye, T.A. Daschle, A. Diaz et al. 2017b. Vital Directions for Health and Health Care: A policy initiative of the National Academy of Medicine. National Academy of Medicine. March 21, 2017. [https://nam.edu/initiatives/vital-directions-for-health-and-health-care/] accesses June 5, 2018.

[IOM] Institute of Medicine. 2014. Graduate Medical Education that Meets the Nation's Health Needs. July 29, 2014. [http://www.nationalacademies.org/hmd/Reports/2014/Graduate-Medical-Education-That-Meets-the-Nations-Health-Needs.aspx] accessed June 28, 2018.

Papanicolas, I., L. Woskie and A. Jha. 2018. Health Care Spending in the United States and Other High-Income Countries, JAMA March 13.

Richardson, M. 2013. Funding trends. A funding profile of the NIH. Research Trends. September 2013. Website. [https://www.researchtrends.com/issue-34-september-2013/a-funding-profile-of-the-nih/] accessed May 18, 2018.

Shi, L. and D. Singh. 2018. Delivering Health Care in America, Seventh Edition, Jones and Bartlett. Burlington, MA.

Wartman, S. 2015. Transformation of Academic Health Centers: Meeting the challenges of healthcare's changing landscape. Academic Press, London, UK.

Wartman, S. 2016. Graduate Medical Education for 21st Century. Health Affairs. August 16, 2016. [https://www.healthaffairs.org/do/10.1377/hblog20160816.056164/full/] accessed June 28, 2018.

Wartman, S. and D. Combs. 2018. Medical education must move from the information age to the age of artificial intelligence. Acad Med 93: 1107–1109.

QUESTIONS

1. What are the most important incentives for collaboration between medicine and engineering in academic health centers?
2. What impact on engineering and medicine collaboration results from the design of educational programs for medical students, residents, and graduate students?
3. How do academic health centers affect the economic impact of engineering and medicine collaboration?

PROBLEM

Design the elements of the educational experience that will create an interface between the medical school curriculum and graduate education in bioengineering.

- Create specific collaborative projects in which bioengineering and medical students will partner.
- Determine literature review materials addressing the interface of medicine and engineering.
- Provide descriptions of opportunities for students to observe disruptive innovation examples of applications to clinical health care delivery resulting from in medicine-engineering collaboration.

26

Environmental Protection

Lawrence S. Chan

QUOTABLE QUOTES

"God put the human race in charge of managing the resources of the entire planet for the benefit of all life. Therefore, we, of all people on this planet, should be concerned about environmental issues and doing what we can to enhance the beauty and productivity of the natural realm."

Hugh N. Ross, Canadian Astrophysicist and Christian Apologist (BRAINY QUOTE 2018a)

"As stewards of God's creation, we are called to make the earth a beautiful garden for the human family. When we destroy our forests, ravage our soil and pollute our seas, we betray that noble calling."

Pope Francis (McAfee 2017)

"Environmental protection doesn't happen in a vacuum. You can't separate the impact on the environment from the impact on our families and communities."

Jim Clyburn, United States Congressman from South Carolina (BRAINY QUOTE 2018b)

Learning Objectives

The learning objectives of this chapter are to familiarize students diseases caused by environmental exposure, the impact environment on our health and the impact of medical practice on our environment, and the roles of biomedical engineering and medical professions in protecting our environment.
After completing this chapter, the students should:

- Understand the importance of environmental protection for our future planet.
- Able to identify diseases and health conditions triggered in part or in full by environmental conditions.
- Understand the impact of medicine-related waste on our environment.
- Understand the huge energy consumptions by medical facilities.
- Understand the roles of bioengineering and medical professions in protecting our environment.
- Able to think ways of waste reduction in medicine practice.
- Able to apply the principle of "Reduce, Reuse, and Recycle" in medicine practice.

University of Illinois College of Medicine, 808 S. Wood Street, R380, Chicago, IL 60612; larrycha@uic.edu

Environmental Protection in Engineering

Engineering has a major role in environmental protection. Our water treatment system and our waste management system are all derived from engineering processes. Many engineering schools have environmental engineering programs (CORNELL 2018, MIT 2018, PENN 2018, PRINCETON 2018, UIUC 2018, YALE 2018). In one such program, it defines the goals of environmental engineering this way, "Environmental engineers are called upon to understand, arrange, and manipulate the biological, chemical, ecological, economic, hydrological, physical, and social processes that take place in our environment in an effort to balance our materials needs with the desire for sustainable environmental quality" (CORNELL 2018). In another program, it states its major aims of "prepares students for the increasingly critical role of addressing the world's challenges of air, land and water pollution" (UIUC 2018). At Massachusetts Institute of Technology, the Department of Civil and Environmental Engineering has this teaching focus, "Here you will learn how to understand the Earth's biomes; design benign materials and structures; model air, water and climate; discover new energy resources; develop quantitative systems thinking to understand and design complex infrastructure; and build sustainable infrastructure and cities" (MIT 2018).

At Yale University, two bachelor degrees relating to environment are offered: B.S. degree in Environmental Engineering is a technically comprehensive program that provides rigorous preparation for graduate study and a career in environmental engineer in consulting firm or industry. B.A. degree in Engineering Sciences (Environmental) is designed for students aiming for careers in law, business, medicine, or public service (YALE 2018).

Based on the Association of Environmental Engineering and Science Professors, environmental engineering education programs teach students on various subjects including the followings (AEESP 2018):

- Water and waste water treatment
- Air pollution control
- Water and air resource management
- Industrial and hazardous waste management
- Solid waste management
- Contaminated site investigation and remediation
- Waste repositories
- Pollution prevention
- Environmental chemistry
- Aquatic ecology
- Environmental toxicology
- Public health engineering
- Environmental policy management

According to Bureau of Labor Statistics, environmental engineers' main activities are defined as the followings (BLS 2018):

- Utilize the principles of engineering, soil science, biology, and chemistry to solve environmental problems.
- Make efforts to improve recycling, waste disposal, public health, and air and water pollution control.

Viewing from the perspectives of both the educational goals and aims of environmental engineering programs and the main activities of environmental engineers (as stated by the Bureau of Labor Statistics), there is a strong connection between environmental engineers and medical professionals in that they both encounter the same medicine-related environmental problems that affect us all. Thus, collaboratively, environmental engineers and medical professionals could help solving the environmental problems impacted by medical waste and help preventing environment-related diseases.

Environmental Protection in Engineering-Medicine

Rationale

Environmental protection is also an important part of engineering-medicine. The significance is two folds: First, our environment affects health and medicine in that some diseases are actually causes by environmental problems. Second, the very practice of medicine generates wastes that affect our environment. The link of environmental engineering to health and medicine is so strong that some environmental engineering programs are actually situated within the College of Agriculture and Life Sciences (CORNELL 2018).

Having delineated the rationale of environmental protection in engineering-medicine, we now look at two areas of critical issues involving environmental protection and medicine, namely the diseases that are environment-related and medicine-related waste that affect our environment.

Environment-Related Diseases

There are many diseases that are generally recognized as environment-related. In certain diseases or condition we can state with confident that environmental situation directly causes the diseases or the condition to occur. In the followings, three examples of diseases or health condition are reviewed in relationship to environment as the causes.

Asthma

Asthma, the most common chronic childhood disease, is the leading cause of childhood morbidity with respect to school absences, emergency department visits, and hospital admissions (Ferrante and La Grutta 2018, Naja et al. 2018). Globally, asthma accounts for 1.1% of the overall estimate of "Disability-adjusted life years" (DALYs)/100,000 for all causes. Moreover, it is troubling that the prevalence of asthma in children and adolescents has increased in recent years worldwide, especially in low-middle income countries (Ferrante and La Grutta 2018). Asthma is generally recognized as chronic allergic/inflammatory disease where environment condition plays a critical factor, but is not the only factor (Gardeux et al. 2018, Gray et al. 2018, Kantor et al. 2018, Orellano et al. 2017, Papi et al. 2018, Scherzer and Grayson 2018, Soyiri and Alcock 2018). Asthma is likely a disease that has the influencing factors of genetic predisposition and an exposure to environmental triggers (Gardeux et al. 2018, Jeong et al. 2018, Kantor et al. 2018, Papi et al. 2018). The link between environmental conditions and asthma has been validated not only by epidemiological studies, but also supported by laboratory-generated experimental data (Jeong et al. 2018).

As a complex disease, asthma not only has various clinical manifestations but also variable responses to treatment. In the future, utilizing the precision medicine principle, particularly with pharmacogenomic data to optimize treatment, should be considered (Matera et al. 2017). In addition, precision medicine method can help predicting disease exacerbation by examination of personal transcriptome response to environmental challengers with the assistance of artificial intelligence (Gardeux et al. 2017).

Lead Poisoning

Lead poisoning is a clear example of abnormal health condition or disease in which lead exposure is the direct cause. Although lead poisoning posts the greatest risk to children who are at the developmental stage of our lives, it can also cause health problems for adults who are exposed to it (Schnur and John 2014). Lead poisoning is essentially a man-made problem as the metal was contained in many commercial products such as paint, gasoline, water pipe, pottery, batteries, roofing, bullets, some cosmetics, certain herbal medicines, and toys for many years. Since the removal of lead from gasoline and paint products in the 1970s, the prevalence and severity of lead poisoning in children have been reduced substantially in the US, but it is still a health risk as 300,000 US children younger than 5 years have been found to have elevated blood lead levels, a limit defined by US Center for Disease Control and Prevention as higher than 10 microgram per dL (Warniment et al. 2010). Lead poisoning affects many organs and can result in many serious medical conditions. The common problems encountered in children including the followings (MAYO 2018, Raihan et al. 2018, Talbot et al. 2018):

- *Developmental*: developmental delay, learning difficulties, stunting, underweight, premature birth and low birth weight (in newborns).
- *Neurological*: irritability, hearing loss, seizure, encephalopathy, loss of consciousness.
- *Gastrointestinal*: loss of appetite, weight loss, abdominal pain, vomiting, constipation.

For adults, the usual symptoms are the followings (MAYO 2018):

- *Neurological*: difficulties in memory and concentration, headache, mood disorders.
- *Gastrointestinal*: abdominal pain.
- *Male Reproductive*: reduced and abnormal sperm production.
- *Female Reproductive*: miscarriage, stillbirth.
- *Cardiovascular*: high blood pressure.
- *Musculoskeletal*: joint and muscle pain.

In addition, wild and companion animals living near lead-contaminated areas can also be affected by the lead exposure (Langlois et al. 2017, Isomursu et al. 2018). In addition to epidemiological link, many studies have documented on the cellular and molecular levels how lead poisoning could cause serious damages in human and animals (Duan et al. 2017, Mitra et al. 2017, Chwalba et al. 2018, de Souza et al. 2018, Ge et al. 2018, Gu et al. 2018, Yang et al. 2018).

Arsenic Toxicity

Arsenic is a heavy metal and its contamination in drinking water has affected millions of people worldwide (Ratnaike 2003). The sources of arsenic contamination and toxicity are from both natural geological presence and industrial products, including an anti-neoplastic medicine termed arsenic trioxide that is used for the treatment of promyelocytic leukemia. Absorbed through small intestine following oral ingestion, arsenic exerts its toxicity to humans through inactivation of multiple enzymes (up to 200), targeting particularly those involving DNA synthesis and repair and cellular energy pathways. Two areas of the world that are particularly affected by arsenic toxicity are located in West Bengal, India and Southern Bangladesh, where a total of more than 120 millions of people are exposed to ground water containing arsenic concentration greater than 50 microgram per liter, the maximum permissive limit by the World Health Organization (Ratnaike 2003). In the US, a geographic location of major public health concern is the Millard County, Utah, where residents have been chronically exposed to low to medium concentrations of arsenic in drinking water with associated diseases and cancers (Lewis et al. 1999, Ratnaike 2003).

For patients suffered from acute arsenic toxicity, defined as arsenic level 1–3 mg/kg detected on hair sample, usually manifested the followings (Ratnaike 2003):

- *Death*: lethal dose is around 100 mg to 300 mg (about 0.6 mg/kg/day).
- *Gastrointestinal*: nausea, vomiting, abdominal pain, severe diarrhea, excessive salivation, esophagitis, gastritis, hepatic steatosis.
- *Neurological*: encephalopathy, peripheral neuropathy, seizure, acute psychosis.
- *Cardiovascular*: cardiomyopathy, blood volume loss from diarrhea, circulatory collapse from blood volume loss.
- *Hematological*: hematuria, intravascular coagulation, bone marrow depression, pancytopenia.
- *Renal*: failure.
- *Pulmonary*: edema, respiratory failure.
- *Cutaneous*: rash.

Patients with chronic arsenic toxicity, documented by 0.1–0.5 mg/kg arsenic concentration detected on hair sample, commonly suffer the followings (Ratnaike 2003):

- *Cardiovascular*: hypertension, ischemic heart disease, cardiac arrhythmia, cardiomyopathy, gangrene.
- *Gastrointestinal*: diarrhea, vomiting, hepatomegaly, liver fibrosis, liver cirrhosis.
- *Neurological*: peripheral neuropathy, anesthesia, confusion, memory loss, cognitive impairment, cerebral infarction.
- *Genitourinary*: nephritis, prostate cancer, bladder cancer, kidney cancer.
- *Endocrine*: diabetes.
- *Cutaneous*: palmar and solar keratosis, skin cancer.

Medicine-Related Waste

Having considered diseases or abnormal health conditions that are fully or partially caused by environmental factors, we should now turn our attention to the impact of activities in practicing medicine onto our environment. This issue is very essential for discussion. Every single day, health facilities in the US generate about 6,600 tons of waste (Kaplan et al. 2012). In 2013, the healthcare sector contributed a significant fraction of the national air pollution emissions and impact: 12% in acid rain, 10% in greenhouse gas emission, 10% in smog formation, 9% in criteria air pollution, 1% in stratospheric ozone depletion, and 1–2% in carcinogenic and non-carcinogenic air toxics (Eckelman and Sherman 2016). Physicians took the Hippocratic oath of doing no harm to the patients in our caring mission, likewise physicians should also examine carefully if medical practice itself would generate wastes affecting our environment negatively. To be sure, why would medicine practice creates substances to harm our environment and by extension harm our health, if the very mission of medicine is to provide better health. The follow up question is, therefore, how should negative impact on our environment by those waste be minimized if present. Better yet, if we can find way to eliminate any negative impact on our environment.

Infectious Waste

Since medical professionals care for patients who have various infectious diseases, including viral, bacterial, fungal, and parasitic diseases, it is inevitable that medical practices, be it in the hospital setting or in the physician's office, naturally encounter infectious materials that require proper disposal. Furthermore, any medical supplies that came in contact with blood or body fluid are considered to be potentially infectious in nature. These infectious and potentially infectious wastes are usually embedded with cotton gauze, Q-tips, absorbing pads, bandages, and other cotton and paper products.

Radioactive Waste

Whether it is for the diagnostic purpose or for the treatment purpose, medical professionals utilize radioactive materials commonly in both the physician's office and the hospital. Radioactive materials are needed for many diagnostic tests such as nuclear medicine imaging, cardiovascular imaging, and arterial imaging. In addition, radioactive materials are used for the treatment of hyperthyroidism and certain cancers. Therefore, the radioactive wastes are routinely generated. Since radioactive isotopes require a long time for radioactivity decay, these radioactive wastes need to be kept in protective container for proper disposal during a long decay process.

Instrument Waste

In carrying out their caring missions, medical professionals utilize many instruments. Surgical procedures require blade, curettes, forceps, scissors, clamps, needle holders, retractors, hemostats, and many other instruments. For those instruments that came in direct contact with patients' body fluid or blood, they are considered to be infectious. These contaminated instruments need to be treated in one of the two methods. For the expensive and easily cleaned instruments, they are thoroughly disinfected, cleaned, and autoclaved for future use. For the inexpensive and hard to clean items, they are generally disposed as instrument wastes after only one use. Many of these disposable items are sharp instruments, including surgical blades, needles, and curettes. The other disposable items are primarily plastic in nature, such as syringes, suction tubes, urinary draining tubes and bags, central lines, and intravenous cannula. Since these plastic items have been contacted with human fluid, they also need to disposed in biohazard containers.

Chemical Waste

Cleaning is a big activity in medical practice. In a profession that takes care of sick people, the practice of medicine is very keen on preventing cross-contamination between patients. Naturally, cleaning the room between patients is a routine activity and by doing so many chemical wastes are generated. Worse yet, many of these chemical waste are indeed toxic to the environment.

Medication Waste

All medications have clearly indicated "shelf life" which is a federal government regulation mandated by the Food and Drug Administration (FDA 2018). The requirement of shelf life indication is based on the fact that medication, once manufactured, is only stable for certain period of time when stored at certain temperature environment. After that period of shelf life time is passed, the medication may either loss its active ingredient function or be degraded. So the shelf life indication is to ensure that patients will indeed receive functional medication when they need them. Shelf life is tested rigorously during manufactural process to validate the proper function of the medication remain throughout the entire shelf life time span (Puglielli 2014). When medications expire, hospital and outpatient pharmacies will dispose them in a regulated manner to prevent environmental pollution. However, the same expired or unused medications at the hands of patients may not end up in proper disposal pathway. Some of these expired or unused medications may be found in our trash and subsequently in land fill, sewage, surface and ground water, and even in drinking water (Heberer 2002, Bound and Voulvoulis 2005, Sui et al. 2015).

Healthcare-related Energy Consumption

The huge amount of energy consumed by healthcare facilities is a little known fact that has significant implication. A 2012 government survey has ranked hospital as the second highest energy-intensive commercial building, only slightly less than food service (EIA 2012). A major contribution of this sizable energy consumption (18%) is estimated to be from medical equipment in the hospital setting by the US Department of Energy, which has identified medical equipment as a major area for energy-saving opportunity (DOE 2011, CANADA 2016). Proportionally, more energy consumed, more environmental pollution will be generated (WHO 2018).

Medical Waste Management Applications

Having delineated the rationale of environmental protection in engineering-medicine, we now turn to the applications of engineering-medicine in protecting our environment (Quinn et al. 2015, Windfeld and Brooks 2015). This kind of effort requires a close cooperation between medical and engineering professionals, as we should discuss the following important aspects of medicine-related environmental protection. The general strategy of "Reduce, Reuse, Recycle", promoted by the US Environmental Protection Agency, would be a sensible approach for handling medicine-related waste (EPA 2018).

Waste Reduction

It would be great for the environment and for the finance if the medicine-related waste could be reduced, since less waste will translate into less waste being burden to our environment and the overall cost of healthcare will be lower. There are several possible ways that medicine waste could be decreased:

Prescription Drug Reduction. Studies have been pointing out that as a society we are over medicated and sometimes received unnecessary treatments. In some unfortunate individuals, overuse of medication results in a chronic disease called "medication overuse headache" associated with documented abnormal brain structure and function (Symvoulakis et al. 2015, Schwedt and Chong 2017). To achieve this reduction, there is a great need for evidence-based medical publications which provide solid evidences of proper treatments and non-treatments. Such evidence-based medical information will greatly help healthcare professionals in making appropriate medication decision and to avoid prescribing unneeded medications. Artificial intelligence, particularly machine learning, will likely be a great help in this regard.

Unnecessary Surgery Reduction. Studies have indicated that for certain diseases, such as uncomplicated appendicitis, medical treatment would be as effective as surgical treatment, even though surgery has been the traditional "standard" approach (Salminen et al. 2015). Since medical treatment would likely generate less medical waste and cost less than surgical treatment for the same disease, healthcare professionals would be benefited from receiving such information, if these information is provided as evidence-based solid scientific data (Salminen et al. 2015, Sippola et al. 2017). Furthermore, some studies with the assistance of artificial intelligence, have indicated that some slow growing tumors, like specific types of prostate cancer, do not require surgical treatment (Schroder et al. 2009, Flores-Morales and Iglesias-Gato 2017). Such information, provided as evidence-based scientific data, would be greatly helpful to healthcare providers to avoid unnecessary surgery procedures, thus reducing medicine waste and morbidity to the patients.

Reusable Instrumentation. Perhaps one way to reduce medicine-related waste is to convert some of the instruments that are usually disposable into reusable. This is particularly true for surgical instruments. For example, forceps and scissors used for suture removal, biopsy punches, and curettes are instruments that are commonly disposed after just one use. As a result, many infectious metal waste are generated. Converting these instruments to reusable form will require more labor in cleaning, disinfecting, and autoclaving process, but will certainly help cut down the amounts of medical waste and perhaps even reducing the overall cost as well. Reusable biopsy punches, inexpensive reusable forceps, scissors, and curettes are commercially available (EMS 2018, GT 2018, MM 2018).

Accurate Waste Sorting

According to a government report, about 85% of medically related waste taken from hospitals are not hazardous in nature (Kaplan et al. 2012). Since the hazardous waste disposal requires more labor-intense, more costly procedures, and more pollution emission-processes (from incinerations) than non-hazardous waste disposal, it makes good sense for accurate waste sorting so as to reduce the burden of hazardous waste disposal (Kaplan et al. 2012). It would be greatly beneficial if healthcare professionals and environmental engineers could work together to find solutions in this regard, as healthcare professionals would be able to delineate their daily operations in relationship to medical waste generation and environmental engineers would be able to utilize their engineering skills to deduce user-friendly and fool-proofed methodology for an accurate waste sorting, in line with healthcare professionals' daily activities.

Green Infection Control

Since cleaning and disinfection is a major task for medicine practice and generates many toxic waste, it would make good sense if the disinfectant, cleaner, and other infection control supplies are products of biodegradable or otherwise non-toxic, so that the resulting waste would not post such negative impact on our environment. Medline company has, in 2010, introduced the first biodegradable surgical drape called "EcoDrape™", which was made almost entirely from natural fibers (96%) without dyes, bleach, chemical binders or fluorochemicals and is degradable in 5 months in landfill conditions (ECODRAPE 2018). Using nanotechnology, EnviroSystems, Inc. has introduced a few years ago a hospital grade nanoemulsive disinfectant cleaner, named "EcoTru®", that is effective on metal, metal alloy, plastic, synthetic, rubber, glass, and paint surfaces; and is non-corrosive and non-toxic. In addition, EcoTru® has been approved by Environmental Protection Agency for safe use on food contact surfaces (NANOTECH 2018). I am hopeful that environmental engineers would help develop more of these earth-friendly infection control products in the not-to-distant future.

Recycling

Some of the medicine-related waste are paper products and non-infectious. It is certainly beneficial if these non-infectious paper waste could be recycled for future reuse. This recycling effort will not only reduce the amount of waste requiring special treatment, but would also reduce the amount of landfill occupation, both are good for our environment. If optimal results are the goal, however, this recycling process will require a conscious effort by the healthcare professionals to dispose their paper products into designated containers and the environmental engineers to design user-friendly methodology to assist this disposal effort.

Green Medical Machine and Machine Control

Part of equation of reducing healthcare-related environmental impact will be to reduce energy consumption, particularly energy consumed by those large diagnostic machines. Thus a collaboration between medical professions and biomedical engineers would be needed to help reducing this huge environmental footprint by designing and optimizing equipment that are less energy hunger (WHO 2018). In Canada, a classification project initiated to develop ENERGY STAR specification for large medical equipment that do not have such energy usage label, would be a good measure for future energy saving (CANADA 2018). Another energy waste occurred in medical equipment is idle energy consumed, as certain equipment needed to be ready at all time. Thus, Bioengineers and medical professionals could work together to develop programmable control devices to reduce idle energy usage while ensuring fast response and emergency readiness (ENERGY 2018).

Summary

Environmental protection is essential for both engineering and medicine. On one hand, some diseases and health conditions are due to the effects from our environment. On the other hand, healthcare activities generated many wastes that could negatively impact our environment, and by extension, our health. In this chapter, some major environment-related diseases are discussed. To protect our environment from negative impact resulting from our medical practices, health care professionals and environmental engineers would need to work together to counter these negative impact, by following the three "Rs", namely "Reduce" (medical waste reduction), "Reuse" (disposable converted to reusable), and "Recycle" (recycle for repurposing when possible). Some specific suggestions are included for the purpose of stimulating future discussions.

References Cited

[AEESP] 2018. About AEESP. Association of Environmental Engineering & Science Professors. [https://www.aeesp.org/about] accessed July 17, 2018.

[BLS] 2018. Environmental Engineers. Occupational Outlook Handbook. Bureau of Labor Statistics. [https://www.bls.gov/ooh/architecture-and-engineering/environmental-engineers.htm] accessed July 17, 2018.

Bound, J.P. and N. Voulvoulis. 2005. Household disposal of pharmaceuticals as a pathway for aquatic contamination in the United Kingdom. Environ Health Perspect 113: 1705–1711.

[BRAINY QUOTE] 2018a. Hugh Ross Quotes. Brainy Quote. [https://www.brainyquote.com/authors/hugh_ross] accessed July 13, 2018.

[BRAINY QUOTE]. 2018b. Environmental Quotes. Brainy Quote. [https://www.brainyquote.com/topics/environmental] accessed July 13, 2018.

[CANADA] 2016. Medical Imaging Equipment Energy Use Study. Nov. 2, 2016. [http://greenhealthcare.ca/imging/] accessed August 4, 2018.

Chwalba, A., B. Maksym, M. Dobrakowski, S. Kasperczyk, N. Pawlas, E. Birkner et al. 2018. The effects of occupational chronic lead exposure on the complete blood count and the levels of selected hematopoietic cytokines. Toxicol Appl Pharmacol 355: 174–179.

[CORNELL] 2018. Environmental Engineering Program. Department of Biological and Environmental Engineering. Cornell College of Agriculture and Life Sciences. [https://bee.cals.cornell.edu/undergraduate/environmental-engineering-program/] accessed July 13, 2018.

de Souza, I.D., A.S. de Andrade and R.J.S. Dalmolin. 2018. Lead-interacting proteins and their implication in lead poisoning. Crit Rev Toxicol 48: 375–386.

[DOE] 2011. U.S. Department of Energy. Hospital pulling the plug on energy-wasting electric equipment and procedures. July 2011.

Duan, Y.I. Peng, H. Shi and Y. Jiang. 2017. The effects of lead on GABAergic interneurons in rodents. Toxicol Ind Health 33: 867–875.

Eckelman, M.J. and J. Sherman. 2016. Environmental impacts of the U.S. health care system and effects on public health. PLoS One June 9; 11(6):e0157014. Doi: 10.137/journal.pone.0157014.

[ECODRAPE] 2018. [https://vimeo.com/162844828] accessed July 20, 2018.

[EIA] 2012. U.S. Energy Information Administration. 2012 Commercial building energy consumption survey: energy usage summary. [https://www.eia.gov/consumption/commercial/reports/2012/energyusage/] accessed August 4, 2018.

[EMS] 2018. Express Medical Supplies. [https://www.expressmedicalsupplies.com] accessed July 20, 2018.

[ENERGY] 2018. Healthcare energy: spotlight on medical equipment. Office of Energy Efficiency and Renewable Energy. [https://www.energy.gov/eere/buildings/healthcare-energy-spotlight-medical-equipment] accessed August 4, 2018.

[EPA] 2018. Reduce, Reuse, Recycle. US Environmental Protection Agency. [https://epa.gov/recycle] accessed July 20, 2018.

[FDA] 2018. Expiration Dating and Stability Testing for Human Drug Products. US Food and Drug Administration. [https://www.fda.gov/iceci/inspections/inspectionguides/inspectiontechnicalguides/ucm072919.htm] accessed July 20, 2018.

Ferrante, G. and S. La Grutta. 2018. The burden of pediatric asthma. Front Pediatr June 22; 6: 186. Doi: 10.3389/fped.2018.00186.

Flores-Morales, A. and D. Iglesias-Gato. 2017. Quantitative mass spectrometry-based proteomic profiling for precision medicine in prostate cancer. Frontiers Oncology 7: 1–8.

Gardeux, V., J. Berghout, I. Achour, A.G. Schissler, Q. Li, C. Kenost et al. 2017. A geneome-by-environment interaction classifier for precision medicine: personal transcriptome response to rhinovirus identifies children prone to asthma exacerbations. J Am Med Inform Assoc 24: 1116–1126.

Ge, Y., L. Chen, X. Sun, Z. Yin, X. Song, C. Li et al. 2018. Lead-induced changes of cytoskeletal protein is involved in the pathological basis in mice brain. Environ Sci Pollut Res Int 25: 11748–11753.

Gray, C.L., D.T. Lobdell, K.M. Rappazzo, Y. Jian, J.S. Jagal, L.C. Messer et al. 2018. Association between environmental quality and adult asthma prevalence in medical claims data. Environ Res 166: 529–536.

[GT] 2018. George Tiemann & Co. [www.georgetiemann.com] accessed July 20, 2018.

Gu, X., Y. Qi, Z. Feng, L. Ma, K. Gao, Y. Zhang. 2018. Lead (Pb) induced ATM-dependent mitophagy via PINK1/Parkin pathway. Toxicol Lett 291: 92–100.

Heberer, T. 2002. Occurrence, fate, and removal of pharmaceutical residues in the aquatic environment: a review of recent research data. Toxicol Lett 131: 5–17.

Isomursu, M., J. Koivusaari, T. Stiemberg, V. Hirvela-Koski and E.R. Venalaninen. 2018. Lead poisoning and other human-related factors cause significant mortality in white-tailed eagles. Ambio March 29. Doi: 10.1007/s13280-018-1052-9.

Jeong, A., G. Fiorito, P. Keski-Rahkonen, M. Imboden, A. Kiss, N. Robinot et al. 2018. Perturbation of metabolic pathways mediates the association of air pollutants with asthma and cardiovascular diseases. Envrion Int 119: 334–345.

Kantor, D.B., W. Phipatanakul and J.N. Hirschhorn. 2018. Gene-environment interactions associated with the severity of acute asthma exacerbation in children. Am J Respir Crit Care Med 197: 545–547.

Kaplan, S., B. Sadler, K. Little, C. Franz and P. Orris. 2012. Can sustainable hospitals help bend the health care cost curve? November 2, 2012. The Commonwealth Fund. [www.commonwealthfund.org/publications/issue-briefs/2012/nov/can-sustainable-hospitals-help-bend-health-care-cost-curve] accessed July 15, 2018.

Langlois, D.K., J.B. Kaneene, V. Yuzbasiyan-Gurkan, B.L. Daniels, H. Mejia-Abreu, N.A. Frank et al. 2017. Investigation of blood lead concentrations in dogs living in Flint, Michigan. J Am Vet Med Assoc 251: 912–921.

Lewis, D.R., J.W. Southwick, R. Ouellet-Hellstrom, J. Rench and R.L. Calderon. 1999. Drinking water arsenic in Utah. A cohort mortality study. Environ Mealth Perspect 107: 359–365.

Matera, M.G., B. Rinaldi, L. Calzetta and M. Cazzola. 2017. Pharmacogenetic and pharmacogenomics considerations of asthma treatment. Expert Opin Drug Toxicol 13: 1159–1167.

[MAYO] 2018. Lead Poisoning. Mayo Clinic. [https://www.mayoclinic.org/diseases-conditions/lead-poisoning/sumptoms-causes/syc-20354717] accessed July 18, 2018.

McAfee, S. 2017. 10 quotes from Pope Francis about nature. April 14, 2017. [www.wideopenspaces.com/10-quotes-pope-francis-nature/] accessed July 13, 2018.

[MIT] 2018. Civil and Environmental Engineering. MIT. [https://cee.mit.edu/undergraduate/] accessed July 15, 2018.

Mitra, P., S. Sharma, P. Purohit and P. Sharma. 2017. Clinical and molecular aspects of lead toxicity: An update. Crit Rev Clin Lab Sci 54: 506–528.

[MM] 2018. Fox Dermal Curettes. Marne Medical. [https://www.marnemedical.com.au/product/fox-dermal-curettes/] accessed July 20, 2018.

Naja, A.S., P. Permaul and W. Phipatanakul. 2018. Taming asthma in school-aged children: a comprehensive review. J Allergy Clin Immunol Pract 6: 726–735.

[NANOTECH] 2018. EcoTru® Professional. The Project on Emerging Nanotechnologies. [www.nanotechproject.org/cpi/products/ecotru-r-professional/] accessed July 20, 2018.

Orellano, P., N. Quaranta, J. Reynoso, B. Balbi and J. Vasquez. 2017. Association of outdoor air pollution with the prevalence of asthma in children of Latin America and the Caribbean: A systematic review and meta-analysis. J Asthma 6: 1–13.

Papi, A., C. Brightling, S.E. Pedersen and H.K. Reddel. 2018. Asthma. Lancet 391: 783–800.

[PENN] 2018. Undergraduate Program Overview. Penn Engineering. [http://www.cbe.seas.upenn.edu/prospective-students/undergraduates/index.php] accessed July 15, 2018.

[PRINCETON] 2018. Environmental Engineering. Civil and Environmental Engineering. School of Engineering and Applied Sciences. Princeton University. [https://www.princeton.edu/cee/undergraduate/program-tracks/environmental-engineering/] accessed July 13, 2018.

Puglielli, M. What is shelf life testing of a pharmaceutical product? Wellspring Manufacturing Solutions. Nov 24, 2014. [www.wellspringcmo.com/blog/what-is-shelf-lfe-testing-of-pharmaceutical-product] accessed July 20, 2018.

Quinn, M.M., P.K. Henneberger, B. Braun, G.L. Delclos, K. Fagan, V. Huang et al. 2015. Cleaning and disinfecting environmental surfaces in health care: Toward an integrated framework for infection and occupational illness prevention. Am J Infection Control 42: 424–434.

Raihan, M.J., E. Briskin, M. Mahfuz, M.M. Islam, D. Mondal, M.I. Hossain et al. 2018. Examining the relationship between blood lead level and stunting, wasting and underweight: A cross-sectional study of children under 2 years-of-age in a Bangladeshi slum. PLoS One May 24; 13(5): e0197856. Doi: 10.1371/journal.pone.0197856.

Ratnaike, R.N. 2003. Acute and chronic arsenic toxicity. Postgrad Med 79: 391–396.

Salminen, P., H. Paajanen, T. Rautio, P. Nordstrom, M. Aarnio, T. Rantanen et al. 2015. Antibiotic therapy vs. appendectomy for treatment of uncomplicated appendicitis. JAMA 313: 2340–8.

Scherzer, R. and M.H. Grayson. 2018. Heterogeneity and the origins of asthma. Ann Allergy Asthma Immunol June 18, 2018. Doi: 10.1016/j.anal.2018.06.009.

Schnur, J. and R.M. John. 2014. Childhood lead poisoning and the new centers for disease control and prevention guidelines for lead exposure. J Am Assoc Nurse Pract 26: 238–247.

Schroder, F.H., J. Hugosson, M.J. Roobol, T.L. Tammela, S. Ciatto, V. Nelen et al. 2009. Screening and prostate-cancer mortality in a randomized European study. N Eng J Med 360: 1320–1328.

Schwedt, T.J. and C.D. Chong. 2017. Medication overuse headache: pathophysiological insights from structural and functional brain MRI research. Headache 57: 1173–1178.

Sippola, S., J. Gronroos, R. Tuominen, H. Paajanen, T. Rautio, P. Nordstrom et al. 2017. Economic evaluation of antibiotic therapy versus appendicectomy for the treatment of uncomplicated appendicitis from the APPAC randomized clinical trial. Br J Surg 104: 1355–1361.

Soyiri, I.N. and I. Alcock. 2018. Green spaces could reduce asthma admissions. Lancet Respir Med Jan; 6(1):e1. Doi: 10.1016/s2213-2600(17)30441-1.

Sui, Q., X. Cao, S. Lu, W. Zhao, Z. Qiu and G. Yu. 2015. Occurrence, sources and fate of pharmaceuticals and personal care products in the groundwater: a review. Emerging Contaminants 1: 14–24.

Symvoulakis, E.K., D. Anyfantakis and A. Markaki. 2015. Overuse of antibiotics in primary health care: a practitioner—or patient-induced problem? JAMA Internal Med 175: 863.

Talbot, A., C. Lippiatt and A. Tantry. 2018. Lead in a case of encephalopathy. BMJ Case Rep March 9; 2018. Doi: 10.1136/bcr-2017-222388.

[UIUC] 2018. Environmental Engineering and Science. Department of Civil and Environmental Engineering. UIUC. [https://cee.illinois.edu/areas/environmental-engineering-and-science] accessed July 15, 2018.

Warniment, C., K. Tsang and S.S. Galazka. 2010. Lead poisoning in children. Am Acad Family Physicians 81: 751–757.

[WHO] 2018. Energy-efficient medical devices. World Health Organization. [http://www.who.int/sustainable-development/health-sector/strategies/energy-efficient-medical-devices/en/] accessed August 4, 2018.

Windfeld, E.S. and M.S.L. Brooks. 2015. Medical waste management: A review. J Environmental Management 163: 98–108.

[YALE] 2018. Undergraduate Study-Environmental Engineering. Chemical and Environmental Engineering. School of Engineering and Applied Science. Yale University. [https://seas.yale.edu/departments/chemical-and-environmental-engineering/undergraduate-study-environmental/undergradu ate-c] accessed July 13, 2018.

Yang, M., Y. Li, Y. Wang, N. Cheng, Y. Zhang, S. Pang et al. 2018. The effects of lead exposure on the expression of HMGB1 and HO-1 in rats and PC12 cells. Toxicol Lett 288: 111–118.

QUESTIONS

1. What do environmental engineers do?

2. What is the relationship between environmental engineering and medicine?

3. Can you name several diseases that are in part or in full caused by environmental condition? What are the routes of the environmental exposures in these diseases?

4. What do the practice of medicine generate in terms of waste that could negatively impact environment?

5. What is a major area in hospital setting where energy-saving opportunity exists?

6. What are the environmental engineers' roles in medical waste disposal?

7. What are the practice physicians' roles in medical waste disposal?

PROBLEM

- Grouping with at least one medical student and one engineering student in a goup
- Brainstorming to determine a real-life medical situation where large amount of medicine waste are generated
- Develop a detailed methodology for medical waste management so as to reduce the amount of waste
- Present the detailed idea and methodology to the class

Index

About the Editors

Lawrence S. Chan, MD, MHA, a graduate of Massachusetts Institute of Technology with double B.S. degrees in Chemical Engineering and Life Sciences, obtained his medical training at the University of Pennsylvania School of Medicine and a M.D. degree. After an internship in Internal Medicine at the Cooper Hospital/University of Medical Center, he then completed his residency training in Dermatology and a fellowship training in Immunology at the University of Michigan Medical Center. He initially served as a junior faculty member at the Wayne State University and Northwestern University. He subsequently served as the Chief of Dermatology at the University of Illinois Hospital and the Department Head of Dermatology at the University of Illinois College of Medicine at Chicago for 11 years as well as the Program Director of Dermatology Residency for 9 years. His research interests focused on the immune mechanisms of autoimmune and inflammatory skin diseases and had received grant funding from the Dermatology Foundation, Veterans Affairs, and the National Institutes of Health. His academic achievements include more than 110 full-length biomedical publications, 35 book chapters, and 3 textbooks. Professionally Dr. Chan has been recognized as one of the "America's Top Physicians" in *Guide to America's Top Physicians* by Consumers' Research Council of America. Currently he is the Dr. Orville J. Stone Endowed Professor of Dermatology at the University of Illinois College of Medicine and has recently completed a master degree in Health Administration (MHA) under the Clinician Executive Master in Health Administration (CEMHA) Program at University of Illinois at Chicago School of Public Health.
Contact: Lawrence S. Chan, MD, Department of Dermatology, University of Illinois College of Medicine, 808 S. Wood Street, R380, Chicago, IL 60612; Tel. (312) 996-6966; Email: larrycha@uic.edu

William C. Tang, PhD., obtained his B.S., M.S. and Ph.D. degrees in Electrical Engineering and Computer Sciences from the University of California at Berkeley. Dr. Tang has substantial hands-on engineering and leadership experience in both academic and industrial arenas. He has previously served as Member of Technical Staff at TRW, Engineer Senior at IBM Corp., Sensor Research Manager at Ford Microelectronics, Inc., Group Supervisor at Caltech Jet Propulsion Laboratory, Program Manager at the Defense Advanced Research Projects Agency, and subsequently served as the founding Associate Dean for Research of the Henry Samueli School of Engineering at the University of California, Irvine. Currently he is Professor of Biomedical Engineering with joint appointments with the Electrical Engineering and Computer Sciences, Chemical and Biomolecular Engineering, and Materials Science and Engineering Departments at the same institution. His major research interests are in the development of devices and platforms that enable both in-vitro and in-vivo studies of the mechanical aspects of physiological activities at the length-scales that are beyond the reach of traditional biomedical instrumentation. His Microbiomechanics Laboratory has been supported by grants from DARPA, NIH, NSF, and private and foundation donations. His academic achievements include over 100 full-length biomedical publications, 4 book chapters, 8 patents, and multiple Achievement Awards.
Contact: William C. Tang, PhD. Department of Biomedical Engineering, 3120 Natural Sciences II, UC Irvine, Irvine, CA 92697-2715; Tel. (949) 824-9892; Email: wctang@uci.edu

Milton Keynes UK
Ingram Content Group UK Ltd.
UKHW050453071024
449327UK00015B/360